Reviews in Computational Chemistry Volume 24

1807

WILEY

2007

THE WILEY BICENTENNIAL–KNOWLEDGE FOR GENERATIONS

ℰach generation has its unique needs and aspirations. When Charles Wiley first opened his small printing shop in lower Manhattan in 1807, it was a generation of boundless potential searching for an identity. And we were there, helping to define a new American literary tradition. Over half a century later, in the midst of the Second Industrial Revolution, it was a generation focused on building the future. Once again, we were there, supplying the critical scientific, technical, and engineering knowledge that helped frame the world. Throughout the 20th Century, and into the new millennium, nations began to reach out beyond their own borders and a new international community was born. Wiley was there, expanding its operations around the world to enable a global exchange of ideas, opinions, and know-how.

For 200 years, Wiley has been an integral part of each generation's journey, enabling the flow of information and understanding necessary to meet their needs and fulfill their aspirations. Today, bold new technologies are changing the way we live and learn. Wiley will be there, providing you the must-have knowledge you need to imagine new worlds, new possibilities, and new opportunities.

Generations come and go, but you can always count on Wiley to provide you the knowledge you need, when and where you need it!

WILLIAM J. PESCE
PRESIDENT AND CHIEF EXECUTIVE OFFICER

PETER BOOTH WILEY
CHAIRMAN OF THE BOARD

Reviews in Computational Chemistry Volume 24

Edited by

Kenny B. Lipkowitz
Thomas R. Cundari

Editor Emeritus

Donald B. Boyd

WILEY-VCH

Kenny B. Lipkowitz
Department of Chemistry
Howard University
525 College Street, N.W.
Washington, D.C., 20059, U.S.A.
ken.lipkowitz@cox.net

Thomas R. Cundari
Department of Chemistry
University of North Texas
Box 305070,
Denton, Texas 76203-5070, U.S.A.
tomc@unt.edu

Donald B. Boyd
Department of Chemistry and Chemical
   Biology
Indiana University–Purdue University
   at Indianapolis
402 North Blackford Street
Indianapolis, Indiana 46202-3274, U.S.A.
boyd@chem.iupui.edu

Published by John Wiley & Sons, Inc., Hoboken, New Jersey.
Published simultaneously in Canada.

For general information on our other products and services or for technical support, please contact our
Customer Care Department within the United States at (800) 762-2974, outside the United States at
(317) 572-3993 or fax (317) 572-4002.

Wiley also publishes its books in a variety of electronic formats. Some content that appears in print may
not be available in electronic format. For information about Wiley products, visit our web site at
www.wiley.com.

Wiley Bicentennial Logo: Richard J. Pacifico

Library of Congress Cataloging-in-Publication Data:

ISBN   978-0-470-11281-6
ISSN   1069-3599

Printed in the United States of America.

10 9 8 7 6 5 4 3 2 1

Foreword

Reviews in Computational Chemistry is a pedagogically driven review series covering all aspects of computational chemistry relevant to the scientific and engineering communities. Unlike traditional books and journals that are dedicated to reviewing a given subject, this series also provides tutorials covering the topic at hand so that the reader is brought up to speed before the literature is reviewed. And, unlike traditional textbooks that focus exclusively on delivering pedagogy, this series provides an overview of the extant literature, pointing out to the reader what can and what cannot be done with a given computational method, illustrating for the practicing molecular modeler how one method compares with a rival technique to address a given scientific of engineering problem, and highlighting for the novice computational chemist what to do, what not to do, and pitfalls to avoid. The success of this series lies in the fact that *Reviews in Computational Chemistry* is nontraditional; it is a hybrid species that covers what textbooks and review journals accomplish independently of one another.

Textbooks, by their very nature, are limited in what they can provide to the reader in terms of diversity of topic and depth of coverage. The coverage of *Reviews in Computational Chemistry* is very wide in scope. Moreover, we publish in-depth, didactic chapters that are more highly focused on an important computational method than comparable chapters found in encyclopedias or in traditional textbooks. We want this series to be a learning experience for both the novice modeler who has a basic comprehension of physical chemsitry as well as for seasoned professionals who may be working in academia or industry and need to learn a new method quickly to solve a problem.

Many topics covered in this series are sufficiently narrow in scope so that coverage in a single, comprehensive chapter is possible. Some topics, however, are more expansive in nature, cutting extraordinarily wide swaths through the scientific landscape shared by researchers in disciplines as disparate as mathematics, biology, geology, physics, chemistry, and computer science. An example includes the treatment of solvents where in the past we have covered

individual topics like continuum solvation models, computational approaches
to lipophilicity, molecular models of water, cellular automata models of aequous solution systems, and computing hydrophobicity among other related
topics. Even broader in scope is the topic of quantum chemistry. We have
published individual chapters in this book series covering techniques like
semiempirical molecular orbital methods, density functional theory, post-
Hartree-Fock methods, quantum Monte Carlo methods, basis sets, basis set
superposition errors, effective core potentials, relativistic effects, coupled
cluster methods, valence bond theory, and molecular quantum similarity.
Further attesting to the breadth of quantum chemistry are individual chapters covering specific applications of those techniques focusing, for example,
on calculating the properties of hydrogen bonds by ab initio methods, deriving molecular electrostatic potentials and chemical reactivity, obtaining
electron densities from quantum mechanics, calculating vibrational circular
dichroism intensities, ab initio computation of nuclear magnetic resonance
chemical shifts, quantum mechanical methods for predicting nonlinear optical properties, and many other topics.

In this volume of *Reviews in Computational Chemistry*, we deviate somewhat from the past volumes by covering a single topic: confined fluids. In
this volume Drs. Martin Schoen and Sabine H. L. Klapp highlight the theoretical underpinnings of this field of study, and provide the mathematical
derivations and computer implementations of those theories. Drs. Schoen and
Klapp then compare the numerical results of those implementations with analytical solutions and experimental data to illustrate how one should treat
confined fluids under various conditions. Understanding confined fluids and
predicting their complex behaviors is especially important in the disciplines
of science, engineering, and technology. Relevant areas of study that quickly
come to mind include the hydrodynamics and rheology of ground water in
soils, the action of lubricants in piston-driven motors or in automotive trans-
missions, the characteristics of compressed fluids used as refrigerants, and
perhaps more relevant to the traditional chemistry community, the behavior
of confined fluids in chromatography, and, more recently, comprehending and
manipulating the behaviors of fluids in the micro- and nano-fluidic devices
used for both synthesis and analysis.

The behavior and characteristics of confined fluids is more complex than
that of bulk liquids or of simple solvated systems described in the chapters
mentioned above. One must consider the complexities of the interface between confining walls and the fluid along with confinement-induced phase
transitions, critical points, the stratification of the fluid near the confining
walls, the idea that confined fluids may sustain certain shear stress without
exhibiting structural features normally associated with solid-like phases, and

other phenomena that scientists and engineers alike need to be aware of to become proficient modelers. In this book, the authors put all of these issues into perspective and assist the reader with many required mathematical derivations that are needed to comprehend this technologically relevant and scientifically interesting area of research.

Reviews in Computational Chemistry is highly rated and well received by the scientific community at large; the reason for these accomplishments rests firmly on the shoulders of the authors whom we have contacted to provide pedagogically driven reviews that have made this ongoing book series so popular. To those authors we are especially grateful.

We are also glad to note that our publisher now makes our most recent volumes available in an online form through Wiley InterScience. Please consult the Web (http://www.interscience.wiley.com/onlinebooks) or contact reference@wiley.com for the latest information. For readers who appreciate the permanence and convenience of bound books, these will, of course, continue.

We thank the authors of this and previous volumes for their excellent chapters.

Kenny B. Lipkowitz,
Washington

Thomas R. Cundari,
Denton

December 2006

Preface

In classifying condensed-matter systems, it often proves useful to identify relevant length scales if one aims at a theoretical understanding of their properties and behavior. For example, the notion of a mean-free path is a useful concept in the kinetic theory of (dilute) gases. If sufficiently large, it tells us that we may treat the dynamic evolution of a gaseous system (and therefore its transport properties) as a result of isolated collision events involving no more than a pair of gas molecules. In solids the lattice constant poses another such length scale by which different crystallographic structures can be classified and distinguished conveniently.

In dense gases and liquids, which we subsume under the term "fluids" henceforth, the correlation length may be viewed as a key length scale in the above sense because it permits one to classify phenomena in both the non- and near-critical regimes of the fluid. The correlation length is a quantitative indicator of density fluctuations in the fluid. It may thus be viewed as a measure of the length over which fluid structures decay such that they may no longer be distinguished from a random arrangement of fluid molecules. Because of this interpretation the correlation length is comparable with the range of a typical intermolecular interaction potential as long as one stays off the near-critical regime of the fluid where density fluctuations are rather minute; the correlation length may, however, become macroscopic as one penetrates into the near-critical regime. From a theoretical perspective, the concept of a correlation length is the key ingredient in scaling theories that permit one to understand critical phenomena in thermal many-particle systems quantitatively.

The idea that a specific property of a thermal system may be intimately linked to a characteristic length scale immediately raises the question: What might happen if there is more than just one relevant length scale? Moreover, what might happen if different length scales governing different features of a thermal system are comparable in magnitude and what will be the result of a competition between these different length scales, that is, a competition between physical features with which they are intertwined? To answer these questions or, to be a bit more modest, to shed some light on aspects touched upon by them, was at the core of our motivation to write this book.

Although this rather broad exposition is interesting from a purely academic perspective, it would be nice if systems existed in the "real" world where an answer to the above questions would help to better understand the behavior of such a system. The ultimate goal of such a venture would then be to explore new technologies or to optimize certain processes so that in the very long run fundamental research of this kind might be useful to more applied scientific disciplines. Fortunately, an almost ideal realization of a situation meeting all these conditions and provisos does indeed exist in reality. The class of systems to which we are referring in this context are fluids confined by solid surfaces to tiny volumes of nanoscopic dimensions, which we shall term "confined fluids" throughout this book. Confined fluids play an important role in a variety of practical or applied contexts ranging from say, the swelling of clay soils, which is important in understanding aspects of frost heaving, to catalysis, where new nanoporous media are currently being explored as novel nanoreactors.

In a confined fluid the correlation length remains a key length scale that determines the formation of its equilibrium structures. We immediately limit the scope of our discussion to noncritical fluids where the correlation length is in the range of nanometers. However, the degree of confinement imposes a second such length scale comparable with that of the correlation length. One may think, for example, of the width of a nanoporous medium, which can nowadays be produced rather routinely in a physical chemistry laboratory. As we shall demonstrate below, the presence of a solid surface causes the confined fluid to be inhomogeneous; that is, its density depends on the position relative to the confining substrate. This inhomogeneity has profound repercussions for thermophysical properties of the confined fluid.

By means of modern lithographic and related techniques, it is also possible to impose on the confining substrate a chemical or geometrical substructure of nanometer dimensions. These substructures impose yet a third length scale comparable with the two aforementioned ones, which will also have fundamental consequences for the confined fluid. Perhaps the most fascinating of these consequences is that the existence of thermodynamic phases may be triggered, which have no counterpart in the bulk.

Because of these intriguing and complex features of confined fluids, we feel that this text comes timely for several reasons. First, research on confined fluids has been carried out for the past 15–20 years with increasing intensity so that it is still one of the most active fields in soft condensed matter research. During this time, our understanding of many aspects of confined fluids has grown enormously. At the same time, experimental and theoretical "tools" have been designed and applied to investigate these systems so that the time seems right for a more comprehensive introduction to the field. Second, even though quite a few excellent reviews of actual scientific work in this field have already been published over the past years [1–5], there still does not exist any text aiming at a more pedagogic introduction from a statistical physics background. In particular this second aspect was our primary motivation of compiling the material below.

However, looking back at research in this area, it became immediately clear that a conscientious choice of techniques and applications had to be made to preserve the pedagogic impetus on the one hand and a sufficiently broad selection of illustrating material on the other hand. As the current text focuses on theoretical aspects, it goes without saying that mathematics is the appropriate language for such a discussion. In selecting the level of formal (i.e., mathematical) presentation of the necessary and relevant key concepts, we had in mind a reader with knowledge in algebra and calculus typical of a well-educated (physical) chemist. More specifically, we intend to address an audience ranging from advanced students in the physical sciences to experienced scientists interested in beginning research in the field of nanoconfined fluids.

Unfortunately, the subject of this manuscript makes it inevitable to go quite a bit beyond this level at certain points. In these events we summarize the mathematical techniques to an extent necessary to follow our line of arguments. To keep the main text legible, we deferred these discussions to various appendices where each appendix is related to a specific chapter of the main text. An example is a brief introduction to the theory of complex functions in Appendix B.2 for which excellent textbooks are, of course, available. Nevertheless the goal of keeping this manuscript self-contained as much as possible prompted us to include this background material to some minimum extent.

A guiding principle in selecting the topics to be discussed below was that a specific physical problem should be tackled by a combination of different techniques wherever feasible. For example, in Chapter 4, we employ mean-field theory to study the phase behavior of confined fluids. In Chapter 5 we revisit this issue but employ Monte Carlo simulations instead. The latter, being a first-principles numerical method, permit a rigorous test of predictions deduced from mean-field theory. This way we do not only emphasize the importance of applying more than just one approach because of the complexity of the systems under study. Moreover, we can illustrate the mutual limitations and advantages of each one of them. Thereby we hope to provide a broader (and more useful) overview, which is particularly important from the educational or training aspect central to this work. At the same time we intended to keep this text sufficiently focused and therefore made a deliberate choice from the outset in limiting the discussion exclusively to equilibrium properties.

From a theoretical perspective, thermodynamics is then the central theory on which such a discussion must build. Below we use a formulation of thermodynamics usually applied to solid-like systems, because confined fluids have a lot in common with bulk solids in that they are highly inhomogeneous and anisotropic. However, unlike a solid, a confined fluid lacks any long-range spatial order. As we demonstrate in Chapter 1, symmetry considerations play an integral part of the current formulation of equilibrium thermodynamics with which the nonexpert in the field will not necessarily be accustomed.

The link to the molecular level of description is provided by statistical thermodynamics where our focus in Chapter 2 will be on specialized statistical physical ensembles designed specifically for capturing features that make confined fluids distinct among other soft condensed matter systems. We develop statistical thermodynamics from a quantum-mechanical foundation, which has at its core the existence of a discrete spectrum of energy eigenstates of the Hamiltonian operator. However, we quickly turn to the classic limit of (quantum) statistical thermodynamics. The classic limit provides an adequate framework for the subsequent discussion because of the region of thermodynamic state space in which most confined fluids exist.

We immediately apply the concepts of statistical thermodynamics in Chapter 3 to a class of systems that can be handled analytically, namely one-dimensional hard-rod fluids confined between hard walls. Despite its simplicity, this system exhibits key features of confined fluids as we shall demonstrate by comparing the results obtained in Chapter 3 with those for more realistic systems in later chapters.

However, one-dimensional confined fluids with purely repulsive interactions can be expected to be only of limited usefulness, especially if one is interested in phase transitions that cannot occur in any one-dimensional system. In treating confined fluids in such a broader context, a key theoretical tool is the one usually referred to as "mean-field theory." This powerful theory, by which the key problem of statistical thermodynamics, namely the computation of a partition function, becomes tractable, is introduced in Chapter 4 where we focus primarily on lattice models of confined pure fluids and their binary mixtures. In this chapter the emphasis is on features rendering confined fluids unique among other fluidic systems. One example in this context is the solid-like response of a confined fluid to an applied shear strain despite the absence of any solid-like structure of the fluid phase.

The mean-field theory developed in Chapter 4 is, however, plagued by

several simplifying and a priori uncontrollable assumptions, the most prominent one being the complete neglect of intermolecular correlations and, in
our case, the discretization of space by employing lattice models. To demonstrate the somewhat astonishing power and reliability of the mean-field treatment, we amend its discussion by presenting parallel Monte Carlo computer
simulations in Chapter 5. As we emphasize in that chapter and as we already pointed out above, Monte Carlo should be viewed essentially as a *first
principles* approach free of any additional assumption. Monte Carlo simulations thus provide an ideal test ground for the mean-field results presented
in Chapter 4. To make such a comparison we selected applications of the
Monte Carlo computer simulation method in Chapter 5.

In all examples discussed in that chapter, fluid molecules interact with
each other via *short-range* potentials decaying as r^{-6} like dispersion interactions between polarizable molecules (r being the intermolecular distance) or
even faster (as, for example, in the nearest-neighbor lattice models also discussed in Chapter 5). There is, however, an increasing interest in modeling
confined biological, electrochemical, or colloidal fluids where the dominant
interactions are electrostatic in nature. As far as Coulombic interactions between charged molecules or molecules with a permanent dipole are concerned,
these interactions are long-range. In these two cases, the relevant interaction
potentials decay as r^{-1} and r^{-3}, respectively. In Chapter 6 we discuss special
techniques to deal with such long-range interactions in computer simulations
and present selected applications for confined dipolar fluids.

Again we limit the discussion of simulation techniques in Chapter 6 to the
Monte Carlo method as a key numerical technique to stay focused as much
as possible. Molecular dynamics simulations, which are the other simulation
technique one would immediately think of, are explicitly disregarded here
because they are more suitable to study dynamic rather than equilibrium
properties with which we are concerned here. In a similar spirit, off-lattice
density functional theory is also disregarded here because this is already a
vast and flourishing field in its own right to which a separate such text should
be devoted.

For most of this book we consider cases of "ideal" confinement, that is,
situations where the geometry of the confining substrates is simple. The
most prominent example is that of a slit-pore where the confining substrates
are planar and parallel to one another. In Chapter 7 we focus on the opposite extreme, that is, a fluid confined to a randomly disordered porous
matrix. Experimentally this situation is encountered in aerogels. The simultaneous presence of both confinement and (quenched) disorder representing
the nearly-random silica network renders the treatment of such systems quite
challenging from a theoretical perspective. In Chapter 7 we discuss one of the

most powerful techniques to study fluids confined to random porous media, namely the so-called "replica integral equations", which allow one to calculate, at an admittedly approximate level, both structural and thermodynamic properties of confined fluids.

Given the amount of material covered by this manuscript, it is apparent that over the years many people have contributed significantly to the book we are now presenting. In particular, we thank our current and former students and coworkers Henry Bock, Thomas Gruhn, Dirk Woywod, Sophie Sacquin-Mora, Carsten Spöler, Gabriel M. Range, Fabien Porcheron, Jörg R. Silbermann, Holger Bohlen, Jochen Sommerfeld, Matthias Gramzow, Madeleine Kittner, and Vladimir Froltsov. We are also grateful for many discussions, interactions, and most enjoyable and fruitful collaborations we enjoyed over the years with our colleagues Professors John H. Cushman (Purdue University), Dennis J. Diestler (University of Nebraska at Lincoln), Siegfried Dietrich (Max-Planck-Institut für Metallforschung and Universität Stuttgart), Bob Evans (University of Bristol), Gerhard H. Findenegg (Technische Universität Berlin and North Carolina State University), Frank Forstmann (Freie Universität Berlin), Alain H. Fuchs (Université de Paris-Sud), Keith E. Gubbins (North Carolina State University), Enrique Díaz-Herrera (Universidad Autónoma Metropolitana), Siegfried Hess (Technische Universität Berlin), Christos N. Likos and Hartmut Löwen (both at Heinrich-Heine-Universität Düsseldorf), Peter A. Monson (University of Massachusetts), and Gren Patey (University of British Columbia).

Finally, we would like to thank our families and friends for their continued support and patience. Without their appreciation of the contraints on our time at the stage of writing and correcting the manuscript, this book would never have come into existence. One of us (S.H.L.K) would like to express her sincere gratitude to D. Fliegner for help whenever it was needed. In addition, M.S. wishes to thank G. Arnold, K. Behrens, E. Egorov, and M. Wahl for their friendship in difficult times.

Sabine H. L. Klapp and Martin Schoen

Berlin

December 2006

Dimensionless units

At this point we would like to emphasize that throughout this book we are giving all quantities in dimensionless (i.e., "reduced") units. However, these units will vary between lattice and off-lattice models, pure fluids and binary mixtures, and systems with short- and long-range interactions. To avoid specifying the specific dimensionless quantities we are using at any point in the text, which, in our opinion, would reduce the legibility of this text markedly, we compile the basic quantities in Table 1. The reader should note, however, that only basic quantities like lengths or energies are given explicitly; "derived" quantities like density or stress follow by suitable combination of the basic units.

Table 1: Dimensionless (i.e., "reduced") units used for basic physical quantities in various parts of this book.

Portion of text	Quantity	In units of
Chapter 3	length	d
	stress/pressure	$k_B T/d$
Section 4.2	length	σ_f
	energy	ε_{ff}
	temperature	ε_{ff}/k_B
Sections 4.3–4.5	length	ℓ
	energy	ε_{ff}
	temperature	ε_{ff}/k_B
Sections 4.6–4.8	length	ℓ
	energy	ε_{AA}
	temperature	ε_{AA}/k_B
Sections 5.3–5.7	length	σ_f
	energy	ε_{ff}
	temperature	ε_{ff}/k_B
Section 5.8	length	ℓ
	energy	ε_{ff}
	temperature	ε_{ff}/k_B
Section 6.3.3	length	ℓ
	energy	$\sigma^3/\mu^{2\,a}$
Sections 6.4.1 and 6.4.2	temperature	ε/k_B
	dipole moment	$\sqrt{\varepsilon\sigma^3}$
Section 6.5	length	σ
	energy	ε
Fig. 6.12	energy	$\sigma^3/\mu^{2\,a}$
Sections 7.7.2 and 7.7.3	length	σ
	temperature	ε/k_B
	dipole moment	$\sqrt{\varepsilon\sigma^3}$
Fig. 7.4	length	σ
	temperature	$\mu^2/(k_B\sigma^3)^{a}$
	dipole moment[b]	$\sqrt{k_B T_0 \sigma^3}$

[a]μ denotes the dipole moment.
[b]Frozen species.

Contents

Contributors

Martin Schoen, Stranski-Laboratorium für Physikalische und Theoretische Chemie, Sekretariat C7, Institut für Chemie, Fakultät für Mathematik und Naturwissenschaften, Technische Universität Berlin, Straße des 17. Juni 115, 10623 Berlin, Germany
(Electronic mail: martin.schoen@fluids.tu-berlin.de)

Sabine H. L. Klapp, Stranski-Laboratorium für Physikalische und Theoretische Chemie, Sekretariat C7, Institut für Chemie, Fakultät für Mathematik und Naturwissenschaften, Technische Universität Berlin, Straße des 17. Juni 115, 10623 Berlin, Germany
(Electronic mail: sabine.klapp@fluids.tu-berlin.de)

Contributors to previous volumes

Volume 1 (1990)

David Feller and **Ernest R. Davidson**, Basis Sets for Ab Initio Molecular Orbital Calculations and Intermolecular Interactions.
James J. P. Stewart, Semiempirical Molecular Orbital Methods.
Clifford E. Dykstra, Joseph D. Augspurger, Bernard Kirtman and **David. J. Malik**, Properties of Molecules by Direct Calculation.
Ernest L. Plummer, The Application of Quantitative Design Strategies in Pesticide Design.
Peter C. Jurs, Chemometrics and Multivariate Analysis in Analytical Chemistry.
Yvonne C. Martin, Mark G. Bures, and **Peter Willet**, Searching Databases of Three-Dimensional Structures.
Paul G. Mezey, Molecular Surfaces.
Terry P. Lybrand, Computer Simulation of Biomolecular Systems Using Molecular Dynamics and Free Energy Perturbation Methods.
Donald B. Boyd, Aspects of Molecular Modeling.
Donald B. Boyd, Successes of Computer Assisted Molecular Design.
Ernest R. Davidson, Perspectives of Ab Initio Calculations.

Volume 2 (1991)

Andrew R. Leach, A Survey of Methods for Searching the Conformational Space of Small and Medium-Sized Molecules.
John M. Troyer and **Fred E. Cohen**, Simplified Models for Understanding and Predicting Protein Structure.
J. Phillip Bowen and **Norman L. Allinger**, Molecular Mechanics: The Art and Science of Parameterization.
Uri Dinur and **Arnold T. Hagler**, New Approaches to Empirical Force

Fields.

Steve Scheiner, Calculating the Properties of Hydrogen Bonds by Ab Initio Methods.

Donald E. Williams, Net Atomic Charge and Multipole Models fort the Ab Initio Molecular Electric Potential.

Peter Politzer and **Jane S. Murray**, Molecular Electrostatic Potentials and Chemical Reactivity.

Michael C. Zerner, Semiempirical Molecular Orbital Methods.

Lowell H. Hall and **Lemont B. Kier**, The Molecular Connectivity Chi Indexes and Kappa Shape Indexes in Structure-Property Modeling.

I. B. Bersuker and **A. S. Dimoglo**, The Electron-Topological Approach to the QSAR Problem.

Donald B. Boyd, The Computational Chemistry Literature.

Volume 3 (1992)

Tamar Schlick, Optimization Methods in Computational Chemistry.

Harold A. Scheraga, Predicting Three-Dimensional Structures of Oligopeptides.

Andrew E. Torda and **Wilfred F. van Gunsteren**, Molecular Modeling Using NMR Data.

David F. V. Lewis, Computer-Assisted Methods in the Evaluation of Chemical Toxicity.

Volume 4 (1993)

Jerzy Cioslowski Ab Initio Calculations on Large Molecules: Methodology and Applications.

Michael L. McKee and **Michael Page**, Computing Reaction Pathways on Molecular Potential Energy Surfaces.

Robert M. Whitnell and **Kent R. Wilson**, Computational Molecular Dynamics of Chemical Reaction in Solution.

Roger L. DeKock, **Jeffry D. Madura**, **Frank Rioux** and **Joseph Casanova**, Computational Chemistry in the Undergraduate Curriculum.

Volume 5 (1994)

John D. Bolcer and **Robert B. Hermann**, The Development of Computational Chemistry in the United States.

Rodney J. Bartlett and **John F. Stanton**, Applications of Post-Hartree-Fock Methods: A Tutorial.

Steven M. Bachrach, Population Analysis and Electron Densities from Quantum Mechanics.

Jeffry D. Madura, Malcolm E. Davis, Michael K. Gilson, Rebecca C. Wade, Brock A. Luty and **J. Andrew McCammon**, Biological Applications of Electrostatic Calculations and Brownian Dynamics Simulations.

K. V. Damodaran and **Kenneth M. Merz, Jr.**, Computer Simulation of Lipid Systems.

Jeffrey M. Blaney and **J. Scott Dixon**, Distance Geometry in Molecular Modelling.

Lisa M. Balbes, S. Wayne Mascarella and **Donald B. Boyd**, A Perspective of Modern Methods in Computer-Aided Drug Design.

Volume 6 (1995)

Christopher J. Cramer and **Donald G. Truhlar**, Continuum Solvation Models: Classical and Quantum Mechanical Implementations.

Clark R. Landis, Daniel M. Root and **Thomas Cleveland**, Molecular Mechanics Force Field for Modeling Inorganic and Organometallic Compounds.

Vassilios Galiatsatos, Computational Methods for Modeling Polymers: An Introduction.

Rick A. Kendall, Robert J. Harrison, Rik J. Littlefield and **Martyn F. Guest**, High Performance Computing in Computational Chemistry: Methods and Machines.

Donald B. Boyd, Molecular Modeling Software in Use: Publication Trends.

Eiji Ōsawa and **Kenny B. Lipkowitz**, Appendix: Published Force Field Parameters

Volume 7 (1996)

Geoffrey M. Downs and **Peter Willett**, Similarity Searching in Databases of Chemical Structures.

Andrew C. Good and **Jonathan S. Mason**, Three-Dimensional Structure Database Searches.

Jiali Gao, Methods and Applications of Combined Quantum Mechanical and Molecular Mechanical Potentials.

Libero J. Bartolotti and **Ken Flurchick**, An Introduction to Density Functional Theory.

Alain St-Amant, Density Functional Methods in Biomolecular Modeling.
Danya Yang and **Arvi Rauk**, The A Priori Calculation of Vibrational Circular Dichroism Intensities.
Donald B. Boyd, Appendix: Compendium of Software for Molecular Modeling.

Volume 8 (1996)

Zdenek Slanina, Shyi-Long Lee and **Chin-hui Yu**, Computations in Treating Fullerenes and Carbon Aggregates.
Gernot Frenking, Iris Antes, Marlis Böhme, Stefan Dapprich, Andreas W. Ehlers, Volker Jonas, Arndt Neuhaus, Michael Otto, Ralf Stegmann, Achim Veldkamp, and **Sergei F. Vyboishchikov**, Pseudopotential Calculations of Transition Metal Compounds: Scope and Limitations.
Thomas R. Cundari, Michael T. Benson, M. Leigh Lutz and **Shaun O. Sommerer**, Effective Core Potential Approaches to the Chemistry of the Heavier Elements.
Jan Almlöf and **Odd Gropen**, Relativistic Effects in Chemistry.
Donald B. Chesnut, The Ab Inito Computation of Nuclear Magnetic Resonance Chemical Shielding.

Volume 9 (1996)

James R. Damewood, Jr., Peptide Mimetic Design with the Aid of Computational Chemistry.
T. P. Straatsma, Free Energy by Molecular Simulation.
Robert J. Woods, The Application of Molecular Modeling Techniques to the Determination of Oligosaccharide Solution Conformations.
Ingrid Pettersson and **Tommy Liljefors**, Molecular Mechanics Calculated Conformational Energies of Organic Molecules: A Comparison of Force Fields.
Gustavo A. Arteca, Molecular Shape Descriptors.

Volume 10 (1997)

Richard Judson, Genetic Algorithms and Their Use in Chemistry.
Eric C. Martin, David C. Spellmeyer, Roger E. Critchlow, Jr. and **Jeffrey M. Blaney**, Does Combinatorial Chemistry Obviate Computer-Aided Drug Design?
Robert Q. Topper, Visualizing Molecular Phase Space: Nonstatistical

France: A Historical Survey.

Volume 13 (1999)

Thomas Bally and **Weston Thatcher Borden**, Calculations on Open-Shell Molecules: A Beginners Guide.

Neil R. Kestner and **Jaime E. Combariza**, Basis Set Superposition Errors: Theory and Practice.

James B. Anderson, Quantum Monte Carlo: Atoms, Molecules, Clusters, Liquids, and Solids.

Anders Wallqvist and **Raymond D. Mountain**, Molecular Models of Water: Derivation and Description.

James M. Briggs and **Jan Antosiewicz**, Simulation of pH-Dependent Properties of Proteins Using Mesoscopic Models.

Harold E. Helson, Structure Diagram Generation.

Volume 14 (2000)

Michelle Miller Francl and **Lisa Emily Chirlian**, The Pluses and Minuses of Mapping Atomic Charges to Electrostatic Potentials.

T. Daniel Crawford and **Henry F. Schaefer III**, An Introduction to Coupled Cluster Theory for Computational Chemists.

Bastiaan van de Graaf, Swie Lan Njo, and **Konstantin S. Smirnov**, Introduction to Zeolite Modeling.

Sarah. L. Price, Toward More Accurate Model Intermolecular Potentials for Organic Molecules.

Christopher J. Mundy, Sundaram Balasubramanian, Ken Bagchi, Mark E. Tuckerman, Glenn J. Martyna, and **Michael L. Klein**, Nonequilibrium Molecular Dynamics.

Donald B. Boyd and **Kenny B. Lipkowitz**, History of the Gordon Research Conferences on Computational Chemistry.

Mehran Jalaie and **Kenny B. Lipkowitz**, Appendix: Published Force Field parameters for Molecular Mechanics, Molecular Dynamics, and Monte Carlo Simulations.

Volume 15 (2000)

F. Matthias Bickelhaupt and **Evert Jan Baerends**, Kohn-Sham Density Functional Theory: Predicting and Understanding Chemistry.

Michael A. Robb, Marco Garavelli, Massimo Olivucci and **Fernando Bernardi**, A Computational Strategy for Organic Photochemistry.

Larry A. Curtiss, Paul C. Redferm and David J. Frurip, Theoretical Methods for Computing Enthalpies of Formation of Gaseous Compounds.
Russel J. Boyd, The Development of Computational Chemistry in Canada.

Volume 16 (2000)

Richard A. Lewis, Stephen D. Pickett, and David E. Clark, Computer-Aided Molecular Diversity Analysis and Combinatorial Library Design.
Keith L. Peterson, Artificial Neural Networks and Their Use in Chemistry.
Jörg-Rüdiger Hill, Clive M. Freeman and Lalitha Subramanian, Use of Force Fields in Materials Modeling.
M. Rami Reddy, Mark D. Erion and Atul Agarwal, Free Energy Calculations: Use and Limitations in Predicting Ligand Binding Affinities.

Volume 17 (2001)

Ingo Muegge and Matthias Rarcy, Small Molecule Docking and Scoring.
Lutz P. Ehrlich and Rebecca C. Wade, Protein-Protein Docking.
Christel M. Marian, Spin-Orbit Coupling in Molecules.
Lemont B. Kier, Chao-Kun Cheng and Paul G. Seybold, Cellular Automata Models of Aqueous Solution Systems.
Kenny B. Lipkowitz and Donald B. Boyd, Appendix: Books Published on the Topics of Computational Chemistry.

Volume 18 (2002)

Geoff M. Downs and John M. Barnard, Clustering Methods and Their Uses in Computational Chemistry.
Hans-Joachim Böhm and Martin Stahl, The Use of Scoring Functions in Drug Discovery Applications.
Steven W. Rick and Steven J. Stuart, Potentials and Algorithms for Incorporating Polarizability in Computer Simulations.
Dmitry V. Matyushov and Gregory A. Voth, New Developments in the Theoretical Description of Charge-Transfer Reactions in Condensed Phases.
George R. Famini and Leland Y. Wilson, Linear Free Energy Relationships Using Quantum Mechanical Descriptors.
Sigrid D. Peyerimhoff, The Development of Computational Chemistry in Germany.
Donald B. Boyd and Kenny B. Lipkowitz, Appendix: Examination of

the Employment Environment for Computational Chemistry.

Volume 19 (2003)

Robert Q. Topper, David L. Freeman, Denise Bergin and **Keirnan R. LaMarche**, Computational Techniques and Strategies for Monte Carlo Thermodynamic Calculation, with Applications to Nanoclusters.
David E. Smith and **Anthony D. J. Haymet**, Computing Hydrophobicity.
Lipeng Sun and **William L. Hase**, Born Oppenheimer Direct Dynamics Classical Trajectory Simulations.
Gene Lamm, The Poisson Boltzmann Equation.

Volume 20 (2004)

Sason Shaik and **Philippe C. Hiberty**, Valence Bond Theory: Its History, Fundamentals and Applications. A Primer.
Nikita Matsunaga and **Shiro Koseki**, Modeling of Spin Forbidden Reactions.
Stefan Grimme, Calculation of the Electronic Spectra of Large Molecules.
Raymond Kapral, Simulating Chemical Waves and Patterns.
Costel Sârbu and **Horia Pop**, Fuzzy Soft-Computing Methods and Their Applications in Chemistry.
Sean Ekins and **Peter Swaan**, Development of Computational Models for Enzymes, Transporters, Channels and Receptors Relevant to ADME/Tox.

Volume 21 (2005)

Roberto Dovesi, Bartolomeo Civalleri, Roberto Orlando, Carla Roetti and **Victor R. Saunders**, Ab Initio Quantum Simulation in Solid State Chemistry.
Patrick Bultinck, Xavier Gironés and **Ramon Carbó-Dorca**, Molecular Quantum Similarity: Theory and Applications.
Jean-Loup Faulon, Donald P. Visco, Jr. and **Diana Roe**, Enumerating Molecules.
David J. Livingstone and **David W. Salt**, Variable Selection-Spoilt for Choice.
Nathan A. Baker, Biomolecular Applications of Poisson Boltzmann Methods.
Baltazar Aguda, Georghe Craciun and **Rengul Cetin-Atalay**, Data

Sources and Computational Approaches for Generating Models of Gene Regulatory Networks.

Volume 22 (2006)

Patrice Koehl, Protein Structure Classification.
Emilio Esposito, Dror Tobi and **Jeffry Madura**, Comparative Protein Modeling.
Joan-Emma Shea, Miriam Friedel and **Andrij Baumketner**, Simulations of Protein Folding.
Marco Saraniti, Shela Aboud and **Robert Eisenberg**, The Simulation of Ionic Charge Transport in Biological Ion Channels: An Introduction to Numerical Methods.
C. Matthew Sundling, Nagamani Sukumar, Hongmei Zhang, Curt Breneman and **Mark Embrechts**, Wavelets in Chemistry and Chemoinformatics.

Volume 23 (2007)

Christian Ochsenfeld, Jörg Kussmann and **Daniel Lambrecht**, Linear Scaling in Quantum Chemistry.
Spiridoula Matsika, Conical Intersections in Molecular Systems.
Antonio Fernandez-Ramos, Benjamin Ellingson, Bruce C. Garrett and **Donald G. Truhlar**, Variational Transition State Theory with Multidimensional Tunneling.
Roland Faller, Coarse Grain Modeling of Polymers.
Jeffrey Godden and **Jürgen Bajorath**, Analysis of Chemical Information Content using Shannon Entropy.
Ovidiu Ivanciuc, Applications of Support Vector Machines in Chemistry.
Donald B. Boyd, How Computational Chemistry Became Important in the Pharmaceutical Industry.

Nanoconfined Fluids:
Soft Matter Between Two and Three Dimensions

Martin Schoen and Sabine H. L. Klapp

Stranski-Laboratorium für Physikalische
und Theoretische Chemie
Institut für Chemie
Fakultät für Mathematik und Naturwissenschaften
Technische Universität Berlin
Straße des 17. Juni 135, 10623 Berlin, GERMANY

Chapter 1

Thermodynamics of confined phases

1.1 Introductory remarks

Our understanding of phenomena in the nonanimated part of nature (and perhaps to a lesser extent even those in its animated part) is promoted by the four cornerstones of modern theoretical physics: classic mechanics, quantum mechanics, electrodynamics, and thermodynamics. Among these four fields, thermodynamics occupies a unique position in several respects. For example, its mathematical structure is by far the simplest and can be grasped by anyone with knowledge of elementary calculus. Yet, most students and at times even long-time practitioners find it hard to apply its concepts to a given physical situation.

The axiomatic basis of thermodynamics is quite scarce compared with the other three theoretical fields. Just four elementary and completely general principles, the so-called *Laws of Thermodynamics*, are required to lay the axiomatic foundation. They are essentially deduced from everyday experience. Thus, thermodynamics is by far the most self-contained of the four theoretical fields. However, what appears as a particular strength and certainly a source of mathematical beauty also gives rise to perhaps the most serious shortcoming of thermodynamics, namely its almost total lack of any predictive power. This disadvantage is caused by the fact that thermodynamics equips us only with general mathematical relations between its key quantities. It is virtually incapable of quantifying any of them without having to take recourse to additional sources of information such as experimental data or (empirical) equations of state. Thus, there is a substantial price to pay for mathematical rigor, self-containment, and beauty (i.e., structural simplicity) of thermodynamics as a central theoretical cornerstone in the modern physical sciences.

The total lack of predictive power as far as properties of a specific physical system are concerned turns out to be caused by the fact that thermal systems are composed of (a *macroscopic* piece of) matter whose properties depend on the interaction between the *microscopic* constituents (i.e., electrons, phonons, atoms, or molecules) of which it is composed. Thermodynamics, on the other hand, has no concept whatsoever of the underlying microscopic structure of the macroscopic world with which it is dealing; that is, it knows nothing about interactions between microscopic constituents.

To outline the conceptual framework of this chapter, we think that this latter aspect can hardly be overemphasized because it has become a widely accepted but deplorable practice in some physical chemistry courses to introduce students to thermodynamics by taking recourse to inherently molecular concepts that are completely alien to thermodynamics. Although this notion is usually motivated didactically (but in our opinion utterly confused), it must be regarded as ill-founded and conceptually misleading.

To establish a conceptually sound link between molecules as entities forming a macroscopic piece of matter on the one hand and thermodynamics on the other hand, one needs to resort to (quantum) statistical physics, which could not be established in its modern form until after the advent of quantum mechanics in the early twentieth century (see Chapter 2). In other words, an attempt to base thermodynamics on molecular concepts like interacting molecules or, even worse, *molecular chaos*, deliberately ignores the fundamental character of the postulatory basis of thermodynamics among the laws of nature as we know them today [6, 7]. In light of these comments, thermodynamics appears as a typical physical theory of the nineteenth century, with the engineers Watt[1] and Carnot[2] being among its "founding fathers." In the nineteenth century, it was by no means undisputed whether entities like atoms or molecules really existed or whether they were merely a construction of the human mind [8]. Although "the atom" was a well-established but purely philosophical entity around 1800 to which rather bizarre properties were ascribed [9], it did not have precise meaning in a physics or chemistry context. Meaning in the sense of a sound concept in the natural sciences

---

[1] James Watt (1736–1819) significantly improved the heat engine developed by Thomas Newcomen.

[2] In 1824 Sadi Nicolas Léonard Carnot (1796–1832) published an analysis of what became known as the "Carnot cycle" in his book entitled *Réflexion sur la Puissance Motrice du Feu et sur les Machines Propres à Développer Cette Puissance*, where also he introduced the concept of a nonmolecular fluid, the *caloric*, as the working substance of heat engines.

was given to it in experiments reviewed by Lord Kelvin [9] who showed that atoms could be counted and weighed, so that the molecular structure of matter was soon to become an accepted fact among scientists. From there it was only a small step to realize the relation between heat as a key quantity in thermodynamics and kinetic energy of individual molecules [10].

1.2 Deformation of macroscopic bodies

In the subsequent analysis of thermal systems, a key issue will be a transformation of energy between its various forms, namely heat and work. The latter term refers to energetic changes in the state of a thermal system on account of its interaction with the environment through mechanical means (i.e., pistons). In other words, the environment imposes external forces on the system to which it responds by deformation. In the context of this book, two types of forces are relevant: those changing the volume of the body (i.e., compressional/dilational forces) and those changing its shape (i.e., shear forces). The purpose of this section is to develop a rigorous (macroscopic) description of the changes of the state of a macroscopic elastic body under these strains and to calculate the associated stresses. Our treatment follows in spirit the discussion in Chapter 3.6 of Arfken's book [11].

1.2.1 Strain tensor

Consider two mass elements of an elastic body in an unstrained reference state. One of them is located at a point r_0 and the other one at a point $r_0 + \delta r_0$. If the body is strained, the first one changes position according to

$$r = r_0 + u(r_0) \tag{1.1}$$

whereas the other one is displaced simultaneously by $u(r_0 + \delta r_0)$. Therefore, in the strained state, deformation of the body may be described by the quantity

$$\delta u \equiv u(r_0 + \delta r_0) - u(r_0) \tag{1.2}$$

The dependence of the vector *field* u on the position of the mass elements in the unstrained reference state is a necessary prerequisite to describe changes in the shape of the elastic body. Note that if u would be a vector (independent of the position of the mass elements), all mass elements would be displaced in the same direction and by the same amount $|u|$. Hence, we would be dealing with a rigid rather than an elastic body incapable of changing its shape. In other words, the body would only be capable of moving in space as an undeformable entity.

If the two mass elements considered above are separated by an infinitesimally small distance $|\delta r_0|$ in the unstrained reference state, it makes sense to approximate the displacement $u\left(r_0 + \delta r_0\right)$ by a Taylor series expansion of $u\left(r_0\right)$ according to

$$u\left(r_0 + \delta r_0\right) = u\left(r_0\right) + \sum_{n=1}^{\infty} \frac{1}{n!} \left(\delta r_0 \cdot \nabla_{r_0}\right)^n u\left(r_0\right) \tag{1.3}$$

where $\nabla_{r_0}^{\mathrm{T}} \equiv \left(\partial/\partial r_{0x}, \partial/\partial r_{0y}, \partial/\partial r_{0z}\right)$. Retaining in this expansion only terms up to first order in δr_0 (i.e., $n = 1$), we obtain from Eqs. (1.1) and (1.3)

$$\delta u = \nabla_{r_0} u\left(r_0\right) \cdot \delta r_0 \tag{1.4}$$

In Eq. (1.4), the dyad

$$\left(\nabla_{r_0} u\right)_{\alpha\beta} = \frac{\partial u_\alpha\left(r_0\right)}{\partial r_{\beta 0}} = \frac{\partial\left(r_\alpha - r_{\alpha 0}\right)}{\partial r_{\beta 0}} \tag{1.5}$$

is a second-rank tensor in component notation, where r_α and $r_{\alpha 0}$ are the (Cartesian) α-components of the vectors r and r_0, respectively.

To proceed it is convenient to split the tensor $\nabla_{r_0} u$ into a symmetric $\left(\sigma\right)$ and antisymmetric part $\left(\eta\right)$ via

$$\left(\nabla_{r_0} u\right)_{\alpha\beta} = \frac{1}{2}\left(\frac{\partial u_\alpha}{\partial r_{\beta 0}} + \frac{\partial u_\beta}{\partial r_{\alpha 0}}\right) - \frac{1}{2}\left(\frac{\partial u_\beta}{\partial r_{\alpha 0}} - \frac{\partial u_\alpha}{\partial r_{\beta 0}}\right) \equiv \sigma_{\alpha\beta} - \eta_{\alpha\beta} \tag{1.6}$$

At this point it is instructive to analyze η in more detail. To this end it is useful to introduce a vector

$$\begin{aligned}
\xi &\equiv \nabla_{r_0} \times u\left(r_0\right) \\
&= \left(\frac{\partial u_z}{\partial r_{y0}} - \frac{\partial u_y}{\partial r_{z0}}\right)\hat{e}_x + \left(\frac{\partial u_x}{\partial r_{z0}} - \frac{\partial u_z}{\partial r_{x0}}\right)\hat{e}_y + \left(\frac{\partial u_y}{\partial r_{x0}} - \frac{\partial u_x}{\partial r_{y0}}\right)\hat{e}_z
\end{aligned} \tag{1.7}$$

such that [see Eqs. (1.4) and (1.6)]

$$\eta \cdot \delta r_0 = \xi \times \delta r_0 \tag{1.8}$$

where $\hat{e}_x$, $\hat{e}_y$, and $\hat{e}_z$ are unit vectors along the x-, y-, and z-axis of the Cartesian coordinate system, respectively. Equation (1.8) has a lucid physical interpretation. It shows that the antisymmetric part of the displacement tensor $\nabla_{r_0} u\left(r_0\right)$ describes a rotation about an instantaneous axis through the mass element at r_0 in the unstrained state in the direction of ξ by $|\xi|$

radians [11]. As we wish to focus on deformations of the elastic body and have already disregarded mere translations by Eq. (1.1) we shall henceforth ignore the antisymmetric part of the displacement tensor $\boldsymbol{\nabla}_{\boldsymbol{r}_0} \boldsymbol{u}\,(\boldsymbol{r}_0)$. Instead we consider only its symmetric part satisfying [see Eq. (1.6)] the symmetry relation

$$\sigma_{\alpha\beta} = \sigma_{\beta\alpha}, \qquad \forall \alpha \neq \beta \tag{1.9}$$

where we refer to $\boldsymbol{\sigma}$ as the strain tensor henceforth.

Let us illustrate the above analysis by a specific example where

$$\delta \boldsymbol{r}_0 = \delta r_{x0} \widehat{\boldsymbol{e}}_x \tag{1.10}$$

i.e., we consider two mass elements separated by a small distance δr_{x0} along the x-axis in the unstrained reference state. Because

$$\frac{\delta \boldsymbol{u}}{\delta r_{x0}} = \frac{\boldsymbol{\sigma} \cdot \delta \boldsymbol{r}_0}{\delta r_{x0}} = \boldsymbol{\sigma} \cdot \widehat{\boldsymbol{e}}_x = \begin{pmatrix} \sigma_{xx} & \sigma_{xy} & \sigma_{xz} \\ \sigma_{xy} & \sigma_{yy} & \sigma_{yz} \\ \sigma_{xz} & \sigma_{yz} & \sigma_{zz} \end{pmatrix} \cdot \begin{pmatrix} 1 \\ 0 \\ 0 \end{pmatrix} = \begin{pmatrix} \sigma_{xx} \\ \sigma_{xy} \\ \sigma_{xz} \end{pmatrix} \tag{1.11}$$

$\delta \boldsymbol{u}$ has three nonzero components. Therefore, in the strained state, the two mass elements are separated by

$$\delta \boldsymbol{r} = \delta \boldsymbol{r}_0 + \delta \boldsymbol{u} = \delta r_{x0} \begin{pmatrix} \sigma_{xx} + 1 \\ \sigma_{xy} \\ \sigma_{xz} \end{pmatrix} \tag{1.12}$$

so that the two mass elements, which were originally located at the same point in the y–z plane are now displaced also along the y- and z-axis. Therefore, we conclude that diagonal elements of the matrix representing the strain tensor describe compression ($\sigma_{\alpha\alpha} < 0$) or dilatation ($\sigma_{\alpha\alpha} > 0$) of the elastic body, whereas off-diagonal components of $\boldsymbol{\sigma}$ represent shear deformations.

1.2.2   Stress tensor

In the previous section, we focused on deformations of elastic bodies. Clearly, such deformations are the consequence of external forces acting on the body.[3] These forces can be cast in terms of a set of stresses $\{\tau_{\alpha\beta}\}$ acting on the faces of an infinitesimally small parallelepiped, $\tau_{\alpha\beta} dA_\beta$, where A_β is the area of such a face whose normal is pointing in the β-direction (see Fig. 1.1). As we would again like to disregard mere translation of the parallelepiped in space, certain symmetry relations between the stresses must hold, which are illustrated in Fig. 1.1. Moreover, we assume

---

[3]This argument is based on the implicit but plausible assumption of validity of the causality principle.

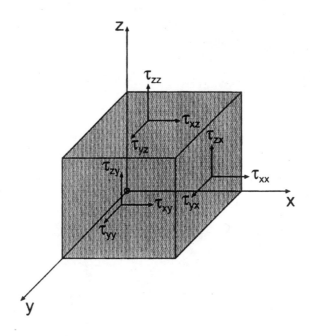

Figure 1.1: Sketch of a parallelepiped (i.e., the fluid lamella) on which various stresses $\tau_{\alpha\beta}$ are acting. The quantity $\tau_{\alpha\beta}dA_\beta$ is the α-component of the force exerted on the β-directed face of the parallelepiped. Its direction is represented by the arrows. For the sake of clarity, only those stresses acting on the front faces are displayed.

1. The stress to be homogeneous throughout the body.

2. A state of mechanical stability to exist.

3. Absence of external body forces (such as gravity) or body torques (such as magnetic fields).

As in Section 1.2.1 we would also like to disregard rotations of the parallelepiped as a result of the external forces acting on it. That is, we wish all torques about the three axes to vanish [11].

Consider as a specific example the torque acting on the parallelepiped about the z-axis. The normal stresses $\tau_{\alpha\alpha}$ do not contribute to this torque. Stresses τ_{zx} and τ_{zy} do not contribute because they point in the z-direction (see Fig. 1.2). Similarly, τ_{xz} and τ_{yz} cannot add to the net torque because they are balanced by stresses that are equal in magnitude but point in the opposite direction on the bottom plane $z = 0$. This then leaves us with two remaining contributions to the net torque, namely $\tau_{yx}\,(\mathrm{d}y\mathrm{d}z)\,\mathrm{d}x$ and $\tau_{xy}\,(\mathrm{d}x\mathrm{d}z)\,\mathrm{d}y$.

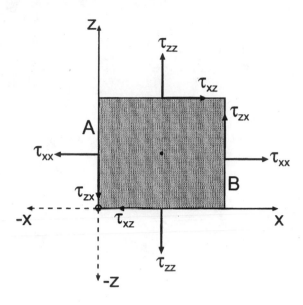

Figure 1.2: Side view of the parallelepiped sketched in Fig. 1.1.

These torques need to satisfy the equation

$$\tau_{yx}\,(\mathrm{d}y\mathrm{d}z)\,\mathrm{d}x = \tau_{xy}\,(\mathrm{d}x\mathrm{d}z)\,\mathrm{d}y \tag{1.13}$$

if we wish the parallelepiped to stay at rest, that is in the absence of rotation around the z-axis. Because in Eq. (1.13), $\mathrm{d}x\mathrm{d}y\mathrm{d}z$ is arbitrary, $\tau_{yx} = \tau_{xy}$ follows without further ado. The above argument may be applied also to rotations around the x- and y-axis so that in general the stresses must satisfy the symmetry relation

$$\tau_{\alpha\beta} = \tau_{\beta\alpha}, \qquad \forall \alpha \neq \beta \tag{1.14}$$

In other words, the set $\{\tau_{\alpha\beta}\}$ can be represented by a symmetric matrix if we consider only external forces that cause deformations (i.e., compression/dilatation or shear) of the unstrained parallelepiped in its reference state. Notice also that, on account of our argumentation above, Eq. (1.14) is not a statement about equality of *direction* but rather one about *magnitude* of stresses.

What remains to be shown is that the *matrix* $\{\tau_{\alpha\beta}\}$ is actually a representation of a second-rank *tensor* $\boldsymbol{\tau}$ to which we shall henceforth refer as the stress tensor. We need to demonstrate that the matrix representing $\boldsymbol{\tau}$ satisfies transformation properties under rotation of the coordinate system that constitute a second–rank tensor. To this end consider an infinitesimally

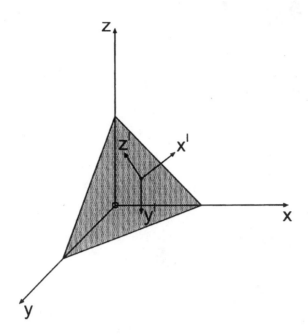

Figure 1.3: Sketch of an infinitesimal tetrahedron whose three faces coincide with the x–y, x–z, and y–z planes of the original (unprimed) Cartesian coordinate system. The third slant face appears to be oriented such that the axis x' is normal to the area of the slant face. The two remaining axes, y' and z' lie in the plane of the slant face but are orthogonal to one another as well as to the x'-axis. Hence, x', y', and z' define a Cartesian coordinate system rotated with respect to the original one.

small tetrahedron plotted in Fig. 1.3. Three of its (triangular) faces coincide with the planes located at $x = 0$, $y = 0$, and $z = 0$, respectively, of the original (i.e., unprimed) coordinate system. The fourth face of area $\mathrm{d}A$ is slanted with respect to the original coordinate system. The fourth face may be employed to define a second (Cartesian) coordinate system whose axis x', say, points in a direction normal to $\mathrm{d}A$; the other two axes, y' and z', lie in the plane of area $\mathrm{d}A$ such that any one axis in the primed coordinate system is orthogonal to the other two just as any axis in the original coordinate system is orthogonal to the other two. Clearly, both coordinate systems are rotated with respect to one another where the orientation of the primed coordinate system relative to the unprimed one may be expressed in terms of the matrix of direction cosines $\mathbf{a}$, which is the cosine of the angle between any pair of axes, one in the primed and the other one in the unprimed coordinate system (see Section 4.3 in [11]).

From this description and the previous discussion in this section, it is clear that in the primed coordinate system forces acting on the slant face of the tetrahedron are given by $\tau_{ij}\mathrm{d}A$. We shall use Latin letters to refer to properties pertaining to the rotated coordinate system and Greek letters for the corresponding original coordinate system. Given the forces $\tau_{ij}\mathrm{d}A$ we obtain the corresponding ones $\tau_{\alpha\beta}a_{j\beta}\mathrm{d}A$ in the original coordinate system, where $a_{j\beta}\mathrm{d}A$ is the area $\mathrm{d}A$ of the slant face projected onto the plane $\beta = 0$. The force $\tau_{\alpha\beta}a_{j\beta}\mathrm{d}A$ points along the α-axis. Its component along the i-axis of the rotated coordinate system is given by $\tau_{\alpha\beta}a_{i\alpha}a_{j\beta}\mathrm{d}A$. Summing these forces over α gives the sum of the i-components of the three forces on the planes $\beta = 0$ in the i-direction. Finally, summing over all three planes $\beta = 0$ the *total* force along the i-axis is given by

$$\tau_{ij}\mathrm{d}A = \sum_{\alpha}\sum_{\beta}\tau_{\alpha\beta}a_{i\alpha}a_{j\beta}\mathrm{d}A \qquad (1.15)$$

and because $\mathrm{d}A$ is arbitrary

$$\tau_{ij} = \sum_{\alpha}\sum_{\beta}\tau_{\alpha\beta}a_{i\alpha}a_{j\beta} \qquad (1.16)$$

which is the transformation that a matrix representing a tensor of second rank must satify [11].

Finally, the reader should appreciate a significant difference between the way in which $\boldsymbol{\sigma}$ and $\boldsymbol{\tau}$ were introduced. In Section 1.2.1 the strain tensor was defined on purely mathematical grounds, whereas the conjugate stress tensor was introduced by purely physical reasoning (i.e., force balance). However, both $\boldsymbol{\sigma}$ and $\boldsymbol{\tau}$ are defined such that mere translations or rotations of a macroscopic elastic body are explicitly excluded. As far as $\boldsymbol{\sigma}$ is concerned, this is effected by introducing the displacement of mass elements $\boldsymbol{u}\,(\boldsymbol{r})$ as a vector field and by defining $\boldsymbol{\sigma}$ as the symmetric part of the displacement tensor $\boldsymbol{\nabla}\boldsymbol{u}\,(\boldsymbol{r})$ [see Eqs. (1.1) and (1.6)]; for the stress tensor $\boldsymbol{\tau}$, the symmetry property stated in Eq. (1.14) serves to eliminate rotations.

Employing the concepts of stress and conjugate strain, and their proper mathematical formulation as second-rank tensors, now enables us to deal with mechanical work in a general anisotropic piece of matter. One realization of such a system are fluids in confinement to which this book is devoted. However, at the core of our subsequent treatment are thermal properties of confined fluids. In other words, we need to understand the relation between mechanical work represented by stress-strain relationships and other forms of energy such as heat or chemical work. This relation will be formally established within the framework of (equilibrium or phenomenological) thermodynamics to which we shall turn now.

1.3   Gibbs fundamental equation

Perhaps the most important concepts of the axiomatic foundation of thermodynamics are the ones referred to as the First and Second Laws dealing with the internal energy $\mathcal{U}$ and the entropy $\mathcal{S}$. They are essentially statements dealing with energy conservation and the transformation of one form of energy (e.g., work) into another one (e.g., heat). If combined, the First and Second Laws give rise to the so-called Gibbs fundamental equation

$$d\mathcal{U} = Td\mathcal{S} + dW \tag{1.17}$$

where T is the absolute temperature and dW is an infinitesimal amount of work exchanged between the thermodynamic system and its environment. Following standard practice we refer to T as "absolute" temperature because it is defined such that any physical properties of the measuring device (e.g., thermal expansivity of a thermometer substance) are completely irrelevant. This reemphasizes our earlier argument that thermodynamics as a physical theory has no concept of matter. Incidentally, the only difficult and challenging step required in the analysis of *any* thermodynamic problem is to identify the system of interest, its environment, and to specify the interaction between the two. The distinction between *system* and *environment* is the key order principle of thermodynamics.

Mathematically speaking, both $d\mathcal{U}$ and $d\mathcal{S}$ are exact differentials; that is, they satisfy the equation

$$\oint_C d\mathcal{X} = 0, \qquad \mathcal{X} = \mathcal{U}, \mathcal{S} \tag{1.18}$$

along any arbitrary closed path in the space spanned by the *natural* variables (see Section 1.4.2) on which $\mathcal{X}$ depends (i.e., state space). Thermodynamic functions $\mathcal{X}$ satisfying Eq. (1.18) are referred to as *state* functions because they depend only on initial and final states of the system but not on the path connecting the two in state space. On the contrary

$$\oint_C dW = f(C) \tag{1.19}$$

is the amount of work exchanged between the system and its environment and depends on *how* this exchange is being effected. In other words, a state function W does not exist.

So far nothing specific has been said about the nature of the work exchanged between the system and its environment. In this book we shall restrict ourselves to two types of work. The first is chemical work

$$dW_{\text{chem}} = \mu dN \qquad (1.20)$$

in which matter can be exchanged between the system and its environment. In Eq. (1.20), μ is the chemical potential and N is the number of molecules accommodated by the system. The somewhat abstract quantity $\mu < 0$ may be viewed as the amount of work required to add a new molecule to the system from some external reservoir of matter (i.e., the environment).[4] Hence, we are dealing with *open* systems in general. Note, that the use of the term *molecule* does not contradict our expository remarks concerning molecular concepts in thermodynamics because nothing is being said about the (microscopic) properties of molecules at this point.

The second type of work to which we shall restrict our discussion is *mechanical* work. As we shall see later, confined phases can be exposed to two types of mechanical work, namely compression (dilation) and shear. In that regard, confined phases have a lot in common with bulk solids in that they are generally inhomogeneous and anisotropic in one or more spatial dimensions. Therefore, it seems sensible to cast the mechanical work term in terms of stress (τ) and dimensionless strain tensors (σ) introduced in Section 1.2 and suggested earlier by Callen for a proper treatment of the thermodynamics of bulk solids [12], by writing

$$dW_{\text{mech}} = V_0 \text{Tr}\left(\tau d\sigma\right) \qquad (1.21)$$

where V_0 is the volume of an undeformed (uncompressed and unsheared) reference system. Thus, in its most general form suitable for this book, the Gibbs fundamental equation may be cast as

$$d\mathcal{U} = Td\mathcal{S} + \mu dN + V_0 \text{Tr}\left(\tau d\sigma\right) \qquad (1.22)$$

which follows from Eqs. (1.17), (1.20), and (1.21). In Eq. (1.22), "Tr" represents the trace of the product of the matrices representing τ and $d\sigma$ (see Appendix A.1). As we saw in Sections 1.2.1 and 1.2.2, τ and $d\sigma$ can be

---

[4]If $dN < 0$, $dW_{\text{chem}} > 0$; i.e., the environment does work *on* the thermodynamic system.

represented by symmetric 3×3 matrices, namely

$$
\boldsymbol{\tau} \;=\; \begin{pmatrix} \tau_{xx} & \frac{1}{2}\tau_{xy} & \frac{1}{2}\tau_{xz} \\ \frac{1}{2}\tau_{xy} & \tau_{yy} & \frac{1}{2}\tau_{yz} \\ \frac{1}{2}\tau_{xz} & \frac{1}{2}\tau_{yz} & \tau_{zz} \end{pmatrix}
\tag{1.23a}
$$

$$
\mathrm{d}\boldsymbol{\sigma} \;=\; \begin{pmatrix} \mathrm{d}\sigma_{xx} & \mathrm{d}\sigma_{xy} & \mathrm{d}\sigma_{xz} \\ \mathrm{d}\sigma_{xy} & \mathrm{d}\sigma_{yy} & \mathrm{d}\sigma_{yz} \\ \mathrm{d}\sigma_{xz} & \mathrm{d}\sigma_{yz} & \mathrm{d}\sigma_{zz} \end{pmatrix}
\tag{1.23b}
$$

As we also showed in Sections 1.2.1 and 1.2.2, both tensors have a lucid physical interpretation. Components $\tau_{\alpha\beta}$ of the stress tensor $\boldsymbol{\tau}$ may be perceived as the β-component of the force acting on an α-directed area; likewise, diagonal components of $\mathrm{d}\boldsymbol{\sigma}$ account for (infinitesimal) compressional/dilational strains acting on the system, whereas off-diagonal components represent various shear strains. By convention, $\tau_{\alpha\beta} < 0$ if the force acting on the α-directed area point outward. In Eq. (1.23a), the factor $\frac{1}{2}$ arises for off-diagonal elements of $\boldsymbol{\tau}$ to give the correct magnitude of shear contributions to the mechanical work.

1.3.1   Bulk fluids

Let us apply the above general formalism to two simple examples that are central to this book chapter, namely that of a bulk fluid and a fluid confined to a slit-pore (see Sections 1.3.2 and 1.3.3). In both cases, we take as the reference system a rectangular prism of volume $V_0 = s_{x0}s_{y0}s_{z0}$, where a body-fixed coordinate system is employed such that the faces of the prism coincide with the planes $x = \pm s_{x0}/2$, $y = \pm s_{y0}/2$, and $z = \pm s_{z0}/2$. If the unstrained system is exposed to an infinitesimally small compressional or shear strain, $V_0 \to V = s_x s_y s_z$. This implies that a mass element originally at a point $\boldsymbol{r}_0$ in the unstrained system changes position to a point $\boldsymbol{r}$ in the strained system.

As most readers may not be familiar with this formulation, let us illustrate the approach by considering a specific situation, namely that of a homogeneous bulk fluid. By definition, at thermodynamic equilibrium, bulk fluids cannot sustain any shear strain, that is, $\tau_{\alpha\beta} = 0$ for all $\alpha \neq \beta$. Thus,

$$
\boldsymbol{r} = \mathbf{D}\boldsymbol{r}_0
\tag{1.24}
$$

where the transformation is effected by the matrix

$$
\mathbf{D} = \begin{pmatrix} s_x/s_{x0} & 0 & 0 \\ 0 & s_y/s_{y0} & 0 \\ 0 & 0 & s_z/s_{z0} \end{pmatrix}
\tag{1.25}
$$

From Eq. (1.1) we then find

$$u_\alpha \left(r_{\alpha 0}\right) = r_\alpha - r_{\alpha 0} = \frac{s_\alpha - s_{\alpha 0}}{s_{\alpha 0}} r_{\alpha 0} \tag{1.26}$$

such that from the definition of $\boldsymbol{\sigma}$ given in Eq. (1.6) we have the explicit expression

$$\boldsymbol{\sigma} = \begin{pmatrix} \left(s_x - s_{x0}\right)/s_{x0} & 0 & 0 \\ 0 & \left(s_y - s_{y0}\right)/s_{y0} & 0 \\ 0 & 0 & \left(s_z - s_{z0}\right)/s_{z0} \end{pmatrix} \tag{1.27}$$

At this point it is important to realize that by definition bulk fluids are homogeneous on account of the absence of any external fields; that is, the density of a bulk fluid is spatially constant. This implies that its properties are translationally invariant in all three spatial dimensions. Hence,

$$\boldsymbol{\tau} = \tau_b \mathbf{1} \tag{1.28}$$

where the scalar τ_b represents the compressional (dilational) stress exerted on the bulk fluid (in any spatial direction) and $\mathbf{1}$ is the unit tensor. Moreover, to preserve the isotropy of the system [see Eq. (1.28)] under compressional (dilational) strains, these cannot be independent but must satisfy

$$\frac{s_x}{s_{x0}} = \frac{s_y}{s_{y0}} = \frac{s_z}{s_{z0}} = \frac{\sigma_b}{3} + 1 \tag{1.29}$$

where σ_b is defined through the expression

$$\boldsymbol{\sigma} = \frac{\sigma_b}{3} \mathbf{1} \tag{1.30}$$

similar to τ_b in Eq. (1.28). From Eqs. (1.21) and (1.29), we therefore have

$$dW_{\text{mech}} = \tau_b V_0 d\sigma_b \equiv -P_b dV \tag{1.31}$$

which also defines the bulk pressure through the relation $\tau_b = -P_b$. From Eqs. (1.17) and (1.31), we finally arrive at the well-known form of the Gibbs fundamental equation for homogeneous, isotropic bulk fluids, namely

$$d\mathcal{U} = Td\mathcal{S} + \mu dN - P_b dV \tag{1.32}$$

1.3.2   Slit-pore with unstructured substrate surfaces

The discussion of bulk phases in the preceding section is, of course, presented solely for didactic reasons because it demonstrates that the somewhat more involved formulation of mechanical work in terms of stresses and conjugate strains [see Eq. (1.21)] leads to well-known textbook results like Eq. (1.32). Hence, it should serve to help the reader get familiar with the current treatment of mechanical work, which turns out to be particularly useful for confined fluids as we shall demonstrate now.

In this section we are considering one of the systems of key interest in this work, namely that of a nanoscopic slit-pore with chemically homogeneous substrate surfaces. For simplicity we disregard the atomic structure of the solid substrate such that the external field exerted on the confined fluid depends only on the distance of a mass element from the substrate surface. In this case, the fluid confined by the solid surfaces is anisotropic on account of the external potential represented by the pore walls. Assuming these walls to be parallel with the x–y plane, the confined fluid is inhomogeneous along the z-direction; that is, its local density depends on the position (z) with respect to the confining planar substrate surfaces. However, the fluid is *homogeneous* across all planes parallel with the substrate surfaces (at different locations z). Hence, in two dimensions (x and y), isotropy is preserved despite confinement. In other words, compared with the previously discussed bulk, the confined phase has lower symmetry. However, note that like the bulk, the currently described confined phase cannot be exposed to a shear strain because the external field depends only on the distance of a mass element from the substrate surfaces.

The reduced symmetry is reflected by the relation

$$\frac{s_x}{s_{x0}} = \frac{s_y}{s_{y0}} = \frac{\sigma_\parallel}{2} + 1 \tag{1.33}$$

which replaces Eq. (1.29) so that

$$\mathsf{d}\left(V_0\sigma_{xx}\right) = s_{z0}\mathsf{d}\left(s_x s_{y0}\right) = \frac{s_{z0}}{2}\mathsf{d}\left(\sigma_\parallel A_0 + 2A_0\right) = \frac{s_{z0}}{2}\mathsf{d}A = \mathsf{d}\left(V_0\sigma_{yy}\right) \tag{1.34a}$$

$$\mathsf{d}\left(V_0\sigma_{zz}\right) = A_0\mathsf{d}s_z \tag{1.34b}$$

where $A_0 = s_{x0}s_{y0}$ is the (constant) area of the z-directed face of the unstrained reference system; likewise, $A = A_0\sigma_\parallel$ represents the area of the z-directed face in the strained system. Because of Eqs. (1.33) and (1.34), the mechanical work term can then be cast as

$$
V_0 \mathrm{Tr}\left(\boldsymbol{\tau} \mathrm{d}\boldsymbol{\sigma}\right) = \frac{1}{2}\mathrm{Tr}\begin{pmatrix} \tau_{xx} & \frac{1}{2}\tau_{xy} & \frac{1}{2}\tau_{xz} \\ \frac{1}{2}\tau_{xy} & \tau_{yy} & \frac{1}{2}\tau_{yz} \\ \frac{1}{2}\tau_{xz} & \frac{1}{2}\tau_{yz} & \tau_{zz} \end{pmatrix}\begin{pmatrix} s_{z0}\mathrm{d}A & 0 & 0 \\ 0 & s_{z0}\mathrm{d}A & 0 \\ 0 & 0 & 2A_0\mathrm{d}s_z \end{pmatrix}
$$
$$
= \tau_\parallel s_{z0}\mathrm{d}A + \tau_\perp A_0\mathrm{d}s_z \tag{1.35}
$$

where we employ the symmetry of the system to define

$$
\tau_\parallel \equiv \tau_{xx} = \tau_{yy} \tag{1.36a}
$$
$$
\tau_\perp \equiv \tau_{zz} \tag{1.36b}
$$

and therefore have [see Eq. (1.22)]

$$
\mathrm{d}\mathcal{U} = T\mathrm{d}\mathcal{S} + \mu\mathrm{d}N + \tau_\parallel s_{z0}\mathrm{d}A + \tau_\perp A_0\mathrm{d}s_z \tag{1.37}
$$

Equation (1.37) is the Gibbs fundamental equation specialized to a fluid confined to a slit-pore with chemically homogeneous, (infinitesimally) smooth substrate surfaces.

1.3.3 Slit-pore with structured substrate surfaces

The next slightly more complicated situation concerns a fluid confined to a nanoscopic slit-pore by *structured* rather than *unstructured* solid surfaces. For the time being, we shall restrict the discussion to cases in which the symmetry of the external field (represented by the substrates) preserves translational invariance of fluid properties in one spatial dimension. An example of such a situation is depicted in Fig. 5.7 (see Section 5.4.1) showing substrates endowed with a chemical structure that is periodic in one direction (x) but quasi-infinite (i.e., macroscopically large) in the other one (y).

A new feature entering the picture is that fluids confined by substrates of the kind illustrated by the sketch in Fig. 5.7 can be sheared in addition to mere compression (dilatation). This is because the relative alignment of the substrates matters on account of the periodicity of their structure in the x-direction. In this case, the transformation matrix assumes the form

$$
\mathbf{D} = \begin{pmatrix} s_x/s_{x0} & 0 & \alpha s_{x0}/s_{z0} \\ 0 & s_y/s_{y0} & 0 \\ \alpha s_{x0}/s_{z0} & 0 & s_z/s_{z0} \end{pmatrix} \tag{1.38}
$$

so that

$$
\boldsymbol{r} = \left(\frac{s_x}{s_{x0}}r_{x0} + \frac{\alpha s_{x0}}{s_{z0}}r_{z0}\right)\widehat{\boldsymbol{e}}_x + \frac{s_y}{s_{y0}}r_{y0}\widehat{\boldsymbol{e}}_y + \left(\frac{\alpha s_{x0}}{s_{z0}}r_{x0} + \frac{s_z}{s_{z0}}r_{z0}\right)\widehat{\boldsymbol{e}}_z \tag{1.39}
$$

and therefore [see Eq. (1.1)]

$$
\begin{aligned}
\boldsymbol{u} &= \left(\frac{s_x - s_{x0}}{s_{x0}} r_{x0} + \frac{\alpha s_{x0}}{s_{z0}} r_{z0} \right) \widehat{\boldsymbol{e}}_x + \frac{s_y - s_{y0}}{s_{y0}} r_{y0} \widehat{\boldsymbol{e}}_y \\
&\quad + \left(\frac{\alpha s_{x0}}{s_{z0}} r_{x0} + \frac{s_z - s_{z0}}{s_{z0}} r_{z0} \right) \widehat{\boldsymbol{e}}_z
\end{aligned}
\tag{1.40}
$$

Thus, $\boldsymbol{\sigma}$ assumes slightly more complicated forms, namely

$$
\boldsymbol{\sigma} = \begin{pmatrix} (s_x - s_{x0})/s_{x0} & 0 & \alpha s_{x0}/s_{z0} \\ 0 & (s_x - s_{x0})/s_{x0} & 0 \\ \alpha s_{x0}/s_{z0} & 0 & (s_z - s_{z0})/s_{z0} \end{pmatrix}
\tag{1.41}
$$

where $0 \leq \alpha \leq \frac{1}{2}$ is a dimensionless parameter specifying the relative alignment of the two substrates in the x-direction. The substrates are perfectly aligned for $\alpha = 0$, whereas their misalignment is maximum for $\alpha = \frac{1}{2}$ because of the periodicity of the chemical pattern. Note also that the symmetry of the confined fluid is once again reduced with respect to the case discussed in Section 1.3.2. This is reflected by the inequality

$$
\tau_{xx} \neq \tau_{yy} \neq \tau_{zz}
\tag{1.42}
$$

With Eqs. (1.41) the specialized Gibbs fundamental equation [see Eq. (1.22)] can be written as

$$
\mathrm{d}\mathcal{U} = T\mathrm{d}\mathcal{S} + \mu \mathrm{d}N + A_{x0}\tau_{xx}\mathrm{d}s_x + A_{y0}\tau_{yy}\mathrm{d}s_y + A_{z0}\tau_{zz}\mathrm{d}s_z + A_{z0}\tau_{xz}\mathrm{d}\left(\alpha s_{x0}\right)
\tag{1.43}
$$

where $A_{\alpha 0}$ denotes the area of the α-directed face of undeformed reference system.

1.4 Equilibrium states and thermodynamic potentials

1.4.1 Conditions for thermodynamic equilibrium

To this point we have been concerned with various expressions for mechanical work depending on the symmetry of the thermodynamic system. A key issue in this regard was the unstrained reference system characterized by quantities like $s_{\alpha 0}$, $A_{\alpha 0}$, and V_0. By external agents (i.e., by exposing the system to mechanical or chemical work), the reference system may be transformed into a strained system that may or may not be in a state of thermodynamic

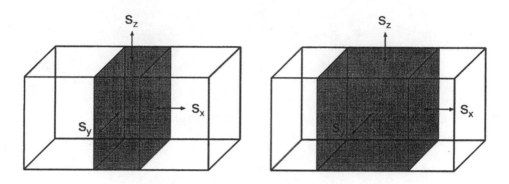

Figure 1.4: Sketch of a fluid lamella (shaded area) whose faces in x-, y-, and z-direction can be moved independently in the direction of the double arrows. As an example we show states of the lamella differing in strain in the x-direction such that the right plot shows an expanded lamella (relative to the plot on the left side) of different width in that direction.

equilibrium. Thus, it is the purpose of this section to develop criteria that allow us to identify such equilibrium states.

A key issue in establishing such criteria is to distinguish between the thermodynamic system and its environment and to specify the interaction mechanism between the two. In this case we shall take as the system a finite lamella of the (infinitely large) confined fluid, regarding the (infinitely large) remainder of the lamella as its environment. The faces separating the lamella from the environment can be moved as indicated in Fig. 1.4 by the double arrows. The lamella is therefore assumed to be coupled to its environment thermally, materially, and mechanically. Criteria for thermodynamic equilibrium of the lamella can be established by viewing it *plus* the environment as an *isolated* supersystem. That is, if temperature, chemical potential, and strains are to be maintained constant by means of external agents (i.e., virtual "pistons"), we wish to find the appropriate thermodynamic function attaining a global minimum when the lamella is in a state of thermodynamic equilibrium.

Because the supersystem is isolated and rigid by supposition, its entropy must be constant in this equilibrium state and its internal energy must be minimum according to the Second and First Laws, respectively. Hence, an infinitesimal virtual transformation δ that would take the supersystem from this state must satisfy

$$\delta \left(\mathcal{U} + \mathcal{U}' \right) \;\geq\; 0 \tag{1.44a}$$

$$\delta \left(\mathcal{S} + \mathcal{S}' \right) \;\geq\; 0 \tag{1.44b}$$

where unprimed and primed quantities refer to the lamella and its environment, respectively, and the equal sign holds if the transformation brings back the system to the original equilibrium state. It is important to note that Eqs. (1.44) constitute nothing but special versions of a general extremum principle applying to many different problems and fields in physics. Examples are Gauss' principle of least constraint [13] or Hamilton's principle leading to the Lagrangian equations of motion in classic mechanics [14].

From Gibbs fundamental equation [see Eq. (1.22)], it follows that

$$\delta S = \frac{1}{T}\delta \mathcal{U} - \frac{\mu}{T}\delta N - \frac{V_0}{T}\text{Tr}\left(\boldsymbol{\tau}\delta\boldsymbol{\sigma}\right) \tag{1.45a}$$

$$\delta S' = \frac{1}{T'}\delta \mathcal{U}' - \frac{\mu'}{T'}\delta N' - \frac{V_0'}{T'}\text{Tr}\left(\boldsymbol{\tau}'\delta\boldsymbol{\sigma}'\right) \tag{1.45b}$$

The supersystem is materially closed such that

$$\delta N = -\delta N' \tag{1.46}$$

and it is assumed rigid so that

$$V_0\delta\boldsymbol{\sigma} = -V_0'\delta\boldsymbol{\sigma}' \tag{1.47}$$

also holds. Because the supersystem is in a state of thermodynamic equilibrium, we have from Eqs. (1.44), (1.46), and (1.47) the expression

$$0 = \left(\frac{1}{T} - \frac{1}{T'}\right)\delta \mathcal{U} - \left(\frac{\mu}{T} - \frac{\mu'}{T'}\right)\delta N - \left(\frac{\boldsymbol{\tau}}{T} - \frac{\boldsymbol{\tau}'}{T'}\right)\delta\boldsymbol{\sigma} \tag{1.48}$$

As the transformations $\delta \mathcal{U}$, δN, and $\delta\boldsymbol{\sigma}$ are independent and arbitrary, Eq. (1.48) can only be satisfied if

$$T = T' \tag{1.49a}$$

$$\frac{\mu}{T} = \frac{\mu'}{T'} \rightarrow \mu = \mu' \tag{1.49b}$$

$$\frac{\boldsymbol{\tau}}{T} = \frac{\boldsymbol{\tau}'}{T'} \rightarrow \boldsymbol{\tau} = \boldsymbol{\tau}' \tag{1.49c}$$

hold simultaneously. These latter expressions constitute conditions of thermal [see Eq. (1.49a)], material [see Eq. (1.49b)], and mechanical equilibrium [see Eq. (1.49c)] that must be satisfied if the confined lamella is in thermodynamic equilibrium.

1.4.2 Thermodynamic potentials

In general, *any* transformation of the thermodynamic state of the confined lamella is associated with a variation of a characteristic function assuming a global minimum if the lamella is in thermodynamic equilibrium. However, there are various ways in which such a transformation may be effected in practice. Hence, the precise form of the characteristic function may vary between different (experimental) situations. Below we shall briefly discuss the characteristic functions that are key to the analysis of confined fluids.

1.4.2.1 Closed system, fixed strains

We begin with a lamella on which no compressional or shear stresses are acting; i.e., no mechanical work is exchanged between the lamella and its environment. Hence, $\delta\boldsymbol{\sigma} = \delta\boldsymbol{\sigma}' = \mathbf{0}$, where $\mathbf{0}$ is the zero tensor. In addition, the lamella is separated from its environment by a fictitious, impermeable, diathermal membrane so that the number of molecules that constitute the lamella remains fixed; that is, $\delta N = \delta N' = 0$. In other words, only heat is exchanged between the lamella and its surroundings. Under these conditions, we have from Eqs. (1.44), (1.45b), and (1.49a)

$$\delta\mathcal{U} + \delta\mathcal{U}' = \delta\mathcal{U} + T'\delta\mathcal{S}' = \delta\left(\mathcal{U} - T\mathcal{S}\right) \equiv \delta\mathcal{F} \geq 0 \qquad (1.50)$$

which defines the (Helmholtz) free energy $\mathcal{F}$. Equation (1.50) states, that under conditions of fixed N and $\boldsymbol{\sigma}$, equilibrium states of the lamella are characterized by a global minimum of $\mathcal{F}$.

1.4.2.2 Open system, fixed strains

Consider now a situation where the lamella is still not subject to any mechanical work but is permitted to exchange heat *and* matter with its environment. Thus, as in the previous example we still have $\delta\boldsymbol{\sigma} = \delta\boldsymbol{\sigma}' = \mathbf{0}$ and it follows from Eqs. (1.44), (1.45b), and (1.49a) that

$$\delta\mathcal{U} + \delta\mathcal{U}' = \delta\mathcal{U} + T'\delta\mathcal{S}' + \mu'\delta N' = \delta\left(\mathcal{U} - T\mathcal{S} - \mu N\right) \equiv \delta\Omega \geq 0 \qquad (1.51)$$

where Ω is the grand potential. Equation (1.51) tells us that, under conditions of fixed compressional (dilational) and shear strains, fixed temperature, and fixed chemical potential, the equilibrium state of the confined lamella is characterized by a global minimum of Ω. Experimental situations where Ω is the relevant thermodynamic potential are frequently encountered in the sorption of gases in nanoporous materials because the pore gas is coupled both thermally and materially to a bulk reservoir (see Section 4.2.1).

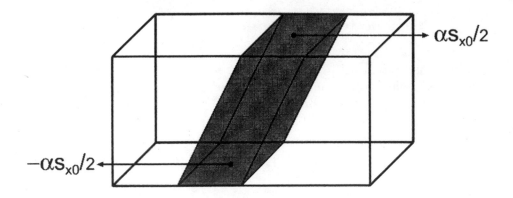

Figure 1.5: Sketch of a fluid lamella (shaded area) sheared in the x-direction, where a shear strain of $\pm\alpha s_{x0}/2$ is applied to the upper and lower substrate, respectively.

1.4.2.3 Closed system, nonvanishing strains

Last but not least we analyze a situation in which the lamella is still separated from its environment by a virtual impermeable, heat conducting membrane. However, this time we allow the confining substrates to move along the z-direction thereby exerting a nonvanishing compressional strain of the environment on the lamella. In addition, the relative alignment of the solid surfaces in the x–y plane may be altered by external agents. For concreteness we assume the solid surfaces to be structured as shown in Fig. 5.7, which shows that only the alignment in the x-direction can be altered. This situation, depicted schematically in Fig. 1.5, is frequently encountered in experiments employing the so-called *surface forces apparatus* by which mechanical properties and the phase behavior of nanoscopic films can be investigated (see Section 5.3.1). Thus, the strain tensor can be expressed as

$$\delta\boldsymbol{\sigma} = \begin{pmatrix} 0 & 0 & \delta\sigma_{xz} \\ 0 & 0 & 0 \\ \delta\sigma_{xz} & 0 & \delta\sigma_{zz} \end{pmatrix} \tag{1.52}$$

It is then easy to verify that the mechanical work term is [see Eq. (1.23a)]

$$\mathrm{Tr}\,(\boldsymbol{\tau}\delta\boldsymbol{\sigma}) = \tau_{xz}\delta\sigma_{xz} + \tau_{zz}\delta\sigma_{zz} \tag{1.53}$$

As we still have $\delta N = \delta N' = 0$, it is immediately clear from Eqs. (1.44), (1.45b), and (1.53) that in this case

$$\begin{aligned} \delta\mathcal{U} + \delta\mathcal{U}' &= \delta\mathcal{U} + T'\delta\mathcal{S}' + \tau'_{xz}\delta\sigma'_{xz} + \tau'_{zz}\delta\sigma'_{zz} \\ &= \delta\left(\mathcal{U} - T\mathcal{S} - \tau_{xz}\sigma_{xz} - \tau_{zz}\sigma_{zz}\right) \equiv \delta\mathcal{G} \geq 0 \end{aligned} \tag{1.54}$$

where $\mathcal{G}$ is a generalized Gibbs potential. If the lamella is in a state of thermodynamic equilibrium for fixed N, T, τ_{xz}, and τ_{zz}, $\mathcal{G}$ attains a global minimum. To arrive at the second line of Eq. (1.54), Eqs. (1.47), (1.49a), and (1.49c) have also been invoked. Note that, depending on the set of strains controlled, several such Gibbs potentials exist. They all have in common that, besides N and T, a subset of stresses is also controlled during a thermodynamic transformation of the confined lamella.

The above examples showed that, for a given set of variables, functions such as $\mathcal{U}$, $\mathcal{F}$, Ω, or $\mathcal{G}$ may be defined complying with the general extremum principle; that is, for a given set of variables, these functions are minimum if the system is in a state of thermodynamic equilibrium. Thus, by analogy with mechanical systems, functions such as $\mathcal{U}$, $\mathcal{F}$, Ω, or $\mathcal{G}$ are frequently referred to as *thermodynamic potentials* and their various sets of parameters are called *natural variables* of the function in question. According to the discussion in Section 1.3, the set of natural variables of the internal energy is $\{\mathcal{S}, N, \boldsymbol{\sigma}\}$.

It needs to be emphasized at this point that one could, of course, express each thermodynamic potential in terms of different sets of (nonnatural) variables. The immediate consequence is that thermodynamic potentials would not necessarily attain a minimum value if the system is in a state of thermodynamic equilibrium. This point is important to realize because it implies that the set of natural variables of a given thermodynamic potential is distinguished and unique among other conceivable sets of variables.

1.5 Legendre transformation

Because of the discussion in Section 1.3 the question arises: Is there a formal way to switch between various thermodynamic potentials by changing the set of natural variables? The answer to this question is particularly important because controlling, for instance, the set $\{\mathcal{S}, N, \boldsymbol{\sigma}\}$ in a laboratory experiment might be cumbersome at best, if not at all impossible, because it requires control of the entropy during a thermodynamic transformation. However, there is no direct way of measuring $\mathcal{S}$ directly nor is there any device to control it. Without elaborating further on this issue, the reader will surely agree (at least intuitively) that it might at least be hard to think of a device that permits one to control $\mathcal{S}$ experimentally. Thus, one would like to transform variables such that in our example $\mathcal{S}$ is no longer a natural variable but instead becomes a dependent function of another set of thermodynamic variables. The formal way of effecting such a change of variables is the Legendre transformation that we introduce formally in Appendix A.2.

Applying the concepts of Legendre transformation to thermodynamic potentials, we realize that Gibbs' fundamental equation [see Eq. (1.22)] can be rewritten as

$$d\mathcal{U} = \left(\frac{\partial \mathcal{U}}{\partial \mathcal{S}}\right)_{\{\cdot\}\backslash \mathcal{S}} d\mathcal{S} + \left(\frac{\partial \mathcal{U}}{\partial N}\right)_{\{\cdot\}\backslash N} dN + \mathrm{Tr}\left[(\nabla_\sigma \mathcal{U})\, d\boldsymbol{\sigma}\right] \qquad (1.55)$$

because $d\mathcal{U}$ is an exact differential. In Eq. (1.55), $\nabla_\sigma \mathcal{U}$ can be represented by the matrix

$$\nabla_\sigma \mathcal{U} = \begin{pmatrix} (\partial \mathcal{U}/\partial \sigma_{xx})_{\{\cdot\}\backslash \sigma_{xx}} & \frac{1}{2}(\partial \mathcal{U}/\partial \sigma_{xy})_{\{\cdot\}\backslash \sigma_{xy}} & \frac{1}{2}(\partial \mathcal{U}/\partial \sigma_{xz})_{\{\cdot\}\backslash \sigma_{xz}} \\ \frac{1}{2}(\partial \mathcal{U}/\partial \sigma_{xy})_{\{\cdot\}\backslash \sigma_{xy}} & (\partial \mathcal{U}/\partial \sigma_{yy})_{\{\cdot\}\backslash \sigma_{yy}} & \frac{1}{2}(\partial \mathcal{U}/\partial \sigma_{yz})_{\{\cdot\}\backslash \sigma_{yz}} \\ \frac{1}{2}(\partial \mathcal{U}/\partial \sigma_{xz})_{\{\cdot\}\backslash \sigma_{xz}} & \frac{1}{2}(\partial \mathcal{U}/\partial \sigma_{yz})_{\{\cdot\}\backslash \sigma_{yz}} & (\partial \mathcal{U}/\partial \sigma_{zz})_{\{\cdot\}\backslash \sigma_{zz}} \end{pmatrix}$$
$$(1.56)$$

where we used shorthand notation "$\{\cdot\}\backslash\sigma_{\alpha\beta}$" to indicate that, upon performing the partial differentiation, the set of variables $\{\mathcal{S}, N\}$ is being held fixed together with all strain tensor elements except for $\sigma_{\alpha\beta}$. Elements of the matrix in Eq. (1.56) may be identified with thermodynamic forces acting on the confined lamella on account of compressional (dilational) and shear strains exerted on it by its environment.

The set of natural variables $\{\mathcal{S}, N, \boldsymbol{\sigma}\}$ of $\mathcal{U}$ in Eq. (1.55) may thus be perceived as the set $\{x_k\}$ in Eq. (A.4) of Appendix A.2. Moreover, by comparison with Eq. (1.17), it is also clear that the analogs of the $\{f'_k\}$ in Eq. (A.4) are given by

$$T \equiv \left(\frac{\partial \mathcal{U}}{\partial \mathcal{S}}\right)_{\{\cdot\}\backslash \mathcal{S}} \qquad (1.57a)$$

$$\mu \equiv \left(\frac{\partial \mathcal{U}}{\partial N}\right)_{\{\cdot\}\backslash N} \qquad (1.57b)$$

$$\tau_{\alpha\beta} \equiv \frac{1}{V_0}\left(\frac{\partial \mathcal{U}}{\partial \sigma_{\alpha\beta}}\right)_{\{\cdot\}\backslash \sigma_{\alpha\beta}} \qquad (1.57c)$$

In addition, we realize from Eqs. (A.4) and (1.57) that the grand potential defined in Eq. (1.51) can be obtained as a Legendre transform of $\mathcal{U}$ expressed as

$$\Omega(T, \mu, \boldsymbol{\sigma}) = \mathcal{U}(\mathcal{S}, N, \boldsymbol{\sigma}) - T\mathcal{S} - \mu N \qquad (1.58)$$

such that its exact differential is given by

$$d\Omega(T, \mu, \boldsymbol{\sigma}) = d(\mathcal{U} - T\mathcal{S} - \mu N) = -\mathcal{S}dT - Nd\mu + V_0 \mathrm{Tr}(\boldsymbol{\tau} d\boldsymbol{\sigma}) \qquad (1.59)$$

where Eqs. (1.55), (1.57a), and (1.57b) have also been employed. These considerations can easily be extended to other kinds of thermodynamic potentials of interest. If the Legendre transformation is to be performed in order to replace a (sub)set of strains by conjugate stresses, one needs to realize that the operators "d" and "Tr" commute.

1.6 Homogeneity of confined phases

1.6.1 Mechanical expressions for the grand potential

At an elementary level, one of the dogmas taught to almost every chemist is that in thermodynamics only differences between thermodynamic potentials at various state points matter. This is essentially a consequence of the discussion in Section 1.3 where we emphasized that exact differentials exist for thermodynamic potentials such as $\mathcal{U}$, $\mathcal{S}$, $\mathcal{F}$, $\mathcal{G}$, or Ω. These potentials therefore satisfy Eq. (1.18). However, one is frequently confronted with the problem of calculating *absolute* values of thermodynamic potentials theoretically. An example is the determination of phase equilibria, which is one of the key issues in this book chapter. In this context a theorem associated with the Swiss mathematician Leonhard Euler[5] is quite useful. We elaborate on Euler's theorem in Appendix A.3 where we also introduce the notion of homogeneous functions of degree k.

1.6.1.1 Bulk phases

Applying Euler's ideas to the thermodynamic potentials introduced in Section 1.4.2, one realizes that homogeneous functions of degree 1 are of particular interest in the context of equilibrium thermodynamics [see Eq. (A.10)]. For example, consider the grand potential whose exact differential is given by Eq. (1.59). For the special case of a homogeneous bulk phase, it follows that at constant T and μ

$$d\Omega\left(V\right) = \tau_b dV \tag{1.60}$$

because of the special form of the mechanical work term [see Eq. (1.31)]. On account of the homogeneity of bulk phases, it is immediately clear that if we

---

[5]Leonhard Euler (1707–1783), Professor of Mathematics in St. Petersburg (Russia), contributed in an outstanding way to a variety of fields in mathematics. His philosophical writings had a great impact on the German philosopher Immanuel Kant (1724–1804) who was Professor of Logic and Metaphysics in his hometown Königsberg, which he never left during his entire lifetime. With his book *Kritik der Reinen Vernunft*, Kant laid one of the foundations of modern philosophy.

change the volume of our system by a factor $\lambda > 0$ at constant T and μ, the expression

$$\Omega\left(T, \mu, \lambda V\right) = \lambda \Omega\left(T, \mu, V\right) \tag{1.61}$$

must be satisfied, which tells us that Ω is a homogeneous function of degree 1 in V according to the definition given in Eq. (A.8). Transforming variables in Eq. (1.60) according to $V \to \widetilde{V} = \lambda V$ ($\lambda > 0$) and replacing in Eq. (A.10), $x_i = V$ and $f = \Omega$, we obtain

$$V \tau_{\mathrm{b}} = \Omega\left(T, \mu, V\right) \tag{1.62}$$

as a "mechanical" expression for the grand potential in terms of the bulk stress (pressure). Equation (1.62) can be found in most standard texts on bulk phases. However, almost always it is not clearly stated that Ω still depends on T and μ, which is important later on when we discuss the Gibbs-Duhem equation.

1.6.1.2   Slit-pore with unstructured substrate surfaces

A somewhat more complicated situation is encountered for slit-pores with (infinitesimally) smooth (homogeneous) substrates. As we explained in Section 1.3.2, the confined fluid is *homogeneous* across each x–y plane located at different positions z relative to the confining surfaces. Thus, from Eqs. (1.37) and (1.59), we find

$$\mathrm{d}\Omega\left(T, \mu, A, s_z\right) = -\mathcal{S}\mathrm{d}T - N\mathrm{d}\mu + \tau_{\|}s_{z0}\mathrm{d}A + \tau_{\perp}A_0\mathrm{d}s_z \tag{1.63}$$

Applying the homogeneity argument, we realize that

$$\Omega\left(T, \mu, \lambda A, s_z\right) = \lambda \Omega\left(T, \mu, A, s_z\right) \tag{1.64}$$

where, of course, T, μ, and s_z need to be held constant. From Eqs. (A.10) and (1.63), one obtains

$$A s_{z0} \tau_{\|} = \Omega\left(T, \mu, A, s_z\right) \tag{1.65}$$

as the analog of Eq. (1.62). In Eq. (1.65) it is important to realize, that despite s_z being an extensive variable, Ω is not a homogeneous function of s_z because of the inhomogeneous nature of the confined fluid in the z-direction. The notion of extensivity of a thermodynamic function does not necessarily imply that a thermodynamic potential depending on this variable is homogeneous of degree 1 in that variable. In other words, extensivity of a thermodynamic variable is a *necessary* condition but *not sufficient* to conclude that a thermodynamic potential is a homogeneous function of degree 1 of that variable. Consequently, increasing s_z by a factor of λ does not cause nanoscopic slits to imbibe λ times the amount of matter it originally accommodated. We thus readily conclude that, for Ω to be a homogeneous function in any of its extensive variables, it is necessary that the system is *homogeneous* in at least one spatial direction. In other words, system properties must be *translationally invariant* in at least one spatial dimension.

1.6.1.3   Slit-pore with structured substrate surfaces

Consider next the system depicted schematically in Fig. 5.7, namely a fluid confined to a slit–pore with chemically striped walls. From Eqs. (1.59) and (1.43), we obtain

$$\mathrm{d}\Omega = -\mathcal{S}\mathrm{d}T - N\mathrm{d}\mu + A_{x0}\tau_{xx}\mathrm{d}s_x + A_{y0}\tau_{yy}\mathrm{d}s_y + A_{z0}\tau_{zz}\mathrm{d}s_z + A_{z0}\tau_{xz}\mathrm{d}\left(\alpha s_{z0}\right) \quad (1.66)$$

On account of the infinite length of the chemical stripes in the y-direction, system properties are translationally invariant in that direction. Hence, Ω is a homogeneous function of degree 1 only in s_y and therefore

$$\Omega\left(T, \mu, s_x, \lambda s_y, s_z, \alpha s_{x0}\right) = \lambda\Omega\left(T, \mu, s_x, s_y, s_z, \alpha s_{x0}\right) \quad (1.67)$$

Equation (A.10) then gives

$$A_{y0}s_y\tau_{yy} = \Omega\left(T, \mu, s_x, s_y, s_z, \alpha s_{x0}\right) \quad (1.68)$$

as the desired mechanical expression for the grand potential in the current case.

Moreover, the reader may realize that, because of the periodicity of the substrate structure in the x-direction one may write

$$\Omega\left(T, \mu, \widetilde{\lambda}s_x, s_y, s_z, \alpha s_{x0}\right) = \widetilde{\lambda}\Omega\left(T, \mu, s_x, s_y, s_z, \alpha s_{x0}\right) \quad (1.69)$$

for $\widetilde{\lambda} \in \mathbb{N}$, which at first glance may look as if Ω may also be perceived as a homogeneous function of degree 1 in s_x. However, on account of the structural periodicity of the substrates, $\widetilde{\lambda}$ is restricted to integer values unlike its counterpart λ in Eq. (1.67). In fact, contemplating the definition of partial derivatives, the relationship

$$\frac{\partial f}{\partial\left(\lambda x\right)} = \lim_{\Delta\lambda \to 0} \frac{f\left[\left(\lambda + \Delta\lambda\right)x\right] - f\left(\lambda x\right)}{\Delta\lambda x} \quad (1.70)$$

cannot be made arbitrarily small for arbitrary values of x. This is because, for integer values of $\lambda = \widetilde{\lambda}$, $\Delta\lambda = \delta\widetilde{\lambda}$ cannot be made arbitrarily small and therefore the right side of Eq. (1.70) does not exist. Hence, in this case the grand potential $\Omega\left(T,\mu,\widetilde{\lambda}s_x,s_y,s_z,\alpha s_{x0}\right)$ does not satisfy Eq. (A.10) because the partial derivative of Ω with respect to $\widetilde{\lambda}s_x$ is ill–defined.

Thus, whether or not a closed mechanical expression for Ω exists for a confined fluid (as well as for any other system exposed to an external field) depends critically on the nature of the external potenial. It must be such that the confined fluid is homogeneous in at least one spatial direction. In other words, the conclusion that Ω is a homogeneous function of degree 1 in one or more of its extensive variables is ineluctably coupled to considerations of the symmetry of the external potential representing the confining substrate.

1.6.2   Gibbs-Duhem equations and symmetry

The existence of mechanical expressions for the grand potential introduces an additional equation for Ω. Take as an example Eq. (1.62) whose exact differential may be cast as

$$d\Omega\left(T,\mu,V\right) = \tau_b dV + V d\tau_b \qquad (1.71)$$

If this equation, which is valid for bulk systems, is combined with Eq. (1.59), we arrive at a so-called Gibbs-Duhem equation, namely

$$0 = SdT + Nd\mu + Vd\tau_b \qquad (1.72)$$

which states that the three intensive variables T, μ, and τ_b cannot be varied independently but are related to each other through an (*a priori* unknown) equation of state $\tau_b = f(\mu, T)$. This is standard textbook knowledge.

However, as we shall demonstrate shortly, Eq. (1.72) is by no means general as far as confined systems (or any system exposed to an external field) are concerned. Take as an example a fluid confined to a slit-pore with homogeneous (infinitesimally) smooth substrates. For this system, we derived an expression for Ω in Eq. (1.65). Differentiating it we obtain

$$d\Omega\left(T,\mu,A,s_z\right) = As_{z0}d\tau_\parallel + \tau_\parallel s_{z0}dA \qquad (1.73)$$

Combining this latter expression with Eq. (1.63) yields yet another Gibbs-Duhem equation of the form

$$0 = SdT + Nd\mu + As_{z0}d\tau_\parallel - \tau_\perp A_0 ds_z \qquad (1.74)$$

that tells us that an equation of state $\tau_\parallel = f(T,\mu,s_z)$ exists in which one intensive variable (e.g., $\tau_\parallel$) can be expressed in terms of the other two (T,μ) and one extensive variable (s_z).

If the external potential is nonconstant across the x–y plane but varies periodically along the x-axis, say, as the one describing the chemically striped substrate surfaces depicted in Fig. 5.7, the symmetry of the fluid is reduced even further. This causes the equation of state to depend on even more parameters as in the previously discussed case. This can be realized from Eqs. (1.66) and (1.68), which permit us to derive yet another Gibbs-Duhem equation, namely

$$0 = \mathcal{S}dT + Nd\mu - A_{x0}\tau_{xx}ds_x - A_{y0}s_yd\tau_{yy} - A_{z0}\tau_{zz}ds_z - A_{x0}\tau_{xz}d\left(\alpha s_{x0}\right) \quad (1.75)$$

which suggests the existence of an equation of state of the form $\tau_{yy} = \tau_{yy}\left(T, \mu, s_x, s_z, \alpha s_{x0}\right)$ that now depends on the periodicity of the chemical structure (i.e., the widths of the chemically distinct stripes in the x-direction) through s_x as well as on their lateral alignment (i.e., the shear strain αs_{x0}) (see Fig. 1.5). Hence, symmetry considerations play an important rôle in the thermodynamics of confined *fluids* similar to bulk *solids* (see Chapter 13 in the book of Callen [12]).

We shall return to this issue in Section 5.5 where the symmetry of the external potential representing a confining solid surface is such that the grand potential is not a homogeneous function of degree 1 in *any* of its extensive variables. The reason in this particular case is that the surface is decorated with a chemical nanopattern of finite extent, which together with the mere presence of the substrates, abolishes the translational invariance of the local density in all three spatial directions (i.e., the homogeneity of the confined fluid). As a consequence, a Gibbs-Duhem equation does not exist, which precludes the existence of an equation of state in the above sense as well.

1.7 Phase transitions

Within the scope of this book, phase transitions play a prominent role. From a thermodynamic perspective, phase transitions can be discussed most conveniently on the basis of the grand potential Ω introduced in Section 1.4. There is a twofold reason for this distinguished position of the grand among other thermodynamic potentials:

1. With regard to confined systems, one is often confronted with situations in which the confined phase is in thermodynamic equilibrium with a bulk reservoir with which it exchanges heat and matter. Under these conditions, it was shown in Section 1.4.2 [see Eq. (1.51)] that Ω is the relevant thermodynamic potential to identify equilibrium states of the system of interest.

2. As was demonstrated in Section 1.6.1, a mechanical expression for Ω can be derived in many cases of interest. This permits access to (absolute values of) Ω from stress tensor components that can be calculated[6] or controlled experimentally. This also holds for other thermodynamic potentials where N has been replaced by μ as a thermodynamic state variable via a Legendre transformation (see Section 1.5).

The objective then is to identify stable phases in the context of phase transitions on the basis of variations of Ω. We will concentrate mostly on *discontinuous* (i.e., *first-order*) phase transitions where in addition the participating phases will always be fluid (i.e., gas- or liquid-like). In general, two phases α and β undergo a discontinuous phase transition at some fixed temperature if their grand-potential *density* ω (see below) satisfies the conditions

$$\omega^i\left(\mu^{ij}, T\right) = \omega^j\left(\mu^{ij}, T\right) \tag{1.76a}$$

$$\left(\frac{\partial \omega^i}{\partial \mu}\right)_T\bigg|_{\mu=\mu^{ij}} \neq \left(\frac{\partial \omega^j}{\partial \mu}\right)_T\bigg|_{\mu=\mu^{ij}} \tag{1.76b}$$

where μ^{ij} denotes the chemical potential at coexistence between phases i and j at a given temperature T. In other words, at a discontinuous phase transition, grand-potential density curves of different slopes intersect. The following discussion is therefore devoted to an investigation of conditions for the existence of such intersections and their relation to measurable thermodynamic quantities.

The simplest case that we shall be discussing here in some detail is that of a fluid confined to a nanoscopic slit-pore with homogeneous (infinitesimally) smooth substrate surfaces. For this prototypical model, it was shown in Section 1.6.1 that a mechanical expression for the grand potential exists. However, in what follows, it is more convenient to focus on the grand-potential *density* rather than on Ω itself. The former is defined through the relation

$$\omega\left(T, \mu, A, s_z\right) \equiv \frac{\Omega}{V} = \tau_\| \tag{1.77}$$

which has a number of important properties. For example, from Eqs. (1.63) and (1.77), it follows that

$$\left(\frac{\partial \omega}{\partial \mu}\right)_{T,A,s_z} = -\rho < 0 \tag{1.78}$$

where ρ is the (mean) density of the confined fluid in the strained system. Hence, for a given temperature and geometry of the fluid lamella, ω is a monotonically decreasing function of the chemical potential because $\rho > 0$.

---

[6]By, for example, statistical mechanical methods (see Chapter 2).

Another important quantity in the context of discontinuous phase transitions is the isothermal compressibility $\kappa_\parallel$. For this system, $\kappa_\parallel$ may be defined starting from the relevant Gibbs-Duhem equation [see Eq. (1.74)], which reduces to

$$N \mathrm{d}\mu = -A s_{z0} \mathrm{d}\tau_\parallel \tag{1.79}$$

because T and s_{z0} are supposed to be constant. Under these conditions, both μ and $\tau_\parallel$ are solely functions of N. Hence,

$$\mathrm{d}\mu = \left(\frac{\partial \mu}{\partial N}\right)_{T,A,s_z} \mathrm{d}N \tag{1.80a}$$

$$\mathrm{d}\tau_\parallel = \left(\frac{\partial \tau_\parallel}{\partial N}\right)_{T,A,s_z} \mathrm{d}N \tag{1.80b}$$

Substituting these expressions into Eq. (1.79) one obtains

$$
\begin{aligned}
N \left(\frac{\partial \mu}{\partial N}\right)_{T,A,s_z} &= -A s_{z0} \left(\frac{\partial \tau_\parallel}{\partial N}\right)_{T,A,s_z} = -s_{z0} \left(\frac{\partial \tau_\parallel}{\partial (N/A)}\right)_{T,A,s_z} \\
&= \frac{A^2 s_{z0}}{N} \left(\frac{\partial \tau_\parallel}{\partial A}\right)_{T,N,s_z} \equiv \frac{A s_{z0}}{N} \frac{1}{\kappa_\parallel}
\end{aligned}
\tag{1.81}
$$

because $\mathrm{d}N$ is arbitrary. This expression can be rearranged to give

$$\left(\frac{\partial^2 \omega}{\partial \mu^2}\right)_{T,A,s_z} = -\left(\frac{\partial \rho}{\partial \mu}\right)_{T,A,s_z} = -\rho^2 \kappa_\parallel \leq 0 \tag{1.82}$$

because both ρ and the isothermal transverse compressibility $\kappa_\parallel$ are positive definite. Together, Eqs. (1.78) and (1.82) allow us to conclude that the function $\omega(\mu)$ (for fixed T, A, and s_z) is monotonic and concave; that is, $\omega(\mu)$ satisfies the inequality

$$\omega(\mu) \geq \frac{\omega(\mu_1)(\mu - \mu_0) + \omega(\mu_0)(\mu_1 - \mu)}{\mu_1 - \mu_0}, \qquad \mu_0 \leq \mu \leq \mu_1 \tag{1.83}$$

where μ_0 and μ_1 are two arbitrary chemical potentials for which $\omega(\mu)$ exists (see below). The right side of Eq. (1.83) represents the secant to $\omega(\mu)$ between μ_0 and μ_1.

For fluid phases i and j differing in density, one realizes from Eqs. (1.78) and (1.82) that the associated curves $\omega(\mu)$ will have different slopes and curvatures. Assuming $\rho^i < \rho^j$, monotonicity of $\omega(\mu)$ suggests that one and only one intersection μ^{ij} exists, which may be obtained as a solution of Eq. (1.76a) for each fixed value of T. In the thermodynamic limit, both curves (ω^i and

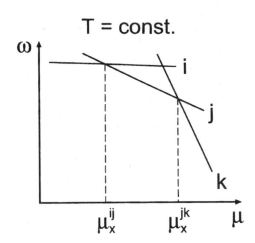

Figure 1.6: Schematic plot of the grand-potential density ω as a function of chemical potential μ under isothermal conditions. The plot shows grand–potential density curves for a situation where three different phases i, j, and k are (meta)stable over certain ranges of μ. Because of Eq. (1.78) their mean densities satisfy the inequality $\overline{\rho}^i < \overline{\rho}^j < \overline{\rho}^k$. Notice that the concavity of the curves $\omega(\mu)$ as predicted by Eq. (1.83) has been deliberately ignored. Chemical potentials μ_x^{ij} and μ_x^{jk} refer to values at coexistence between phases i and j and j and k, respectively.

ω^j) would have to end at μ^{ij} such that, for $\mu < \mu^{ij}$, phase i would be thermodynamically stable, whereas this is the case for phase j over the range $\mu > \mu^{ij}$. However, as we shall demonstrate below, metastable states may exist for both phases such that the curves $\omega^{i,j}$ exist above and below μ^{ij} for a certain finite range of chemical potentials as indicated in Fig. 1.6. Thus, over a certain range of chemical potentials, the grand-potential density at fixed temperature may turn out to be a double-valued function of μ. The globally stable phase is then identified as the one satisfying

$$\omega^i\left(\mu^{ij}, T\right) = \min_k \omega^k\left(\mu^{ij}, T\right) \tag{1.84}$$

in addition to Eq. (1.76a) where we tacitly assume that several phases are conceivable.

Notice also that

$$\rho^j\left(\mu^{ij}, T\right) = \rho^i\left(\mu^{ij}, T\right) + \Delta\rho\left(T\right) \tag{1.85}$$

where $\Delta\rho\left(T\right) \geq 0$ according to our definition of the relative magnitude of ρ^i and ρ^j. Thus, at the intersection between ω^i and ω^j, the density changes discontinuously. This is characteristic of a so-called "discontinuous" or first-order phase transition.

If, in the above equation, $\Delta\rho(T) = 0$, then phases i and j are indistinguishable. This is indicative of a critical point (T_c^{ij}, μ_c^{ij}). Because of the equality of ρ^i and ρ^j, the slope of ω^i and ω^j is the same at the critical point [see Eq. (1.78)]. In other words, at μ_c^{ij} and for $T = T_c^{ij}$, ω^i changes continuously to ω^j. Even though critical phenomena are not at the core of our discussion we note in passing that, in the immediate vicinity of the critical point, a power law applies

$$\lim_{t \to 0^-} \Delta\rho(t) \propto t^\beta \tag{1.86}$$

where β is a critical exponent in standard notation [15], $t \equiv (T_c^{ij} - T)/T_c^{ij}$, and the notation "$0^-$" is used to indicate that t vanishes as T approaches T_c^{ij} from *below*.

It is then convenient to introduce the notion of a coexistence line $\mu_x^{ij}(T)$ as the set of points $\{\mu^{ij}, T\}$ obtained as a solution of Eqs. (1.76a) and (1.84). We also note for later reference that one may encounter situations where a pair of coexistence lines intersects, that is, where

$$\omega^i\left(\mu^{ijk}, T^{ijk}\right) = \omega^j\left(\mu^{ijk}, T^{ijk}\right) = \omega^k\left(\mu^{ijk}, T^{ijk}\right) \tag{1.87}$$

is satisfied, thus defining a so-called triple point $\left(\mu^{ijk}, T^{ijk}\right)$ at which three phases i, j, and k coexist. It therefore seems sensible to introduce the notion of a phase diagram as the union of all coexistence lines

$$\mu_x(T) \equiv \bigcup_{i,j} \mu_x^{ij}(T) \tag{1.88}$$

With Eq. (1.88), we conclude our discussion of phenomenological thermodynamics of confined fluids. In Chapter 2, we shall turn to an interpretation of the various thermodynamic quantities introduced above in terms of interactions between the microscopic constituents forming the *system* at a molecular level of description (i.e., atoms and molecules).

Chapter 2

Elements of statistical thermodynamics

2.1 Introductory remarks

In Chapter 1 we introduced thermodynamics as the central *macroscopic* physical theory that allows us to deal with thermophysical phenomena in confined fluids. However, as we mentioned at the outset, thermodynamics as such does not permit us to draw any quantitative conclusions about a specific physical system without taking recourse to additional sources of information such as experimental data or (empirical) equations of state based on these data. Instead thermodynamics makes rigorous statements about the *relation* among its key quantities such as temperature, internal energy, entropy, heat, and work. It does not permit one to calculate any numbers for these quantities.

As we pointed out in Chapter 1, the reason for this lack of predictive power is that thermodynamics as a theory of the macroscopic world does not have any concept of the underlying microscopic world governed by entities like electrons, atoms, or molecules, say. In fact, it was one of the great intellectual achievements of the nineteenth century to realize that, for example, heat as a thermodynamic quantity is intimately coupled to the motion of molecules (see, for example, Ref. 10).

However, if one accepts the hypothesis that all phenomena perceived by our physical senses are a result of the interaction and consequently the motion of microscopic constituents, we are immediately confronted with three central problems:

1. Any macroscopic piece of matter is composed of an astronomically large number of molecules, which is usually of the order of 10^{23}. How do we follow the spatio-temporal evolution of such a large number of objects

given the overwhelming amount of information?

2. Suppose we could, at least *in principle*, solve the (classic or quantum mechanical) equations of motion for such a huge number of microscopic entities, we would still need to specify initial conditions of the system, which, again, is a task of overwhelming complexity in practice.

3. In Chapter 1 we saw that only a rather small set of variables is required to completely specify the thermodynamic state of a system. How is the enormous reduction of information taking place as we go from a very large number of microscopic constituents and their motion in space and time to the macroscopic equilibrium behavior of matter?

Because of the substantial reduction of information associated with the transition between the micro- and macroworlds, one may readily conclude that the overwhelming amount of information buried in the detailed description of the spatio-temporal evolution of a many-particle system must be largely irrelevant for the thermodynamics of an equilbrium system. In fact, one may suspect that it is the *on-average* behavior of the many-particle system what matters for its macroscopic properties, which, in turn, immediately suggests to employ statistical concepts.

The extent to which statistical concepts enter the picture as we go from the micro- to the macroworld is not at all at our disposal. For example, quantum mechanics as we know it today teaches us that it is impossible *in principle* to obtain complete information about a microscopic entity (i.e., the *precise* and *simultaneous* knowledge of an electron's location and momentum, say) at any instant in time. On account of Heisenberg's Uncertainty Principle, conjugate quantities like, for instance, position and momentum can only be known with a certain maximum precision. Quantum mechanics therefore already deals with *averages* only (i.e., expectation values) when it comes to actual measurements.

The key then is to somehow calculate the probability with which a specific quantum state contributes to the average values. As far as thermal systems in thermodynamic equilibrium are concerned, this is the central problem addressed by statistical thermodynamics. We therefore begin our discussion of some core elements of statistical thermodynamics at the quantum level but will eventually turn to the classic limit because the phenomena addressed by this book occur under conditions where a classic description turns out to be adequate. We shall see this at the end of this chapter in Section 2.5 where we introduce a quantitative criterion for the adequacy of such a classic description.

Two additional comments apply at this point. First, there is a conceptual difference between the probabilistic element in quantum and classic statistical physics. For instance, in quantum mechanics, the outcome of a measurement of properties even of a *single* particle can be known *in principle* only with a certain probability. In classic mechanics, on the other hand, a probabilistic element is usually introduced for many-particle systems where we would *in principle* be able to specify the state of the system with absolute certainty; however, *in practice*, this is not possible because we are dealing with too many degrees of freedom. Recourse to a probabilistic description within the framework of classic mechanics must therefore be regarded a matter of mere convenience. The reader should appreciate this less fundamental meaning of probabilistic concepts in classic as opposed to quantum mechanics.

However, once this fundamental difference in the role played by probabilistic elements in quantum and classic mechanics is accepted, there is essentially no operational difference in the way in which one would compute any property of a thermal many-particle system regardless of whether the system is treated at the quantum level. Our discussion below will reveal that in any case the key quantity required is the so-called partition function, which conveys information about the probability with which a particular (quantum or classic) microstate is realized.

The second comment applies to the use of quantum statistical physics in this chapter. As is well known there are two versions of quantum statistics, namely Bose-Einstein and Fermi-Dirac statistics depending on whether the spin of the quantum system is integer or half-integer. This difference, which manifests itself in symmetry properties of the wave function (see Section 2.5.2), is of importance if one wishes to calculate thermal properties of a specific many-particle quantum system. For most of the treatment here, the different statistics are not an issue. What is important, however, is that, regardless of the specific version of quantum statistics, we are always dealing with a discrete spectrum of eigenstates. The discreteness of this spectrum simplifies the formulation of statistical thermodynamics considerably as we shall see below. Classic systems do normally not have a discrete spectrum of microstates. However, there are exceptions like the celebrated Ising model of a magnet. In the Ising model, one is dealing with a classic spin system where each spin can attain one of two discrete values ("spin up" or "spin down"). The Hamiltonian governing the interaction between those spins is, however, purely classic.

We mention the Ising model at this point because it is intimately linked to another model, which is central to the discussion in Chapter 4, namely that referred to as "lattice fluid." In the lattice fluid, molecules are not permitted to move continuously along classic trajectories but are restricted to discrete positions on some lattice. In this model we may interpret the "spin-up" configuration in the Ising model as a site on the lattice occupied by a fluid molecule, whereas the "spin-down" configuration may be interpreted as an empty lattice site. As one would anticipate, it is possible to translate between the "magnetic" language used for the Ising model and the one appropriate for the lattice fluid [16].

2.2   Concepts of quantum statistical thermodynamics

2.2.1   The most probable distribution

For the sake of concreteness of the following developments, we consider a fluid confined to a slit-pore such that the solid surfaces representing the pore walls are planar, parallel to one another, and perpendicular to the z-axis of a Cartesian coordinate system. The separation between the pore walls will be denoted s_z. In addition, the two solid surfaces can be manipulated by external agents normal to the fluid-solid interface such that s_z may be altered. Eventually, these planar surfaces will come to rest at some equilibrium separation s_z. As we shall see later in Section 5.3.1, the situation just described is akin to laboratory experiments in which the rheology of confined fluids is investigated by means of the so-called surface forces apparatus (SFA).

Consider now an astronomically large number of $N - 1$ identical virtual replicas[1] of the original slit–pore. Each replica is capable of exchanging heat with the slit-pore through diathermal walls to maintain the confined fluid at constant T. In addition, matter may be exchanged between the slit-pore and its environment formed by the replicas. As the confining substrates may move on account of external manipulation, the slit-pore may also exchange (normal) compressional stress with the replicas. Together the N systems are forming a supersystem (i.e., a statistical physical *ensemble*) that is assumed to be closed in the thermodynamic sense, meaning that the supersystem is insulated against its own surroundings such that it has fixed energy $\mathcal{E}$, a fixed number of molecules $\mathcal{N}$, and fixed volume $\mathcal{V}$.

Moreover, we assume that, for the slit-pore (and therefore for all its $N - 1$ identical replicas), we can somehow solve the time-independent (i.e.,

---

[1]Use of the term "replica" for a virtual system in statistical physical ensembles should not be confused with the same term used in conjunction with integral equation theories in Chapter 7.

stationary) N-particle Schrödinger equation

$$\widehat{H} \left| \psi_{j(N,s_z)} \right\rangle = E_{j(N,s_z)} \left| \psi_{j(N,s_z)} \right\rangle \tag{2.1}$$

where

$$\widehat{H} = \sum_{i=1}^{N} \frac{\widehat{\boldsymbol{p}}_i^2}{2m} + \widehat{U} \left(\widehat{\boldsymbol{r}}^N; N, s_z \right) \tag{2.2}$$

is the Hamiltonian operator. For simplicity we assume our fluid to be composed of particles of mass m that have only translational degrees of freedom.[2] In Eq. (2.2), $\widehat{\boldsymbol{p}}_i$ and $\widehat{\boldsymbol{r}}_i$ are operators pertaining to momentum and position of a particle, respectively, and $\widehat{U} \left(\widehat{\boldsymbol{r}}^N; N, s_z \right)$ is the configurational-energy operator that can be formally split into an intrinsic (i.e., fluid-fluid) and an external (i.e., fluid-substrate) contribution. In Eq. (2.1), $\left| \psi_{j(N,s_z)} \right\rangle$ are the quantum mechanical wave functions in the well-known representation-independent Dirac notation.

The reader should realize that this treatment of the fluid-substrate interaction assumes that we may perceive the substrate as an external field superimposed onto the intermolecular interactions. This is possible if we disregard thermodynamic equilibrium between the fluid molecules and the substrate. In other words, the substrates are not thermally coupled to the confined fluid. Such a coupling may be incorporated but would make the statistical physical analysis much more involved than is needed for our current purposes.

We use the notation $E_{j(N,s_z)}$ to indicate that the spectrum of energy eigenstates depends $\left\{ E_{j(N,s_z)} \right\}$ on the number of molecules N accommodated by the slit-pore and the actual substrate separation s_z, which may, of course, vary such that at any instant in time $n_{i(N,s_z)}$ systems of the ensemble are in a quantum state characterized by $E_{i(N,s_z)}$, $n_{j(N',s_z')}$ in a quantum state characterized by $E_{j(N',s_z')}$ and so on. As the spectrum of energy eigenstates obtained by solving Eq. (2.1) is discrete, we may order them such that

$$E_{1(N,s_z)} \leq E_{2(N,s_z)} \leq \ldots \leq E_{i(N,s_z)} \leq \ldots E_{j(N',s_z')} \leq \ldots \tag{2.3}$$

Henceforth, we shall tacitly assume the set of orthonormal functions $\left| \psi_{j(N,s_z)} \right\rangle$ to be complete. Therefore, we know from Eq. (2.1) the complete spectrum of energy eigenstates $\left\{ E_{j(N,s_z)} \right\}$ (i.e., energy eigenvalues of the N-particle Hamiltonian). We note in passing that this is, of course, a formidable if not entirely impossible task *in practice* if one is dealing with a many-particle

---

[2]The treatment can be extended to nonspherical particles (see, for example, Chapters 6, 7, and Ref. 17).

system of interacting microscopic constituents because of the many degrees of freedom one needs to consider explicitly. One of the simplest cases for which a solution of Eq. (2.1) turns out to be possible (and with which each chemist should be familiar) is the particle in a box, which we shall briefly discuss to get a feel for the number of quantum states accessible to a macroscopic piece of matter under conditions relevant to the current discussion.

Consider a cubic box of side length L containing N noninteracting particles of mass m. At the boundaries of the box, that is, at planes $x = 0, L$, $y = 0, L$, and $z = 0, L$, the particles are exposed to a potential $\widehat{U} = \widehat{U}_{\text{ext}} = \infty$, which keeps them inside the volume $V = L^3$ of the box. By solving Eq. (2.1) for this model, the energy eigenstates of each particle are given by

$$E = \frac{h^2}{8mL^2}\left(k_x^2 + k_y^2 + k_z^2\right), \qquad k_x, k_y, k_z = 0, 1, 2, \ldots \qquad (2.4)$$

For sufficiently large values of the quantum numbers k_x, k_y, and k_z, these are distributed quasi-continuously on one octant of a sphere of radius $R \equiv \sqrt{k_x^2 + k_y^2 + k_z^2}$. Hence, the number of quantum states with an energy less than or equal to E maybe calculated from

$$\phi(E) = \frac{1}{8}\frac{4\pi R^3}{3} = \frac{\pi}{6}\left(\frac{8mL^2}{h^2}E\right)^{3/2} \qquad (2.5)$$

where the factor of $\frac{1}{8}$ arises because the quantum numbers are positive semidefinite such that only one octant of the sphere needs to be considered. From Eq. (2.5), we may estimate the number of quantum states within a small energy interval ΔE from the expression

$$\begin{aligned}
\Delta\phi(E, \Delta E) &\equiv \phi(E + \Delta E) - \phi(E) \\
&= \frac{\pi}{6}\left(\frac{8mL^2}{h^2}E\right)^{3/2}\left[\left(1 + \frac{\Delta E}{E}\right)^{3/2} - 1\right] \\
&\approx \frac{\pi}{4}\left(\frac{8mL^2}{h^2}\right)^{3/2}\sqrt{E}\Delta E \qquad (2.6)
\end{aligned}$$

where we assumed $\Delta E/E \ll 1$ such that we can expand $(1 + \Delta E/E)^{3/2}$ in a Taylor series truncated after the first-order term. To estimate $\Delta\phi$ we take as an example a particle having thermal energy $E = \frac{3}{2}k_{\text{B}}T$ at room temperature $T = 300\text{K}$, where k_{B} is Boltzmann's constant. The particle has a mass $m = 10^{-25}\text{kg}$ typical of a rare-gas atom like Ar. Assuming $L = 0.1\text{m}$ and a typical thermal "noise" of $\Delta E = 0.01E$, we obtain $\Delta\phi = \mathcal{O}(10^{28})$ for each particle in our box. As we are considering a macroscopic piece of matter,

$N = \mathcal{O}(10^{23})$. Thus, our entire system has of the order of $\mathcal{O}(10^{51})$ accessible quantum states that are compatible with the specified conditions. This is indeed an astronomically large number of states, which becomes even larger by many more orders of magnitude if we also include interactions between all these particles as we should for more realistic situations.

Notice also that, if we take our system as a quantum mechanical model for an ideal gas, the thermal state of the gas would be completely specified by the density and temperature of the gas (which would permit us to compute its pressure). Hence, in going from the microscopic to the macroscopic level of description, an enormous reduction of information from $\mathcal{O}(10^{51})$ down to only 2 degrees of freedom has taken place.

Let us now return to our original situation, namely that we can solve Eq. (2.1) for the systems of interest in this book and that we can arrange energy eigenstates according to the inequality given in Eq. (2.3). For the subsequent development, it will turn out to be useful to introduce the concept of a distribution of quantum states through the vector $\boldsymbol{n} \equiv |\boldsymbol{n}\rangle = |n_{1(N,s_z)}, n_{2(N,s_z)}, \ldots\rangle$, where $n_{j(N,s_z)}$ is the occupation number of quantum state j for all those replicas having the same number of molecules N and substrate separation s_z. Thus, we can interpret $\boldsymbol{n}$ as a quantity that tells us how the available quantum states are *distributed* over the ensemble of replicas. Therefore, we shall henceforth refer to the vector $\boldsymbol{n}$ as a "distribution" of quantum states. Because the supersystem is thermodynamically closed against its own environment, the distribution $\boldsymbol{n}$ is subject to four constraints, namely

$$\varphi_1(\boldsymbol{n}) \equiv \mathsf{N} - \sum_{N,s_z \to j} n_{j(N,s_z)} = 0 \tag{2.7a}$$

$$\varphi_2(\boldsymbol{n}) \equiv \mathcal{E} - \sum_{N,s_z \to j} n_{j(N,s_z)} E_{j(N,s_z)} = 0 \tag{2.7b}$$

$$\varphi_3(\boldsymbol{n}) \equiv \mathcal{N} - \sum_{N,s_z \to j} n_{j(N,s_z)} N = 0 \tag{2.7c}$$

$$\varphi_4(\boldsymbol{n}) \equiv \mathcal{S}_z - \sum_{N,s_z \to j} n_{j(N,s_z)} s_z = 0 \tag{2.7d}$$

Equation (2.7a) expresses the fact that the number of systems forming the ensemble is fixed and finite but may be made arbitrarily large at the end of our calculation. Equation (2.7b) states that the energy of the isolated supersystem is fixed as well as its total volume $\mathcal{V} = \mathsf{N}^2 s_x s_y \mathcal{S}_z$ [see Eq. (2.7d)] and total number of molecules $\mathcal{N}$ [see Eq. (2.7c)]. The expression for $\mathcal{V}$ implicitly assumes all systems of the ensemble to be exposed to the same compressional strains proportional to s_x and s_y as well as shear strain proportional to αs_{x0}.

In Eqs. (2.7) we are using " $\sum\limits_{N,s_z \to j} \ldots$ " as shorthand notation for the triple sum " $\sum\limits_{N} \sum\limits_{s_z} \sum\limits_{j(N,s_z)} \ldots$ " where the arrow is used to emphasize that within the set of quantum states $\{\psi_{j(N,s_z)}\}$ *depends* on the actual number of particles N and the specific substrate separation s_z, which may differ between the individual systems of the ensemble.

Let us now assume N to be finite but overwhelmingly large. Because we are imposing only four constraints on the construction of $\boldsymbol{n}$, consisting themselves of an overwhelmingly large number of elements $n_{j(N,s_z)}$ as well, it seems reasonable to expect many distributions $\{\boldsymbol{n}\}$ to exist, all of which are complying with Eqs. (2.7). From the preceding discussion, it should be obvious that at any instant it is the specific quantum state occupied by one of the (many-particle) replica systems of the ensemble that makes it distinct and distinguishable from any other. In other words, the replicas can only be discriminated on an energetic basis. However, it is conceivable that two or more replicas occupy the same quantum state and are thus indistinguishable. There are several such groups in which $n_{i(N,s_z)}$ replicas are characterized by the same energy $E_{i(N,s_z)}$, $n_{j(N',s'_z)}$ having energy $E_{j(N',s'_z)}$ and so on and so forth. Hence, a specific distribution $\{\boldsymbol{n}\}$ can be realized in a number of

$$W\left(\boldsymbol{n}\right) = \frac{\mathsf{N}!}{\prod\limits_{N,s_z \to j} n_{j(N,s_z)}!} \tag{2.8}$$

different ways by assigning $n_{i(N,s_z)}$ randomly chosen systems of the ensemble to quantum state $E_{i(N,s_z)}$, $n_{j(N',s''_z)}$ to quantum state $E_{j(N',s''_z)}$ and so forth. Hence, if we pick any one system of the ensemble at random and ask for the probability to find it *on average* in a quantum state $E_{j(N,s_z)}$ accommodating N molecules at a substrate separation s_z, this probability is given by

$$p_{j(N,s_z)} = \frac{\overline{n}_{j(N,s_z)}}{\mathsf{N}} = \frac{1}{\mathsf{N}} \frac{\sum\limits_{\{\boldsymbol{n}\}} W\left(\boldsymbol{n}\right) n_{j(N,s_z)}}{\sum\limits_{\{\boldsymbol{n}\}} W\left(\boldsymbol{n}\right)} \tag{2.9}$$

where $\overline{n}_{j(N,s_z)}$ is the number of times a quantum state j is realized on average with N molecules present and substrate separation s_z. The sums in Eq. (2.9) extend over all a priori equally probable distributions. We assume these distributions to be equally likely because they all comport with a state of the isolated supersystem characterized by fixed $\mathcal{N}$, $\mathcal{V}$, and $\mathcal{E}$. The assumption of equal likelihood of all distributions $\{\boldsymbol{n}\}$ complying with $\mathcal{E} = \text{constant}$ is known as the *Principle of Equal A Priori Probability*. It is part of the postulatory basis of statistical thermodynamics.

As we pointed out, N can be made arbitrarily large and may, in fact, become infinite. Suppose now that in the limit $N \to \infty$ a distribution $\boldsymbol{n}^\star$ exists such that $W(\boldsymbol{n}^\star)$ overwhelms all other values $\{W(\boldsymbol{n})\}$. An immediate consequence of this assumption is that, in the sums on the right-hand side of Eq. (2.9), all summands except those involving $\boldsymbol{n}^\star$ become negligible so that Eq. (2.9) may be replaced by a single term, namely

$$p_{j(N,s_z)} = \frac{1}{N} \frac{W(\boldsymbol{n}^\star) n^\star_{j(N,s_z)}}{W(\boldsymbol{n}^\star)} = \frac{n^\star_{j(N,s_z)}}{N} \qquad (2.10)$$

2.2.2 Justification of the most probable distribution

The existence of a most probable distribution $\boldsymbol{n}^\star$ with properties reflected by Eq. (2.10) is by no means obvious, however, or even plausible despite the fact that N is at our disposal and may therefore be increased beyond limits. A formal way of *proving* the existence of $\boldsymbol{n}^\star$ was suggested by Darwin and Fowler [18]. Their method is based on function-theoretical arguments that are mathematically a bit involved. The advantage of Darwin and Fowler's argument is that it is mathematically rigorous and completely general. We have therefore decided to include it in this book. Moreover, we were prompted to do so because the Darwin-Fowler approach can hardly be found elsewhere in the literature [19].

Because of the somewhat more formal character of Darwin and Fowler's analysis, however, less interested readers may skip this portion of the current section. Nonetheless, we feel that these readers should at least be equipped with an argument suggesting that the assumption of a most probable distribution in the sense of Section 2.2.1 is justified and physically sensible. However, unlike the Darwin-Fowler approach, this alternative reasoning, which is based on a Taylor expansion of $\ln W(\boldsymbol{n})$, amounts only to a plausibility argument and is by no means rigorous.

2.2.2.1 Taylor expansion of Eq. (2.8)

We begin by rewriting Eq. (2.8) as

$$
\begin{aligned}
\ln W(\boldsymbol{n}) &\simeq N \ln N - N - \sum_{N,s_z \to j} n_{j(N,s_z)} \ln n_{j(N,s_z)} + \sum_{N,s_z \to j} n_{j(N,s_z)} \\
&= N \ln N - \sum_{N,s_z \to j} n_{j(N,s_z)} \ln n_{j(N,s_z)} \qquad (2.11)
\end{aligned}
$$

where we employed Eq. (2.7a) and Stirling's approximation [see Eq. (B.7)]. Expanding $\ln W(\boldsymbol{n})$ in a Taylor series around the most probable distribution, we obtain

$$\ln W(\boldsymbol{n}) \simeq \ln W(\boldsymbol{n}^\star) - \frac{1}{2} {\sum_{N,s_z \to j}}' \frac{\left(n^\star_{j(N,s_z)} - n_{j(N,s_z)}\right)^2}{n^\star_{j(Ns_z)}} + \ldots \qquad (2.12)$$

where we retain in the expansion only terms up to second order in the deviation of $\boldsymbol{n}$ from $\boldsymbol{n}^\star$. The prime attached to the summation sign emphasizes that the summation is to be carried out subject to the constraints posed by Eqs. (2.7). Under these conditions, the first-order term vanishes because it is the necessary condition for the existence of the most probable distribution. Assuming $|\boldsymbol{n}^\star - \boldsymbol{n}|$ to be sufficiently small, higher-order terms may be neglected in Eq. (2.12). We may then rewrite Eq. (2.12) as

$$\begin{aligned} W(\boldsymbol{n}) &= W(\boldsymbol{n}^\star) \prod_{N,s_z \to j} \exp\left[-\frac{1}{2} \frac{\left(n^\star_{j(N,s_z)} - n_{j(N,s_z)}\right)^2}{n^\star_{j(N,s_z)}}\right] \\ &= \begin{cases} W(\boldsymbol{n}^\star), & \text{if} \quad \boldsymbol{n} = \boldsymbol{n}^\star \\ 0, & \text{if} \quad \boldsymbol{n} \neq \boldsymbol{n}^\star \end{cases} \end{aligned} \qquad (2.13)$$

This follows by noting that, for $\boldsymbol{n} = \boldsymbol{n}^\star$, the exponential factors are exactly one, whereas they are always less than one otherwise. However, because the number of such factors is astronomically large as $\mathsf{N} \to \infty$ the product in Eq. (2.13) goes to zero regardless of how small the deviation between $\boldsymbol{n}$ and $\boldsymbol{n}^\star$ becomes. In other words, compared with $W(\boldsymbol{n}^\star)$, $W(\boldsymbol{n}) \approx 0$ for all other distributions $\{\boldsymbol{n}\} \backslash \boldsymbol{n}^\star$, which is the plausibility argument justifying the subsequent development detailed in Section 2.2.3.

2.2.2.2 Darwin-Fowler analysis of the most probable distribution of microstates

To justify the concept of existence of a most probable distribution of microstates overwhelming all other possible distributions [i.e., all other distributions consistent with the constraints posed by Eqs. (2.7)] a treatment originally due to Darwin and Fowler may be employed[3] [18]. It starts by defining the zero of the energy scale and energy unit such that the discrete energy spectrum $\{E_{j(N,s_z)}\}$ can be represented by a set of nonnegative

---

[3]This section may be skipped by less interested readers.

integers $\{E\}$. Let us then define an auxiliary function

$$Q(z) \equiv \sum_{E=0}^{\infty} z^E t(E) = \sum_{E=0}^{\infty} z^E \sum_{\{n\}} W(n)$$

$$= \sum_{E=0}^{\infty} \sum_{\{n\}}{}' \frac{N!}{\prod_{N,s_z \to j} n_{j(N,s_z)}!} z^{n_{0(N,s_z)} E_{0(N,s_z)} + n_{1(N,s_z)} E_{1(N,s_z)} + \cdots} \quad (2.14)$$

where $z \in \mathbb{C}$ is complex [see Eq. (B.8b)] and the prime attached to the second summation sign signifies that this summation has to be carried out such that, for each energy E, the constraint in Eq. (2.7a) is satisfied. Keeping this in mind Eq. (2.14) can be recast as

$$Q(z) = \sum_{N} \sum_{s_z} \sum_{n_{0(N,s_z)}=0}^{\infty} \sum_{n_{1(N,s_z)}=0}^{\infty} \cdots \sum_{n_{m(N,s_z)}=0}^{\infty} \cdots \frac{N!}{n_{0(N,s_z)}! n_{1(N,s_z)}! \cdots}$$

$$\times \left(z^{E_{0(N,s_z)}} \right)^{n_{0(N,s_z)}} \left(z^{E_{1(N,s_z)}} \right)^{n_{1(N,s_z)}} \cdots \left(z^{E_{m(N,s_z)}} \right)^{n_{m(N,s_z)}} \cdots$$

$$(2.15)$$

One immediately realizes that the previous expression can be rewritten more compactly using the multinomial expansion [17]

$$Q(z) = \left[\sum_{N}{}' \sum_{s_z}{}' \left(z^{E_{0(N,s_z)}} + z^{E_{1(N,s_z)}} + \cdots \right) \right]^N \equiv [q(z)]^N \quad (2.16)$$

where

$$q(z) = \sum_{N,s_z \to j}{}' z^{E_{j(N,s_z)}} \quad (2.17)$$

One member of the set of (nonnegative integer) energies $\{E\}$ is the fixed energy $\mathcal{E}$ of the ensemble [see Eq. (2.7b)]. Therefore, by definition, $t(\mathcal{E})$ is the coefficient associated with $z^{\mathcal{E}}$ in the expansion of Q in terms of a power series in z in Eq. (2.14). As the complex quantity z may become zero, $Q(z)/z^{\mathcal{E}+1}$ has a singularity for this value of z. Because of our definition of the zero and unit of the energy scale, $\mathcal{E} \geq 0$. To avoid an undetermined expression of the form 0^0, we consider $z^{\mathcal{E}+1}$ rather than $z^{\mathcal{E}}$ henceforth and expand $Q(z)/z^{\mathcal{E}+1}$ in a Laurent series around $z_0 = 0$ [see Appendix B.2.4, Eq. (B.40)] according to

$$\frac{Q(z)}{z^{\mathcal{E}+1}} = \sum_{k=-\infty}^{\infty} a_k z^k \quad (2.18)$$

where the $\{a_k\}$ are unknown coefficients of the Laurent expansion. Terms in the last expression may be rearranged slightly to give

$$Q(z) = \sum_{k=-\infty}^{\infty} a_k z^{k+\mathcal{E}+1} \qquad (2.19)$$

Comparing the term for $k = -1$ in the Laurent expansion with our starting expression for $Q(z)$ in Eq. (2.14), it is clear that $a_{-1} = t(\mathcal{E})$ because a_{-1} is the expansion coefficient associated with $z^{\mathcal{E}}$. This coefficient, the so-called residue [see Eqs. (B.33), (B.41)], may be obtained from

$$t(\mathcal{E}) = a_{-1} = \frac{1}{2\pi i} \oint_C dz \frac{Q(z)}{z^{\mathcal{E}+1}} \qquad (2.20)$$

where C is some yet to be specified closed path in the complex plane enclosing the singularity of the integrand at $z = 0$.

Let us now define the complex auxiliary function [see Eqs. (B.8)]

$$f(z) \equiv \ln q(z) - \frac{\mathcal{E}+1}{N} \ln z \qquad (2.21)$$

which permits us to rewrite Eq. (2.20) as

$$t(\mathcal{E}) = \frac{1}{2\pi i} \oint_C dz \exp[N f(z)] \qquad (2.22)$$

where we used Eq. (2.16). The integral in Eq. (2.22) is identical with the one in Eq. (B.17) and can therefore be evaluated using the method of steepest descent detailed in Appendix B.2.2.

It requires $f(z)$ to assume a maximum at some $z = z_0$. Expanding $f(z)$ in a Taylor series and retaining terms to second order only, it is shown in Appendix B.2.2 that Eq. (2.22) can be rewritten as

$$\lim_{N\to\infty} \ln t(\mathcal{E}) = -\frac{1}{2} \lim_{N\to\infty} \{\ln[2\pi N f''(x_0)] + N f(x_0)\} \simeq N f(x_0) \qquad (2.23)$$

where $x_0 = \mathrm{Re} z_0$ is the real part of z_0. These expressions permit us now to calculate the mean occupation number $\overline{n}_{j(N,s_z)}$ of quantum state $j(N, s_z)$ and its variance $\overline{n^2}_{j(N,s_z)} - \overline{n}^2_{j(N,s_z)}$. To proceed we introduce

$$W(\boldsymbol{n}) \quad \to \quad W_\xi(\boldsymbol{n}) \equiv \frac{N!}{\displaystyle\prod_{N,s_z\to j} n_{j(N,s_z)}!} \prod_{N,s_z\to j} \xi_{j(N,s_z)}^{n_{j(N,s_z)}} \qquad (2.24a)$$

$$t(\mathcal{E}) \quad \to \quad t_\xi(\mathcal{E}) \equiv \sum_{\{\boldsymbol{n}\}} W_\xi(\boldsymbol{n}) \qquad (2.24b)$$

$$q(z) \quad \to \quad q_\xi(z) \equiv \sum_{N,s_z\to j} \xi_{j(N,s_z)} z^{E_{j(N,s_z)}} \qquad (2.24c)$$

where $\{\xi_{j(N,s_z)}\}$ is a set of arbitrary parameters that will be set equal to one at the end of the calculation. From the definition in Eq. (2.9), it follows that

$$
\begin{aligned}
\overline{n}_{j(N,s_z)} &= \xi_{j(N,s_z)} \frac{\partial}{\partial \xi_{j(N,s_z)}} \ln t_\xi \left(\mathcal{E} \right) \Bigg|_{\{\xi_{j(N,s_z)}\}=1} \\
&= N\xi_{j(N,s_z)} \frac{\partial}{\partial \xi_{j(N,s_z)}} \ln q_\xi \left(x_0 \right) \Bigg|_{\{\xi_{j(N,s_z)}\}=1} \\
&= N \frac{x_0^{E_{j(N,s_z)}}}{q_\xi \left(x_0 \right)} \Bigg|_{\{\xi_{j(N,s_z)}\}=1} = N \frac{x_0^{E_{j(N,s_z)}}}{\sum_{N,s_z \, ,j}' x_0^{E_{j(N,s_z)}}} \qquad (2.25)
\end{aligned}
$$

In Eq. (2.25) we employed Eq. (2.23) and the fact that [see Eq. (2.21)]

$$
\lim_{N \to \infty} f \left(z \right) = \ln q \left(z \right) \qquad (2.26)
$$

In a similar fashion we calculate

$$
\begin{aligned}
\overline{n^2}_{j(N,s_z)} &= \frac{\sum_{\{n\}} W_\xi \left(\boldsymbol{n} \right) n^2_{j(N,s_z)}}{\sum_{\{n\}} W_\xi \left(\boldsymbol{n} \right)} \Bigg|_{\{\xi_{j(N,s_z)}\}=1} \\
&= \frac{1}{t_\xi \left(\mathcal{E} \right)} \xi_{j(N,s_z)} \frac{\partial}{\partial \xi_{j(N,s_z)}} \left(\xi_{j(N,s_z)} \frac{\partial}{\partial \xi_{j(N,s_z)}} t_\xi \left(\mathcal{E} \right) \right) \Bigg|_{\{\xi_{j(N,s_z)}\}=1} \\
&= \xi_{j(N,s_z)} \frac{\partial}{\partial \xi_{j(N,s_z)}} \left(\xi_{j(N,s_z)} \frac{\partial}{\partial \xi_{j(N,s_z)}} \ln t_\xi \left(\mathcal{E} \right) \right) \Bigg|_{\{\xi_{j(N,s_z)}\}=1} \\
&\quad + \underbrace{\xi^2_{j(N,s_z)} \left(\frac{\partial}{\partial \xi_{j(N,s_z)}} \ln t_\xi \left(\mathcal{E} \right) \right)^2 \Bigg|_{\{\xi_{j(N,s_z)}\}=1}}_{\overline{n}^2_{j(N,s_z)}} \qquad (2.27)
\end{aligned}
$$

In Eq. (2.27), we can rewrite

$$
\xi_{j(N,s_z)} \frac{\partial}{\partial \xi_{j(N,s_z)}} \left(\xi_{j(N,s_z)} \frac{\partial}{\partial \xi_{j(N,s_z)}} \ln t_\xi \left(\mathcal{E} \right) \right) \Bigg|_{\{\xi_{j(N,s_z)}\}=1}
$$

$$
= N x_0^{E_{j(N,s_z)}} \xi_{j(N,s_z)} \frac{\partial}{\partial \xi_{j(N,s_z)}} \left(\frac{\xi_{j(N,s_z)}}{q_\xi \left(x_0 \right)} \right) \Bigg|_{\{\xi_{j(N,s_z)}\}=1}
$$

$$
= N \frac{x_0^{E_{j(N,s_z)}}}{\sum\limits_{N,s_z \to j} {}' x_0^{E_{j(N,s_z)}}} - N \left(\frac{x_0^{E_{j(N,s_z)}}}{\sum\limits_{N,s_z \to j} {}' x_0^{E_{j(N,s_z)}}} \right)^2
$$

$$
= \overline{n}_{j(N,s_z)} \left(1 - \frac{\overline{n}_{j(N,s_z)}}{N} \right) \tag{2.28}
$$

Hence, from Eqs. (2.27) and (2.28), we realize that

$$
\overline{n^2}_{j(N,s_z)} - \overline{n}^2_{j(N,s_z)} = \overline{n}_{j(N,s_z)} \left(1 - \frac{\overline{n}_{j(N,s_z)}}{N} \right) = \overline{n}_{j(N,s_z)} \left[1 - \mathcal{O}\left(1 \right) \right] \tag{2.29}
$$

where the term in brackets is constant because $\overline{n}_{j(N,s_z)} \propto N$. Because of this proportionality, we may conclude that

$$
\lim_{N \to \infty} \sqrt{\frac{\overline{n^2}_{j(N,s_z)} - \overline{n}^2_{j(N,s_z)}}{\overline{n}^2_{j(N,s_z)}}} \propto \lim_{N \to \infty} \frac{1}{\sqrt{N}} = 0 \tag{2.30}
$$

which proves the existence of a most probable distribution $\boldsymbol{n}^\star = \overline{\boldsymbol{n}}$ with vanishing variance in the limit $N \to \infty$ so that indeed $\boldsymbol{n}^\star$ overwhelms all other distributions of quantum states consistent with the constraints [see Eqs. (2.7)] imposed on the (astronomically large number of) virtual systems of the ensemble.

2.2.3   The Schrödinger-Hill approach

The problem we are now confronted with consists of determining the most probable distribution $\boldsymbol{n}^\star$ discussed in the preceding Sections 2.2.1 and 2.2.2. This approach was first suggested by Schrödinger [20] and later extended and reformulated by Hill [21]. We already noted that, mathematically speaking, $\boldsymbol{n}^\star$ is that distribution maximizing $W\left(\boldsymbol{n} \right)$ subject to the constraints specified in Eqs. (2.7). Thus, $\boldsymbol{n}^\star$ can be determined using the method of Lagrangian multipliers detailed in Appendix B.3. However, we apply this technique to Eq. (2.11) rather than directly to $W\left(\boldsymbol{n} \right)$. This is possible because both $W\left(\boldsymbol{n} \right)$

and $\ln W\,(\boldsymbol{n})$ are monotonic functions of N, which permits us to prefer one over the other on the grounds of mere computational convenience.

By analogy with the treatment developed in Appendix B.3 we begin by introducing

$$F\,(\boldsymbol{n}) \equiv \ln W\,(\boldsymbol{n}) + \boldsymbol{\lambda} \cdot \boldsymbol{\varphi}\,(\boldsymbol{n}) \tag{2.31}$$

where [see Eq. (2.7)]

$$\boldsymbol{\varphi}\,(\boldsymbol{n}) = \begin{pmatrix} \varphi_1\,(\boldsymbol{n}) \\ \varphi_2\,(\boldsymbol{n}) \\ \varphi_3\,(\boldsymbol{n}) \\ \varphi_4\,(\boldsymbol{n}) \end{pmatrix} \tag{2.32}$$

and in the current situation $\boldsymbol{\lambda}^{\mathrm{T}} \equiv (\lambda_1, \lambda_2, \lambda_3, \lambda_4)$ is a four-dimensional vector of undetermined Lagrangian multipliers. From Eq. (2.11), it is easy to verify that

$$\frac{\partial \ln W\,(\boldsymbol{n})}{\partial n_{j(N,s_z)}} = -\ln n_{j(N,s_z)} - 1 \tag{2.33}$$

because N is constant (but can be made arbitrarily large). Moreover, it follows from Eqs. (2.7) that a matrix $\boldsymbol{\varphi}'\,(\boldsymbol{n})$ can be formed whose elements in each row are given by

$$\varphi'_1\,(\boldsymbol{n}) \equiv \frac{\partial \varphi_1\,(\boldsymbol{n})}{\partial n_{j(N,s_z)}} = 1 \tag{2.34a}$$

$$\varphi'_2\,(\boldsymbol{n}) \equiv \frac{\partial \varphi_2\,(\boldsymbol{n})}{\partial n_{j(N,s_z)}} = E_{j(N,s_z)} \tag{2.34b}$$

$$\varphi'_3\,(\boldsymbol{n}) \equiv \frac{\partial \varphi_3\,(\boldsymbol{n})}{\partial n_{j(N,s_z)}} = N \tag{2.34c}$$

$$\varphi'_4\,(\boldsymbol{n}) \equiv \frac{\partial \varphi_4\,(\boldsymbol{n})}{\partial n_{j(N,s_z)}} = s_z \tag{2.34d}$$

Hence, using Eqs. (2.33) and (2.34) in Eq. (B.47), we readily obtain

$$n^{\star}_{j(Ns_z)} = \exp\left(-\lambda_1 - 1\right) \exp\left(-\lambda_2 E_{j(N,s_z)}\right) \exp\left(-\lambda_3 N\right) \exp\left(-\lambda_4 s_z\right) \tag{2.35}$$

after a trivial rearrangement of terms. We can immediately eliminate one of the four Lagrangian multipliers by summing both sides of the previous expression over j, N, and s_z, which leads to the more compact expression

$$n^{\star}_{j(Ns_z)} = \frac{\mathsf{N}}{\chi} \exp\left(-\lambda_2 E_{j(N,s_z)}\right) \exp\left(-\lambda_3 N\right) \exp\left(-\lambda_4 s_z\right) \tag{2.36}$$

where

$$\chi \equiv \sum_{N,s_z \to j} \exp\left(-\lambda_2 E_{j(N,s_z)}\right) \exp\left(-\lambda_3 N\right) \exp\left(-\lambda_4 s_z\right)$$

$$= \sum_N \exp\left(-\lambda_3 N\right) \sum_{s_z} \exp\left(-\lambda_4 s_z\right) \mathcal{Q}\left(N, s_z\right) \tag{2.37}$$

is the partition function of a so-called *grand mixed isostress isostrain* ensemble[4] and

$$\mathcal{Q}\left(N, s_z\right) \equiv \sum_j \exp\left(-\lambda_2 E_{j(N,s_z)}\right) \tag{2.38}$$

is the canonical partition function, which of course, still depends on the substrate separation and the number of particles. It is termed "canonical" because of the distinguished role played by this ensemble, as we shall demonstrate below (see also Section 4.1).

An alternative formulation of Eq. (2.38) is obtained by realizing that, with the aid of Schrödinger's equation, one may write [see Eq. (2.1)]

$$\mathcal{Q}\left(N, s_z\right) = \sum_{j(N,s_z)} \sum_{k=0}^{\infty} \frac{\left(\pm\lambda_2 E_{j(N,s_z)}\right)^k}{k!}$$

$$= \sum_{j(N,s_z)} \sum_{k=0}^{\infty} \frac{\left(\pm\lambda_2\right)^k}{k!} \left\langle \psi_{j(N,s_z)} \left| \widehat{H}^k \right| \psi_{j(N,s_z)} \right\rangle$$

$$= \sum_{j(N,s_z)} \left\langle \psi_{j(N,s_z)} \left| \exp\left(-\lambda_2 \widehat{H}\right) \right| \psi_{j(N,s_z)} \right\rangle$$

$$\equiv \mathrm{Tr}\left[\exp\left(-\lambda_2 \widehat{H}\right)\right] \tag{2.39}$$

where the expression in $[\ldots]$ is a matrix element of the operator $\exp\left(-\lambda_2 \widehat{H}\right)$ defined via its MacLaurin-series expansion on the second line of Eq. (2.39). The trace operation is defined in Eq. (A.1). Provided one can expand an arbitrary complete set of orthonormal functions ϕ_i in terms of the eigenfunctions $\left\{\psi_{j(N,s_z)}\right\}$ of $\widehat{H}$ according to

$$\phi_i = \sum_{j=0}^{\infty} a_{ji} \psi_j \tag{2.40}$$

---

[4]The term *grand mixed isostress isostrain* ensemble is used to indicate that the slit-pore is materially coupled to its environment and that its thermodynamic state depends on the control of a set of stresses and strains.

where we changed the notation $j\,(N, s_z) \rightarrow j$ temporarily for the sake of clarity. We emphasize that the functions $\{\phi_i\}$ are not eigenfunctions of $\widehat{H}$. One can then show that

$$\sum_j \left\langle \psi_j \left| \exp\left(-\lambda_2 \widehat{H}\right) \right| \psi_j \right\rangle = \sum_j \left\langle \phi_j \left| \exp\left(-\lambda_2 \widehat{H}\right) \right| \phi_j \right\rangle \qquad (2.41)$$

which expresses the fact that, in order to calculate the canonical partition function $\mathcal{Q}$, one does not need to know the complete set of energy eigenfunctions. In fact, *any* complete set of orthonormal functions permits one to calculate $\mathcal{Q}$. A proof of Eq. (2.41) based on Parseval's equation [see Eq. (B.65)] is presented in Appendix B.5.2.

2.3 Connection with thermodynamics

From the previous discussion it should be apparent that, in the limit of an astronomically large number of systems, expressions such as the ones given in Eqs. (2.37) or (2.38) for χ or $\mathcal{Q}$, respectively, are exact. The correctness of this statement is, of course, intimately linked to the existence of a most probable distribution $\boldsymbol{n}^\star$ (see Section 2.2.2).

However, as expressions for both partition functions have been derived quantum mechanically, they do not permit us per se to calculate equilibrium properties of thermal systems because quantum (as well as classic) mechanics as such does not have any concept of temperature. However, an inspection of Eqs. (2.37) and (2.38) reveals that both partition functions still contain one or more yet-to-be-determined Lagrangian multipliers. These need to be calculated in a way that the resulting expressions become consistent with thermodynamics as it was introduced in Chapter 1.

2.3.1 Determination of Lagrangian multipliers

We begin by noticing from Eqs. (2.10) and (2.36) that the probability of observing (on average) our slit-pore in quantum state j with N particles and substrate separation s_z is given by

$$p_{j(N, s_z)} = \chi^{-1} \exp\left(-\lambda_2 E_{j(N, s_z)} - \lambda_3 N - \lambda_4 s_z\right) \qquad (2.42)$$

Consequently the average energy of the slit-pore (i.e., its mean energy) can be written as

$$\langle E \rangle = \sum_{N, s_z \to j} E_{j(N, s_z)} p_{j(N, s_z)} \qquad (2.43)$$

such that its exact differential may be cast as

$$\text{d}\langle E \rangle = \sum_{N,s_z \to j} \left[E_{j(N,s_z)}\text{d}p_{j(N,s_z)} + p_{j(N,s_z)}\text{d}E_{j(N,s_z)} \right] \tag{2.44}$$

As we pointed out in Section 2.2.1, all systems in the ensemble are subject to the same fixed compressional strains proportional to s_x and s_y; in addition, they are all exposed to the same fixed shear strain αs_{x0}. Hence, the energy eigenvalues $E_{j(N,s_z)} = E_{j(N,s_z)}(s_x, s_y, \alpha s_{x0})$ so that

$$
\begin{aligned}
\text{d}E_{j(N,s_z)} &= \left(\frac{\partial E_{j(N,s_z)}}{\partial s_x} \right)_{\{\cdot\}\backslash s_x} \text{d}s_x + \left(\frac{\partial E_{j(N,s_z)}}{\partial s_y} \right)_{\{\cdot\}\backslash s_y} \text{d}s_y \\
&+ \left(\frac{\partial E_{j(N,s_z)}}{\partial (\alpha s_{x0})} \right)_{\{\cdot\}\backslash(\alpha s_{x0})} \text{d}(\alpha s_{x0})
\end{aligned}
\tag{2.45}
$$

and from Eq. (2.42)

$$E_{j(N,s_z)} = -\lambda_2^{-1}\left(\ln p_{j(N,s_z)} + \ln \chi + \lambda_3 N + \lambda_4 s_z \right) \tag{2.46}$$

Inserting now Eqs. (2.45) and (2.46) into Eq. (2.44), we obtain

$$
\begin{aligned}
\text{d}\langle E \rangle &= -\frac{1}{\lambda_2}\text{d}\left(\sum_{N,s_z \to j} p_{j(N,s_z)} \ln p_{j(N,s_z)} \right) - \frac{\lambda_3}{\lambda_2}\text{d}\langle N \rangle - \frac{\lambda_4}{\lambda_2}\text{d}\langle s_z \rangle \\
&+ \left\langle \left(\frac{\partial E_{j(N,s_z)}}{\partial s_x} \right)_{\{\cdot\}\backslash s_x} \right\rangle \text{d}s_x + \left\langle \left(\frac{\partial E_{j(N,s_z)}}{\partial s_y} \right)_{\{\cdot\}\backslash s_y} \right\rangle \text{d}s_y \\
&+ \left\langle \left(\frac{\partial E_{j(N,s_z)}}{\partial (\alpha s_{x0})} \right)_{\{\cdot\}\backslash(\alpha s_{x0})} \right\rangle \text{d}(\alpha s_{x0})
\end{aligned}
\tag{2.47}
$$

where, of course, $\langle \ldots \rangle$ stands for $\sum_{N,s_z \to j} \ldots p_{j(N,s_z)}$. To arrive at Eq. (2.47), we invoked the *Principle of Probability Conservation*, which is

$$
\begin{aligned}
\sum_{N,s_z \to j} \ln p_{j(N,s_z)}\text{d}p_{j(N,s_z)} &= \sum_{N,s_z \to j} \ln p_{j(N,s_z)}\text{d}p_{j(N,s_z)} + \sum_{N,s_z \to j} \text{d}p_{j(N,s_z)} \\
&= \text{d}\left(\sum_{N,s_z \to j} p_{j(N,s_z)} \ln p_{j(N,s_z)} \right)
\end{aligned}
\tag{2.48}
$$

This follows because

$$\sum_{N,s_z \to j} p_{j(N,s_z)} = 1 \tag{2.49}$$

and therefore

$$d \left(\sum_{N,s_z \to j} p_{j(N,s_z)} \right) = \sum_{N,s_z \to j} dp_{j(N,s_z)} = 0 \qquad (2.50)$$

We now *postulate* that $\langle E \rangle = \mathcal{U}$. Hence, comparing Eq. (2.47) with the thermodynamic expression given in Eq. (1.43), we readily conclude that

$$\lambda_3 = -\lambda_2 \mu \qquad (2.51a)$$
$$\lambda_4 = -\lambda_2 \tau_{zz} A_{z0} \qquad (2.51b)$$

Moreover, we identify

$$\tau_{xx} A_{x0} = \left\langle \left(\frac{\partial E_{j(N,s_z)}}{\partial s_x} \right)_{\{\cdot\}\backslash s_x} \right\rangle \qquad (2.52a)$$

$$\tau_{yy} A_{y0} = \left\langle \left(\frac{\partial E_{j(N,s_z)}}{\partial s_y} \right)_{\{\cdot\}\backslash s_y} \right\rangle \qquad (2.52b)$$

$$\tau_{xz} A_{x0} = \left\langle \left(\frac{\partial E_{j(N,s_z)}}{\partial (\alpha s_{x0})} \right)_{\{\cdot\}\backslash(\alpha s_{x0})} \right\rangle \qquad (2.52c)$$

so that we are still left with λ_2 as the remaining undetermined Lagrangian multiplier.

2.3.2   Statistical expression for the entropy

Comparing Eqs. (2.47) and (1.43), one may speculate that physical meaning can be assigned to λ_2 by relating it to the entropy $\mathcal{S}$, i.e., by making the identification

$$d\mathcal{S} = \frac{1}{\lambda_2 T} dw \equiv \phi(w) \, dw \qquad (2.53)$$

where

$$w \equiv - \sum_{N,s_z \to j} p_{j(N,s_z)} \ln p_{j(N,s_z)} \qquad (2.54)$$

Moreover, as pointed out in Section 1.3, $d\mathcal{S}$ is an exact differential such that the factor $1/(\lambda_2 T)$ in Eq. (2.53) must be an integrating factor because dw cannot be expected to be an exact differential per se. Hence, we can write

$$\phi(w) \, dw \equiv df(w) \qquad (2.55)$$

where df must be an exact differential because of Eq. (2.53). Thus, we may integrate $d\mathcal{S}$ along an arbitrarily chosen path in thermodynamic state space to obtain (see Section 1.3)

$$\mathcal{S} = f(w) + c \qquad (2.56)$$

where c is an integration constant.

Two additional properties of $\mathcal{S}$ turn out to be important for the current discussion.

1. As we saw in Section 1.3, $\mathcal{S}$ is a state function such that

$$\int_C d\mathcal{S} = \mathcal{S}(C_a) - \mathcal{S}(C_b) \tag{2.57}$$

   where $C_a = (\mathcal{U}_a, N_a, \boldsymbol{\sigma}_a)$ and $C_b = (\mathcal{U}_b, N_b, \boldsymbol{\sigma}_b)$ denote the start and end point of some thermodynamic transformation, respectively, along an open path C in thermodynamic state space. Without loss of generality, we can therefore set $c = 0$ in Eq. (2.56).

2. $\mathcal{S}$ is additive in the sense that, if we were given two systems A and B both in the same thermodynamic equilibrium state, $\mathcal{S}_{A \cup B} = \mathcal{S}_A + \mathcal{S}_B$. This prompts us to conclude from Eq. (2.56) that

$$f(w_{A \cup B}) = f(w_A) + f(w_B) \tag{2.58}$$

   should also hold.

Let A and B be systems pertaining to our ensemble. We may then ask what is the probability of finding A in a quantum state characterized by $E_{i(N^A, s_z^A)}$, whereas B is simultaneously in a quantum state represented by an energy eigenvalue $E_{j(N^B s_z^B)}$? As we tacitly assumed the interactions between systems in the ensemble to be very weak, so that the spectrum of energy eigenstates is essentially that of an isolated system, we conclude that the answer to the question is given by

$$p_{A \cup B} = p_{i(N^A s_z^A)} p_{j(N^B s_z^B)} \equiv p_A p_B \tag{2.59}$$

so that, from Eq. (2.54), we may write

$$
\begin{aligned}
w_{A \cup B} &= -\sum_A \sum_B p_{A \cup B} \ln p_{A \cup B} \\
&= -\sum_A p_A \ln p_A \underbrace{\sum_B p_B}_{=1} - \sum_B p_B \ln p_B \underbrace{\sum_A p_A}_{=1} \\
&= -\sum_A p_A \ln p_A - \sum_B p_B \ln p_B = w_A + w_B
\end{aligned}
\tag{2.60}
$$

where the summation extends over all quantum states accessible to systems A and B, respectively. In other words, because of Eqs. (2.58) and (2.60), our function f must satisfy the relation

$$f(w_{A\cup B}) = f(w_A + w_B) = f(w_A) + f(w_B) \tag{2.61}$$

Differentiating both sides of the previous expression with respect to w_A and w_B, respectively, we obtain

$$\frac{df(w_A + w_B)}{d(w_A + w_B)} \frac{\partial(w_A + w_B)}{\partial w_A} = \frac{df(w_A + w_B)}{d(w_A + w_B)} = \frac{df(w_A)}{dw_A} \tag{2.62a}$$

$$\frac{df(w_A + w_B)}{d(w_A + w_B)} \frac{\partial(w_A + w_B)}{\partial w_B} = \frac{df(w_A + w_B)}{d(w_A + w_B)} = \frac{df(w_B)}{dw_B} \tag{2.62b}$$

Equations (2.62) result because

$$\frac{\partial(w_A + w_B)}{\partial w_A} = \frac{\partial(w_A + w_B)}{\partial w_B} = 1 \tag{2.63}$$

or, alternatively, because w_A and w_B are mutually independent of one another. This is a direct consequence of our initial assumption of a negligibly weak interaction between various systems of the ensemble such that in each system the spectrum of energy eigenstates is identically the same.

Equations (2.62) can only be satisfied simultaneously if and only if

$$\frac{df(w_A)}{dw_A} = \frac{df(w_B)}{dw_B} = k_B \tag{2.64}$$

where k_B is some constant henceforth referred to as *Boltzmann's constant*. It is then immediately clear from Eq. (2.55) that the integrating factor is given by

$$\phi(w) = \frac{df}{dw} = k_B = \frac{1}{\lambda_2 T} \tag{2.65}$$

where the far right side follows by comparing the right side of Eq. (2.53) with the left side of Eq. (2.55). Equation (2.65) determines the remaining Lagrangian multiplier λ_2 such that from Eqs. (1.43) and (2.47) we have

$$S = -k_B \sum_{N,s_z \to j} p_{j(N,s_z)} \ln p_{j(N,s_z)} \tag{2.66}$$

as the expression for the entropy in the grand mixed isostress isostrain ensemble. Similar expressions may be derived in other statistical physical ensembles [21].

Replacing in Eq. (2.66) $p_{j(N,s_z)}$ through the expression given in Eq. (2.10), another interesting expression may be derived. We readily obtain

$$S = -\frac{k_B}{N} \sum_{N,s_z \to j} n^\star_{j(N,s_z)} \ln n^\star_{j(N,s_z)} + \frac{k_B \ln N}{N} \sum_{N,s_z \to j} n^\star_{j(N,s_z)} \qquad (2.67)$$

Noting also that $S_{\text{sup}} \equiv NS$ is the entropy of the isolated supersystem we may write from the previous expression

$$S_{\text{sup}} = -k_B \left[\sum_{N,s_z \to j} n^\star_{j(N,s_z)} \ln n^\star_{j(N,s_z)} - N \ln N \right] = -k_B \ln W(\boldsymbol{n}^\star) \qquad (2.68)$$

where Eqs. (2.7a) and (2.11) have also been employed. As energy $\mathcal{E}$ [see Eq. (2.7b)], substrate separation S_z [see Eq. (2.7d)], and the number of molecules $\mathcal{N}$ [see Eq. (2.7c)] of the isolated supersystem are fixed, we can interpret $W(\boldsymbol{n}^\star)$ as the number of quantum states accessible to the supersystem $\Omega(\mathcal{N}, \mathcal{V}, \mathcal{E})$ subject to the constraints spelled out in Eqs. (2.7). Hence, we may write

$$S_{\text{sup}} = -k_B \ln \Omega(\mathcal{N}, \mathcal{V}, \mathcal{E}) \qquad (2.69)$$

which is nothing but the well-known expression for the entropy in the microcanonical ensemble.[5] In Eq. (2.69), S_{sup} may be interpreted as the thermodynamic potential associated with the microcanonical ensemble.

2.3.3 Statistical physical averages and thermodynamics

Inserting now Eqs. (2.51a), (2.51b), and (2.65) into the expression for the partition function χ in Eq. (2.37), we eventually arrive at

$$\chi = \sum_{N,s_z \to j} \exp\left(\frac{\tau_{zz} A_{z0} s_z}{k_B T}\right) \exp\left(\frac{\mu N}{k_B T}\right) \exp\left(\frac{-E_{j(Ns_z)}}{k_B T}\right) \qquad (2.70a)$$

$$= \sum_{Ns_z} \exp\left(\frac{\tau_{zz} A_{z0} s_z}{k_B T}\right) \exp\left(\frac{\mu N}{k_B T}\right) \mathcal{Q} \qquad (2.70b)$$

$$= \sum_{s_z} \exp\left(\frac{\tau_{zz} A_{z0} s_z}{k_B T}\right) \Xi \qquad (2.70c)$$

---

[5] An expression of the form of Eq. (2.69) was first proposed by the Austrian physicist Ludwig Boltzmann (1844–1906) within the framework of classic physics. To honor Boltzmann's contributions to statistical physics, Eq. (2.69) is engraved on his tombstone at the Zentralfriedhof in Vienna (Austria).

where we define the partition function in the grand canonical ensemble Ξ in Eq. (2.70c) for later reference.

Another important relation can be derived from the expression for the entropy in the grand mixed isostress isostrain ensemble. Substituting in Eq. (2.66), $p_{j(N,s_z)}$ via Eq. (2.42) together with Eqs. (2.51a), (2.51b), and (2.65) permits us to write

$$
\begin{aligned}
-k_{\mathrm{B}}T \ln \chi &= \langle E \rangle - TS - \mu \langle N \rangle - \tau_{zz} A_{z0} \langle s_z \rangle \\
&= \mathcal{U} - TS - \mu N - \tau_{zz} A_{z0} s_z \equiv \Phi
\end{aligned}
\tag{2.71}
$$

where the second line follows from the equivalence between statistical and thermodynamic expressions in the thermodynamic limit. Equation (2.71) also defines the thermodynamic potential Φ of the grand mixed isostress isostrain ensemble. Taking the exact differential of Φ, we realize that as

$$
\begin{aligned}
\mathrm{d}\Phi = {}& -\mathcal{S}\mathrm{d}T - N\mathrm{d}\mu \\
& + \tau_{xx} A_{x0}\mathrm{d}s_x + \tau_{yy} A_{y0}\mathrm{d}s_y - A_{z0}s_{z0}\mathrm{d}\tau_{zz} + A_{x0}\tau_{xz}\mathrm{d}\left(\alpha s_{x0}\right)
\end{aligned}
\tag{2.72}
$$

$\Phi = \Phi\left(T, \mu, s_x, s_y, \tau_{zz}, \alpha s_{x0}\right)$ depends on a mixed set of stress $\left(\tau_{zz}\right)$ and strains $\left(s_x, s_y, \alpha s_{x0}\right)$ as natural parameters. Therefore, the ensemble characterized by the partition function χ has been termed the grand mixed isostress isostrain ensemble in Section 2.2.3. Note that, in the derivation of Eq. (2.72), Eq. (1.43) has also been employed.

A comparison between Eqs. (2.71) and (2.72) reveals that statistical averages are related to first-order derivatives of the relevant thermodynamic potential. For example, from Eqs. (2.70a) and (2.71), we have

$$
\begin{aligned}
\left(\frac{\partial \Phi}{\partial \mu}\right)_{\{\cdot\}\backslash\mu} &= -k_{\mathrm{B}}T \left(\frac{\partial \ln \chi}{\partial \mu}\right)_{\{\cdot\}\backslash\mu} \\
&= -\frac{1}{\chi} \sum_{N, s_z \to j} N \exp\left(\frac{\tau_{zz} A_{z0} s_z}{k_{\mathrm{B}}T}\right) \exp\left(\frac{\mu N}{k_{\mathrm{B}}T}\right) \exp\left(-\frac{E_{jN s_z}}{k_{\mathrm{B}}T}\right) \\
&= -\langle N \rangle
\end{aligned}
\tag{2.73}
$$

Another example is the average substrate separation that the system will attain if exposed to a fixed external (normal) stress. From Eqs. (2.70a) and

(2.71), one immediately sees by a similar token that

$$
\begin{aligned}
\left(\frac{\partial \Phi}{\partial \tau_{zz}}\right)_{\{\cdot\}\backslash \tau_{zz}} &= -k_{\mathrm{B}}T\left(\frac{\partial \ln \chi}{\partial \tau_{zz}}\right)_{\{\cdot\}\backslash \tau_{zz}} \\
&= -\frac{1}{\chi}\left[\sum_{N,s_z \to j} s_z \exp\left(\frac{\tau_{zz}A_{z0}s_z}{k_{\mathrm{B}}T}\right)\exp\left(\frac{\mu N}{k_{\mathrm{B}}T}\right)\right. \\
&\qquad\qquad \left. \times \exp\left(-\frac{E_{j(N,s_z)}}{k_{\mathrm{B}}T}\right)\right] = -\langle s_z\rangle
\end{aligned}
\tag{2.74}
$$

2.3.4   Fluctuations

By virtue of their nature, thermal averages like $\langle N\rangle$ [see Eq. (2.73)] or $\langle s_z\rangle$ [see Eq. (2.74)] are associated with a certain variance of their instantaneous values N or s_z. The variance is associated with second-order partial derivatives of the relevant thermodynamic potential(s). For example, from Eq. (2.73), we find

$$
\begin{aligned}
\left(\frac{\partial^2 \Phi}{\partial \mu^2}\right)_{\{\cdot\}\backslash \mu} &= -\left(\frac{\partial \langle N\rangle}{\partial \mu}\right)_{\{\cdot\}\backslash \mu} \\
&= -\frac{1}{\chi k_{\mathrm{B}}T}\left[\sum_{N,s_z \to j} N^2 \exp\left(\frac{\tau_{zz}A_{z0}s_z}{k_{\mathrm{B}}T}\right)\exp\left(\frac{\mu N}{k_{\mathrm{B}}T}\right)\right. \\
&\qquad\qquad \left. \times \exp\left(-\frac{E_{j(N,s_z)}}{k_{\mathrm{B}}T}\right)\right] \\
&\quad + \frac{1}{k_{\mathrm{B}}T}\left[\frac{1}{\chi}\sum_{N,s_z \to j} N \exp\left(\frac{\tau_{zz}A_{z0}s_z}{k_{\mathrm{B}}T}\right)\exp\left(\frac{\mu N}{k_{\mathrm{B}}T}\right)\right. \\
&\qquad\qquad \left. \times \exp\left(-\frac{E_{j(N,s_z)}}{k_{\mathrm{B}}T}\right)\right]^2 \\
&= -\frac{1}{k_{\mathrm{B}}T}\left[\langle N^2\rangle - \langle N\rangle^2\right] \equiv -\frac{\sigma_{\mathrm{N}}^2}{k_{\mathrm{B}}T}
\end{aligned}
\tag{2.75}
$$

where σ_{N} is the variance of $\langle N\rangle$. Noting from Eq. (2.75) that $\Phi \propto \langle N\rangle \propto \sigma_{\mathrm{N}}^2$, it follows that

$$
\lim_{\langle N\rangle \to \infty}\frac{\sigma_{\mathrm{N}}}{\langle N\rangle} = \lim_{\langle N\rangle \to \infty}\frac{\mathcal{O}\left(\sqrt{\langle N\rangle}\right)}{\mathcal{O}\left(\langle N\rangle\right)} = \lim_{\langle N\rangle \to \infty}\mathcal{O}\left(1/\sqrt{\langle N\rangle}\right) = 0
\tag{2.76}
$$

In other words, as we let the system size grow, fluctuations about the average value of a thermal quantity become negligible. However, there are exceptions.

For example, near a critical point or at a spinodal (i.e., at the stability limit of a fluid phase in the metastable regime), the above is no longer true. However, it should also be noted that, at a discontinuous phase transition (i.e., *before* entering the metastable regime), Eq. (2.76) still holds.

To arrive at the far right side of Eq. (2.76), we employed the *additivity* of Φ. That is, if we prepare two identical equilibrium systems characterized by Φ_A and Φ_B, then the composite system has twice the number of particles than any one of the two separate systems and $\Phi_{A \cup B} = \Phi_A + \Phi_B = 2\Phi$. It is important to realize that *additivity* in this sense does not necessarily imply *extensivity* in the sense of Section 1.6, where we showed that thermodynamic potentials may be homogeneous functions of degree one in only a few of their extensive variables depending on the symmetry of the external field represented by the (structured) substrates.

Expressions similar to the one given in Eq. (2.75) result for other thermal averages as well. In most cases, a relation like Eq. (2.76) also holds, which is essentially the reason why one measures "sharp" values of thermal properties in macroscopic systems (except for statistical errors). A result peculiar to confined fluids is, however, obtained by considering thermal fluctuations of s_z about $\langle s_z \rangle$. By manipulations precisely parallel to those leading to the expression for σ_N in Eq. (2.75), one can show that

$$\left(\frac{\partial^2 \Phi}{\partial \tau_{zz}^2} \right)_{\{ \cdot \} \setminus \tau_{zz}} = -\frac{A_{z0}}{k_B T} \left[\langle s_z^2 \rangle - \langle s_z \rangle^2 \right] \equiv -\frac{A_{z0}}{k_B T} \sigma_{s_z}^2 \qquad (2.77)$$

Consider now the *relative* variance $\sigma_{s_z} / \langle s_z \rangle$. Clearly, it does not decrease if we make the system larger in the x- or y-directions but will only decrease as we approach the bulk limit by letting s_z approach infinity. However, this inevitably changes the physical nature of the confined fluid.

Notice that the same problem would be encountered for the ratio $\sigma_N / \langle N \rangle$ in cases where the external potential prevents the thermodynamic potential to be a homogeneous function of degree one in either s_x or s_y and these two quantities would be increased by noninteger factors. These examples show that for confined fluids one has to be cautious to approach the thermodynamic limit properly.

These considerations also have an immediate practical consequence for the experimental determination of $\langle s_z \rangle$ (in, say, the SFA experiment; see Section 5.3.1) because these average values would inevitably be plagued by a certain systematic error on account of thermal fluctuations. In other words, there will be a distribution of substrate separations s_z about the average value that is characteristic of the physical nature of the confined fluid. Therefore, this error cannot be reduced by any more sophisticated device but is characteristic of the experimental situation. However, in practice the error may still be small enough to be inconsequential from a purely practical perspective.

2.4   Equivalence of ensembles

So far our discussion has focused on the grand mixed isostress isostrain ensemble, which was devised to mimic at a microscopic level operating conditions encountered in the SFA. However, this ensemble is by no means distinguished among other conceivable ensembles. In other words, it was defined merely out of convenience. The close relation between the grand mixed isostress isostrain ensemble and the more conventional canonical or grand canonical ensembles is already suggested by Eqs. (2.70). However, one can demonstrate that the grand mixed isostress isostrain ensemble is not only linked to others but also formally *equivalent* to the other two.

The demonstration of equivalence departs from the observation that partition functions in *any* statistical physical ensemble comply with the general form given in Eq. (B.52) with individual summands as given in Eq. (B.54). Thus, according to the discussion in Appendix B.4, we may approximate the grand mixed isostress isostrain ensemble partition function χ in Eq. (2.70a) by the maximum terms in the sums appearing on the right side of that equation

$$\Phi = -k_{\mathrm{B}}T \ln \chi \simeq -\mu N^{\star} - \tau_{zz} A_{z0} s_z^{\star} - k_{\mathrm{B}}T \ln \mathcal{Q}\left(N^{\star}, T, s_{\mathrm{x}}, s_{\mathrm{y}}, s_z^{\star}, \alpha s_{\mathrm{x}0}\right) \quad (2.78)$$

where $N^{\star}$ and $s_z^{\star}$ denote values of N and s_z, respectively, corresponding to the maximum term in the sums in Eq. (2.70a). The previous expression may be recast with the aid of Eq. (2.71) as

$$\mathcal{F}\left(N^{\star}, T, s_{\mathrm{x}}, s_{\mathrm{y}}, s_z^{\star}, \alpha s_{\mathrm{x}0}\right) = -k_{\mathrm{B}}T \ln \mathcal{Q}\left(N^{\star}, T, s_{\mathrm{x}}, s_{\mathrm{y}}, s_z^{\star}, \alpha s_{\mathrm{x}0}\right) \quad (2.79)$$

where the definition of the (Helmholtz) free energy given in Eq. (1.50) has also been employed. These expressions show that the canonical ensemble for fixed $N = N^{\star}$ and $s_z = s_z^{\star}$ is equivalent to the grand mixed isostress isostrain ensemble, which is solely a consequence both of the infinitely large number of terms contributing to the partition function χ in Eq. (2.70a) and of the functional form of these terms [cf., Eq. (B.54)].

Likewise, starting from Eq. (2.70c) and employing the maximum term approximation (see Section B.4), we may write

$$\Phi = -k_{\mathrm{B}}T \ln \chi \simeq -\tau_{zz} A_{z0} s_z^{\star} - k_{\mathrm{B}}T \ln \Xi\left(T, \mu, s_{\mathrm{x}}, s_{\mathrm{y}}, s_z^{\star}, \alpha s_{\mathrm{x}0}\right) \quad (2.80)$$

which together with Eq. (2.71) and the definition of the grand potential given in Eq. (1.51) yields

$$\Omega\left(T, \mu, s_{\mathrm{x}}, s_{\mathrm{y}}, s_{\mathrm{z}}^{\star}, \alpha s_{\mathrm{x}0}\right) = -k_{\mathrm{B}}T \ln \Xi\left(T, \mu, s_{\mathrm{x}}, s_{\mathrm{y}}, s_{\mathrm{z}}^{\star}, \alpha s_{\mathrm{x}0}\right) \qquad (2.81)$$

after some straightforward algebraic manipulations. This expression states that the grand mixed isostress isostrain ensemble is equivalent to the grand canonical one at fixed $s_{\mathrm{z}} = s_{\mathrm{z}}^{\star}$. Similar considerations can be employed to demonstrate equivalence of any two (physically sensible) statistical physical ensembles.

In closing, we note that fluctuations in statistical physics arise in two separate contexts. As we saw in Section 2.2.2, fluctuations around the most probable distribution of quantum states are completely suppressed on account of the astronomically large number of systems of which an ensemble is composed in the thermodynamic limit. However, *within* this most probable distribution, *thermal* fluctuations arise. Their magnitude can be quite large depending on the specific thermodynamic conditions, as we pointed out above.

2.5   The classic limit

Until now, our formulation of statistical thermodynamics has been based on quantum mechanics. This is reflected by the definition of the canonical ensemble partition function Q, which turns out to be linked to matrix elements of the Hamiltonian operator $\widehat{H}$ in Eq. (2.39). However, the systems treated below exist in a region of thermodynamic state space where the *exact* quantum mechanical treatment may be abandoned in favor of a *classic* description. The transition from quantum to classic statistics was worked out by Kirkwood [22, 23] and Wigner [24] and is rarely discussed in standard texts on statistical physics. For the sake of completeness, self-containment, and as background information for the interested readers we summarize the key considerations in this section.

In essence, the classic approximation will enable us to replace sums over discrete quantum states in the various partition functions [see, for example, Eqs. (2.70)] by integrals over the so-called phase space spanned by the $6N$ coordinates r^N and momenta p^N that fully specify a classic microstate of a system in which the N particles have only translational degrees of freedom. Thus, $\left(r^N\right)^{\mathrm{T}} = (r_1, r_2, \ldots, r_N)$ and $\left(p^N\right)^{\mathrm{T}} = (p_1, p_2, \ldots, p_N)$ are (the transposes of) $3N$-dimensional vectors. We refer to $\Gamma = r^N \otimes p^N$ as a point in (classic) phase space and to r^N as a configuration of particles.

At the core of the following development are symmetry properties of the wave function. In quantum mechanics, one is concerned with two general types of particles, namely, Bosons (e.g., photon, π-meson, and ^{4}He) and Fermions (e.g., electron, proton, neutron, and ^{3}He). Fermions are characterized by half-integer spin whereas for Bosons the spin is integer. This difference has an immediate consequence for the wave function describing the state of a many-particle Fermion as opposed to a Boson system. If we are dealing with identical particles, then the wave function must be antisymmetric with respect to the exchange of any two particles as far as a fermionic system is concerned; for a bosonic system, the wave function remains symmetric during such a permutation. This is a consequence of Pauli's Principle, which can only be fully appreciated within the framework of quantum electrodynamics.

2.5.1 Symmetry considerations

Consider a system of N particles whose quantum state can be described by a wave function $\psi(r_1, r_2, \ldots, r_N)$, where r_i is the location of particle i in space. Here and below we deliberately ignore spin coordinates. This seems justified because including those additional degrees of freedom would not affect our final conclusions concerning conditions that render adequate the desired classic treatment of microscopic systems, with which we shall be concerned almost exclusively[6][19].

Let us then define an operator $\widehat{\mathcal{P}}_r$ that serves to permute pairs of coordinates r_i and r_j. If this operator is applied to the wave function, we may thus write

$$(\pm 1)^{\left|\widehat{\mathcal{P}}_r\right|} \widehat{\mathcal{P}}_r \psi_m (r_1, r_2, \ldots, r_N) = \psi_m (r_1, r_2, \ldots, r_N) \qquad (2.82)$$

where $\left|\widehat{\mathcal{P}}_r\right|$ is the total number of pair-wise exchanges into which the permutation can be decomposed and the set $\{\psi_m\}$ is taken to be energy eigenfunctions [see Eq. (2.1)]. However, here and below we shall be working in the canonical ensemble for convenience. Hence, N and s_z are assumed constant such that we may replace the label $j(N, s_z)$ on the wave function (and its associated energy eigenvalues) by an integer m, say (which is independent of both N and s_z). In Eq. (2.82), the prefactor (i.e., the parity of the wave function) is $+1$ for a system of Bosons because of their *symmetric* wave function. In other words, regardless of whether $\left|\widehat{\mathcal{P}}_r\right|$ is even or odd, the sign of the wave function remains unaltered. For a system of Fermions, however, the prefactor is -1. The wave function changes sign if $\left|\widehat{\mathcal{P}}_r\right|$ is odd so that the wave function

---

[6]The only exception is the confined ideal quantum gas discussed in Section 5.7.4.3.

of a system of Fermions is *antisymmetric* with respect to the permutation of a *single* pair of particles. These symmetry properties give rise to different types of statistics associated with the names Bose-Einstein and Fermi-Dirac [17]. Moreover, in a system of N particles, $N!$ different permutations are possible so that from Eq. (2.82)

$$\psi_m\left(\boldsymbol{r}^N\right) = \frac{1}{N!}\sum_{\hat{\mathcal{P}}_r}(\pm 1)^{|\hat{\mathcal{P}}_r|}\,\hat{\mathcal{P}}_r\psi_m\left(\boldsymbol{r}^N\right) \tag{2.83}$$

follows.

To proceed, we now expand the wave function in terms of eigenfunctions $\exp\left(i\boldsymbol{k}^N\cdot\boldsymbol{r}^N\right)$ of the momentum operator,

$$\psi_m\left(\boldsymbol{r}^N\right) = \int a_m\left(\boldsymbol{k}^N\right)\exp\left(i\boldsymbol{k}^N\cdot\boldsymbol{r}^N\right)\mathrm{d}\boldsymbol{k}^N \tag{2.84}$$

where the N-dimensional vector $\boldsymbol{k}^N$ whose elements are $\boldsymbol{k}_i = \boldsymbol{p}_i/\hbar$, where $\boldsymbol{p}_i$ is the momentum of particle i and $\hbar = h/2\pi$. That

$$\exp\left(i\boldsymbol{k}^N\cdot\boldsymbol{r}^N\right) = \prod_{j=1}^{N}\exp\left(i\boldsymbol{k}_j\cdot\boldsymbol{r}_j\right) \tag{2.85}$$

is an eigenfunction of the momentum operator as can easily be verified in space representation by considering

$$\hat{\boldsymbol{p}}_j\exp\left(i\boldsymbol{k}^N\cdot\boldsymbol{r}^N\right) = \frac{h}{i}\frac{\partial}{\partial\boldsymbol{r}_j}\prod_{j=1}^{N}\exp\left(i\boldsymbol{k}_j\cdot\boldsymbol{r}_j\right) = \boldsymbol{p}_j\exp\left(i\boldsymbol{k}^N\cdot\boldsymbol{r}^N\right) \tag{2.86}$$

which also shows that $\boldsymbol{p}_j$ is the eigenvalue of the momentum operator $\hat{\boldsymbol{p}}_j$. Mathematically, Eq. (2.84) constitutes a Fourier inversion of the wave function from momentum ($\boldsymbol{k}$) to real ($\boldsymbol{r}$) space. The corresponding Fourier transformation [i.e., the transformation conjugate to Eq. (2.84)] is then given by

$$a_m\left(\boldsymbol{k}^N\right) = \frac{1}{(2\pi)^{3N}}\int\psi_m\left(\boldsymbol{r}^N\right)\exp\left(-i\boldsymbol{k}^N\cdot\boldsymbol{r}^N\right)\mathrm{d}\boldsymbol{r}^N \tag{2.87}$$

Thus, inserting the expansion given in Eq. (2.84) into the right side of Eq. (2.83), we obtain

$$\psi_m\left(\boldsymbol{r}^N\right) = \frac{1}{\sqrt{N!}}\int a_m\left(\boldsymbol{k}^N\right)\theta\left(\boldsymbol{k}^N,\boldsymbol{r}^N\right)\mathrm{d}\boldsymbol{k}^N \tag{2.88}$$

where

$$\theta\left(\boldsymbol{k}^N, \boldsymbol{r}^N\right) = \frac{1}{\sqrt{N!}} \sum_{\widehat{\mathcal{P}}_r} (\pm 1)^{|\widehat{\mathcal{P}}_r|} \widehat{\mathcal{P}}_r \exp\left(i\boldsymbol{k}^N \cdot \boldsymbol{r}^N\right)$$

$$= \frac{1}{\sqrt{N!}} \sum_{\widehat{\mathcal{P}}_k} (\pm 1)^{|\widehat{\mathcal{P}}_k|} \widehat{\mathcal{P}}_k \exp\left(i\boldsymbol{k}^N \cdot \boldsymbol{r}^N\right) \qquad (2.89)$$

and the permutation operator $\widehat{\mathcal{P}}_k$ acts on elements of the vector $\boldsymbol{k}^N$. The second line of Eq. (2.89) follows because any permutation of a pair of elements of the vector $\boldsymbol{r}^N$ is equivalent to a permutation of the corresponding elements in the vector $\boldsymbol{k}^N$ as far as the scalar product $\boldsymbol{k}^N \cdot \boldsymbol{r}^N$ is concerned. Note that the function $\theta\left(\boldsymbol{k}^N, \boldsymbol{r}^N\right)$ introduced in Eqs. (2.88) and (2.89) is an eigenfunction of the permutation operator to an eigenvalue of ± 1.

Eventually, the current considerations will serve to express the partition function $\mathcal{Q}$ as a (multidimensional) integral over configuration $\boldsymbol{r}^N$ and momentum space $\boldsymbol{p}^N$. It is therefore necessary to investigate the effect of $\widehat{\mathcal{P}}_k$ on integrals of the general type

$$I = \int F\left(\ldots, \boldsymbol{k}_{n-1}, \boldsymbol{k}_n, \boldsymbol{k}_{n+1}, \ldots\right) \ldots d\boldsymbol{k}_{n-1}, d\boldsymbol{k}_n, d\boldsymbol{k}_{n+1} \ldots \quad (2.90a)$$

$$= \int F\left(\ldots, \boldsymbol{k}_n, \boldsymbol{k}_{n-1}, \boldsymbol{k}_{n+1}, \ldots\right) \ldots d\boldsymbol{k}_{n-1}, d\boldsymbol{k}_n, d\boldsymbol{k}_{n+1} \ldots \quad (2.90b)$$

$$= \int F\left(\ldots, \boldsymbol{k}'_{n-1}, \boldsymbol{k}'_n, \boldsymbol{k}'_{n+1}, \ldots\right) \ldots d\boldsymbol{k}'_n, d\boldsymbol{k}'_{n-1}, d\boldsymbol{k}'_{n+1} \ldots \quad (2.90c)$$

$$= \int F\left(\ldots, \boldsymbol{k}'_{n-1}, \boldsymbol{k}'_n, \boldsymbol{k}'_{n+1}, \ldots\right) \ldots d\boldsymbol{k}'_{n-1}, d\boldsymbol{k}'_n, d\boldsymbol{k}'_{n+1} \ldots \quad (2.90d)$$

where we exchanged variables $\boldsymbol{k}_{n-1} \leftrightarrow \boldsymbol{k}_n$ [see Eqs. (2.90a) and (2.90b)], changed variables according to $\boldsymbol{k}_i \to \boldsymbol{k}'_i$ ($i \leq n - 2 \wedge i \geq n + 2$), $\boldsymbol{k}_{n-1} \to \boldsymbol{k}'_n$, and $\boldsymbol{k}_n \to \boldsymbol{k}'_{n-1}$. Finally, the order of integration is changed between Eqs. (2.90c) and (2.90d), thus bringing us back to the original expression in Eq. (2.90a), which leads us to conclude that permutations of the arguments of the function F are irrelevant for the integral I. Hence, the operations detailed in Eqs. (2.90) prompt us to write

$$\int F\left(\boldsymbol{k}^N\right) d\boldsymbol{k}^N = \frac{1}{N!} \int \sum_{\widehat{\mathcal{P}}_k} \widehat{\mathcal{P}}_k F\left(\boldsymbol{k}^N\right) d\boldsymbol{k}^N \qquad (2.91)$$

Applying these considerations to the right side of Eq. (2.88), we obtain

$$
\begin{aligned}
\psi_m \left(\boldsymbol{r}^N \right) &= \frac{1}{\left(N! \right)^{3/2}} \int \sum_{\widehat{\mathcal{P}}_k} \widehat{\mathcal{P}}_k \left[a_m \left(\boldsymbol{k}^N \right) \theta \left(\boldsymbol{k}^N, \boldsymbol{r}^N \right) \right] \mathrm{d}\boldsymbol{k}^N \\
&= \frac{1}{\left(N! \right)^{3/2}} \int \sum_{\widehat{\mathcal{P}}_k} \left(\pm 1 \right)^{\left| \widehat{\mathcal{P}}_k \right|} \theta \left(\boldsymbol{k}^N, \boldsymbol{r}^N \right) \widehat{\mathcal{P}}_k a_m \left(\boldsymbol{k}^N \right) \\
&= \frac{1}{N!} \int b_m \left(\boldsymbol{k}^N \right) \theta \left(\boldsymbol{k}^N, \boldsymbol{r}^N \right) \mathrm{d}\boldsymbol{k}^N \qquad (2.92)
\end{aligned}
$$

Notice that $\theta \left(\boldsymbol{k}^N, \boldsymbol{r}^N \right)$ is introduced as the *complete* set of permutations caused by the action of the set of permutation operators $\left\{ \widehat{\mathcal{P}}_k \right\}$ on the function $\exp \left(i \boldsymbol{k}^N \cdot \boldsymbol{r}^N \right)$ [see Eq. 2.85)]. Hence, if any one of the members of $\left\{ \widehat{\mathcal{P}}_k \right\}$ acts on $\theta \left(\boldsymbol{k}^N, \boldsymbol{r}^N \right)$, it is impossible by definition to generate new permutations that are not already accounted for by the original $\theta \left(\boldsymbol{k}^N, \boldsymbol{r}^N \right)$. However, the action of a specific operator $\widehat{\mathcal{P}}_k$ may change the parity of $\theta \left(\boldsymbol{k}^N, \boldsymbol{r}^N \right)$ such that the factor $\left(\pm 1 \right)^{\left| \widehat{\mathcal{P}}_k \right|}$ appears as the eigenvalue of the operator $\widehat{\mathcal{P}}_k$ acting on its eigenfunction $\theta \left(\boldsymbol{k}^N, \boldsymbol{r}^N \right)$. Coefficients $b_m \left(\boldsymbol{k}^N \right)$ in Eq. (2.92) are then given by

$$
\begin{aligned}
b_m \left(\boldsymbol{k}^N \right) &= \frac{1}{\sqrt{N!}} \sum_{\widehat{\mathcal{P}}_k} \left(\pm 1 \right)^{\left| \widehat{\mathcal{P}}_k \right|} \widehat{\mathcal{P}}_k a_m \left(\boldsymbol{k}^N \right) \\
&= \frac{1}{\left(2\pi \right)^{3N}} \int \psi_m \left(\boldsymbol{r}^N \right) \theta^* \left(\boldsymbol{k}^N, \boldsymbol{r}^N \right) \mathrm{d}\boldsymbol{r}^N \qquad (2.93)
\end{aligned}
$$

where the asterisk denotes the complex conjugate and Eqs. (2.87) and (2.89) have also been used.

2.5.2 Kirkwood-Wigner theory

We now realize from Eqs. (2.39) and (2.65) that

$$
Q = \mathrm{Tr} \left[\exp \left(-\widehat{H}/k_{\mathrm{B}} T \right) \right] = \sum_{j(N, s_z)} \left\langle \psi_j \left| \exp \left(-\widehat{H}/k_{\mathrm{B}} T \right) \right| \psi_j \right\rangle \qquad (2.94)
$$

in the representation-independent Dirac ("bra"–"ket") notation. As we already emphasized in Section 2.2.3 (see also Appendix B.5.2), the trace is independent of the choice of a specific basis set. However, the subsequent discussion will benefit from turning immediately to a specific representation in which we express the set $\{\psi_j\}$ in the basis of eigenfunctions of the permutation operator $\theta\left(\boldsymbol{k}^N, \boldsymbol{r}^N\right)$ [see Eq. (2.92)]. Moreover, writing the Hamiltonian operator defined in Eq. (2.2) as

$$\widehat{H} = -\frac{\hbar^2}{2m} \sum_{i=1}^{N} \Delta_i + \widehat{U}\left(\widehat{\boldsymbol{r}}^N; N, s_z\right) \tag{2.95}$$

in space representation, we may recast Eq. (2.94) more explicitly as

$$\begin{aligned}
\mathcal{Q} &= \frac{1}{N!} \sum_m \iint \psi_m^*\left(\boldsymbol{r}^N\right) b_m\left(\boldsymbol{k}^N\right) \exp\left(-\frac{\widehat{H}}{k_{\mathrm{B}}T}\right) \theta\left(\boldsymbol{k}^N, \boldsymbol{r}^N\right) \mathrm{d}\boldsymbol{k}^N \mathrm{d}\boldsymbol{r}^N \\
&= \frac{1}{N!} \frac{1}{(2\pi)^{3N}} \iiint \left[\sum_m \psi_m^*\left(\boldsymbol{r}^N\right) \psi_m\left(\boldsymbol{r}^{N'}\right)\right] \theta^*\left(\boldsymbol{k}^N, \boldsymbol{r}^N\right) \\
&\quad \times \exp\left(-\frac{\widehat{H}}{k_{\mathrm{B}}T}\right) \theta\left(\boldsymbol{k}^N, \boldsymbol{r}^N\right) \mathrm{d}\boldsymbol{r}^{\prime N} \mathrm{d}\boldsymbol{r}^N \mathrm{d}\boldsymbol{k}^N
\end{aligned} \tag{2.96}$$

In Eq. (2.95), we use the shorthand $\Delta_i \equiv \nabla_i \cdot \nabla_i$ [$\widehat{\boldsymbol{p}}_i = (\hbar/i)\nabla_i$, $\nabla_i^{\mathrm{T}} \equiv (\partial/\partial x_i, \partial/\partial y_i, \partial/\partial z_i)$ in Cartesian coordinates] to simplify the notation. The second line of Eq. (2.96) follows by inserting Eq. (2.93) and noticing that the two integrations over spatial coordinates have to be independent. In Eq. (2.96), we also used the fact that the Hamiltonian operator as defined in Eq. (2.95) acts only on particle coordinates and not on their momenta. Because the $\{\psi_j\}$ form a complete orthonormal basis, the term in brackets can be replaced through the celebrated completeness relation

$$\sum_m \left[\psi_m^*\left(\boldsymbol{r}^N\right) \psi_m\left(\boldsymbol{r}^{\prime N}\right)\right] = \delta\left(\boldsymbol{r}^N - \boldsymbol{r}^{\prime N}\right) \tag{2.97}$$

where δ denotes Dirac's δ-function. A proof of Eq. (2.97) is deferred to Appendix B.6.2. The integration over $\boldsymbol{r}^{N'}$ in Eq. (2.96) can then be carried out in closed form to yield

$$\begin{aligned}
\mathcal{Q} &= \frac{1}{N!} \frac{1}{(2\pi)^{3N}} \iint \theta^*\left(\boldsymbol{k}^N, \boldsymbol{r}^N\right) \exp\left(-\frac{\widehat{H}}{k_{\mathrm{B}}T}\right) \theta\left(\boldsymbol{k}^N, \boldsymbol{r}^N\right) \mathrm{d}\boldsymbol{k}^N \mathrm{d}\boldsymbol{r}^N \\
&= \frac{1}{(N!)^2} \frac{1}{(2\pi)^{3N}} \sum_{\widehat{\mathcal{P}}'_k} \sum_{\widehat{\mathcal{P}}''_k} (\pm 1)^{|\widehat{\mathcal{P}}'_k + \widehat{\mathcal{P}}''_k|} \iint \left[\widehat{\mathcal{P}}''_k \exp\left(-i\boldsymbol{k}^N \cdot \boldsymbol{r}^N\right)\right] \\
&\quad \times \exp\left(-\frac{\widehat{H}}{k_{\mathrm{B}}T}\right) \left[\widehat{\mathcal{P}}'_k \exp\left(i\boldsymbol{k}^N \cdot \boldsymbol{r}^N\right)\right] \mathrm{d}\boldsymbol{k}^N \mathrm{d}\boldsymbol{r}^N
\end{aligned} \tag{2.98}$$

where the second and third lines follow from Eq. (2.89). In Eq. (2.98), $\widehat{\mathcal{P}}'_k$ and $\widehat{\mathcal{P}}''_k$ are permutation operators acting on elements of $\boldsymbol{k}^N$ (see Section 2.5.1).

To proceed, let us define the inverse permutation operator $\widehat{\mathcal{P}}_k{}'^{-1}$, which reverses the permutation effected by $\widehat{\mathcal{P}}'_k$ (see Section 2.5.1). We may then start with an expression like the one given in Eq. (2.90d), change the order of integration [see Eq. (2.90c)], and change variables to arrive at Eq. (2.90b). Finally, let $\widehat{\mathcal{P}}_k{}'^{-1}$ act on the integral in Eq. (2.90b) to recover the original expression given in Eq. (2.90a). Therefore, like its counterpart $\widehat{\mathcal{P}}'_k$, the inverse operator $\widehat{\mathcal{P}}_k{}'^{-1}$ does not affect integrals like the one given in Eq. (2.98). Thus, we may multiply both terms in brackets in Eq. (2.98) by $\widehat{\mathcal{P}}_k{}'^{-1}$ to obtain

$$
\mathcal{Q} = \frac{1}{N!}\frac{1}{(2\pi)^{3N}}\sum_{\widehat{\mathcal{P}}_k}(\pm 1)^{|\hat{P}_k|}\iint \left[\widehat{\mathcal{P}}_k \exp\left(-i\boldsymbol{k}^N\cdot\boldsymbol{r}^N\right)\right]\exp\left(-\frac{\widehat{H}}{k_{\mathrm{B}}T}\right)
$$
$$
\times \widehat{1}\exp\left(i\boldsymbol{k}^N\cdot\boldsymbol{r}^N\right)\mathrm{d}\boldsymbol{r}^N\mathrm{d}\boldsymbol{p}^N \tag{2.99}
$$

where $\widehat{\mathcal{P}}_k = \widehat{\mathcal{P}}_k{}'^{-1}\widehat{\mathcal{P}}''_k$ and $\widehat{1} \equiv \widehat{\mathcal{P}}_k{}'^{-1}\widehat{\mathcal{P}}'_k$. In Eq. (2.99), $\widehat{1}$ is the unit operator. We also used the fact that $\sum_{\hat{P}'_k}\sum_{\hat{P}''_k}\dots \longrightarrow N!\sum_{\hat{P}_k}$ because each of the two independent sums generates $N!$ permutations.

2.5.3 The canonical partition function in the classic limit

Turning now to the classic limit, we assume that for sufficiently high temperatures

$$
\widehat{H} \rightarrow H\left(\boldsymbol{r}^N,\boldsymbol{p}^N\right) = \sum_{i=1}^{N}\frac{p_i^2}{2m} + U\left(\boldsymbol{r}^N\right) \tag{2.100}
$$

where H is the Hamiltonian function of classic mechanics. This is the simplest approximation that will turn out to give rise to correction terms arising solely from symmetry properties of the wave function. For a more refined treatment, the interested reader is referred to Appendix B.6.4 where we show that Eq. (2.100) is the lowest-order approximation in our semiclassic treatment of the quantum mechanical Hamiltonian operator. However, for current purposes Eq. (2.100) will be sufficient (see the discussion in Section 5.7.4.3).

Replacing in Eq. (2.99) the Hamiltonian operator by its classic analog and noting that $\mathrm{d}\boldsymbol{k}^N = \hbar^{-3N}\mathrm{d}\boldsymbol{p}^N$ ($\hbar = h/2\pi$, h is Planck's constant), we can rewrite Eq. (2.99) as

$$
\begin{aligned}
Q \;=\; & \frac{1}{h^{3N}\,N!} \iint \exp\left[-\frac{H\left(\boldsymbol{r}^N,\boldsymbol{p}^N\right)}{k_{\mathrm B}T}\right] \mathrm{d}\boldsymbol{r}^N \mathrm{d}\boldsymbol{p}^N \\
& \pm \frac{1}{(2\pi)^{3N}\,N!} \sum_{\widehat{\mathcal P}_k\neq\widehat{1}} \iint \left[\widehat{\mathcal P}_k \exp\left(-i\boldsymbol{k}^N\cdot\boldsymbol{r}^N\right)\right] \exp\left(i\boldsymbol{k}^N\cdot\boldsymbol{r}^N\right) \\
& \times \exp\left[-\frac{H\left(\boldsymbol{r}^N,\boldsymbol{p}^N\right)}{k_{\mathrm B}T}\right] \mathrm{d}\boldsymbol{r}^N \mathrm{d}\boldsymbol{k}^N \qquad (2.101)
\end{aligned}
$$

where we split the sum over permutations $\widehat{\mathcal P}_k$ such that the identity permutation $\widehat{\mathcal P}_k = \widehat{1}$ is treated separately, thus giving rise to the first summand in Eq. (2.101). Note that in front of the second summand of Eq. (2.101) the operator "$\pm$" instead of "$(\pm 1)^{|\widehat{\mathcal P}_k|}$" arises where the "$+$" sign refers to Bosons and the "$-$" sign to Fermions. This becomes possible by realizing that the sum over permutations in Eq. (2.99) generates $N!$ terms. The integer $N!$ is inevitably an *even* number regardless of N. If we treat the identity permutation separately, only $N! - 1$ terms are generated by the sum over permutations in Eq. (2.101), which turns out to be an *odd* number independent of N. Therefore, symmetry properties of the wave function dictate that in a fermionic system the factor in front of the second term in Eq. (2.101) must always be negative, whereas a positive contribution results for a system of Bosons.

One may derive a more explicit expression for the second summand in Eq. (2.101) by realizing that the sum over all permutations (except the identity permutation) can be rearranged such that terms resulting from a permutation of pairs of particle momenta, triplets, quadruplets, and so on are grouped together. Each of these sums contains products of two, three, four, and so on factors of the general form

$$
\pm\frac{1}{h^{3N}\,N!} \iint \exp\left(\frac{i\boldsymbol{p}_k\cdot\boldsymbol{r}_{ij}}{\hbar}\right) \exp\left(-\frac{p_k^2}{2mk_{\mathrm B}T}\right) \mathrm{d}\boldsymbol{p}_k = \pm\frac{1}{\Lambda^3}\exp\left(-\frac{\pi r_{ij}^2}{\Lambda^2}\right)
$$
$$(2.102)$$

where

$$
\Lambda \equiv \sqrt{\frac{h^2}{2\pi m k_{\mathrm B}T}} \qquad (2.103)
$$

is the thermal de Broglie wavelength, $r_{ij} = |\boldsymbol{r}_{ij}| = |\boldsymbol{r}_i - \boldsymbol{r}_j|$ is the distance between particles i and j, and "$+$" and "$-$" refer to Bosons and Fermions, respectively.

To rationalize Eq. (2.102), consider as an example an operator $\widehat{\mathcal{P}}_k$ that exchanges the momenta of molecules i and j. From Eq. (2.101), one realizes that

$$
\begin{aligned}
\widehat{\mathcal{P}}_k \exp\left(-i\boldsymbol{k}^N \cdot \boldsymbol{r}^N\right) &= \widehat{\mathcal{P}}_k \prod_{l=1}^{N} \exp\left(-i\boldsymbol{k}_l \cdot \boldsymbol{r}_l\right) \\
&= \exp\left(-i\boldsymbol{k}_j \cdot \boldsymbol{r}_i\right)\exp\left(-i\boldsymbol{k}_i \cdot \boldsymbol{r}_j\right) \\
&\quad \times \prod_{l=1 \neq i,j}^{N} \exp\left(-i\boldsymbol{k}_l \cdot \boldsymbol{r}_l\right)
\end{aligned}
\tag{2.104}
$$

It is then clear that, in Eq. (2.101), $N-2$ of the N factors in the integrand survive, namely

$$
\begin{aligned}
\left[\widehat{\mathcal{P}}_k \exp\left(-i\boldsymbol{k}^N \cdot \boldsymbol{r}^N\right)\right]\exp\left(\boldsymbol{k}^N \cdot \boldsymbol{r}^N\right) &= \exp\left(-i\boldsymbol{k}_j \cdot \boldsymbol{r}_i\right)\exp\left(-i\boldsymbol{k}_i \cdot \boldsymbol{r}_j\right) \\
&\quad \times \exp\left(i\boldsymbol{k}_i \cdot \boldsymbol{r}_i\right)\exp\left(i\boldsymbol{k}_j \cdot \boldsymbol{r}_j\right) \\
&= \exp\left[i\boldsymbol{k}_i \cdot \boldsymbol{r}_{ij}\right]\exp\left[i\boldsymbol{k}_j \cdot \boldsymbol{r}_{ji}\right]
\end{aligned}
\tag{2.105}
$$

because

$$
\prod_{l=1 \neq i,j}^{N} \exp\left(-i\boldsymbol{k}_l \cdot \boldsymbol{r}_l\right)\exp\left(i\boldsymbol{k}_l \cdot \boldsymbol{r}_l\right) = 1
\tag{2.106}
$$

Thus, the surviving terms have precisely the form of the complex exponential function in the integrand of Eq. (2.102) if we replace the wave vector of a molecule by its momentum. The only difference between Eq. (2.105) and Eq. (2.102) is that for general permutations of three or more wave vectors the indices on momentum and on the distance vector may be different. The general form of the resulting terms remains unaltered as the reader may verify, which is a bit tedious but straightforward. The Gaussian in the integrand of Eq. (2.102) represents the kinetic part of the Hamiltonian function because of [see Eq. (2.100)]

$$
\exp\left(-\frac{H\left(\boldsymbol{r}^N, \boldsymbol{p}^N\right)}{k_\mathrm{B}T}\right) = \exp\left(-\frac{U\left(\boldsymbol{r}^N\right)}{k_\mathrm{B}T}\right)\prod_{k=1}^{N}\exp\left(-\frac{p_k^2}{2mk_\mathrm{B}T}\right)
\tag{2.107}
$$

The right side of Eq. (2.102) is then obtained as detailed in Appendix B.6.3 [see Eq. (B.103)].

Focusing from now on only on pair-wise permutations, it is immediately clear from the previous discussion that, for each permutation $\boldsymbol{k}_i \leftrightarrow \boldsymbol{k}_j$, $N-2$

integrals of the form

$$\frac{1}{h^3} \int_0^{2\pi} \int_0^\pi \int_0^\infty p^2 \exp\left(-\frac{p^2}{2mk_BT}\right) dp \sin\varphi \, d\varphi \, d\vartheta = \frac{1}{\Lambda^3} \tag{2.108}$$

arise because the permutation does not affect $N-2$ wave vectors and therefore Eq. (2.106) holds for those wave vectors. Hence, as far as the momentum integration is concerned in Eq. (2.101), one is left only with the kinetic part of the Hamiltonian function as the integrand. Realizing that momentum space is isotropic and homogeneous for a system in thermodynamic equilibrium, one may conveniently use spherical coordinates, which gives rise to a product of $N-2$ terms of the form presented in Eq. (2.108). In addition, a factor

$$\pm\frac{1}{\Lambda^6} \exp\left(-\frac{\pi r_{ij}^2}{\Lambda^2}\right) \exp\left(-\frac{\pi r_{ji}^2}{\Lambda^2}\right) = \pm\frac{1}{\Lambda^6} \exp\left(-\frac{2\pi r_{ij}^2}{\Lambda^2}\right) \tag{2.109}$$

appears, which is caused by the pair-wise permutation as discussed above.

Hence, we can recast Eq. (2.101) as

$$\mathcal{Q} = \mathcal{Q}_{cl} \pm \frac{1}{N!\Lambda^{3N}} \int \left[\sum_{i=1}^{N-1}\sum_{j=i+1}^N \exp\left(-\frac{2\pi r_{ij}^2}{\Lambda^2}\right) + \ldots\right]$$

$$\times \exp\left[-\frac{U\left(\mathbf{r}^N\right)}{k_BT}\right] d\mathbf{r}^N \tag{2.110}$$

where $\mathcal{Q}_{cl}$ is the "classic" partition function given by

$$\mathcal{Q}_{cl} = \frac{1}{h^{3N}N!} \iint \exp\left[-\frac{H\left(\mathbf{r}^N,\mathbf{p}^N\right)}{k_BT}\right] d\mathbf{r}^N d\mathbf{p}^N = \frac{Z}{N!\Lambda^{3N}} \tag{2.111}$$

In Eq. (2.111)

$$Z = \int \exp\left[-\frac{U\left(\mathbf{r}^N\right)}{k_BT}\right] d\mathbf{r}^N \tag{2.112}$$

is the configuration integral. From Eq. (2.110), it is clear that $\mathcal{Q} \simeq \mathcal{Q}_{cl}$ in the limit $r_{kl}/\Lambda \to \infty$. For most fluids, $\Lambda = \mathcal{O}\left(10^{-1}\sigma\right)$ or smaller where σ is the "diameter" of a spherical fluid molecule. As $r_{kl} = \mathcal{O}\left(\sigma\right)$, the second term in Eq. (2.110) vanishes to a good approximation provided m and T are not too small [see Eq. (2.103)].

However, it is important to realize that this does not hold for ideal gases. Because there is no interaction between the molecules of an ideal gas, the

Boltzmann factor $\exp\left[-U\left(\boldsymbol{r}^N\right)/k_BT\right]$ does not prevent molecules from approaching one another closely. In fact, because in an ideal gas $U\left(\boldsymbol{r}^N\right) = 0$ configurations are conceivable in which two or more molecules occupy the same point in space. This, in turn, implies that in an ideal gas the separation between any pair of molecules may and will become much smaller than the thermal de Broglie wavelength *regardless* of T and m. In other words, quantum effects are *maximized* in a semiclassic ideal gas compared with nonideal fluids.

In Eq. (2.110), we deliberately ignored additional terms arising from the permutation of triplets of momenta, quadruplets, and so on because we will be interested mostly in situations in which $r_{ij} \gg \Lambda$. Because these higher-order terms involve sums over products of three and more factors of the form $\exp\left(-\pi r_{ij}^2/\Lambda^2\right)$, their contribution to the semiclassic correction to $\mathcal{Q}_{\text{cl}}$ vanishes rapidly. We note in passing that a simple graphical method can be devised to derive explicit forms for the contributions from triplet, quadruplet, and so on permutations. A detailed discussion of this technique is, however, beyond the scope of this chapter.

It is instructive to summarize the above analysis, which is a bit involved at certain points, in a more qualitative manner. The argument is based on the well-known fact that in quantum mechanics one may associate a wave length

$$\widetilde{\lambda} = \frac{h}{|\boldsymbol{p}|} \tag{2.113}$$

with a free particle of mass m and momentum $\boldsymbol{p}$. Using [see Eq. (2.100)]

$$|\boldsymbol{p}| = \sqrt{2mE_{\text{kin}}} = \sqrt{3mk_BT} \tag{2.114}$$

we realize that

$$\widetilde{\lambda} = \sqrt{\frac{2\pi}{3}}\Lambda \tag{2.115}$$

where Eq. (2.103) has also been used. In Eq. (2.114) the far right side is obtained by invoking the equipartition theorem, which states that each of the (three) translational degrees of freedom of the particle contributes an amount of $k_BT/2$ to the total kinetic energy E_{kin}. The important point about Eq. (2.115) is that apparently the thermal de Broglie wavelength Λ is a measure of the size of the quantum mechanical wave packet $\widetilde{\lambda}$ associated with the (free) particle. Hence, one may argue that a classic description is adequate whenever the mean distance between the particles $\propto 1/\sqrt[3]{\rho}$ (ρ density) is larger than the size of the wave packet, i.e., larger than Λ. Typical examples of fluids where, on the contrary, quantum corrections are important arc H_2 or (liquid) He. In both cases, the fluid molecules have a small mass m and exist at quite low temperatures T.

Therefore, in the classic limit, thermal averages in the grand mixed isostress isostrain ensemble may be cast as

$$\langle O \rangle = \sum_{Ns_z} \int \mathrm{d}\boldsymbol{r}^N O\left(\boldsymbol{r}^N; N, s_z\right) p\left(\boldsymbol{r}^N; N, s_z\right) \tag{2.116}$$

where $O\left(\boldsymbol{r}^N; N, s_z\right)$ is a microscopic analog of the macroscopic thermal average $\langle O \rangle$. For example, taking $O\left(\boldsymbol{r}^N; N, s_z\right) = U\left(\boldsymbol{r}^N; N, s_z\right)$, $\langle O \rangle = \langle U \rangle$ would be the configurational contribution to the internal energy, which is to say that $\mathcal{U} = \frac{3}{2}Nk_BT + \langle U \rangle$. In Eq. (2.116)

$$p\left(\boldsymbol{r}^N; N, s_z\right) = \frac{1}{N!\Lambda^{3N}\chi_{cl}} \exp\left[\frac{\mu N}{k_BT}\right] \exp\left[\frac{\tau_{zz}A_{z0}s_z}{k_BT}\right] \exp\left[-\frac{U\left(\boldsymbol{r}^N; N, s_z\right)}{k_BT}\right] \tag{2.117}$$

is the probability density in the grand mixed isostress isostrain ensemble replacing its quantum statistical counterpart p_{jNs_z} in the classic limit where

$$\chi_{cl} = \sum_{Ns_z} \frac{1}{N!\Lambda^{3N}} \exp\left[\frac{\mu N}{k_BT}\right] \exp\left[\frac{\tau_{zz}A_{z0}s_z}{k_BT}\right] Z\left(N, s_z\right) \tag{2.118}$$

is the classic analog of the quantum statistical partition function defined in Eq. (2.37) and $Z\left(N, s_z\right)$ is the configuration integral already introduced in Eq. (2.112).

2.5.4 Laplace transformation of probability densities

Equations (2.116)–(2.118) can be rewritten in a slightly different way, which permits to derive a general relation between partition functions in various mixed isostress isostrain ensembles. Notice, for example, that we may define

$$\langle O\left(s_z\right) \rangle = \frac{1}{\Xi_{cl}} \sum_{N=0}^{\infty} \frac{1}{N!\Lambda^{3N}} \exp\left[\frac{\mu N}{k_BT}\right]$$
$$\times \int \mathrm{d}\boldsymbol{r}^N O\left(\boldsymbol{r}^N; N, s_z\right) \exp\left[-\frac{U\left(\boldsymbol{r}^N; N, s_z\right)}{k_BT}\right] \tag{2.119}$$

where

$$\Xi_{cl} = \sum_{N=0}^{\infty} \frac{1}{N!\Lambda^{3N}} \exp\left[\frac{\mu N}{k_BT}\right] \int \mathrm{d}\boldsymbol{r}^N \exp\left[-\frac{U\left(\boldsymbol{r}^N; N, s_z\right)}{k_BT}\right] \tag{2.120}$$

is the partition function of the grand canonical ensemble in the classic limit. Noticing that in the classic limit s_z is continuous on the interval $[0, \infty]$, we may thus rewrite Eq. (2.116) as

$$\langle O\left(\tau_{zz}\right)\rangle = \frac{\Xi_{cl}}{\chi_{cl}} \int\limits_0^\infty ds_z \exp\left[\frac{\tau_{zz} A_{z0} s_z}{k_B T}\right] \langle O\left(s_z\right)\rangle \qquad (2.121)$$

replacing in Eqs. (2.116) and (2.118), $\sum\limits_{s_z} \ldots \rightarrow \int\limits_{s_z} ds_z \ldots$. Comparing the previous expression with the Laplace transform of a function $f(t)$, namely

$$\mathcal{L}f(t) = F(s) \equiv \int\limits_0^\infty dt \exp\left(-st\right) f(t) \qquad (2.122)$$

we notice that except for the prefactor in Eq. (2.121) both expressions are formally equivalent if we make the identifications

$$t = s_z \qquad (2.123a)$$

$$s = -\frac{\tau_{zz} A_{z0}}{k_B T} \geq 0 \qquad (2.123b)$$

where s is positive semidefinite because $\tau_{zz} \leq 0$ on account of mechanical stability (see Section 1.3).

If the variance of the distribution of $\langle O\left(s_z\right)\rangle$ around its maximum $\langle O\left(s_z^\star\right)\rangle$ vanishes so that one may replace $\langle O\left(s_z\right)\rangle$ in Eq. (2.121) by $\langle O\left(s_z\right)\rangle \delta\left(s_z - s_z^\star\right)$, the integration in Eq. (2.121) may be carried out and one obtains

$$\langle O\left(\tau_{zz}\right)\rangle = \frac{\Xi_{cl}}{\chi_{cl}} \exp\left[\frac{\tau_{zz} A_{z0} s_z^\star}{k_B T}\right] \langle O\left(s_z^\star\right)\rangle = \langle O\left(s_z^\star\right)\rangle \qquad (2.124)$$

where Eq. (2.80) has also been employed. The equivalence between $\langle O\left(s_z^\star\right)\rangle$ and $\langle O\left(\tau_{zz}\right)\rangle$ may be interpreted as a reformulation of the equivalence between statistical physical ensembles demonstrated in Section 2.4.

Chapter 3

A first glimpse: One-dimensional hard-rod fluids

3.1 Introductory remarks

In Chapter 2, we saw that the configuration integral is the key quantity to be calculated if one seeks to compute thermal properties of classical (confined) fluids. However, it is immediately apparent that this is a formidable task because it requires a calculation of Z, which turns out to involve a $3N$-dimensional integration of a horrendously complex integrand, namely the Boltzmann factor $\exp\left[-U\left(r^N\right)/k_BT\right]$ [see Eq. (2.112)]. To evaluate Z we either need additional simplifying assumptions (such as, for example, *mean-field* approximations to be introduced in Chapter 4) or numerical approaches [such as, for instance, Monte Carlo computer simulations (see Chapters 5 and 6), or integral-equation techniques (see Chapter 7)].

There is an alternative, however. It consists of employing a sufficiently simple model for which the configuration integral can be computed *analytically* without having to take recourse to additional simplifying *assumptions*. The immediate disadvantage of such models, on the one hand, is a certain unavoidable lack of realism as far as experimental systems are concerned; they may therefore seem to be of little or no use to the practitioner. On the other hand, if not oversimplified, these models sometimes permit a surprisingly deep insight into the fundamental physics governing in a qualitatively similar fashion a more complex model or even experimental systems.

Based on this notion and reemphasizing the pedagogical impetus behind this work, we find it instructive to begin a deeper discussion of thermal properties of confined fluids by considering one of the simplest nontrivial models still capable of embracing the basic physics characteristic of these systems. This model consists of one-dimensional rods of length d without internal (e.g., spin-like) degrees of freedom, that is "molecules" that cannot orient themselves as, for example, in the one-dimensional Ising model. However, in the model considered below, molecules are not restricted to discrete sites on a one-dimensional lattice but may move continuously in space. In addition, a pair of rods is not allowed to overlap on account of "hard" repulsive interactions, that is, we are dealing with a one-dimensional model fluid whose properties are completely determined by entropic effects. We defer a more detailed discussion of this latter issue to Section 3.2.1 where we consider statistical thermodynamical aspects of our model system.

As we shall also see below, the confined hard-rod fluid exerts a stress on the confining "surfaces" that decays to the (negative) bulk pressure as the distance between the surfaces increases. Moreover, the stress oscillates as a function of substrate separation with a period roughly equal to the rod length. The oscillations may be interpreted as fingerprints of a confinement-induced structure (i.e., inhomogeneity) of the hard-rod fluid. As we shall demonstrate later in Section 5.3.4, this structure really is a stratification of the confined fluid. That is, in confinement, fluid molecules tend to arrange their centers of mass in individual layers. Stratification is perhaps the most prominent structural feature caused by solid surfaces that are separated by a distance comparable with the range of intermolecular interaction potentials.

However, we should also emphasize at the outset of this chapter that the confined fluid is not suitable for the study of yet another feature of central importance to us, namely confinement-induced phase transitions. This is because one can rigorously prove that, in general, one-dimensional systems cannot undergo discontinuous phase changes [16]. However, this apparent lack of realism is outweighed by the analyticity of the current model system and its capability to reproduce other important features of more sophisticated models or even experimental systems sufficiently realistically as we pointed out above. Our analysis in this chapter is based upon the original work by Vanderlick et al. [25] and has in part been adopted from the book of Davis [26].

3.2   Pure hard-rod bulk fluid

3.2.1   Statistical thermodynamics of hard-rod fluids

Let us begin by introducing a system of one-dimensional rods of length d where the interaction between a pair of rods is described by the intermolecular potential

$$u_{\text{ff}}^{\text{hr}}(z_{ij}) = \begin{cases} \infty, & z_{ij} \leq d \\ 0, & z_{ij} > d \end{cases} \tag{3.1}$$

That is to say the potential just prevents any pair of rods from interpenetrating. For convenience, we treat the fluid as a thermodynamically open system such that its equilibrium properties are determined by the grand potential [cf., Eqs. (1.32) and (1.51)]

$$d\Omega(T, \mu, L) = -S dT - N d\mu + \tau_{\text{b}} dL \tag{3.2}$$

where the last term expresses the mechanical work exchanged between the one-dimensional fluid of "volume" L and its surroundings. On account of the dimension of our system the bulk stress has dimensions of energy per unit *length* rather than unit *volume* as in a corresponding three-dimensional system.

The connection to the microscopic level of description is then provided by the standard relation [cf., Eq. (2.81)]

$$\Omega(T, \mu, L) = -k_{\text{B}} T \ln \Xi_{\text{cl}}(T, \mu, L) \tag{3.3}$$

where Ξ_{cl} is defined as in Eq. (2.120) replacing, however, Λ^{3N} by Λ^{N} because of the dimension of the system and because our molecules have only translational degrees of freedom. For the same reason, the configuration integral is given here by

$$Z_{\text{1d}} = \prod_{i=1}^{N} \int_{-L/2}^{L/2} dz_i \exp\left(-\frac{U(z^N)}{k_{\text{B}} T}\right) \tag{3.4}$$

where the configurational potential energy is given by [see Eq. (3.1)]

$$U(z^N) = \frac{1}{2} \sum_{i=1}^{N} \sum_{j \neq i=1}^{N} u_{\text{ff}}^{\text{hr}}(z_{ij}) \tag{3.5}$$

Thus, it is apparent from Eq. (3.5) that U depends on the hard-rod configuration z^N only through intermolecular distances $\{z_{ij}\}$. Thus, we can apply the analysis developed in Appendix C.2 and rewrite the configuration integral as [see Eq. (C.24)]

$$Z_{1d} = N! \int_{-L/2}^{L/2} dz_N \prod_{i=2}^{N} \int_{-L/2}^{z_i} dz_{i-1} \exp\left(-\frac{U\left(z^N\right)}{k_B T}\right) \tag{3.6}$$

Because of the form of the intermolecular interaction potential introduced in Eq. (3.1) we realize that the Boltzmann factor in the integrand of Eq. (3.6) can be zero if any pair of hard rods overlap; it will be equal to one, however, if this is not the case. Hence, we can readjust the integration limits in Eq. (3.6) to restrict the range of integration to those regions in which the Boltzmann factor does not vanish and rewrite Eq. (3.6) as

$$Z_{1d} = N! \int_{[-L+(2N-1)d]/2}^{(L-d)/2} dz_N \int_{[-L+(2N-3)d]/2}^{z_N - d} dz_{N-1} \cdots \int_{(-L+3d)/2}^{z_3 - d} dz_2 \int_{(-L+d)/2}^{z_2 - d} dz_1 \tag{3.7}$$

At this point, it is convenient to introduce a transformation of variables

$$z_i \rightarrow \tilde{z}_i = z_i + \frac{1}{2}\left[L - (2i - 1)d\right] \tag{3.8}$$

so that we can rewrite Eq. (3.7) as

$$\begin{aligned} Z_{1d} &= N! \int_0^{L-Nd} d\tilde{z}_N \int_0^{\tilde{z}_N} d\tilde{z}_{N-1} \cdots \int_0^{\tilde{z}_3} d\tilde{z}_2 \int_0^{\tilde{z}_2} d\tilde{z}_1 = \frac{N!}{(N-1)!} \int_0^{L-Nd} d\tilde{z}_N \tilde{z}_N^{N-1} \\ &= (L - Nd)^N \end{aligned} \tag{3.9}$$

Equation (3.9) permits us to verify that properties of our model system are completely determined by entropy. This becomes apparent by considering the statistical expression for the internal energy, namely

$$\begin{aligned} \mathcal{U} &= k_B T^2 \frac{\partial}{\partial T} \ln \mathcal{Q}_{1d} = k_B T^2 \frac{\partial}{\partial T} \ln \frac{Z_{1d}}{N! \Lambda^N} \\ &= \frac{1}{2} N k_B T + \frac{1}{Z_{1d}} \prod_{i=1}^{N} \int_{-L/2}^{L/2} dz_i U\left(z^N\right) \exp\left(-\frac{U\left(z^N\right)}{k_B T}\right) \\ &= \frac{1}{2} N k_B T + \underbrace{\langle U \rangle}_{=0} = \frac{1}{2} N k_B T \end{aligned} \tag{3.10}$$

which shows that $\mathcal{U}$ consists of only a kinetic contribution. Equation (3.10) is consistent with the equipartition theorem assigning a kinetic energy of $k_BT/2$ to each of the degrees of freedom of the N molecules. In any permissible configuration, no pair of hard rods is permitted to overlap on account of the infinitely hard repulsion between both rods [see Eq. (3.1)]. Therefore, the configurational potential energy $\langle U \rangle$ vanishes in Eq. (3.10). Hence, it follows from Eq. (1.50) that

$$\mathcal{F} = \frac{1}{2}Nk_BT - T\mathcal{S} \qquad (3.11)$$

is completely determined by entropy $\mathcal{S}$ apart from temperature, which appears to be a trivial scaling variable.

It also follows from Eqs. (2.120) and (3.9) that

$$\Xi_{cl} = \sum_{N=0}^{\infty} \frac{1}{N!\Lambda^N} \exp\left[\frac{\mu N}{k_BT}\right] (L - Nd)^N \qquad (3.12)$$

such that we obtain

$$
\begin{aligned}
\tau_b &= -P_b = \left(\frac{\partial\Omega}{\partial L}\right)_{\{\cdot\}\backslash L} \\
&= -\frac{k_BT}{\Xi} \sum_{N=0}^{\infty} \frac{N}{L - Nd} \frac{1}{N!\Lambda^N} \exp\left[\frac{\mu N}{k_BT}\right] (L - Nd)^N \\
&= -k_BT \left\langle \frac{N}{L - Nd} \right\rangle
\end{aligned}
\qquad (3.13)
$$

from Eqs. (3.2) and (3.3) for the bulk equation of state (i.e., the negative bulk pressure P_b).

3.2.2 Virial equation of state

Assuming that the density of the hard-rod fluid is below the density of a close-packed configuration, that is $Nd/L < 1$, we may expand

$$\frac{1}{d}\left\langle \frac{Nd/L}{1 - Nd/L} \right\rangle = \sum_{k=1}^{\infty} \left(\frac{\langle N \rangle}{L}\right)^k d^{k-1} = \sum_{k=1}^{\infty} \overline{\rho}^k d^{k-1} \qquad (3.14)$$

in a power series, which may be reinserted into Eq. (3.13) to give

$$\tau_b = -k_BT\overline{\rho} - k_BT \sum_{k=2}^{\infty} \overline{\rho}^k d^{k-1} \equiv \tau_b^{id} - k_BT \sum_{k=2}^{\infty} B_k\overline{\rho}^k \qquad (3.15)$$

In Eq. (3.15) we introduce the equation of state of the ideal gas of hard rods as

$$\tau_b^{id} = -k_B T \overline{\rho} \tag{3.16}$$

where the mean density is given by

$$\overline{\rho} \equiv \frac{\langle N \rangle}{L} \tag{3.17}$$

and

$$B_k \equiv d^{k-1} \tag{3.18}$$

is the k-th virial coefficient of the hard-rod gas. Because the members of the set $\{B_k\}$ are all positive semidefinite, it is clear that τ_b is a monotonically decreasing and continuous function of the density for all $\overline{\rho} d < 1$. Not unexpectedly, the hard-rod gas cannot undergo any phase transitions at any density as we already pointed out in Section 3.1.

The expression for the virial coefficients can also be obtained in a different fashion (see also Sec. 5.7.4). In this case, the derivation departs from

$$\Xi_{cl} = \exp\left(-\frac{\Omega}{k_B T}\right) = \exp\left(-\frac{\tau_b L}{k_B T}\right) \tag{3.19}$$

where the last equality follows because Ω is a homogeneous function of degree one in L (see Section 1.6.1). Expanding the exponential function, we may rewrite the previous expression as

$$\Xi_{cl} = \sum_{k=0}^{\infty} \frac{(-1)^k}{k!} \left(\frac{\tau_b L}{k_B T}\right)^k = \sum_{N=0}^{\infty} \frac{(L-Nd)^N}{N!} z^N \tag{3.20}$$

where the far right side follows directly from Eq. (3.12) and the definition of the activity

$$z \equiv \exp\left(\mu/k_B T\right)/\Lambda \tag{3.21}$$

Following Rowlinson [27] and McQuarrie and Rowlinson [28], we write as an *ansatz*

$$\tau_b = k_B T \sum_{j=1}^{\infty} b_j z^j \tag{3.22}$$

Inserting Eq. (3.22) into Eq. (3.20), we obtain

$$\begin{aligned} \Xi &= 1 - L\left(b_1 z + b_2 z^2 + b_3 z^3 + \ldots\right) + \frac{L^2}{2}\left(b_1^2 z^2 + 2b_1 b_2 z^3 + \ldots\right) \\ &\quad - \frac{L^3}{6}\left(b_1^3 z^3 + \ldots\right) + \mathcal{O}\left(z^4\right) \\ &= 1 + (L-d)z + \frac{(L-2d)^2}{2} z^2 + \frac{(L-3d)^3}{6} z^3 + \mathcal{O}\left(z^4\right) \end{aligned} \tag{3.23}$$

Comparing in this expression coefficients of equal power in z, it follows after straightforward but somewhat tedious algebraic manipulations that

$$b_1 = \frac{d}{L} - 1 \tag{3.24a}$$

$$b_2 = d\left(1 - \frac{3}{2}\frac{d}{L}\right) \tag{3.24b}$$

$$b_3 = d^2\left(\frac{13}{3}\frac{d}{L} - \frac{3}{2}\right) \tag{3.24c}$$

However, following the discussion at the beginning of this section, we wish to express τ_b in a power series in $\overline{\rho}$ rather than one in terms of the activity. To accomplish this we notice from Eq. (3.20) that

$$\overline{\rho} = \underbrace{\frac{1}{L}z\frac{\partial \ln \Xi}{\partial z}}_{\langle N \rangle} = -\frac{z}{k_{\mathrm{B}}T}\frac{\partial \tau_b}{\partial z} = -\sum_{j=1}^{\infty}jb_j z^j \tag{3.25}$$

where we also used Eqs. (3.20) and (3.22). We now make another *ansatz* expressing

$$z = a_1\overline{\rho} + a_2\overline{\rho}^2 + a_3\overline{\rho}^3 + \mathcal{O}\left(\overline{\rho}^4\right) \tag{3.26}$$

which we insert into the far right side of Eq. (3.25) to yield

$$\begin{aligned}\overline{\rho} =\ & -b_1\left(a_1\overline{\rho} + a_2\overline{\rho}^2 + a_3\overline{\rho}^3 + \ldots\right) \\ & -2b_2\left(a_1^2\overline{\rho}^2 + 2a_1a_2\overline{\rho}^3 + \ldots\right) \\ & -3b_3\left(a_1^3\overline{\rho}^3 + \ldots\right) + \mathcal{O}\left(\overline{\rho}^4\right)\end{aligned} \tag{3.27}$$

Equating in this expression terms of equal power in $\overline{\rho}$ on both sides, we obtain

$$a_1 = -\frac{1}{b_1} \tag{3.28a}$$

$$a_2 = -\frac{2b_2 a_1^2}{b_1} = -\frac{2b_2}{b_1^3} \tag{3.28b}$$

$$a_3 = -\frac{4b_2 a_1 a_2}{b_1} - \frac{3b_3 a_1^3}{b_1} = -\frac{8b_2^2}{b_1^5} + \frac{3b_3}{b_1^4} \tag{3.28c}$$

Inserting Eqs. (3.28) into Eq. (3.26) allows us to reexpress the activity in terms of the set of the original expansion coefficients $\{b_j\}$. Using the resulting expression for z and inserting it into Eq. (3.22) eventually gives us the desired expansion of τ_b in terms of a power series in $\bar{\rho}$,

$$
\begin{aligned}
\tau_b &= k_B T b_1 \left(a_1 \bar{\rho} + a_2 \bar{\rho}^2 + a_3 \bar{\rho}^3 + \ldots \right) \\
&\quad + k_B T b_2 \left(a_1^2 \bar{\rho}^2 + 2 a_1 a_2 \bar{\rho}^3 + \ldots \right) \\
&\quad + k_B T b_3 \left(a_1^3 \bar{\rho}^3 + \ldots \right) + \mathcal{O} \left(\bar{\rho}^4 \right) \\
&= -k_B T \bar{\rho} + k_B T \left(a_2 b_1 + a_1^2 b_2 \right) \bar{\rho}^2 \\
&\quad + k_B T \left(a_3 b_1 + 2 a_1 a_2 b_2 + a_1^3 b_3 \right) \bar{\rho}^3 \mathcal{O} \left(\bar{\rho}^4 \right) \\
&\equiv \tau_b^{\mathrm{id}} - k_B T B_2 \bar{\rho}^2 - k_B T B_3 \bar{\rho}^3 + \mathcal{O} \left(\bar{\rho}^4 \right)
\end{aligned}
\tag{3.29}
$$

so that the second and third virial coefficients are easily identified as

$$
\begin{aligned}
B_2 (L) &\equiv - \left(a_2 b_1 + a_1^2 b_2 \right) = \frac{b_2}{b_1^2} \\
&= \frac{d}{2} \frac{2 - 3d/L}{\left(1 - d/L \right)^2}
\end{aligned}
\tag{3.30a}
$$

$$
\begin{aligned}
B_3 (L) &\equiv - \left(a_3 b_1 + 2 a_1 a_2 b_2 + a_1^3 b_3 \right) = \frac{4 b_2^2}{b_1^4} - \frac{2 b_3}{b_1^3} \\
&= d^2 \frac{\left(2 - 3d/L \right)^2}{\left(1 - d/L \right)^4} - \frac{d^2}{3} \frac{9 - 26d/L}{\left(1 - d/L \right)^3} .
\end{aligned}
\tag{3.30b}
$$

Comparing these expressions with the equivalent ones in Eq. (3.18), it is apparent that the latter are independent of the system size, whereas the former still depend on the ratio d/L. This is because Eq. (3.18) was obtained directly from the equation of state, that is, from Eq. (3.13), which involves a summation over all particle numbers. In other words, we took the thermodynamic limit $N/L = \mathrm{const}$, $N, L \to \infty$ prior to expanding τ_b in a power series in $\bar{\rho}$. The coefficients $B_2 (L)$ and $B_3 (L)$, on the other hand, were obtained from the first few terms of the expansion in Eq. (3.23). In other words, starting from Eq. (3.23), we arrive at the final expressions in Eqs. (3.30) without taking the thermodynamic limit anywhere during the entire derivation. However, we recover B_2 and B_3 given by Eq. (3.18) by noticing that for the bulk L can be made arbitrary large so that $d/L \ll 1$ and therefore

$$
B_2 = \lim_{L \to \infty} B_2 (L) = \lim_{L \to \infty} \frac{d}{2} \frac{2 - 3d/L}{\left(1 - d/L \right)^2} = d
\tag{3.31a}
$$

$$
\begin{aligned}
B_3 &= \lim_{L \to \infty} B_3 (L) = d^2 \lim_{L \to \infty} \left[\frac{\left(2 - 3d/L \right)^2}{\left(1 - d/L \right)^4} - \frac{1}{3} \frac{9 - 26d/L}{\left(1 - d/L \right)^3} \right] \\
&= 4 d^2 - 3 d^2 = d^2
\end{aligned}
\tag{3.31b}
$$

which are in complete agreement with the previously obtained results presented in Eq. (3.18).

3.2.3 Bulk isothermal compressibility

It is also instructive to consider fluctuations in the one-dimensional hard-rod fluid. Focusing on density fluctuations one realizes from Eqs. (1.81) and (2.75) that the isothermal compressibility is a quantitative measure of such fluctuations. For the one-dimensional fluid considered in this section, we may define

$$\kappa_{\rm b} \equiv \frac{1}{L}\left(\frac{\partial L}{\partial \tau_{\rm b}}\right)_{T,N} \tag{3.32}$$

by analogy with Eq. (1.81). Identifying in this expression N with the average number of fluid molecules $\langle N \rangle$, we may employ the virial equation of state to obtain [see Eqs. (3.15) and (3.18)]

$$
\begin{aligned}
L\left(\frac{\partial \tau_{\rm b}}{\partial L}\right)_{T,\langle N \rangle} &= -\tau_{\rm b}^{\rm id} + k_{\rm B}T\sum_{k=2}^{\infty}kB_k\overline{\rho}^k = -\tau_{\rm b}^{\rm id} + k_{\rm B}T\overline{\rho}\sum_{k=2}^{\infty}k\left(\overline{\rho}d\right)^{k-1} \\
&= -\tau_{\rm b}^{\rm id} + k_{\rm B}T\overline{\rho}\left[\frac{1}{\left(1-\overline{\rho}d\right)^2}-1\right] - \frac{k_{\rm B}T\overline{\rho}}{\left(1-\overline{\rho}d\right)^2} \tag{3.33}
\end{aligned}
$$

Alternatively we may use the definition [see Eq. (1.81)]

$$\kappa_{\rm b} \equiv \frac{L}{\langle N \rangle^2}\left(\frac{\partial \langle N \rangle}{\partial \mu}\right)_{T,L} \tag{3.34}$$

for the isothermal compressibility of the one-dimensional hard-rod bulk fluid. From Eqs. (3.2), (3.3), and (3.12), we find that

$$\langle N \rangle = \frac{1}{\Xi\left(\mu\right)}\sum_{N=0}^{\infty}\frac{N}{N!\Lambda^N}\exp\left[\frac{\mu N}{k_{\rm B}T}\right]\left(L-Nd\right)^N \tag{3.35}$$

Remembering that Ξ also depends on μ, we may differentiate the previous expression one more time to obtain

$$
\begin{aligned}
\left(\frac{\partial \langle N \rangle}{\partial \mu}\right)_{T,L} &= \frac{1}{k_{\rm B}T\Xi}\sum_{N=0}^{\infty}\frac{N^2}{N!\Lambda^N}\exp\left[\frac{\mu N}{k_{\rm B}T}\right]\left(L-Nd\right)^N \\
&\quad - \frac{1}{k_{\rm B}T}\left(\frac{1}{\Xi}\sum_{N=0}^{\infty}\frac{N}{N!\Lambda^N}\exp\left[\frac{\mu N}{k_{\rm B}T}\right]\left(L-Nd\right)^N\right)^2 \\
&= \frac{1}{k_{\rm B}T}\left[\langle N^2 \rangle - \langle N \rangle^2\right] \tag{3.36}
\end{aligned}
$$

so that we finally arrive at

$$\kappa_{\rm b} = \frac{L}{k_{\rm B}T}\left[\frac{\langle N^2 \rangle - \langle N \rangle^2}{\langle N \rangle^2}\right] \tag{3.37}$$

where Eq. (3.34) has also been used. Comparing Eq. (3.37) with its counterpart that is valid for a three-dimensional system [see, for example, Eq. (5.78)], it turns out that the two expressions are identical except for the volume V, which, in the three-dimensional system, replaces the variable L in Eq. (3.37).

Equations (3.32), (3.33), and (3.37) may be combined to give

$$\sigma_N^2 = \overline{\rho}d \left(1 - \overline{\rho}d\right)^2 \frac{L}{d} \tag{3.38}$$

where we also used the definition of the variance of the mean particle number introduced in Eq. (2.75). Because the physically sensible density range of the hard-rod fluid complies with the inequality

$$0 \leq \overline{\rho}d \leq 1 \tag{3.39}$$

it follows from Eq. (3.38) that

$$\lim_{\substack{\overline{\rho}d \to 0 \\ \overline{\rho}d \to 1}} \sigma_N = 0 \tag{3.40}$$

That is to say, density fluctuations (relative to the mean density) vanish as one approaches the limits of either vanishing density (i.e., the hard-rod ideal gas) or the density of a close-packed hard-rod fluid (i.e., $\overline{\rho}d = 1$). Notice that at $\overline{\rho}d = \frac{1}{3}$ density fluctuations assume a maximum of

$$\sigma_N^2 \big|_{\overline{\rho}d=1/3} = \frac{4}{27} \frac{L}{d} \tag{3.41}$$

but remain finite over the entire density range defined by the inequality in Eq. (3.39). Hence, density fluctuations do not diverge to infinity, which implies the absence of a critical point.

3.2.4   Density distribution

Despite the absence of capillary condensation, the one-dimensional hard-rod fluid is still so useful because we have an analytic expression for its partition function [see Eq. (3.12)] that permits us to derive closed expressions for any thermophysical property of interest. One such quantity that is closely related to the isothermal compressibility discussed in the preceding section is the particle-number distribution $P(N)$, which one may also employ to compute thermomechanical properties [see, for example, Eqs. (3.65) and (3.68)]. Moreover, in a three-dimensional system $P(N)$ is useful to investigate the system-size dependence of density fluctuations as we shall demonstrate in Section 5.4.2 [see Eq. (5.80)].

As we demonstrate in Appendix C.3, $P(N)$ should be Gaussian in the thermodynamic limit. In Appendix C.3.1, we present an argument showing that $P(N)$ should always approach the Gaussian limit regardless of the specific form of the partition function (see also Section 5.4.2). Moreover, even if we do not know the partition function, the discussion in Appendix C.3.2 gives us the Gaussian limit of $P(N)$ by applying the Moivre-Laplace theorem to the Bernoulli distribution characterizing a general measurement process [see Eq. (C.37)].

Hence, as we are given an analytic expression, it seems worthwhile to apply the analysis detailed in Appendix C.3 to the one-dimensional hard-rod fluid. Our starting point is the probability to find N hard rods in a (one-dimensional) bulk system of "volume" L given by [see Eq. (3.12)]

$$
\begin{aligned}
P(N) &= \frac{1}{\Xi} \exp\left(\frac{\mu N}{k_B T}\right) Q \\
&= \frac{1}{\Xi} \exp\left(\frac{\mu N}{k_B T}\right) \frac{(L - Nd)^N}{N! \Lambda^N} \\
&= \exp\left(\frac{\tau_b L}{k_B T}\right) \exp\left(\frac{\mu N}{k_B T}\right) \frac{(L - Nd)^N}{N! \Lambda^N}
\end{aligned}
\tag{3.42}
$$

where the second line follows with the aid of Eq. (3.9) and the last line is a direct consequence of Eq. (3.19). Rewriting in Eq. (3.42), $N!$ according to Eq. (B.6), one realizes that $P(N)$ may be recast formally as

$$
P(N) = \frac{1}{\Xi} P_\Uparrow(N) P_\Downarrow(N)
\tag{3.43}
$$

where

$$
P_\Downarrow(N) \equiv \frac{1}{\sqrt{2\pi N}} \left(1 - \frac{Nd}{L}\right)^N \exp\left(\frac{\mu N}{k_B T}\right)
\tag{3.44}
$$

is a monotonically decreasing function of N because $\mu < 0$ and $Nd/L < 1$. The function

$$
P_\Uparrow(N) \equiv \left(\frac{d}{\Lambda}\right)^N \left(\frac{L}{Nd}\right)^N \exp(N)
\tag{3.45}
$$

on the other hand, increases monotonically with N because the current analysis is based on the explicit assumption of validity of the classical limit characterized by the inequality

$$
\frac{d}{\Lambda} \gg 1
\tag{3.46}
$$

according to our discussion in Section 2.5.3.

The quantities $P_\Uparrow(N)$ and $P_\Downarrow(N)$ are introduced only to split $P(N)$ into monotonically *increasing* and *decreasing* contributions, respectively, such that the product $P_\Downarrow(N)\,P_\Uparrow(N)$ must have a maximum at some value $\overline{N}$, which we seek to determine now. That the extremum of $P_\Uparrow(N)\,P_\Downarrow(N)$ must be a maximum follows from the fact that both $P_\Uparrow(N)$ and $P_\Downarrow(N)$ are positive semidefinite over the entire physically sensible parameter range.

Clearly, the necessary condition for the existence of this maximum may be stated as $dP(N)/dN = 0$. However, because the functions $P(N)$ and $\ln P(N)$ share the same monotony, it turns out to be more convenient to determine the maximum of $\ln P(N)$ rather than that of $P(N)$ itself. From Eq. (3.42), we obtain

$$
\begin{aligned}
\ln P(N) \;=\; & -\ln\Xi + \frac{\mu N}{k_\mathrm{B}T} + N\ln(L - Nd) \\[4pt]
& -N\ln N + N - \frac{1}{2}\ln(2\pi N) - N\ln\Lambda
\end{aligned}
\tag{3.47}
$$

which we may differentiate with respect to N. Solving then

$$
\left.\frac{d\ln P(N)}{dN}\right|_{N=\overline{N}} = 0
\tag{3.48}
$$

gives us an expression for the chemical potential as a function of $\overline{N}$, namely

$$
\frac{\mu}{k_\mathrm{B}T} = \ln\frac{\overline{N}}{L - \overline{N}d} + \frac{\overline{N}d}{L - \overline{N}d} + \frac{1}{2\overline{N}} + \ln\Lambda
\tag{3.49}
$$

Inserting this equation into Eq. (3.47), we obtain

$$
\begin{aligned}
\ln P(N) \;=\; & -\ln\Xi + N\ln(L - Nd) - N\ln N + N - \frac{1}{2}\ln(2\pi N) \\[4pt]
& +N\ln\frac{\overline{N}}{L - \overline{N}d} + \frac{N\overline{N}d}{L - \overline{N}d} + \frac{N}{2\overline{N}}
\end{aligned}
\tag{3.50}
$$

which satisfies Eq. (3.48) as it must and as the reader may verify for himself.

For the second-order derivative, we obtain

$$
\begin{aligned}
\left.\frac{d^2\ln P(N)}{dN^2}\right|_{N=\overline{N}} &= \left.-\frac{2d}{L - Nd} - \frac{Nd^2}{(L - Nd)^2} - \frac{1}{N} + \frac{1}{2N^2}\right|_{N=\overline{N}} \\[6pt]
&= -\frac{L^2}{\overline{N}\,(L - \overline{N}d)^2} + \frac{1}{2\overline{N}^2}
\end{aligned}
\tag{3.51}
$$

From Eq. (3.51) one can also verify that in general

$$
\left.\frac{d^n\ln P(N)}{dN^n}\right|_{N=\overline{N}} = \mathcal{O}\!\left[\left(\frac{d}{L}\right)^{n-1},\frac{1}{\overline{N}^n}\right], \qquad n \geq 1
\tag{3.52}
$$

in a straightforward fashion that turns out to be algebraically a bit tedious. Hence, if we consider these expressions in the thermodynamic limit, it follows that the leading term in Eq. (3.52) is

$$\frac{d^2 \ln P(N)}{dN^2}\bigg|_{N=\overline{N}} = -\frac{L}{\overline{N}d\left(1 - \overline{N}d/L\right)^2}\frac{d}{L} \tag{3.53}$$

because all others vanish more rapidly as $\overline{N}, L \to \infty$ ($\overline{N}/L = \text{const}$).

Applying the above considerations to the Taylor expansion of $\ln P(N)$ around $\overline{N}$, we realize that in the thermodynamic limit

$$
\begin{aligned}
\ln P(N) &= \ln P\left(\overline{N}\right) + \sum_{n=1}^{\infty}\frac{1}{n!}\frac{d^n \ln P(N)}{dN^n}\bigg|_{N=\overline{N}}\left(N - \overline{N}\right)^n \\
&\overset{\substack{L,\overline{N}\to\infty \\ \overline{N}/L=\text{const}}}{\simeq} \ln P\left(\overline{N}\right) + \frac{1}{2}\frac{d^2 \ln P(N)}{dN^2}\bigg|_{N=\overline{N}}\left(N - \overline{N}\right)^2 \\
&= \ln P\left(\overline{N}\right) - \frac{1}{2}\frac{L^2}{\overline{N}\left(L - \overline{N}d\right)^2}\left(N - \overline{N}\right)^2 \tag{3.54}
\end{aligned}
$$

or, equivalently, if we take the antilogarithm of the previous expression we obtain

$$P(N) = P\left(\overline{N}\right)\exp\left[-\frac{1}{2}\frac{L^2}{\overline{N}\left(L - \overline{N}d\right)^2}\left(N - \overline{N}\right)^2\right] \tag{3.55}$$

The previous expression should be compared with Eq. (C.29a) where we emphasize that, unlike Eq. (3.55), Eq. (C.29a) was derived without employing a specific form of the canonical ensemble partition function Q.

Moreover, the discussion in Appendix C.3.1 reveals that for a Gaussian distribution like the one given in Eq. (3.55), $\overline{N} = \langle N\rangle$ [see Eq. (C.32)]. Hence, we can rewrite the argument of the exponential function in Eq. (3.55) [see also Eq. (3.38)]

$$\frac{L^2}{\overline{N}\left(L - \overline{N}d\right)^2} = \frac{1}{\overline{\rho}d\left(1 - \overline{\rho}d\right)^2}\frac{d}{L} = \frac{1}{\sigma_N^2} \tag{3.56}$$

using the definitions of $\overline{\rho}$ and σ_N given in Eqs. (3.17) and (2.75), respectively. Hence, we see from Eqs. (3.55) and (3.56) that

$$P(N) = P(\langle N\rangle)\exp\left[-\frac{(N - \langle N\rangle)^2}{2\sigma_N^2}\right] \tag{3.57}$$

and determine $P(\langle N \rangle)$ such that $P(N)$ is properly normalized (see Appendix C.3.1). This approach eventually yields

$$P(N) = \frac{1}{\sqrt{2\pi}\sigma_N} \exp\left[-\frac{(N - \langle N \rangle)^2}{2\sigma_N^2} \right] \tag{3.58}$$

according to the arguments given in Appendix C.3.1.

In closing this section the reader should also appreciate the fact that $P(\overline{N})$, as it may be determined from Eq. (3.50), does not equal $P(\langle N \rangle) = 1/\sqrt{2\pi}\sigma_N$, which we obtain from the normalization condition. This is because in reaching Eq. (3.58) we took the thermodynamic limit and truncated the Taylor expansion of $P(N)$ after the quadratic term in Eq. (3.54).

3.3 Hard rods confined between hard walls

3.3.1 Aspects of statistical thermodynamics

The analysis of the virial expansion of the bulk stress in the preceding section showed that the system-size dependence of the virial coefficients in Eqs. (3.30) was an artifact because the thermodynamic limit was not taken properly in deriving those expressions. In other words, the ratio d/L does not have any physical meaning as far as the bulk fluid is concerned. Turning our attention now to a hard-rod fluid confined between hard walls, this situation changes because now the system boundaries become physically significant in that they define the space of a one-dimensional pore accommodating the fluid molecules. To emphasize this we replace L by the distance between the pore walls s_z. The evaluation of the configuration integral proceeds in identically the same fashion as in Section 3.2.1 so that we obtain the equivalent expression

$$Z_{1d} = (s_z - Nd)^N \tag{3.59}$$

from the Analysis in Appendix C.2.

However, we now have to amend this expression by the condition that Nd must not exceed s_z for Eq. (3.59) to be meaningful because the pore is completely filled if $Nd = s_z$. To implement this additional constraint into our statistical thermodynamic treatment, we replace the grand canonical partition function derived in Eq. (3.12) for the (infinitely large) bulk fluid by

$$\Xi_{cl} = \sum_{N=0}^{\infty} \frac{1}{N!\Lambda^N} \exp\left[\frac{\mu N}{k_B T} \right] (s_z - Nd)^N \Theta(s_z - Nd) \tag{3.60}$$

where

$$\Theta\left(x\right) = \begin{cases} 0, & x < 0 \\ 1, & x \geq 0 \end{cases} \tag{3.61}$$

is the Heaviside function.

The link to thermodynamics is provided by Eq. (3.3), where, however, the (exact differential of the) grand potential is now given by

$$d\Omega\left(T, \mu, L\right) = -\mathcal{S}dT - Nd\mu + \tau_\perp ds_z \tag{3.62}$$

and $\tau_\perp$ is the stress exerted by the fluid on the confining substrates (i.e., the pore walls) [cf., Eq. (1.63)]. Hence,

$$
\begin{aligned}
\tau_\perp &= \left(\frac{\partial\Omega}{\partial s_z}\right)_{\{\cdot\}\backslash s_z} = -\frac{k_BT}{\Xi}\left(\frac{\partial\Xi}{\partial s_z}\right)_{\{\cdot\}\backslash s_z} \\
&= -\frac{k_BT}{\Xi}\sum_{N=0}^{\infty}\frac{N}{s_z - Nd}\frac{\left(s_z - Nd\right)^N z^N}{N!}\Theta\left(s_z - Nd\right) \\
&\quad - \frac{k_BT}{\Xi}\sum_{N=0}^{\infty}\frac{\left(s_z - Nd\right)^N z^N}{N!}\delta\left(s_z - Nd\right) \tag{3.63}
\end{aligned}
$$

where the activity z was defined in Eq. (3.21) and we used the fact that

$$\delta\left(x - b\right) = \frac{d}{dx}\Theta\left(x - b\right) \tag{3.64}$$

and the Dirac δ-"function" is defined in Eq. (B.75). From that definition, we conclude that the second summand in the above expression does not contribute to $\tau_\perp$ so that we may rewrite it as

$$\tau_\perp = \sum_{N=0}^{\infty}\tau_\perp^{(N)}P\left(N; s_z\right) \tag{3.65}$$

where

$$\tau_\perp^{(N)} \equiv -\frac{Nk_BT}{s_z - Nd} \tag{3.66}$$

is the stress exerted on the substrates by a confined fluid accommodating N particles and

$$P\left(N; s_z\right) \equiv \frac{z^N}{\Xi}\frac{\left(s_z - Nd\right)^N}{N!}\Theta\left(s_z - Nd\right) \tag{3.67}$$

is the probability of finding N particles in a pore of width s_z.

At this point it seems worthwhile to point out that $P\left(N; s_z\right)$, unlike its bulk counterpart $P\left(N\right)$, does not comply with a Gaussian distribution like the one given in Eqs. (3.55) or (3.58), say. The reason is that we cannot take s_z to infinity because it represents the degree of confinement. Hence, a variation of s_z inevitably changes the physical nature of the confined fluid, whereas the properties of the bulk fluid *must not* depend on a corresponding variation of L. The cutoff represented by the Heaviside function in Eq. (3.67) prevents $P(N; s_z)$ from becoming Gaussian except in the bulk limit where $s_z \to \infty$. In other words, for the confined hard-rod fluid, the thermodynamic limit does not exist in the sense of the second line of Eq. (3.54). Therefore, the one-dimensional confined hard-rod fluid must be considered a somewhat pathological model.

Another quantity of interest is the mean pore density $\overline{\rho} = \langle N \rangle / s_z$. From Eqs. (3.62) and (3.60), we find

$$
\begin{aligned}
\overline{\rho} &= -\frac{1}{s_z} \left(\frac{\partial \Omega}{\partial \mu} \right)_{\{\cdot\}\backslash\mu} = \frac{k_B T}{\Xi s_z} \left(\frac{\partial \Xi}{\partial \mu} \right)_{\{\cdot\}\backslash\mu} = \frac{1}{s_z} \sum_{N=0}^{\infty} N P(N; s_z) \\
&\equiv \sum_{N=0}^{\infty} \rho^N P(N; s_z)
\end{aligned}
\tag{3.68}
$$

3.3.2 "Stratification" of confined one-dimensional fluids

On the basis of the previous theoretical treatment of one-dimensional fluids, in both the bulk and the confined state, we now discuss some key features of these systems. Specifically, we shall consider the confined fluid to be thermally and materially coupled to the (infinitely large) bulk so that in thermodynamic equilibrium both systems are maintained at the same chemical potential μ and temperature T. However, in the absence of any attractive interactions between either fluid molecules or between a fluid molecule and the hard substrate, the latter becomes a more or less trivial parameter that does not affect thermal properties of the hard-rod fluid. Because of Eqs. (1.50) and (2.79), we have

$$
\mathcal{F} = -k_B T \ln \mathcal{Q} = -k_B T \ln \frac{Z_{1d}}{N! \Lambda^N} = -k_B T \ln \frac{(L - Nd)^N}{N! \Lambda^N} \tag{3.69a}
$$

$$
\mu_b = \left(\frac{\partial \mathcal{F}}{\partial N} \right)_{\{\cdot\}\backslash\mu} = \frac{N k_B T d}{L - Nd} + k_B T \ln \frac{N \Lambda}{L - Nd} \tag{3.69b}
$$

where we used Stirling's approximation [see Eq. (B.7)]. With the definition of the bulk stress given in Eq. (3.13), Eq. (3.69b) can be rearranged to give

$$
zd = -\frac{\tau_b d}{k_B T} \exp\left(-\frac{\tau_b d}{k_B T} \right) \tag{3.70}
$$

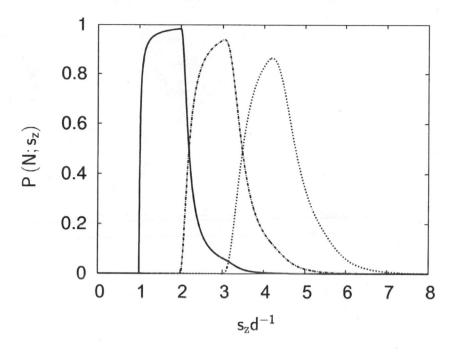

Figure 3.1: Probability density $P(N; s_z)$ as a function of pore "width" s_z. Curves are plotted for $N = 1$ (—), $N = 2$ (– · –·), and $N = 3$ (· · ·).

for the activity [see Eq. (3.21)] of both the bulk and the confined fluid because both are assumed to be in thermodynamic equilibrium. Fixing the bulk density to a sufficiently high fluid density $\bar{\rho}d = 0.75$, we calculate a corresponding bulk stress $\tau_b d/k_B T = -3$ (i.e., a bulk pressure $P_b d/k_B T = 3$) from the equation of state given in Eq. (3.13) using also the definition of the mean density [see Eq. (3.14)]. With these numbers, we calculate a value of $zd = 60.26$ for the activity, which we shall use in the calculations presented in this chapter. In addition, we fix the pore "width" to a nanoscopic range of $1 \leq s_z/d \leq 10$, which is small enough to illustrate confinement effects as well as the onset of ordinary bulk behavior.

We begin with a brief discussion of $P(N; s_z)$, which represents the probability of finding N molecules in a pore of "width" s_z. Plots in Fig. 3.1 show that this quantity is zero as long as s_z is not large enough to accommodate N molecules as one would have guessed. If s_z exceeds this threshold, $P(N; s_z)$ increases quite rapidly until it assumes a maximum at some characteristic value of s_z at which the pore is just becoming large enough to accommodate $N + 1$ molecules. Because of the competition with larger pore occupancies,

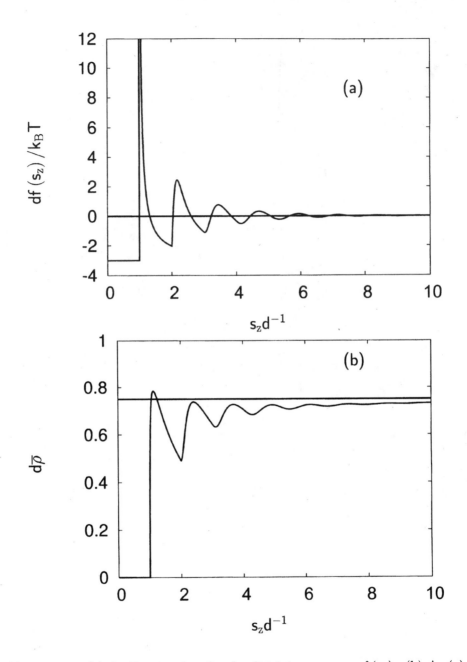

Figure 3.2: (a) As Fig. 3.1, but for the disjoining pressure $f(s_z)$. (b) As (a), but for the mean pore density $d\bar{\rho}$. The horizonal line demarcates the bulk density $d\bar{\rho} = 0.75$.

$p(N; s_z)$ decreases if s_z exceeds this threshold until it vanishes for a sufficiently large pore width. If s_z is not too large (i.e., for typical nanoscopic pore widths), Ξ is completely determined by just a few of the probability distributions $\{p(N; s_z)\}$'s.

Hence, for nanoscopic one-dimensional pores, we are in a position to calculate the so-called disjoining pressure defined as [cf., Eq. (5.57)]

$$f(s_z) \equiv -\tau_\perp(s_z) - P_b \tag{3.71}$$

which is a measure of the excess pressure exerted by the confined fluid on the substrates. For the current system, we calculate $\tau_\perp$ from Eq. (3.65). Clearly, as the distance between the substrates becomes *macroscopic* in magnitude, that is, in the limit

$$\lim_{s_z \to \infty} f(s_z) = 0 \tag{3.72}$$

the impact of confinement diminishes under these conditions and therefore fluid properties become indistinguishable from those of the corresponding bulk system with which it is in thermodynamic equilibrium. If, on the other hand, for sufficiently small s_z, $f(s_z) < 0$ the mechanical state of the fluid is such that it tends to pull the confining substrates together, whereas if $f(s_z) > 0$, the tendency is to push the substrates apart. As we shall explain later in Section 5.3.1, $f(s_z)$ is *in principle* accessible in experiments employing the *surface forces apparatus* (SFA). Plots in Fig. 3.2 show that in general $f(s_z)$ is a nonmonotonic function of s_z oscillating with a period that is slightly larger than the rod length d.

These oscillations reflect the inhomogeneous structure of the confined fluid. In fact, as we shall see below in Section 5.3.4, oscillations in the disjoining pressure are fingerprints of stratification of three-dimensional confined fluids, which is the tendency of fluid molecules to form individual layers parallel with the confining substrate. This structural interpretation is somewhat indirect, however, unless one correlates it with variations in the local density. We shall establish this correlation later in Section 5.3.4. For our current purposes, it suffices to conclude that the confined fluid is apparently highly inhomogeneous if the degree of confinement is sufficiently large (i.e., if s_z is sufficiently small). This notion is supported by plots of the mean pore density in Fig. 3.2(b), which we calculated from Eq. (3.68). The plot in Fig. 3.2(b) indicates that, like $f(s_z)$, $\overline{\rho}$ oscillates as a function of the pore "width" s_z where we notice again that

1. The period of the oscillations corresponds roughly to the rod length d

2. This period increases with s_z.

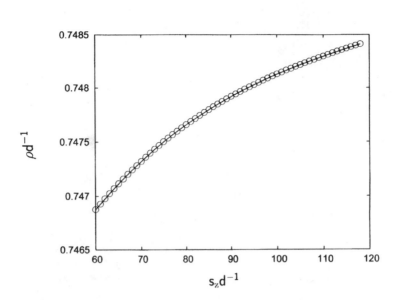

Figure 3.3: Mean pore density $\bar{\rho}$ as function of substrate separation s_z. The full line is a plot of the right side of Eq. (3.74) where $\rho_b d^{-1} = 0.75$.

The increase of the period of oscillations visible in the plot in Fig. 3.2(b) reflects that for larger pore widths the hard rods can pack more comfortably because of the larger space available to them.

Comparing plots in Figs. 3.2(a) and 3.2(b) reveals that minima in $f(s_z)$ and $\bar{\rho}$ coincide, as far as the substrate separation is concerned, at which they occur. At these values of s_z, a confined fluid of n "layers" appears to be strained minimally in the sense that the pore space is not large enough to accommodate $n+1$ such layers. Hence, over the associated range of substrate separations, the mean pore density decreases until s_z eventually becomes large enough to accommodate $n+1$ layers. As s_z increases, "layering" becomes less and less distinct as indicated by the damping of the oscillations in the plots of Figs. 3.2(a) and 3.2(b), which is a direct consequence of the diminishing influence of the substrate surfaces.

However, we notice that another subtle confinement effect prevails up to the largest substrate separations considered in Fig. 3.2(b). As the plot clearly shows, the mean pore density approaches the bulk density from below. This approach is rather weak with increasing substrate separation so that one expects this phenomenon to prevail up to substrate separations exceeding the largest one considered in the plot in Fig. 3.2(b) by more than an order of magnitude. This slow decay of the mean density can be rationalized as follows. Consider a bulk system at some density $\overline{\rho}_{\text{b}}$. Confining this system to a pore with hard walls is equivalent to putting the bulk fluid between two immobile hard rods whose centers are separated by a distance s_z. Hence, the space accessible to the fluid is smaller by some excluded volume on account of the presence of the two "wall particles." The magnitude of the excluded volume is $d/2$ for each pair of wall-fluid particles so that

$$s_z^{\text{eff}} = s_z - \frac{1}{2}\frac{d}{2} \tag{3.73}$$

and therefore

$$\overline{\rho} = \overline{\rho}_{\text{b}}\frac{s_z^{\text{eff}}}{s_z} = \overline{\rho}_{\text{b}}\left(1 - \frac{d}{4s_z}\right) \tag{3.74}$$

where we assign half of the *total* excluded volume to each particle (i.e., wall and fluid) of the interacting pair. A plot of $\overline{\rho}$ in Fig. 3.3 shows that the simple excluded-volume argument presented above is capable of explaining the slow decay of the mean pore density toward its bulk value. In other words, the slow decay is nothing but a trivial effect that could be eliminated by properly rescaling the pore volume according to Eq. (3.74).

Chapter 4

Mean-field theory

4.1 Introductory remarks

In Chapter 2, we developed statistical thermodynamics as the central theory that enables us *in principle* to calculate thermophysical properties of macroscopic confined fluids. A key feature of statistical thermodynamics is an enormous reduction of information that takes place as one goes from the microscopic world of electrons, photons, atoms, or molecules to the macroscopic world at which one performs measurements of thermophysical properties of interest. This information reduction is effected by statistical concepts such as the most probable distribution of quantum states (see Section 2.2.1).

By introducing the notion of various statistical physical ensembles in Section 2.2.1, we saw that we can make the quantum mechanical treatment consistent with several constraints imposed at the macroscopic level of description. That way we obtain an understanding of a thermal system at the microscopic level; that is, we can interpret thermodynamic properties in terms of the interaction between the microscopic constituents forming a macroscopic system.

The notion of an ensemble was first suggested by Gibbs[1] in a remarkably insightful manner. In the preface of his book *Elementary Principles in Statistical Mechanics Developed with Special Reference to the Rational Foundation of Thermodynamics* Gibbs writes [29]:

> "We consider especially ensembles of systems in which the index (or logarithm) of probability of phase is a linear function of the energy. This distribution on account of its unique importance in the theory of statistical equilibrium, I have ventured to call

---

[1]Josiah Willard Gibbs (1839–1903), professor of mathematical physics at Yale University and one of the "founding fathers" of statistical mechanics and vector calculus.

canonical, and the divisor of the energy, the *modulus* of the distribution. The moduli of distributions have properties analogous to temperature ...."[2]

In his writings Gibbs based statistical thermodynamics on entirely classical concepts when, for example, he writes about thermodynamics as pertaining to the "department of rational mechanics" [29]. Nevertheless he knew that classical physics was not entirely adequate. In fact, Gibbs expresses a deep understanding of the status of statistical mechanics of his era in writing:

"In the present state of science it seems hardly possible to frame a dynamic theory of molecular action which shall embrace the phenomena of thermodynamics, of radiation, and of the electrical manifestations which accompany the union of atoms. Yet any theory is obviously inadequate which does not take account of all these phenomena."

Being a contemporary of the nineteenth century, Gibbs could obviously not have had any concept of quantum mechanics and its role in laying a sound foundation of modern statistical thermodynamics. In this modern formulation, the classic Gibbsian version of statistical thermodynamics does, however, emerge as a limiting case as our discussion in Section 2.5 reveals.

However, regardless of whether we base our treatment on classical or quantum statistics, the development of statistical thermodynamics in Chapter 2 shows that the partition function is a key ingredient of the theory. This is because we may deduce from it explicit expressions for the thermophysical properties of equilibrium systems that may be of interest. At its core (and irrespective of the specific ensemble employed), the partition function is determined by the Boltzmann factor $\exp\left[-U\left(r^{N}\right)/k_{\mathrm{B}}T\right]$, where the total configurational potential energy $U\left(r^{N}\right)$ turns out to be a horrendously complex function of the configuration r^{N} on account of the interaction between the microscopic constituents.

Because of these interactions certain spatial arrangements of the microscopic constituents will turn out to be more likely than others. This is immediately apparent from a purely energetic perspective because it will be more likely to find a pair of atoms or molecules at separations from one another corresponding to the minimum of the interaction potential rather than at very short distances where the partial overlap of their electron clouds gives rise to more or less strong repulsion. On account of the interactions, particle

---

[2]See, for example, Eqs. (2.46), (2.51), and (2.65) of this work.

positions appear to be *correlated* and one would need to know the correlations in configuration space to eventually evaluate the configuration integral [see Eq. (2.112)] (and with it the classical partition function Q_{cl}).

From the discussion up to this point the reader will surely appreciate that a rigorous, first-principles calculation of the partition function for a macroscopic system is generally precluded even in the classical limit with the exception of rather simple models of limited usefulness (see Chapter 3). However, the problem of calculating the partition function (or the configuration integral) in closed form becomes tractable if we introduce as a key assumption that correlations between molecules are entirely negligible.

In effect, each molecule is then exposed to a *mean field* exerted on it by all other molecules and external fields such as confining substrate surfaces. Hence, the same mean-field can represent a large number of different configurations which we no longer have to worry about explicitly. The introduction of a mean-field approximation reduces the problem of calculating the configuration integral in Eq. (2.112) greatly because the complex N-dimensional integral then factorizes into single-particle contributions that are obviously far easier to handle computationally. This remarkable reduction of the computational problem is particularly important in the case of confined fluids as we shall demonstrate in this chapter.

4.2   Van der Waals theory of adsorption

Correlations in confined fluids essentially originate from two sources. On account of fluid-fluid interactions one would immediately anticipate short-range order to exist in confined fluids similar to the bulk. This short-range order manifests itself in, say, pair (or higher-order) correlation functions [30]. However, because of the external potential representing the confining substrates, the fluid in their vicinity is highly inhomogeneous (see Section 5.3.4 for a comprehensive discussion). This inhomogeneity may also be viewed as a manifestation of correlations in the fluid phase.

One may, for example, regard the (planar) substrate(s) of a slit-pore as the surface of a spherical particle of infinite radius. The confined fluid *plus* the substrates may then be perceived as a binary mixture in which macroscopically large (i.e., colloidal) particles (i.e., the substrates) are immersed in a "sea" of small solvent molecules. The local density of the confined fluid may then be interpreted as the mixture (A-B) pair correlation function representing correlations of solvent molecules (A) caused by the presence of the solute (B).

As we shall demonstrate in this section, a simple mean-field theory of

confined fluids may be developed based on the assumption that both types
of the aforementioned correlations (i.e., fluid-fluid and fluid-substrate) can
be disregarded altogether. As a result one obtains an analytic equation of
state of the van der Waals type for the confined fluid that permits one to
understand some very basic features of sorption experiments. As an illus-
trative example, we discuss below the volumetric determination of the phase
behavior of a pure fluid confined to a mesoporous silica glass (see, for example, Fig. 4.1) carried out by Thommes and Findenegg [31].

If one is interested only in properties of the pore phase as a whole, such
as the excess adsorption and the phase behavior, and not in properties that
depend explicitly on local density, or on intermolecular correlations, then it
may be sufficient to neglect entirely variations in the local density. It is in
this spirit that we present a simple model for the adsorbed phase that yields
closed expressions for the free energy and for the equation of state. The model
is a direct extension of van der Waals' model for the bulk fluid. For simplicity
we adopt the slit-pore geometry, although the significant conclusions of the
study are not altered for pores of other shapes. As we shall demonstrate
below, some features of the Thommes Findenegg experiment [31] can indeed
be understood in terms of a simple van der Waals equation of state.

4.2.1   Sorption experiments

Using a volumetric technique, Thommes and Findenegg [31] have measured
the excess coverage Γ of SF_6 in controlled pore glasses (CPG, see Fig. 4.1) as a
function of T along subcritical isochoric paths in bulk SF_6. The experimental
apparatus, fully described in Ref. 31, consists of a reference cell filled with
pure SF_6 and a sorption cell containing the adsorbent in thermodynamic
equilibrium with bulk SF_6 gas at a given initial temperature T_i of the fluid
in both cells. The pressure P in the reference cell and the pressure difference
ΔP between sorption and reference cell are measured. The density of (pure)
SF_6 at T_i is calculated from P via an equation of state.

At the beginning of an experimental scan, the reference-cell volume is
adjusted such that $\Delta P(T_i) = 0$; that is, the thermodynamic state of SF_6
is the same in both cells. The temperature is then lowered from T_i to a
new temperature $T_{i+1} = T_i - \Delta T$, at which $\Delta P \neq 0$ because more SF_6 is
adsorbed. The volume of the sorption cell is then adjusted to reestablish
the original condition $\Delta P = 0$ at the new temperature T_{i+1}. The change in
the excess coverage is given by $\Delta\Gamma \propto \rho\Delta V$, where ΔV is the change in the
volume of the sorption cell between T_i and T_{i+1}. Measurements are repeated
by lowering the temperature in a stepwise fashion until the bulk coexistence
temperature T_{xb} of SF_6 for the given isochore is reached and the gas in the

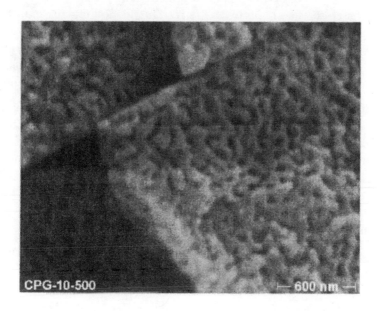

Figure 4.1: The sponge-like structure of a typical sample of controlled-pore glass used in sorption experiments. The silica matrix in lighter gray surrounds the mesopores appearing in darker gray.

reference cell begins to condense. By means of a high-pressure microbalance technique [32], the absolute value of $\Gamma(T_i)$ is determined in an independent experiment so that $\Gamma(T)$ can be calculated from $\Delta\Gamma$ for each temperature $T_i \leq T \leq T_{xb}$.

From a theoretical perspective, these experiments are particularly appealing for two reasons. First, CPG is characterized by a very narrow pore-size distribution. As pointed out in Ref. 31, 80% of all pores have a diameter within 5% of the average radius of the (approximately) cylindrical pores. If connections between individual pores are disregarded, the phase behavior of the adsorbate should therefore closely resemble that of an adsorbate in a single pore (see Section 4.2.4). Second, the CPG employed by Thommes and Findenegg [31] is mesoporous, as reflected by the nominal average pore radii of 24 nm (CPG-240) and 35 nm (CGP-350). As these values are large compared with the range of fluid-substrate intermolecular forces, the inhomogeneous region of the pore fluid is much smaller than the homogeneous region. Therefore the shape of the pores should not matter greatly. This notion is corrobrated by the fact that the structure of CPG is largely bi-continuous having a nearly vanishing mean curvature. The characteristic pore widths of CPG, on the other hand, are still small enough such that confinement effects can be expected to prevail to a significant extent.

In the meantime it has also become feasible to synthesize other meso-porous materials that differ from CPG in that they consist of individual, disconnected cylindrical pores. These so-called SBA-15 or MCM-41 silica pores can be synthesized using a technical-grade triblock copolymer as the structure directing template in aequous H_2SO_4 solution and tetraethyl or-thosilicate as the silica source [33, 34]. After calcination [35], one obtains a regular array of individual cylindrical pores as illustrated by the transmission electron micrographs (TEMs) shown in Fig. 4.2.

A key result of the sorption experiments conducted by Thommes and Findenegg concerns the pore condensation line $T_{xp}(\rho_b) > T_{xb}(\rho_b)$ at which pore condensation occurs along a subcritical isochoric path $\rho_b/\rho_{cb} < 1$ in the bulk (ρ_b and ρ_{cb} are the density of this isochore and the bulk critical density, respectively). Experimentally, $T_{xp}(\rho_b)$ is directly inferred from the temperature dependence of $\Gamma(T)$, which changes discontinuously at $T_{xp}(\rho_b)$ (see Ref. 31 for details). The pore condensation line ends at the pore critical temperature T_{cp} (rigorously defined only in the ideal single slit-pore case) [31]. Because of confinement T_{cp} is shifted to lower values with decreasing pore size. If, on the other hand, the pore becomes large, $T_{cp} \to T_{cb}^-$ (the bulk critical temperature) and $T_{xp} \to T_{xb}$ (see Fig. 6 of Ref. 31).

4.2.2   An equation of state for pure confined fluids

The Thommes Findenegg experiment [31] can be analyzed theoretically via an equation of state for the pore fluid, which can be calculated from the (Helmholtz) free energy of the pore fluid $\mathcal{F}$, given formally by Eq. (2.79). From a molecular perspective, $\mathcal{F}$ is linked to the configuration integral via Eqs. (2.111) and (2.112) assuming that the experiment is carried out in a tem-perature regime where the classical treatment is adequate according to the discussion in Section 2.5 [see Eqs. (2.110) and (2.103)]. Moreover, Eq. (2.112) implies that molecules possess only translational degrees of freedom, which seems justified for SF_6 given its molecular structure.

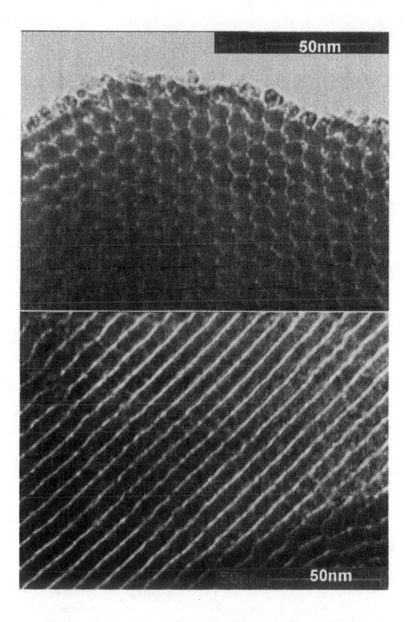

Figure 4.2: Transmission electron micrographs of mesoscopic SBA-15 silica pores. Upper: along the pore axes; lower: perpendicular to the pore axes [36].

4.2.2.1   Perturbation theory

As we see from Eq. (2.112), the key quantity is the configurational energy $U\left(r^N\right)$, which we henceforth separate into the potential energy of an unperturbed (reference) system, $U_0\left(r^N\right)$, and its perturbation represented by $U_1\left(r^N\right)$. We may then rewrite Eq. (2.112) as

$$Z = \int dr^N \exp\left[-\frac{(U_0 + U_1)}{k_B T}\right] = Z_0 \left\langle \exp\left(-\frac{U_1}{k_B T}\right)\right\rangle_0 \qquad (4.1)$$

where the angular brackets signify the ensemble average over the unperturbed probability distribution $Z_0^{-1}\exp\left[-\beta U_0\left(r^N\right)\right]$. In Eq. (4.1)

$$Z_0 \equiv \int dr^N \exp\left[-\frac{U_0\left(r^N\right)}{k_B T}\right] \qquad (4.2)$$

is the configuration integral for the reference system. Assuming the perturbation to be sufficiently small over the temperature range of interest, we approximate the ensemble average in Eq. (4.1) by

$$\left\langle \exp\left(-\frac{U_1}{k_B T}\right)\right\rangle_0 = \sum_{k=0}^{\infty}\left(-\frac{1}{k_B T}\right)^k \frac{\langle U_1^k\rangle_0}{k!} = 1 - \frac{\langle U_1\rangle_0}{k_B T} + \mathcal{O}\left(T^{-2}\right) \qquad (4.3)$$

For sufficiently high temperatures (i.e., for sufficiently low values of $1/T$), we may truncate the expansion in Eq. (4.3) after the linear term. Combining then Eqs. (2.79), (2.111), (4.1), and (4.3) yields

$$\mathcal{F} \simeq \mathcal{F}_0 + \mathcal{F}_1 = -\frac{1}{k_B T}\ln\frac{Z_0}{N!\Lambda^{3N}} + \langle U_1\rangle_0 \qquad (4.4)$$

where we used the fact that $x \equiv \langle U_1\rangle_0 /k_B T \ll 1$ such that $\ln\left(1 - x\right)$ can be expanded in a MacLaurin series[3] to give

$$\ln\left(1 - x\right) = -x + \mathcal{O}\left(x^2\right) \approx -\beta\langle U_1\rangle_0 \qquad (4.5)$$

Henceforth, we consider a Lennard–Jones(12,6) (LJ) fluid between two plane parallel solid substrates. The basic setup of our model is schematically depicted in Fig. 4.3. We assume the fluid-substrate interaction to be a pair-wise additive sum of LJ potentials. As the reference system we take a hard-sphere fluid (diameter σ_f) between hard-sphere substrates (diameter σ_s). Moreover, we "smear" the hard spheres of the surface layer of each

---

[3]That is, a Taylor series around $x = 0$.

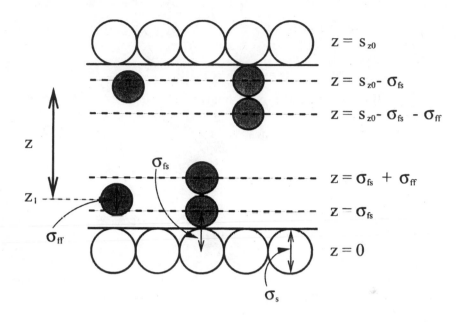

Figure 4.3: Side view of model slit-pore showing wall atoms (white) and fluid molecules (gray).

substrate uniformly over the plane of the layer to obtain a hard wall that is infinitesimally smooth in transverse (x, y) directions. The smeared fluid-wall interaction thus depends only on the distance of the fluid molecule from the wall. The potential energy of the reference system can then be expressed as

$$U_0 = \frac{1}{2} \sum_{i=1}^{N} \sum_{j \neq i}^{N} u_{\text{ff}}^{\text{hs}} (r_{ij}) + \sum_{i=1}^{N} u_{\text{fs}}^{\text{hs}} (z_i) \tag{4.6}$$

where the fluid-fluid (ff) contribution is given by

$$u_{\text{ff}}^{\text{hs}} (r_{ij}) = \begin{cases} 0, & r_{ij} > \sigma_{\text{f}} \\ \infty, & r_{ij} \leq \sigma_{\text{f}} \end{cases} \tag{4.7}$$

and that of the fluid-substrate by

$$u_{\text{fs}}^{\text{hs}} (z_i) = \begin{cases} 0, & \sigma_{\text{fs}} < z_i < s_{z0} - \sigma_{\text{fs}} \\ \infty, & z_i \leq \sigma_{\text{fs}} \text{ or } z_i \geq s_{z0} - \sigma_{\text{fs}} \end{cases} \tag{4.8}$$

In Eq. (4.6), $r_{ij} \equiv |\mathbf{r}_i - \mathbf{r}_j|$ is the distance between a pair of hard spheres located at $\mathbf{r}_i$ and $\mathbf{r}_j$, $\sigma_{\text{fs}} \equiv (\sigma_{\text{f}} + \sigma_{\text{s}})/2$ is the distance between a pore molecule and a substrate atom in contact, and s_{z0} is the distance between the walls (see Fig. 4.3). We approximate the configuration integral of the reference system by

$$Z_0 = \left[Z_0^{(1)} \right]^N \tag{4.9}$$

where $Z_0^{(1)}$ is the effective single-molecule configuration integral. Notice that Eq. (4.9) neglects correlations between molecules in the reference system and may therefore be considered a mean-field approximation in itself. We take $Z_0^{(1)}$ to be equal to the volume accessible to any given molecule, as dictated by Eqs.(4.6)–(4.8):

$$Z_0^{(1)} = A \left(s_{z0} - 2\sigma_{fs} \right) - Nb \tag{4.10}$$

where A is the area of the wall and

$$b \equiv \frac{2\pi\sigma_{ff}^3}{3} \tag{4.11}$$

is (half) the volume excluded to one molecule by another. Hence, we have for the reference free energy

$$Z_0 = \left[A \left(s_{z0} - 2\sigma_{fs} \right) - Nb \right]^N \tag{4.12}$$

The perturbation is similarly given by

$$U_1 = \frac{1}{2} \sum_{i=1}^{N} \sum_{j \neq i}^{N} u_{ff}^a \left(r_{ij} \right) + \sum_{i=1}^{N} u_{fs}^a \left(z_i \right) \tag{4.13}$$

where

$$u_{ff}^a \left(r_{ij} \right) = -4\varepsilon_{ff} \left(\frac{\sigma_{ff}}{r_{ij}} \right)^6, \qquad r_{ij} > \sigma_f \tag{4.14}$$

and

$$u_{fs}^a \left(z_i \right) = -\frac{2\pi\rho_s\varepsilon_{fs}\sigma_{fs}^3}{3d} \left[\left(\frac{\sigma_{fs}}{z_i} \right)^3 + \left(\frac{\sigma_{fs}}{s_{z0} - z_i} \right)^3 \right], \qquad \sigma_{fs} < z_i < s_{z0} - \sigma_{fs} \tag{4.15}$$

In Eq. (4.15), ρ_s is the areal density of the solid substrate. The fluid-fluid perturbation is just the attractive term of the LJ pair potential. Likewise, the (original) fluid-substrate perturbation is the LJ attraction $-4\varepsilon_{fs} \left(\sigma_{fs}/r \right)^6$. To be consistent with the smooth-wall approximation to the reference potential, we average the fluid-substrate attractions over the (x, y) positions of the substrate atoms in the planes in which they lie. We suppose each substrate to comprise an infinite half-space of atomic planes, separated successively by distance d. Approximating the sum over these planes by the Euler-MacLaurin formula [37] yields the expression in Eq. (4.15).

4.2.2.2 Mean-field approximation

Based on Eqs. (4.13)–(4.15) we have

$$\langle U_1 \rangle_0 = \int_V d\mathbf{r}_1 \rho_0^{(1)}(\mathbf{r}_1)\, u_{\mathrm{fs}}^{\mathrm{a}}(z_1) + \frac{1}{2} \int_V d\mathbf{r}_1 \int_V d\mathbf{r}_2 \rho_0^{(2)}(\mathbf{r}_1, \mathbf{r}_2)\, u_{\mathrm{ff}}^{\mathrm{a}}(r_{12}) \quad (4.16)$$

where $\rho_0^{(1)}(\mathbf{r}_1)$ and $\rho_0^{(2)}(\mathbf{r}_1, \mathbf{r}_2)$ are, respectively, the local density and the pair distribution function in the reference system [21], which are related to one another through the pair correlation function by

$$\rho_0^{(2)}(\mathbf{r}_1, \mathbf{r}_2) \equiv \rho_0^{(1)}(\mathbf{r}_1)\, \rho_0^{(1)}(\mathbf{r}_2)\, g(\mathbf{r}_1, \mathbf{r}_2) \quad (4.17)$$

Equation (4.16) may be rationalized by recalling that

$$\rho_0^{(1)}(\mathbf{r}_1) \equiv \left\langle \sum_{i=1}^N \delta(\mathbf{r}_1 - \mathbf{r}_i) \right\rangle \quad (4.18\mathrm{a})$$

$$\rho_0^{(2)}(\mathbf{r}_1, \mathbf{r}_2) \equiv \left\langle \sum_{i=1}^N \sum_{j=1 \neq i}^N \delta(\mathbf{r}_1 - \mathbf{r}_i)\, \delta(\mathbf{r}_2 - \mathbf{r}_j) \right\rangle \quad (4.18\mathrm{b})$$

where we use the definition of the Dirac δ-function (see Appendix B.6.1) and the angular brackets denote an average in the canonical ensemble. Because we are seeking an equation of state at the mean-field level, we ignore intermolecular (fluid-fluid) correlations and set

$$g(\mathbf{r}_1, \mathbf{r}_2) = \begin{cases} 0, & r_{12} < \sigma_{\mathrm{ff}} \\ 1, & r_{12} \geq \sigma_{\mathrm{ff}} \end{cases} \quad (4.19)$$

Moreover, neglecting fluid-substrate correlations, we take the fluid to be homogeneous throughout the pore volume; that is, we approximate the local density by

$$\rho_0^{(1)}(\mathbf{r}_1) = \rho_0^{(1)}(\mathbf{r}_2) = \rho \equiv \frac{N}{A(s_{z0} - 2\sigma_{\mathrm{fs}})} \quad (4.20)$$

It is well established [38] that a fluid confined to a slit-pore is stratified (i.e., the fluid molecules order themselves in strata parallel with the substrates; see Section 5.3.4 for a more detailed discussion). Stratification diminishes rapidly with increasing distance from the substrates, because of the decay of the fluid-substrate interaction; beyond a few molecular diameters σ_{f}, the fluid is essentially homogeneous, as computer simulations have repeatedly demonstrated [39–46]. We therefore expect Eq. (4.20) to be reasonable for mesoscopic pores ($s_{z0} \gtrsim 10\sigma_{\mathrm{fs}}$). From Eq. (4.16) we then obtain

$$
\begin{aligned}
\langle U_1 \rangle_0 &= -\frac{2\pi \rho_{\mathrm{s}} \varepsilon_{\mathrm{fs}} \sigma_{\mathrm{fs}}^3 \rho A}{3d} \int_{\sigma_{\mathrm{fs}}}^{s_{z0}-\sigma_{\mathrm{fs}}} \mathrm{d}z \left[\left(\frac{\sigma_{\mathrm{fs}}}{z} \right)^3 + \left(\frac{\sigma_{\mathrm{fs}}}{s_{z0}-z} \right)^3 \right] \\
&\quad - 2\varepsilon_{\mathrm{ff}} \rho^2 \int \mathrm{d}\boldsymbol{r}_1 \int \mathrm{d}\boldsymbol{r}_2 \left(\frac{\sigma_{\mathrm{f}}}{r_{ij}} \right)^6 \\
&\equiv \Psi(\xi) N - a_{\mathrm{p}}(\xi) N\rho
\end{aligned}
\tag{4.21}
$$

The one-dimensional integral defining $\Psi(\xi)$ in Eq. (4.21) can be readily performed to yield

$$
\Psi(\xi) = \frac{\Psi_0}{\xi - 2} \left[\frac{1}{(\xi - 1)^2} - 1 \right], \qquad \xi > 2
\tag{4.22}
$$

where the energy scale is set by

$$
\Psi_0 \equiv \frac{2\pi \rho_{\mathrm{s}} \varepsilon_{\mathrm{fs}} \sigma_{\mathrm{fs}}^3}{3d}
\tag{4.23}
$$

and $\xi \equiv s_{z0}/\sigma_{\mathrm{fs}}$. Because of constraints enforced by the hard interactions, the evaluation of the double integral defining $a_{\mathrm{p}}(\xi)$ is algebraically more demanding. Its evaluation is therefore deferred to Appendix D.1.1. The final result of this calculation is

$$
a_{\mathrm{p}}(\xi) = a_{\mathrm{b}} \left[1 - \frac{3\sigma_{\mathrm{f}}/\sigma_{\mathrm{fs}}}{4(\xi - 2)} + \frac{\sigma_{\mathrm{f}}/\sigma_{\mathrm{fs}}}{8(\xi - 2)^3} \right], \qquad 2(\sigma_{\mathrm{ff}}/\sigma_{\mathrm{fs}} + 1) < \xi < \infty
\tag{4.24}
$$

where a_{b} refers to the homogeneous bulk fluid, which can readily be obtained by transforming the variables in the second term of Eq. (4.21) according to $\{\boldsymbol{r}_1, \boldsymbol{r}_2\} \rightarrow \{\boldsymbol{r}_1, \boldsymbol{r}_{12}\}$ and reexpressing the three-dimensional integral over $\boldsymbol{r}_{12}$ in spherical polar coordinates. This yields

$$
a_{\mathrm{b}} \equiv 2\varepsilon_{\mathrm{ff}} \sigma_{\mathrm{ff}}^6 \int_0^{2\pi} \mathrm{d}\varphi \int_0^\pi \mathrm{d}\vartheta \sin\vartheta \int_{\sigma_{\mathrm{f}}}^\infty \mathrm{d}r \, r^{-4} = \frac{8\pi \varepsilon_{\mathrm{ff}} \sigma_{\mathrm{ff}}^3}{3} = 4\varepsilon_{\mathrm{ff}} b
\tag{4.25}
$$

A comparison of Eqs. (4.24) and (4.25) shows that $a_{\mathrm{p}}(\xi) \leq a_{\mathrm{b}}$, where the equality applies in the limit $\xi = \infty$. The inequality is a direct consequence of the weakened attractive field "experienced" by a molecule in confinement due to the presence of the substrates, which may, of course, compensate this loss of attractivity through the term $\Psi(\xi)/k_{\mathrm{B}}T$, that is, by virtue of its chemical nature (i.e., the constant Ψ_0) at least as far as the equation of state is concerned as we shall see shortly.

Inserting now Eqs. (4.12) and (4.21) into Eq. (4.4) gives

$$-\frac{\mathcal{F}}{k_{\mathrm{B}}T} = N\left[\ln\frac{A\left(s_{z0} - 2\sigma_{\mathrm{fs}}\right) - Nb}{N\Lambda^{3N}} + 1 + \frac{\Psi\left(\xi\right)}{k_{\mathrm{B}}T} - \frac{a_{\mathrm{p}}\left(\xi\right)N}{Ak_{\mathrm{B}}T\left(s_{z0} - \sigma_{\mathrm{fs}}\right)}\right] \tag{4.26}$$

For a fluid confined between smooth walls, the exact differential of the free energy can be expressed as

$$\mathrm{d}\mathcal{F} = -\mathcal{S}\mathrm{d}T + \mu_p\mathrm{d}N + \tau_{\|}\left(s_{z0} - 2\sigma_{\mathrm{fs}}\right)\mathrm{d}A + \tau_{zz}A_0\mathrm{d}s_z \tag{4.27}$$

where we used the definition of $\mathcal{F}$ given in Eq. (1.50) as well as the expression for the (exact differential of the) internal energy of a fluid confined between structureless, planar substrates surfaces displayed in Eq. (1.37). From Eqs. (4.26) and (4.27), we obtain the mean-field equation of state of the confined fluid as

$$-\tau_{\|} \equiv P_{\|} = -\frac{1}{s_{z0} - 2\sigma_{\mathrm{fs}}}\left(\frac{\partial\mathcal{F}}{\partial A}\right)_{T,N,s_z} = \frac{\rho k_{\mathrm{B}}T}{1 - \rho b} - a_{\mathrm{p}}\left(\xi\right)\rho^2 \tag{4.28}$$

where we also used Eq. (4.20). Equation (4.28) turns out to be independent of the fluid-substrate contribution $N\Psi\left(\xi\right)$ to the perturbation. In Eq. (4.28) we introduce the transverse pressure $P_{\|}$. In the limit $\xi \to \infty$, $a_p\left(\xi\right) = a_b$ and Eq. (4.28) reduces to the well-known van der Waals equation of state for the bulk fluid

$$\lim_{\xi\to\infty} P_{\|} = P_{\mathrm{b}} = \frac{\rho k_{\mathrm{B}}T}{1 - \rho b} - a_b\rho^2 \tag{4.29}$$

4.2.3 Critical behavior and gas-liquid coexistence

It is well known that the van der Waals equation of state is qualitatively correct in the sense that it is capable of predicting gas-liquid phase equilibria as well as critical phenomena. Mathematically speaking this is because the van der Waals equation may be perceived as a third-order polynomial in ρ regardless of whether we consider the bulk or a confined fluid. As for the bulk, the location of the critical point of the confined fluid is determined by the conditions

$$\rho^2\left(\frac{\partial P_{\|}}{\partial\rho}\right)\bigg|_{T=T_{cp}} = \rho^2\frac{\partial}{\partial\rho}\left[\rho^2\left(\frac{\partial P_{\|}}{\partial\rho}\right)\bigg|_{T=T_{cp}}\right] \overset{!}{=} 0, \quad \text{fixed } s_{z0} \tag{4.30}$$

which together with Eq. (4.28) yields

$$k_B T_{cp}(\xi) = \frac{8a_p(\xi)}{27b} \leq k_B T_{cb} = \frac{8a_b}{27b} \tag{4.31a}$$

$$P_{cp}(\xi) = \frac{a_p(\xi)}{27b^2} \leq P_{cb} = \frac{a_b}{27b^2} \tag{4.31b}$$

$$\rho_{cp} = \frac{1}{3b} = \rho_{cb} \tag{4.31c}$$

where T_{cb}, P_{cb}, and ρ_{cb} denote critical values. In the inequalities given in Eqs. (4.31), the equal sign pertains to the limit $\xi = \infty$ [see Eq. (4.24)]. The shift of $T_{cp}(\xi)$ to a value lower than T_{cb} is well established by more sophisticated theories as well as experimentally.

Introducing "reduced" variables through the relations

$$\widetilde{P} \equiv \frac{P}{P_{cp}(\xi)} \tag{4.32a}$$

$$\widetilde{T} \equiv \frac{T}{T_{cp}(\xi)} \tag{4.32b}$$

$$\widetilde{\rho} \equiv \frac{\rho}{\rho_{cp}} \tag{4.32c}$$

it is possible to rewrite Eq. (4.28) as

$$\widetilde{P} = \frac{8\widetilde{T}\widetilde{\rho}}{3 - \widetilde{\rho}} - 3\widetilde{\rho}^2 \tag{4.33}$$

Expanding $\widetilde{P}$ in a Taylor series around the critical point (i.e., around $\widetilde{T} = \widetilde{\rho} = 1$) and retaining terms up to third-order derivatives, the resulting (reduced) equation of state can be cast as

$$\widetilde{P} \simeq 1 + \left(\widetilde{T} - 1\right)\left[4 + 6\left(\widetilde{\rho} - 1\right) + 3\left(\widetilde{\rho} - 1\right)^2\right] + \frac{3}{2}\left(\widetilde{\rho} - 1\right)^3 + \ldots \tag{4.34}$$

Consider now coexisting gas and liquid phases at densities ρ_x^g and ρ_x^l, respectively. If we take these densities to be sufficiently close to the critical density, we may write

$$\widetilde{\rho}_{xp}^l = 1 + \epsilon \tag{4.35a}$$

$$\widetilde{\rho}_{xp}^g = 1 - \epsilon \tag{4.35b}$$

where $\epsilon \ll 1$ is a small number. We may then recast Eq. (4.34) for gas and liquid phases in coexistence as

$$\widetilde{P}_{\parallel}^l = 1 + \left(\widetilde{T} - 1\right)\left[4 + 6\epsilon + 3\epsilon^2\right] + \frac{3}{2}\epsilon^3 \tag{4.36a}$$

$$\widetilde{P}_{\parallel}^g = 1 + \left(\widetilde{T} - 1\right)\left[4 - 6\epsilon + 3\epsilon^2\right] - \frac{3}{2}\epsilon^3 \tag{4.36b}$$

However, if gas and liquid are in coexistence $\widetilde{P}_\parallel^g = \widetilde{P}_\parallel^l$ by definition and the previous expression can be rearranged to give

$$\Delta\rho = \rho_{xp}^l - \rho_{xp}^g = 2\epsilon = 4\rho_{cp}\sqrt{\frac{T_{cp}(\xi) - T}{T_{cp}(\xi)}} \propto \sqrt{t(\xi)} \qquad (4.37)$$

where $t(\xi)$ is defined as in Eq. (1.86) from which we conclude that the critical exponent $\beta = \frac{1}{2}$ for the order parameter $\Delta\rho$ irrespective of ξ. In other words, in the immediate vicinity of the critical point, the coexistence curve for the van der Waals fluid is symmetric, but its location does, of course, depend on the degree of confinement, that is, on ξ.

To illustrate the range of temperatures over which the power law in Eq. (4.37) is valid, we plot the order parameter $\rho_{xp}^l - \rho_{xp}^g$ as a function of T/T_{cb} in Fig. 4.4. The comparison shows that the critical behavior as predicted by Eq. (4.37) prevails for temperatures that are about 10% lower than the (bulk) critical temperature. However, as one would have guessed, the scaling law does not hold for much lower temperatures as one can also see from Fig. 4.4.

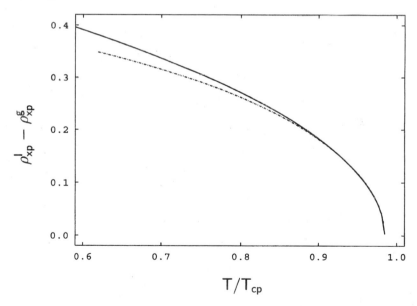

Figure 4.4: Plot of the order parameter $\rho_p^l - \rho_p^g$ as a function of T/T_{cp} calculated from the pore coexistence curve plotted in Fig. 4.5 (—) and from the power law (– · –) [see Eq. (4.37)].

A somewhat more detailed discussion of fluid phase behavior becomes possible on the basis of the full coexistence curves shown in Fig. 4.5 for both the bulk van der Waals fluid and one confined to a slit-pore where $s_z = 50$. Figure 4.5 serves to illustrate the impact of spatial confinement on the phase diagram of a "simple" fluid (i.e., a fluid in which molecules possess only translational degrees of freedom). These curves are obtained by a numerical solution of Eq. (4.43) detailed in Appendix D.1.3. As is evident from the plots in Fig. 4.5, confinement causes a depression of the critical temperature, whereas the critical density remains unaltered. These features are consistent with the analytic expressions for T_{cp} and ρ_{cp} given in Eq. (4.31).

However, the reader should note that the latter feature is not correct with respect to corresponding experimental observations where the critical density is usually shifted to higher values and the coexistence curve of the confined fluid turns out to be narrower with respect to its bulk counterpart [31]. This reflects the fact that, with regard to mean densities, gas- and liquid-like confined phases are more alike than in the bulk. The absence of a shift in critical density in the theoretical curves is caused by the fact that within the context of the current perturbational approach the density dependence of the free energy remains the same in both confined and bulk fluids [see, for example, Eq. (4.26)], which shows that confinement effects are solely restricted to the density-independent van der Waals parameter $a_p(\xi)$. However, on the positive side, we are now equipped with equations of state for both the confined fluid [see Eq. (4.28)] and its bulk counterpart [see Eq. (4.29)]. Together these equations of state enable us to revisit the Thommes Findenegg experiment at mean-field level.

4.2.4 Gas sorption in mesoscopic slit-pores

From a theoretical perspective, the Thommes Findenegg experiment [31] can be represented by the equation

$$\mu_b(T, \rho_b) - \mu_p(T, \rho_p) \stackrel{!}{=} 0, \quad \rho_b = \text{const}, \quad T \to T_{xb}^+ \tag{4.38}$$

where T_{xb} is the bulk liquid-gas coexistence temperature. To derive expressions for μ_b and μ_p, we differentiate Eq. (4.26) [see Eq. (4.27)]

$$\mu_p(T, \rho_p) = \left(\frac{\partial \mathcal{F}}{\partial N}\right)_{T, A, s_z} = k_B T \ln\left(\frac{\Lambda^3 \rho_p}{1 - b\rho_p}\right) + \frac{k_B T b \rho_p}{1 - b\rho_p} + \Psi(\xi) - 2a_p(\xi)\rho_p \tag{4.39}$$

from which

$$\lim_{\xi \to \infty} \mu_p(T, \rho_p) = \mu_b(T, \rho_b) = k_B T \ln\left(\frac{\Lambda^3 \rho_b}{1 - b\rho_b}\right) + \frac{k_B T b \rho_b}{1 - b\rho_b} + 2a_b\rho_b \tag{4.40}$$

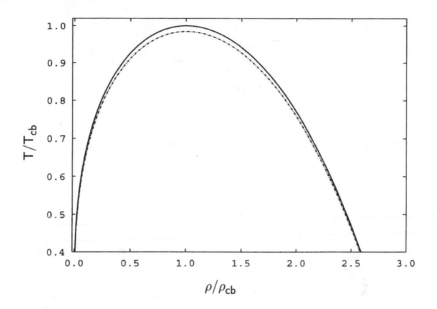

Figure 4.5: Phase diagram of bulk (—) and confined ($s_z = 50$) van der Waals fluid ($- \cdot - \cdot$).

follows. Equation (4.40) is obtained from Eq. (4.39) using the fact that [see Eqs. (4.22), (4.24)]

$$\lim_{\xi \to \infty} \Psi(\xi) = 0 \tag{4.41a}$$

$$\lim_{\xi \to \infty} a_p(\zeta) = a_b \tag{4.41b}$$

Combining now Eqs. (4.38), (4.39), and (4.40), we get

$$0 \stackrel{!}{=} \ln \frac{\rho_p (1 - \rho_b b)}{\rho_b (1 - \rho_p b)} + \frac{(\rho_p - \rho_b) b}{(1 - \rho_p b)(1 - \rho_b b)}$$
$$+ \frac{\Psi(\xi)}{k_B T} - 2 \frac{a_p(\xi) \rho_p - a_b \rho_b}{k_B T} \tag{4.42}$$

As Eq. (4.42) cannot be solved explicitly for ρ_p in terms of ρ_b and T, we must have recourse to a numerical method, which is detailed below. In the remainder of this section, we express all quantities in the customary dimensionless (i.e., "reduced") units, where length is given in units of σ_f and energy in units of ε_{ff}. In addition, for simplicity we set $\sigma_f = \sigma_s$ and $d = \sigma_s/\sqrt{2}$ assuming the (100) configuration of the face-centered cubic lattice for substrate atoms. Then in dimensionless units we have $a_b = 8\pi/3$ and $\rho_s = 1$.

To solve Eq. (4.42) for subcritical bulk isochores in the temperature range $T_{cb} \geq T \geq T_{xb}$, we need to determine T_{xb} first by solving [see Eq. (4.29)]

$$\rho_b^3 - \frac{1}{b}\rho_b^2 + \left(\frac{P_b}{a_b} + \frac{k_B T}{a_b b}\right)\rho_b - \frac{P_b}{a_b b} \overset{!}{=} 0 \qquad (4.43)$$

which has three real roots $\rho_{b1}(T)$, $\rho_{b2}(T)$, and $\rho_{b3}(T)$ for $T \leq T_{cb}$. These are given analytically by the (Cardanic) formulas [37]

$$\rho_{b,n} = 2\sqrt{3}u\cos\left[\phi + 2(n-1)\right] - \frac{r}{3}, \quad n = 1, 2, 3 \qquad (4.44)$$

where $\cos\phi = -q/2u$, $u = \sqrt{-p^3/27}$, $q = 2r^3/27 - rs/3 + t$, $p = s - r^2/3$, $r \equiv -1/b$, $s \equiv P_b/a_b + 1/\beta a_b b$, and $t \equiv -P_b/a_b b$.

The densities of the coexisting gas $[\rho_{xb}^g(T_{xb}) \equiv \min\{\rho_{b1}, \rho_{b2}, \rho_{b3}\}]$ and liquid $[\rho_{xb}^l(T_{xb}) \equiv \max\{\rho_{b1}, \rho_{b2}, \rho_{b3}\}]$ must satisfy Maxwell's constraint (see Appendix D.1.3)

$$0 \overset{!}{=} P(T_{xb})\left(V_x^g - V_x^l\right) - \int_{V_x^l}^{V_x^g} dV\, P(T_{xb})$$

$$= P(T_{xb})\frac{\rho_{xb}^l - \rho_{xb}^g}{\rho_{xb}^l \rho_{xb}^g} + k_B T_{xb}\ln\frac{\rho_{xb}^g\left(1 - \rho_{xb}^l b\right)}{\rho_{xb}^l\left(1 - \rho_{xb}^g b\right)} + a_b\left(\rho_{xb}^l - \rho_{xb}^g\right) \qquad (4.45)$$

As only the gas density $\rho_b^g \equiv \rho_b$ is fixed, and the density of coexisting liquid and the coexistence temperature are unknown, we must solve Eqs. (4.43) and (4.45) simultaneously for T_{xb} and ρ_b^l. We accomplish this numerically by a procedure also detailed in Appendix D.1.3.

Equation (4.42) can now be solved under experimentally relevant conditions [31], that is, for bulk isochoric paths ($\rho_b = $ const) and $T \to T_{xb}^+$. Again we defer a detailed description of the numerical procedure to Appendix D.1.3. Once the numerical solution has been found, we are in a position to calculate the excess coverage for the thermodynamically stable pore phase *via*

$$\Gamma(T, \rho_b) = (s_{z0} - 2\sigma_{wp})(\rho_p - \rho_b) \qquad (4.46)$$

which is the primary experimental quantity [31]. Results are plotted in Figs. 4.6–4.7 for various values of ε_{fs} and $s_z = 50$ corresponding closely to the experimental pore width of CPG-240 where we assume $\sigma_p = 0.47$ nm for SF_6 [47]. If the walls are purely repulsive ($\varepsilon_{fs} = 0$), $\Gamma(T, \rho_b) < 0$ (see Fig. 4.6). This situation corresponds to "drying" because $\rho_p < \rho_b$ regardless of T and ρ_b.

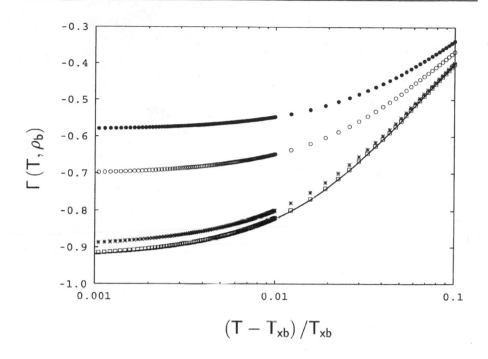

Figure 4.6: Excess coverage $\Gamma(T, \rho_b)$ as a function of temperature T for fluid confined by hard substrate ($\varepsilon_{wp} = 0$) and for representative bulk isochore $\rho_b/\rho_{cb} = 0.640$. ($\bullet$), $s_z = 50$; (O), $s_z = 10^2$; ($*$), $s_z = 10^3$; ($\square$), $s_z = 10^4$; (——), $\xi \to \infty$ limit.

If, on the other hand, the substrates are sufficiently attractive, one notices from the plots in Fig. 4.7 that $\Gamma(T, \rho_b)$ may either vary continuously or discontinuously depending on whether the (bulk) isochoric path is super- or subcritical, respectively, with regard to the critical point of the *confined* fluid. Hence, discontinuities in the plots in Fig. 4.7 indicate capillary condensation (evaporation) in the model pore *prior* to condensation in the bulk, which would, of course, occur at bulk gas-liquid coexistence, i.e., at $(T - T_{xb})/T_{xb} = 0$.

Notice also that, at temperatures higher than that corresponding to capillary condensation, $\Gamma(T, \rho_b)$ increases with decreasing T. This is indicative of a regime where one would observe growth of a wetting film, which at a mean-field level, manifests itself as an increase in overall density of an otherwise homogeneous low-density phase adsorbed all across the slit pore. For temperatures lower than that at which capillary condensation sets in, $\Gamma(\rho_b)$ turns out to be nearly independent of T the lower T becomes. This reflects that, when sufficiently close to bulk gas-liquid coexistence, the pore is com-

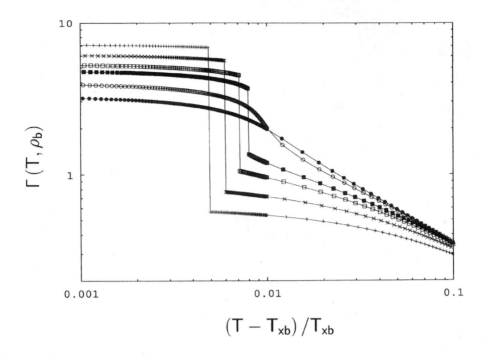

Figure 4.7: As in Fig. 4.6, but for a fluid confined by strongly attractive substrate ($\varepsilon_{\mathrm{fs}} = 1.0$), $s_z = 50$, and bulk isochores $\rho_{\mathrm{b}}/\rho_{\mathrm{cb}} = 0.545$ (+), $\rho_{\mathrm{b}}/\rho_{\mathrm{cb}} = 0.595$ (*), $\rho_{\mathrm{b}}/\rho_{\mathrm{cb}} = 0.645$ ($\square$), $\rho_{\mathrm{b}}/\rho_{\mathrm{cb}} = 0.675$ ($\blacksquare$), $\rho_{\mathrm{b}}/\rho_{\mathrm{cb}} = 0.730$ ($\bigcirc$), and $\rho_{\mathrm{b}}/\rho_{\mathrm{cb}} = 0.780$ ($\bullet$). Solid lines are intended to guide the eye.

pletely filled with a (homogeneous) liquid-like phase resisting compression.

An additional feature one notices from Fig. 4.7 concerns the change

$$\Delta\Gamma\left(T\right) = \left(s_{\mathrm{z}0} - 2\sigma_{\mathrm{wp}}\right)\left(\rho_{\mathrm{xp}}^{\mathrm{l}} - \rho_{\mathrm{xp}}^{\mathrm{g}}\right) \tag{4.47}$$

at the discontinuity where $\rho_{\mathrm{xp}}^{\mathrm{l}}$ and $\rho_{\mathrm{xp}}^{\mathrm{g}}$ are liquid and gas densities in the pore at coexistence. The plots in Fig. 4.7 clearly show that $\Delta\Gamma\left(T\right)$ becomes larger when the density of the subcritical isochoric path in the bulk is lower. The magnitude of $\Delta\Gamma\left(T\right)$ is a direct measure of the difference in density of coexisting phases in the pore, which apparently are becoming more alike the greater the proximity of the bulk isochore is to its own critical point. This is another feature also present in the parallel experiments of Thommes and Findenegg [31].

4.3 Lattice model of confined pure fluids

In the preceding section, we derived a mean-field equation of state for confined fluids based on the assumptions of

1. Irrelevance of fluid-fluid correlations

2. Homogeneity of the confined fluid

Within the approach developed in the previous section, neither of these assumptions can be replaced easily by a more realistic one. However, it turns out that, if one abandons the continuous model fluid in favor of a discrete model in which molecules are restricted to positions on a rigid lattice, the second of the above assumptions is no longer necessary to derive an analytic expression for the partition function of the fluid.

Lattice models are of great importance in a variety of contexts in statistical physics (see, for example, Ref. 48). The charm and usefulness of lattice models lies primarily in their discrete nature and their symmetry. The discreteness reduces greatly the number of configurations that need to be considered in evaluating the partition function. Their inherent symmetry, if combined with short-range interaction potentials, may be exploited to derive a fairly small set of equations describing the location of minima of the relevant thermodynamic potential that can be solved numerically by standard techniques. A key ingredient of this approach is that we deliberately neglect intermolecular correlations as before in Section 4.2.2. The current section is therefore devoted to discussing these features in some detail and illustrate the power of lattice models within the scope of this book.

4.3.1 The model system

Here we consider a lattice model of a "simple" pure confined fluid, that is, a fluid composed of molecules having only translational degrees of freedom. The positions of these molecules are restricted to $\mathcal{N} = n_x n_y n_z$ sites of a simple cubic lattice of lattice constant ℓ. Each site on the lattice can be occupied by one molecule at most which accounts for the infinitely repulsive hard core of each molecule. In addition to repulsion, pair-wise additive attractive interactions between the molecules exist. They are modeled according to square-well potentials where ε_{ff} is the depth of the attractive well whose width equals ℓ.

In addition, the fluid is confined between two planar solid surfaces (slit-pore) exerting an external field on the fluid molecules. Specifically, these solid surfaces are decorated with alternating strongly (width n_s) and weakly attractive stripes (width n_w) that mimic different chemical materials. Over the last decade, researchers have made substantial progress in controlling surface heterogeneities even down to the nanometer length scale. It is now possible to imprint specific geometric or energetic patterns on a surface experimentally at such a small length scale [49–57]. The chemically striped substrate is the most complex model substrate that we wish to consider in this section. It incorporates simpler substrate models such as the purely repulsive one ($\varepsilon_{fs} = \varepsilon_{fw} = 0$) or the chemically homogeneous substrate ($\varepsilon_{fs} = \varepsilon_{fw}$, $\alpha = 0$) as special cases. The latter are also of interest as reference systems.

For the chemically striped substrate, the external field representing the composite solid material can be cast as

$$\Phi_i^{[2]} \equiv \Phi^{[2]}(x_i, z_i) = \begin{cases} \infty & z_i > n_z \\ \begin{cases} -\varepsilon_{fs} & x_i - \alpha n_x \le n_s \\ -\varepsilon_{fw} & x_i - \alpha n_x > n_s \end{cases} & z_i = n_z \\ 0 & z_i \le n_z \end{cases} \quad (4.48a)$$

$$\Phi_i^{[1]} \equiv \Phi^{[1]}(x_i, z_i) = \begin{cases} \infty & z_i > 1 \\ \begin{cases} -\varepsilon_{fs} & x_i \le n_s \\ -\varepsilon_{fw} & x_i > n_s \end{cases} & z = 1 \\ 0 & 2 \le z_i \le n_z \end{cases} \quad (4.48b)$$

where superscripts "[1]" and "[2]" refer to lower and upper substrates, respectively, and $\alpha = \Delta n_x / n_x$ is a (discrete) dimensionless parameter specifying the misalignment of the upper relative to the lower substrate surface in the x-direction. Clearly, if $\alpha = 0$, the two surfaces are perfectly aligned, whereas misalignment is maximum for $\alpha = \frac{1}{2}$ on account of the periodicity that we assume for $\Phi_i^{[k]}$ in the x-direction. By misaligning the two substrate surfaces, the confined lattice fluid can be exposed to a shear strain represented by a strain tensor

$$\boldsymbol{\sigma} = \begin{pmatrix} 0 & 0 & \sigma_{xz} \\ 0 & 0 & 0 \\ \sigma_{xz} & 0 & 0 \end{pmatrix} \quad (4.49)$$

In Eqs. (4.48), we model the fluid-substrate interaction according to a square-well potential; ε_{fs} and ε_{fs} represent the depths of the attractive wells associated with strongly and weakly adsorbing portions of the solid substrates, respectively. Hence, our current model is very similar to the one depicted in Fig. 5.7 in the next chapter. Because of Eqs. (4.48) the lattice fluid at any site i in the x-z plane is exposed to a *total* external field

$$\Phi_i \equiv \Phi(x_i, z_i) = \Phi^{[1]}(x_i, z_i) + \Phi^{[2]}(x_i, z_i) \quad (4.50)$$

The Hamiltonian function of the lattice fluid may then be written as

$$H\left(s^{\mathcal{N}}\right) = -\frac{\varepsilon_{\text{ff}}}{2}\sum_{i=1}^{\mathcal{N}}\sum_{j\neq i}^{\nu(i)} s_i s_j + \sum_{i=1}^{\mathcal{N}} \Phi_i s_i = \sum_{i=1}^{\mathcal{N}} h\left(s_i\right) \qquad (4.51)$$

where $h\left(s_i\right)$ is the single-particle Hamiltonian function that we introduce for future reference in Section 5.8.2. In Eq. (4.51),

$$s_i \equiv s\left(x_i, z_i\right) = \begin{cases} 1, & \text{lattice site occupied} \\ 0, & \text{lattice site empty} \end{cases} \qquad (4.52)$$

is the occupation number; $s^{\mathcal{N}} = (s_1, s_2, \ldots, s_{\mathcal{N}})$ is a specific occupation-number pattern, which is a lattice-fluid configuration (and therefore the *discrete* analog of r^N introduced in Section 2.5), and $\nu(i)$ is the number of nearest-neighbor sites of site i. Depending on whether this site is located at the solid substrate $z_i = 1, n_z$ or elsewhere on the lattice, the number of nearest neighbors is 5 or 6, respectively, because of the simple-cubic symmetry of our lattice.

The confined lattice fluid is coupled to an (infinitely large) external reservoir of heat and matter. Hence, from an equilibrium perspective, the grand potential is the relevant thermodynamic potential [see Eq. (1.58)] where N is equivalent to the number of occupied lattice sites. The exact differential of Ω is given in Eq. (1.59), and the link to the microscopic level is provided by Eq. (2.81) where we need to replace the continuous variables s_α by their discrete analogs n_α ($\alpha =$ x, y, or z). Hence, the discrete analog of Eq. (2.120) may be cast as

$$\Xi = \sum_{\{s^{\mathcal{N}}\}} \exp\left\{-\frac{1}{k_{\text{B}}T}\left[H\left(s^{\mathcal{N}}\right) - \mu\sum_{i=1}^{\mathcal{N}} s_i\right]\right\} \equiv \sum_{\{s^{\mathcal{N}}\}} \exp\left[-\frac{\widetilde{H}\left(s^{\mathcal{N}}\right)}{k_{\text{B}}T}\right] \qquad (4.53)$$

where we dropped the subscript "cl" of Ξ to simplify the notation and define

$$\widetilde{H}\left(s^{\mathcal{N}}\right) \equiv H\left(s^{\mathcal{N}}\right) - \mu\sum_{i=1}^{\mathcal{N}} s_i \qquad (4.54\text{a})$$

$$\widetilde{h}\left(s_i\right) \equiv h\left(s_i\right) - \mu s_i \qquad (4.54\text{b})$$

Because of Eqs. (2.81) and (4.53) we have

$$\Omega\left(T, \mu, \boldsymbol{\sigma}\right) = -k_{\text{B}}T\ln\Xi = -k_{\text{B}}T\ln\sum_{\{s^{\mathcal{N}}\}} \exp\left[-\frac{\widetilde{H}\left(s^{\mathcal{N}}\right)}{k_{\text{B}}T}\right] \qquad (4.55)$$

as the exact expression for the grand potential of the lattice fluid where we note in passing that the dependence of Ω on the strain tensor $\boldsymbol{\sigma}$ arises through the external potential Φ_i [see Eqs. (4.48)–(4.50)].

4.3.2   High-temperature expansion

In the limit of high temperatures one may derive approximate expressions for thermodynamic properties of the lattice fluid introduced in Section 4.3.1. Here we are particularly interested in the energy. Our interest in this quantity is motivated by a later comparison between results obtained for the mean-field lattice fluid to be discussed in Section 4.5 and corresponding Monte Carlo simulations to be presented in Section 5.8. As we shall describe in Section 5.8.3, in these simulations, one has access to the grand potential as the quantity of key interest only through thermodynamic integration along suitably chosen paths in thermodynamic state space. In this regard it will turn out to be important to consider the limit of high temperatures as a reference such that an analytic expression for the energy in this limit is highly desirable.

To simplify its derivation we begin by transforming variables according to

$$s_i \rightarrow \widetilde{s}_i \equiv 2s_i - 1 \tag{4.56}$$

such that the occupation-number vector $\boldsymbol{s}^{\mathcal{N}}$ is replaced by a spin vector $\widetilde{\boldsymbol{s}}^{\mathcal{N}}$ where the elements $\widetilde{s}_i = \pm 1$ ("spin down" $\widetilde{s}_i = -1$; "spin up" $\widetilde{s}_i = +1$). Replacing in Eq. (4.51) occupation numbers by spins permits us to rewrite the Hamiltonian function in so-called "magnetic" language as

$$H_1\left(\widetilde{\boldsymbol{s}}^{\mathcal{N}}\right) \equiv \widetilde{H}\left(\widetilde{\boldsymbol{s}}^{\mathcal{N}}\right) - E_0 = -\frac{J}{2}\sum_{i=1}^{\mathcal{N}}\sum_{j\neq i}^{\nu(i)}\widetilde{s}_i\widetilde{s}_j - \sum_{i=1}^{\mathcal{N}}t_i\widetilde{s}_i \tag{4.57}$$

where

$$J = \frac{\varepsilon_{\mathrm{ff}}}{4} \tag{4.58}$$

is the coupling constant,

$$t_i = \frac{1}{2}\left[\mu + \frac{\varepsilon_{\mathrm{ff}}\nu\left(i\right)}{2} - \Phi_i\right] \tag{4.59}$$

is a (local) magnetic field

$$e_0 \equiv \frac{E_0}{\mathcal{N}} = -\frac{\varepsilon_{\mathrm{ff}}}{4}\left(3 - \frac{5}{n_z}\right) - \frac{\mu}{2} - \frac{1}{n_x n_z}\left(n_{\mathrm{s}}\varepsilon_{\mathrm{fs}} + n_{\mathrm{w}}\varepsilon_{\mathrm{fw}}\right) \tag{4.60}$$

and E_0 is a trivial constant that only shifts the zero of the energy scale. Except for the local magnetic field t_i, Eq. (4.57) represents the Hamiltonian function of the well-known Ising magnet.

From Eq. (4.53) we obtain the corresponding partition function of the Ising magnet by replacing $\widetilde{H}$ by its counterpart H_1 given in Eq. (4.57), where, however, the summation over occupation-number patterns has to be replaced by a summation over spin patterns (configurations). We may cast the expectation value of H_I by differentiating Eq. (4.53) for the Ising magnet according to

$$\langle H_I \rangle = k_B T^2 \frac{\partial}{\partial T} \ln \Xi_I = \Xi_I^{-1} \sum_{\{\tilde{s}^N\}} H_I \exp\left(-\frac{H_I}{k_B T}\right) \qquad (4.61)$$

In the limit of sufficiently high temperature ($\beta \to 0$) this expression may be evaluated approximately using the expansion

$$\exp(-x) = 1 - x + \mathcal{O}\left(x^2\right) \qquad (4.62)$$

such that Ξ_I in the denominator may be approximated by

$$\Xi_I = \sum_{\{\tilde{s}^N\}} 1 + \frac{1}{k_B T} \sum_{\{\tilde{s}^N\}} \sum_{i=1}^{N} \tilde{s}_i \left[\frac{J}{2} \sum_{j \neq i}^{\nu(i)} \tilde{s}_j + t_i\right] + \mathcal{O}\left(T^{-2}\right) \simeq \sum_{\{\tilde{s}^N\}} 1 = \iota \quad (4.63)$$

where ι is the total number of configurations. One realizes fairly easily that the term proportional to $1/T$ must vanish. To see this notice that the expression in brackets has a fixed value for each i. Now, when summing over all configurations, there will be one other configuration where all $\nu(i)$ neighbors of spin i have the same value, but $\tilde{s}_i$ has changed its sign. Therefore each of the terms in brackets arises twice when the sum over all configurations is performed: once with a positive and once with a negative sign so that all these contributions cancel each other in a pair-wise fashion. According to our initial assumption of sufficiently high temperatures, we neglect terms of higher than linear order in $1/T$.

By a similar token we have in the numerator of Eq. (4.61)

$$\sum_{\{\tilde{s}^N\}} H_I \exp(-\beta H_I) = \sum_{\{\tilde{s}^N\}} H_I - \frac{1}{k_B T} \sum_{\{\tilde{s}^N\}} H_I^2 + \mathcal{O}\left(T^{-2}\right) \simeq -\frac{1}{k_B T} \sum_{\{\tilde{s}^N\}} H_I^2 \qquad (4.64)$$

which follows because we wish to retain only the leading term in the temperature expansion and because

$$\sum_{\{\tilde{s}^N\}} H_I = 0 \qquad (4.65)$$

for reasons just explained. Hence, we may write more explicitly as

$$
\sum_{\{\tilde{\mathbf{s}}^N\}} H_{\mathrm{I}}^2 = \sum_{\{\tilde{\mathbf{s}}^N\}} \sum_{i=1}^{N} \tilde{s}_i \sum_{j=1}^{N} \tilde{s}_j \left(\frac{J}{2} \sum_{k \neq i}^{\nu(i)} \tilde{s}_k - t_i \right) \left(\frac{J}{2} \sum_{l \neq j}^{\nu(i)} \tilde{s}_l - t_j \right)
$$
$$
= \iota \sum_{i=1}^{N} \tilde{s}_i^2 \left(\frac{J^2}{4} \sum_{j \neq i}^{\nu(i)} \tilde{s}_j^2 + t_i^2 \right) \tag{4.66}
$$

because of the symmetry between spin-up and spin-down states that we exploited already in Eq. (4.63). From Eqs. (4.61), (4.63), (4.64), and (4.66) we finally obtain

$$
\langle H_{\mathrm{I}} \rangle \simeq -\frac{1}{k_{\mathrm{B}}T} \sum_{i=1}^{N} \tilde{s}_i^2 \left[\frac{J^2}{4} \sum_{j \neq i}^{\nu(i)} \tilde{s}_j^2 + t_i^2 \right] \tag{4.67}
$$

Transforming back to the original lattice-fluid parameters $\varepsilon_{\mathrm{ff}}$, μ, and Φ_i, we may write the last expression as

$$
\langle h_{\mathrm{I}} \rangle \equiv \frac{\langle H_{\mathrm{I}} \rangle}{\mathcal{N}} = -\frac{1}{32} \frac{\varepsilon_{\mathrm{ff}}^2}{k_{\mathrm{B}}T} \left(3 - \frac{1}{n_z} \right) - \frac{1}{4} \frac{(\mu + 3\varepsilon_{\mathrm{ff}})^2}{k_{\mathrm{B}}T} \left(1 - \frac{2}{n_z} \right)
$$
$$
- \frac{1}{2} \frac{\left(\mu + \frac{5}{2}\varepsilon_{\mathrm{ff}} + \varepsilon_{\mathrm{fs}} \right)^2}{k_{\mathrm{B}}T} \frac{n_s}{n_x n_z} - \frac{1}{2} \frac{\left(\mu + \frac{5}{2}\varepsilon_{\mathrm{ff}} + \varepsilon_{\mathrm{fw}} \right)^2}{k_{\mathrm{B}}T} \frac{n_w}{n_x n_z} \tag{4.68}
$$

where we used the fact that $n_x n_y (n_z - 2)$ and $n_x n_y 2$ sites have 6 and 5 nearest neighbors, respectively. Similarly, the external potential Φ_i vanishes at $n_x n_y (n_z - 2)$ sites; at the remaining $n_s n_y 2$ and $n_w n_y 2$ sites, it assumes a value of $-\varepsilon_{\mathrm{fs}}$ and $-\varepsilon_{\mathrm{fw}}$, respectively [see Eq. (4.48)].

Together, Eqs. (4.60) and (4.68) provide a means to calculate $\left\langle \tilde{H} \right\rangle$ in the limit $T \to \infty$. At sufficiently high chemical potentials one may also expect all lattice sites to be fully occupied such that $s_i = \langle s_i \rangle = 1$ so that in the limit of sufficiently high temperatures one has access to the internal energy of the lattice fluid through the expression

$$
\mathcal{U} = \langle H \rangle = \left\langle \tilde{H} \right\rangle + \mu \mathcal{N} = \langle H_{\mathrm{I}} \rangle + E_0 + \mu \mathcal{N} \tag{4.69}
$$

where H is given in Eq. (4.51) and Eqs. (4.54a) and (4.57) have also been used. Equation (4.69) serves as a suitable starting point for the calculation of the grand potential via thermodynamic integration (see Section 5.8.3).

4.4 Thermodynamics of pure confined lattice fluids

4.4.1 The Bogoliubov variational theorem

As we are ultimately interested in the phase behavior of the confined lattice fluid at arbitrary temperatures we are seeking solutions of Eq. (1.76a), where, of course, $\omega^\alpha = \mathcal{N}^{-1}\Omega^\alpha$ on account of the discreteness of the lattice. Because of Eqs. (2.81) and (4.53) this amounts to finding the chemical potential $\mu^{\alpha\beta}$ corresponding to the intersection between ω^α and ω^β at a given temperature T as we explained in Section 1.7. To identify those configurations we take recourse to thermodynamic perturbation theory [58].

Let us assume that we can split the Hamiltonian function into two contributions according to

$$H\left(s^{\mathcal{N}}; \lambda\right) = H_0\left(s^{\mathcal{N}}\right) + \lambda H_1\left(s^{\mathcal{N}}\right) \tag{4.70}$$

where H_0 and H_1 are the Hamilonians governing some reference system (subscript 0) and a perturbation (subscript 1), respectively; $0 \leq \lambda \leq 1$ is a dimensionless coupling parameter that permits us to switch continuously between the reference system and the system of interest. Because of Eqs. (2.81), (4.53), and (4.70) we may also write

$$-\frac{\Omega\left(T, \mu; \lambda\right)}{k_{\mathrm{B}}T} = \ln \sum_{\{s^{\mathcal{N}}\}} \exp\left\{-\frac{1}{k_{\mathrm{B}}T}\left[H_0\left(s^{\mathcal{N}}\right) + \lambda H_1\left(s^{\mathcal{N}}\right) - \mu\sum_{i=1}^{\mathcal{N}} s_i\right]\right\} \tag{4.71}$$

Assuming that the perturbation is not too large we may expand $\Omega\left(T, \mu; \lambda\right)$ in a MacLaurin series in terms of the coupling parameter λ, that is, as a Taylor expansion in λ around $\lambda = 0$ (i.e., the reference system). Retaining in this expansion terms up to first order, we obtain

$$\Omega\left(T, \mu; \lambda\right) = \Omega\left(T, \mu; 0\right) + \lambda\left.\frac{\mathrm{d}\Omega}{\mathrm{d}\lambda}\right|_{\lambda=0} + \mathcal{O}\left(\lambda^2\right) \tag{4.72}$$

From Eq. (4.71) it is immediately apparent that

$$\left.\frac{\mathrm{d}\Omega}{\mathrm{d}\lambda}\right|_{\lambda=0} = \frac{\sum_{\{s^{\mathcal{N}}\}} H_1\left(s^{\mathcal{N}}\right) \exp\left\{-\left[H_0\left(s^{\mathcal{N}}\right) - \mu\sum_{i=1}^{\mathcal{N}} s_i\right]/k_{\mathrm{B}}T\right\}}{\sum_{\{s^{\mathcal{N}}\}} \exp\left\{-\left[H_0\left(s^{\mathcal{N}}\right) - \mu\sum_{i=1}^{\mathcal{N}} s_i\right]/k_{\mathrm{B}}T\right\}} = \langle H_1\rangle_{\lambda=0} \tag{4.73}$$

which is the *perturbation* Hamiltonian averaged over configurations representing the *unperturbed* system. The reader should notice the similarity between these expressions and the ones derived in Section 4.2.2.1. Thus, setting $\lambda = 1$ in Eq. (4.72) and using the definition of H_1 given in Eq. (4.70), we may write

$$\Omega\left(T, \mu; 1\right) = \Omega\left(T, \mu; 0\right) + \left\langle H - H_0 \right\rangle_{\lambda=0} + \mathcal{O}\left(\lambda^2\right) \tag{4.74}$$

As one can verify by differentiating Eq. (4.73) with respect to λ,

$$\left.\frac{\mathrm{d}^2\Omega}{\mathrm{d}\lambda^2}\right|_{\lambda=0} = -\frac{1}{k_{\mathrm{B}}T} \left\langle \left(H_1 - \left\langle H_1 \right\rangle\right)^2 \right\rangle < 0 \tag{4.75}$$

which is a direct consequence of the concavity of the exponential function arising in the grand canonical ensemble partition function [see Eq. (4.53)] [59]. Equation (4.74) may then be recast as

$$\Omega\left(T, \mu; 1\right) \leq \Omega\left(T, \mu; 0\right) + \left\langle H - H_0 \right\rangle_{\lambda=0} \tag{4.76}$$

which constitutes the celebrated Bogoliubov inequality [59, 60]. It shall serve as a basis for the treatment in Section 4.4.2.

4.4.2   Mean-field approximation

Equation (4.76) states that the grand potential calculated by thermodynamic perturbation theory is an upper bound for its "true" value. Hence, one wishes to minimize the deviation between $\Omega\left(T, \mu; 1\right)$ and $\Omega\left(T, \mu; 0\right) + \left\langle H - H_0 \right\rangle_{\lambda=0}$. In this case, we take as an *ansatz*

$$H_0\left(\boldsymbol{s}^{\mathcal{N}}\right) = \sum_{i=1}^{\mathcal{N}} \Psi_i s_i \tag{4.77}$$

where Ψ_i is an *a priori* unknown external reference potential (i.e., the mean field) to be determined by minimizing the right side of Eq. (4.76) with respect to the set $\{\Psi_i\}$ [58]. Putting together Eqs. (4.51), and (4.77) we obtain [see Eq. (4.76)]

$$\left\langle H - H_0 \right\rangle_{\lambda=0} = -\frac{\varepsilon_{\mathrm{ff}}}{2} \sum_{i=1}^{\mathcal{N}} \sum_{j \neq i}^{\nu(i)} \left\langle s_i s_j \right\rangle + \sum_{i=1}^{\mathcal{N}} \left(\Phi_i - \Psi_i\right) \left\langle s_i \right\rangle \tag{4.78}$$

Moreover, we have

$$
\begin{aligned}
\Xi\left(T, \mu; 0\right) &= \sum_{\{s^{\mathcal{N}}\}} \exp\left\{ -\frac{1}{k_{\mathrm{B}}T} \left[\sum_{i=1}^{\mathcal{N}} \left(\Psi_i - \mu\right) s_i \right] \right\} \\
&= \prod_{i=1}^{\mathcal{N}} \sum_{s_i=0}^{1} \exp\left[-\frac{\left(\Psi_i - \mu\right) s_i}{k_{\mathrm{B}}T} \right] \\
&= \prod_{i=1}^{\mathcal{N}} \left\{ 1 + \exp\left[-\frac{\Psi_i - \mu}{k_{\mathrm{B}}T} \right] \right\}
\end{aligned}
\tag{4.79}
$$

which follows by replacing $H\left(s^{\mathcal{N}}\right)$ in Eq. (4.53) by $H_0\left(s^{\mathcal{N}}\right)$ [see Eq. (4.77)] and noting that $s_i = 0, 1$ is discrete and double-valued. To determine the yet unspecified external potential $\{\Psi_i\}$ in the *ansatz* [see Eq. (4.77)], we realize that

$$
\left[\left(\frac{\partial \Omega\left(T, \mu; 0\right)}{\partial \Psi_i} \right)_{T,\mu} + \frac{\partial}{\partial \Psi_i} \langle H - H_0 \rangle_{\lambda=0} \right] \mathrm{d}\Psi_i = 0
\tag{4.80}
$$

is the necessary condition determining the optimal external field. Because Eq. (4.80) must hold for arbitrary infinitesimal changes $\mathrm{d}\Psi_i$, we find from Eqs. (4.71), (4.76), (4.78), and (4.79)

$$
\rho_i \equiv \langle s_i \rangle = \frac{1}{1 + \exp\left[\left(\Psi_i - \mu\right)/k_{\mathrm{B}}T\right]}
\tag{4.81}
$$

where ρ_i is the *mean* occupation number (equivalent to the dimensionless local density in units of ℓ^3). Because of the ensemble average taken in Eq. (4.81), ρ_i is continuous on the interval $[0, 1]$ unlike s_i itself (see above). Equation (4.81) can be rearranged such that

$$
\frac{\Psi_i}{k_{\mathrm{B}}T} = \frac{\mu}{k_{\mathrm{B}}T} + \ln\frac{1 - \rho_i}{\rho_i}
\tag{4.82}
$$

which shows that Ψ_i can be interpreted as an *effective local* chemical potential that also reflects the symmetry inherent in the lattice model, that is, $\Psi_i = \mu$ for $\rho_i = \frac{1}{2}$. If $\rho_i > \frac{1}{2}$, $\Psi_i < \mu$ and vice versa. Because of this interpretation, Eq. (4.77) constitutes a mean-field approximation, i.e., $H_0\left(s^{\mathcal{N}}\right)$ in Eq. (4.77) disregards all intermolecular correlations.

Using Eqs. (4.78), (4.79), and (4.82), we finally obtain

$$
\Omega \leq \Omega\left[\rho^{\mathcal{N}}\right] \equiv k_{\mathrm{B}}T \sum_{i=1}^{\mathcal{N}} \left[\rho_i \ln \rho_i + \left(1 - \rho_i\right) \ln\left(1 - \rho_i\right)\right]
$$
$$
- \frac{\varepsilon_{\mathrm{ff}}}{2} \sum_{i=1}^{\mathcal{N}} \sum_{j \neq i}^{\nu(i)} \rho_i \rho_j + \sum_{i=1}^{\mathcal{N}} \left(\Phi_i - \mu\right) \rho_i
\tag{4.83}
$$

from Bogoliubov's inequality [see Eq. (4.76)], which defines the grand potential of the lattice fluid at mean-field level where $\boldsymbol{\rho}^{\mathcal{N}} \equiv \{\rho_1, \rho_2, \ldots, \rho_{\mathcal{N}}\}$. To arrive at Eq. (4.83) we also assumed that [see Eq. (4.78)]

$$\langle s_i s_j \rangle = \langle s_i \rangle \langle s_j \rangle = \rho_i \rho_j \tag{4.84}$$

Clearly, Eq. (4.84) ignores correlations in the occupation-number patterns (i.e., the configurations of the lattice fluid). This (independent) assumption is, however, required to be consistent with the mean-field *ansatz* in Eq. (4.77).

Finally, the best estimate of Ω is found by minimizing the functional $\Omega\left[\boldsymbol{\rho}^{\mathcal{N}}\right]$ with respect to $\boldsymbol{\rho}^{\mathcal{N}}$ for fixed values of T and μ; that is, we require the functional derivative [26, 30] to satisfy

$$\frac{\delta\Omega\left[\boldsymbol{\rho}^{\mathcal{N}}\right]}{\delta\rho_i} = 0, \qquad i = 1, \ldots, \mathcal{N} \tag{4.85}$$

in the sense of Callen (see above) [58]. Equations (4.83) and (4.85) eventually lead to the coupled set of Euler-Lagrange equations

$$k_{\mathrm{B}}T \ln \frac{\rho_i}{1 - \rho_i} - \varepsilon_{\mathrm{ff}} \sum_{j \neq i}^{\nu(i)} \rho_j + \Phi_i - \mu = 0, \qquad i = 1, \ldots, \mathcal{N} \tag{4.86}$$

which we solve numerically according to a recipe detailed in Appendix D.2.1.

4.4.3   The limit of vanishing temperature

4.4.3.1   Bulk lattice fluid

In the limit of vanishing temperature, the mean-field treatment of the current model becomes exact. To see this we begin by examining the bulk system, which is obtained as a special case of Eq. (4.86) for $\{\Phi_i\} = 0$. In the absence of an external field, all elements of the vector $\boldsymbol{\rho}^{\mathcal{N}}$ assume the same value ρ and each site on the simple cubic lattice has $\nu(i) = 6$ nearest neighbors. Hence, Eq. (4.86) can be simplified to

$$k_{\mathrm{B}}T \ln \frac{\rho}{1 - \rho} - 6\varepsilon_{\mathrm{ff}}\rho - \mu = 0 \tag{4.87}$$

which has a solution $T_{\mathrm{cb}} = \frac{3}{2}$ and $\mu_{\mathrm{cb}} = -3$ defining the critical point of the bulk lattice gas in the customary dimensionless units (distance in units of ℓ, energy in units of $\varepsilon_{\mathrm{ff}}$, temperature in units of $\varepsilon_{\mathrm{ff}}/k_{\mathrm{B}}$). In the spirit of our discussion in Section 1.7, the set of points

$$\mu_{\mathrm{x}}(T) = \{(\mu, T) \mid \mu = \mu_{\mathrm{cb}}, 0 \leq T \leq T_{\mathrm{cb}}\} \tag{4.88}$$

defines the coexistence line of the bulk lattice fluid along which gaseous and liquid phases coexist.

In dimensionless units, Eq. (4.87) can then be recast as

$$\rho = \frac{1}{6}\left(Tx - \mu\right) \tag{4.89}$$

where x is defined through the expression

$$\rho = \frac{1}{1 + \exp\left(-x\right)} \tag{4.90}$$

and $-\infty \leq x \leq \infty$ because $\rho \in [0,1]$. Thus, plotting the right sides of Eqs. (4.89) and (4.90) along $\mu_x\left(T\right)$ as functions of x, we obtain three intersections located at $-x_0$, 0, and x_0 corresponding to densities $1 - \rho\left(x_0\right)$, $\frac{1}{2}$, and $\rho\left(x_0\right)$, respectively. Evaluating the second derivative [see Eq. (4.83)]

$$\left(\frac{\partial^2 \Omega\left(\rho\right)}{\partial \rho^2}\right)_{T,\mu}\bigg|_{\rho=1/2} = \frac{T}{\rho\left(1-\rho\right)}\bigg|_{\rho=1/2} - 6 = 4T - 6 < 0 \tag{4.91}$$

Hence, Ω has a maximum at $\rho = \frac{1}{2}$ corresponding to an unstable thermodynamic state. The two remaining solutions must therefore correspond to minima of Ω, and in light of the symmetry inherent in the expression in Eq. (4.83) for $\mu_x\left(T\right)$, it is clear that $\Omega\left[\rho\left(x_0\right)\right]$ and $\Omega\left[\rho\left(1-x_0\right)\right]$ are the equal values of the grand potential of coexisting "gas" and "liquid" phases.

Furthermore, it is clear from Eq. (4.90) that in the limit $T = 0$ (i.e., $x_0 = \infty$)

$$\lim_{x_0 \to \infty} \rho\left(x_0\right) = \lim_{x_0 \to \infty} \left[1 - \exp\left(x_0\right)\right]^{-1} = 1 \tag{4.92}$$

and therefore $1 - \rho\left(x_0\right) = 0$. By a similar analysis, we obtain for the limiting stable solutions of Eq. (4.89) ($T = 0$) for $\mu > \mu_c$ and $\mu < \mu_c$, the respective values $\rho = 1$ (liquid) and $\rho = 0$ (gas). We conclude that, regardless of the value of the chemical potential, the only stable solutions of Eq. (4.89) are these two, respectively corresponding to the completely filled and completely empty lattice.

This conclusion, reached on the basis of our mean-field approach, is confirmed by the *exact* expression for Ω. From Eqs. (2.81), (4.51), and (4.53) we may write

$$\Omega = -k_B T \ln \Xi$$

$$= \tilde{H}\left(s_0^{\mathcal{N}}\right) - k_B T \ln \left[1 + \sum_{\{s^{\mathcal{N}}\}}' \exp\left(-\frac{\tilde{H}\left(s^{\mathcal{N}}\right) - \tilde{H}\left(s_0^{\mathcal{N}}\right)}{k_B T}\right)\right] \tag{4.93}$$

where the prime is attached to the summation sign to indicate that in the summation the ground-state configuration $s_0^{\mathcal{N}}$ is omitted. As $T \to 0$, the logarithmic term in Eq. (4.93) vanishes, leaving

$$
\begin{aligned}
\Omega = \widetilde{H}\left(s_0^{\mathcal{N}}\right) &= -\frac{\varepsilon_{\mathrm{ff}}}{2} \sum_{i=1}^{\mathcal{N}} \sum_{j \neq i}^{\nu(i)} s_{i0} s_{j0} + \sum_{i=1}^{\mathcal{N}} \left(\Phi_i - \mu\right) s_{i0} \\
&= -\mathcal{N}\left(3\varepsilon_{\mathrm{ff}} s_0^2 + \mu s_0\right) \equiv \mathcal{N}\omega_{\mathrm{b}}
\end{aligned}
\tag{4.94}
$$

where the second line follows for the special case of a bulk lattice fluid ($\Phi_i = 0$) in which s_0 is discrete and bivalued according to the definition given Eq. (4.52). In Eq. (4.94) we also define the grand-potential density of the bulk lattice fluid ω_{b}. The expression on the second line of Eq. (4.94) is therefore equivalent to the *mean-field* one [see Eq. (4.83)] in the limit $T = 0$ where $\rho = s_0$ as we reasoned above.

4.4.3.2 Lattice fluid confined to a chemically heterogeneous slit-pore

The above considerations may be extended to the situation of primary interest here: The fluid is constrained in one dimension (z) by plane-parallel substrates that are chemically decorated with weakly and strongly adsorbing stripes. These stripes alternate periodically in the x-direction, so that the external potential depends only on x_i and z_i [see Eq. (4.52), Fig. 4.8]. Thus, for a given value of x_i and z_i the occupation numbers do not vary with the y-coordinate of the lattice site. That is, by symmetry all densities along lines parallel with the y-axis are equivalent. Thus, using Eq. (4.86) we can write for a particular site i

$$
\beta^{-1} \ln \frac{\rho_i}{1 - \rho_i} - 2\varepsilon_{\mathrm{ff}} \rho_i - \varepsilon_{\mathrm{ff}} \sum_{j=1}^{\widetilde{\nu}(i)} \rho_j + \Phi_i - \mu = 0
\tag{4.95}
$$

where the factor of 2 comes from the two neighbors in the y-direction and $\{\rho_j\}$ are the $\widetilde{\nu}(i) = 3, 4$ densities at the nearest-neighbor sites of site i in the x–z plane. Using the definition of x given in Eq. (4.89), we can rewrite Eq. (4.95) in dimensionless variables as

$$
\rho_i = \frac{1}{2}\left(Tx - \mu_i^{\mathrm{eff}}\right)
\tag{4.96}
$$

where μ_i^{eff} may be interpreted as an *effective* chemical potential

$$
\mu_i^{\mathrm{eff}} \equiv \mu - \Phi_i + \sum_{j=1}^{\widetilde{\nu}(i)} \rho_j
\tag{4.97}
$$

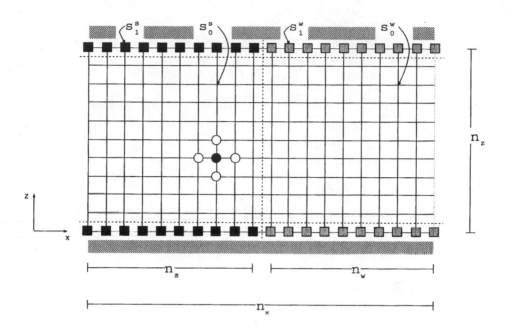

Figure 4.8: Schematic of the model. Dashed lines demarcate four types of energetically distinct regions (slabs): inner (subscript i) and outer slabs (subscript o) in weak (superscript w) and strong (superscript s) (homogeneous) modules. A molecule in the inner region (black circle) interacts with its six nearest neighbors (the four in the x–z plane are depicted as white circles; the two in the y-direction above and below the x–z plane are not shown). Sites at which molecules are subjected to the strongly attractive substrate [$\Phi_i = -\varepsilon_{fs}$, see Eq. (4.48)] are indicated by dark gray squares; those at which molecules are subjected to the weakly attractive substrate [$\Phi_i = -\varepsilon_{fs}$, see Eq. (4.48)] are denoted by light gray squares.

acting on site i in the x–z plane. Because Eq. (4.96) assumes the same functional form as Eq. (4.89) for the bulk lattice fluid, the same reasoning may be applied to allow us to conclude that in the limit $T = 0$, the only stable solutions of Eq. (4.96) are $\rho_i = 0, 1$, irrespective of the value of μ_i^{eff}. Therefore, at $T = 0$, all of the sites along lines parallel with the y-axis are either empty or filled. The stable solution of the overall problem is the set $\{\rho_i\}$ minimizing the functional

$$\Omega\left[\boldsymbol{\rho}\right] = -\frac{\varepsilon_{\text{ff}}}{2} \sum_{i=1}^{N} \sum_{j \neq i}^{\nu(i)} \rho_i \rho_j + \sum_{i=1}^{N} \left(\Phi_i - \mu\right) \rho_i \tag{4.98}$$

where all ρ_i are either 0 or 1. Again, this conclusion is exactly in accord with (the first line of) Eq. (4.94). The specific pattern of $\boldsymbol{\rho}$ minimizing $\Omega\left[\boldsymbol{\rho}\right]$ in

Eq. (4.98) [or, equivalently, the *exact* expression in the first line of Eq. (4.94)] constitutes the "morphologies" of the confined lattice fluid at $T = 0$.

4.5 Phase behavior of pure lattice fluids

Because of the relative complexity of the external field [see Eqs. (4.48)], it is instructive to enumerate possible morphologies in the limit of vanishing temperature. This is particularly so because in the limit $T = 0$ phase equilibria can be determined analytically as our discussion in Section 4.4.3 already indicated. Our recipe for identifying possible morphologies is based on a modular approach in which we construct a hierarchy of increasingly complex modules sequentially from simpler ones, starting from the bulk [61]. Any module, which gives rise to a set of morphologies $\{M^i\}$ consists of a juxtaposition of one or more of the previous (simpler) modules.

The grand potential of a given morphology within a more complex module can be expressed as a sum of grand potentials of the simpler ones, plus corrections accounting for the breaking of "bonds" between nearest neighbors in the simpler modules and the formation of new bonds across the interfaces between modules that make up the composite (more complex) module. The sum of all these contributions determines the number and eventually the stability of the possible morphologies.

4.5.1 Morphologies in the limit $T = 0$

4.5.1.1 Bulk lattice fluid

In the simplest case, $\Phi \equiv 0$ and all sites are equivalent. Thus, as we saw in Section 4.4.3 [see Eq. (4.94)], a closed expression for the grand potential may be derived where the occupation number $s_0 = 0, 1$ is discrete and double-valued. From Eqs. (1.76a) and (4.94), we conclude that

$$\omega_{\mathrm{b}}^{\mathrm{g}} - \omega_{\mathrm{b}}^{\mathrm{l}} = 0 \equiv 3\varepsilon_{\mathrm{ff}} + \mu_{\mathrm{xb}}^{\mathrm{gl}} \qquad (4.99)$$

where $\omega^{\alpha} \equiv \Omega^{\alpha}/\mathcal{N}$, $\omega_{\mathrm{b}}^{\mathrm{g}} = 0$ ($s_0 = 0$) is the grand-potential density of the gaseous and $\omega_{\mathrm{b}}^{\mathrm{l}} = -3\varepsilon_{\mathrm{ff}} - \mu$ ($s_0 = 1$) that of the liquid phase in the bulk. From Eq. (4.99), $\mu_{\mathrm{xb}}^{\mathrm{gl}} = -3\varepsilon_{\mathrm{ff}}$ is readily deduced. Thus, for $\mu < \mu_{\mathrm{xb}}^{\mathrm{gl}}$, gas is the thermodynamically stable bulk phase, whereas for $\mu > \mu_{\mathrm{xb}}^{\mathrm{gl}}$, this is true for the liquid phase.

4.5.1.2 Confinement by "hard" repulsive substrates

The next slightly more complicated situation is one in which the lattice fluid is confined in the z-direction by two planar hard substrates represented by

$$\Phi_i \equiv \Phi\left(z_i\right) \begin{cases} \infty, & z_i < 1 \quad \text{and} \quad z_i > n_z \\ 0, & 1 \le z_i \le n_z \end{cases} \qquad (4.100)$$

According to our modular approach, the lattice fluid confined by "hard" repulsive substrates may be viewed as a bulk system, in which Φ_i serves to introduce "surfaces." We can then express the grand-potential density as

$$\omega_h\left(n_z\right) = \omega_b + \Delta\omega_h\left(n_z\right) \qquad (4.101)$$

where ω_b pertains again to the bulk module defined in Eq. (4.94) and the correction $\Delta\omega_h\left(n_z\right)$ accounts for the interactions that are missing for molecules located in the surface planes $z = 1, n_z$. Because each nearest-neighbor interaction contributes $-\varepsilon_{ff} s_0^2/2$ to the configurational energy of the original bulk module, there are $n_x n_y$ molecules in each surface. The total correction then becomes $n_x n_y \varepsilon_{ff} s_0^2$ and we can rewrite the previous expression more explicitly as

$$\omega_h\left(n_z\right) = \omega_b + \frac{\varepsilon_{ff} s_0^2}{n_z} \qquad (4.102)$$

However, as $s_0 = 0, 1$, no new morphologies arise. The only effect of confinement by hard, repulsive substrates is an upward shift in the chemical potential at gas liquid coexistence. By solving the analog of Eq. (4.99) we obtain

$$\mu_x^{gl} \varepsilon_{ff}^{-1} = -3 + \frac{1}{n_z} \qquad (4.103)$$

which shows that the shift vanishes in the limit $n_z \to \infty$ where we recover the chemical potential at gas liquid coexistence in the bulk.

The upward shift in μ_x^{gl} effected by hard, repulsive walls relative to the bulk value may be interpreted as "drying." This term refers to the fact that a larger chemical potential is needed to initiate condensation of the confined gas relative to its bulk counterpart. This is because the effect of the substrates represented by Eq. (4.100) is to create an energetically less favorable situation by reducing the number of nearest-neighbor attractions from six (bulk) to five in the surface planes of the confined fluid.

4.5.1.3 Confinement by chemically homogeneous substrates

The previously discussed confinement scenario becomes slightly more complex if we allow the substrates to attract molecules in addition to just repelling them. We focus on a chemically homogeneous substrate surface first.

That is, we take in Eq. (4.48) $n_s = n_x$ ($\alpha = 0$) such that

$$\Phi_i \equiv \Phi\left(z_i\right) = \begin{cases} \infty, & z_i < 1 \quad \text{and} \quad z_i > n_z \\ -\varepsilon_{fs}, & z_i = 1 \quad \text{and} \quad z_i = n_z \\ 0, & 2 \le z_i \le n_z - 1 \end{cases} \quad (4.104)$$

The possible morphologies of the lattice fluid can thus be determined by sandwiching a hard-substrate module of "volume" $n_x n_y \left(n_z - 2\right)$ sites between two hard-substrate modules of $n_x n_y \left(n_z = 1\right)$ sites. Using the modular principle in this case, we may express the grand-potential density as

$$\omega_{\text{hom}} = \omega_{\text{h1}} + 2\omega_{\text{h2}} + \Delta\omega_{\text{hom}} \quad (4.105)$$

where

$$\omega_{\text{h1}} = \frac{1}{n_z}\left[\left(n_z - 2\right)\omega_{\text{b1}} + \varepsilon_{ff}s_{01}^2\right] \quad (4.106\text{a})$$

$$\omega_{\text{h2}} = \frac{1}{n_z}\left[\omega_{\text{b2}} + \varepsilon_{ff}s_{02}^2 - \varepsilon_{fs}s_{02}\right] \quad (4.106\text{b})$$

where we use the notation s_{0i} to indicate that the ground-state occupation numbers may assume the values 0 and 1 uniformly but independently in both modules ($i = 1, 2$). The term proportional to ε_{fs} in Eq. (4.106b) arises because of attractive fluid substrate interactions represented by Eq. (4.104). As the expression for $\omega_h \left(n_z\right)$ in Eq. (4.102) already accounts for the breaking of "bonds" to create free surfaces, the correction $\Delta\omega_{\text{hom}}$ in Eq. (4.105) is due solely to the formation of bonds across the two interfaces and is therefore given by $-2\varepsilon_{ff}s_{01}s_{02}/n_z$. Putting all this together we may rewrite Eq. (4.105) as

$$\omega_{\text{hom}} = \omega_{\text{b1}} + \frac{2}{n_z}\left(\omega_{\text{b2}} - \omega_{\text{b1}}\right) + \frac{\varepsilon_{ff}s_{02}^2}{n_z} + \frac{1}{n_z}\left[\varepsilon_{ff}\left(s_{01} - s_{02}\right)^2 - 2\varepsilon_{fs}s_{02}\right] \quad (4.107)$$

Equation (4.107) illustrates the hierarchical character of our modular approach. For example, if the substrates were purely repulsive, $\varepsilon_{fs} = 0$ so that Eq. (4.104) has to be replaced by Eq. (4.100). However, this also implies $\omega_{\text{b1}} = \omega_{\text{b2}}$ and $s_{01} = s_{02} = s_0$ in Eq. (4.107) so that the second and fourth terms vanish and we are left with the expression for ω_h given in Eq. (4.102), which reduces further to ω_b representing the bulk in the limit $n_z \to \infty$.

As s_{01} and s_{02} can assume values 0 and 1 independently, four different morphologies arise for the chemically homogeneous module. These can be identified by sets of occupation numbers $M = \{s_{01}, s_{02}\}$. For example, $M^g = \{0, 0\}$ corresponds to gas, $M^g = \{0, 1\}$ to a monolayer film adsorbed on each substrate. The grand-potential densities $\omega_{\text{hom}}^\alpha$ associated with each of the four morphologies M^α can be determined in closed form from Eq. (4.107).

4.5.1.4 Confinement by chemically heterogeneous substrates

Consider now the situation of ultimate interest, namely a lattice fluid confined between substrates endowed with weakly and strongly attractive stripes that alternate in the x-direction. In this case, the external potential is given by Eqs. (4.48) where we shall focus for simplicity on the case in which the substrates are perfectly aligned, i.e., $\alpha = 0$ in Eqs. (4.48). Following again the modular construction principle, we can enumerate possible morphologies by juxtaposing (in the x-direction) two modules corresponding to the previous, simpler one: the lattice fluid confined between *homogeneous* attractive substrates. Thus, we can write the grand-potential density as

$$\omega_{\text{het}} - \frac{1}{n_x} \left(n_s \omega_{\text{hom}}^{\text{w}} + n_s \omega_{\text{hom}}^{\text{s}} \right) + \Delta\omega_{\text{het}} \qquad (4.108)$$

where $\omega_{\text{hom}}^{\text{s,w}}$ are given by

$$\omega_{\text{hom}}^{\text{s,w}} = \omega_{\text{b1}}^{\text{s,w}} + \frac{2}{n_z} \left(\omega_{\text{h2}}^{\text{s,w}} - \omega_{\text{b1}}^{\text{s,w}} \right) + \frac{\varepsilon_{\text{ff}} s_{02}^{\text{s,w}} s_{02}^{\text{s,w}}}{n_z} + \frac{1}{n_z} \left[\varepsilon_{\text{ff}} \left(s_{01}^{\text{s,w}} - s_{02}^{\text{s,w}} \right)^2 - 2\varepsilon_{\text{fs,w}} s_{02}^{\text{s,w}} \right] \qquad (4.109)$$

by analogy with Eq. (4.107).

The correction in Eq. (4.108) can be derived as follows. We must first create surfaces by breaking bonds between nearest neighbors in the homogeneous modules across a plane plane parallel with the y–z plane. This process increases ω_{het} by the amounts $\varepsilon_{\text{ff}} \left[(n_z - 2) s_{01}^{\text{s,w}} s_{01}^{\text{s,w}} + 2 s_{02}^{\text{s,w}} s_{02}^{\text{s,w}} \right] / n_x n_z$. We must then join the strong and weak homogeneous modules by forming bonds across the interfaces. This joining decreases the grand-potential density by $\varepsilon_{\text{ff}} \left[(n_z - 2) s_{01}^{\text{w}} s_{01}^{\text{s}} + 2 s_{02}^{\text{w}} s_{02}^{\text{s}} \right] / n_x n_z$. Hence,

$$\Delta\omega_{\text{het}} = \frac{\varepsilon_{\text{ff}}}{n_x n_z} \left[2 \left(s_{02}^{\text{s}} - s_{02}^{\text{w}} \right)^2 + (n_z - 2) \left(s_{01}^{\text{s}} - s_{01}^{\text{w}} \right)^2 \right] \qquad (4.110)$$

An immediate consequence of the lower symmetry of the current system compared with the lattice fluid confined between chemically homogeneous substrates is a larger number of possible morphologies for the former. Inspection of Eqs. (4.107)–(4.110) reveals that the grand-potential density is determined by the set $\mathbf{M} = \{ s_{01}^{\text{w}}, s_{01}^{\text{s}}, s_{01}^{\text{w}}, s_{01}^{\text{s}} \}$ where each occupation number can independently assume the value 0 or 1. Thus, 16 different morphologies are possible *in principle*. This fairly large number can be reduced substantially on physical grounds by taking into account the relative magnitudes of ε_{ff}, ε_{fs}, and ε_{fw}. For example, if both ε_{fs} and ε_{fw} are small compared with ε_{ff}, the morphology characterized by $\mathbf{M} = \{0, 0, 1, 1\}$ is physically not sensible because it refers to a situation where energetically less favorable sites in the immediate vicinity of the substrates are occupied, whereas $(n_z - 2)$ energetically more favorable "inner" sites remain empty. By similar considerations, most of the remaining morphologies can be ruled out, without the necessity of calculating their grand potentials. The relevant remaining morphologies and their associated grand potentials are compiled in Table 4.1.

4.5.2   Coexisting phases of the lattice fluid

4.5.2.1   The limit $T = 0$

The remaining morphologies are those where the lattice is completely empty (i.e., "gas") or filled (i.e., "liquid") with molecules. In addition, we have a "bridge" morphology in which fluid spans the gap between the strongly attractive parts of the solid substrates leaving empty those portions of the lattice that are controlled by the weak part of the substrate. Other relevant morphologies in the current context are those listed in Table 4.1 as "droplet" and "vesicle," respectively. The former morphology refers to a situation in which a monolayer is adsorbed by the strongly attractive part of the substrates leaving the entire remainder of the lattice empty. The latter morphology describes in a sense the opposite situation in which the lattice is completely filled with molecules except for a monolayer next to the weakly adsorbing part of the substrate.

At $T = 0$, ω^α is a linear function of μ regardless of the nature of a specific morphologies as Table 4.1 shows. This implies that the equality in Eq. (1.82) holds where, however, on account of the symmetry of the external field, $\kappa_{\parallel}$ has to be replaced by κ_{yy} defined in Eqs. (5.75) and (5.76). Because the

Table 4.1: Possible lattice fluid morphologies M^α and associated grand potential $\Omega^\alpha(\mu)$ for $T = 0$.

α	M^α	Ω/n_y (See Ref. 61)
gas (g)	$\{0,0,0,0\}$	0
droplet (d)	$\{0,0,0,1\}$	$-n_s(\nu - 2 + 2\mu) + 2 - 2n_s\varepsilon_{fs}$
monolayer (m)	$\{1,1,0,0\}$	not stable at $T = 0$
bridge (b)	$\{0,1,0,1\}$	$-\frac{1}{2}n_s n_z(\nu + 2\mu) + n_z + n_s - 2n_s\varepsilon_{fs}$
vesicle (v)	$\{1,1,0,1\}$	$\frac{1}{2}(2n_s - n_x n_z)(\nu + 2\mu) + 2 + n_x - 2n_s\varepsilon_{fs}$
liquid (l)	$\{1,1,1,1\}$	$-\frac{1}{2}n_x n_z(\nu + 2\mu) + n_x - 2n_s\varepsilon_{fs} - 2n_w\varepsilon_{fw}$

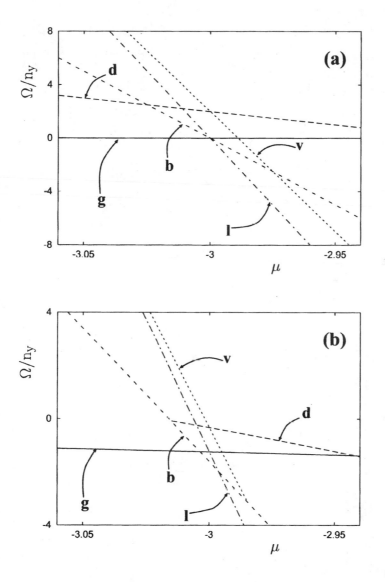

Figure 4.9: Plots of Ω/n_y as a function of μ in mean-field approximation for various temperatures and morphologies of the lattice fluid identified in Table 4.1. Data are shown for $\varepsilon_{fs} = 1$, $\varepsilon_{fs} = 0$, $n_x = 20$, $n_s = n_s = 10$. (a) $T = 0$, (b) $T = 0.6$, (c) $T = 0.9$, (d) $T = 1.2$.

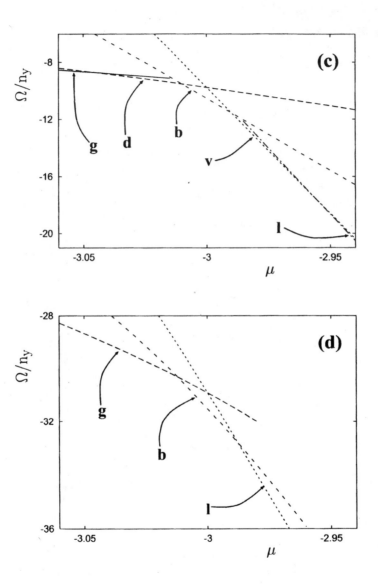

Figure 4.9: Continued.

mean density of a given morphology,

$$\overline{\rho}^\alpha \equiv \frac{1}{\mathcal{N}} \sum_{i=1}^{\mathcal{N}} \rho_i^\alpha \qquad (4.111)$$

does not vanish in general (except for the "gas" morphology, see Table 4.1) we conclude that *in general* $\kappa_\parallel$ must vanish instead. Because the fluid at $T = 0$ is in one of its ground states, the transformation $\{s_i\} \to \{\rho_i\}$ [see Eq. (4.84)] is bijective [62]. This is because s_i is double-valued and assumes one or the other of the two values in every single configuration depending on the actual morphology, that is, $s_i = \langle s_i \rangle = \rho_i$ at $T = 0$. The uniqueness of the transformation reflects the equivalence between the exact expression for the exact [see Eq. (4.55)] and the mean-field expression for the grand potential [see Eq. (4.83)].

Yet another way of looking at Eq. (1.82) is to conclude that at $T = 0$ density fluctuations are completely suppressed. To realize this, consider

$$\frac{\partial^2 \Omega}{\partial \mu^2} = -k_b T \left[\frac{1}{\Xi} \frac{\partial^2 \Xi}{\partial \mu^2} - \left(\frac{1}{\Xi} \frac{\partial \Xi}{\partial \mu} \right)^2 \right] = \frac{1}{k_B T} \left[\sum_{i=1}^{N} \sum_{j=1}^{N} \langle s_i s_j \rangle - \sum_{i=1}^{N} \langle s_i \rangle^2 \right] \qquad (4.112)$$

where Eqs. (4.54a) and (4.55) have also been employed. Hence, the equality in Eq. (1.82) implies that

$$\sum_{i=1}^{N} \sum_{j=1}^{N} \langle s_i s_j \rangle = \sum_{i=1}^{N} \sum_{j=1}^{N} \langle s_i \rangle \langle s_j \rangle \, \delta_{ij} = \sum_{i=1}^{N} \langle s_i \rangle^2 \qquad (4.113)$$

must hold at $T = 0$ where δ_{ij} is the Kronecker symbol. In other words, at $T = 0$ there are no correlations between molecules.

We are thus in a position to calculate the set of chemical potentials $\{\mu^{\alpha\beta}\}$ at which phases α and β coexist at $T = 0$ from Eq. (1.76a) and entries for Ω in Table 4.1.

4.5.2.2 Nonvanishing temperatures

In the limit of nonvanishing temperatures, simple analytic forms for the ω^α's, such as the ones compiled in Table 4.1, do not exist. Hence we need to resort to a numerical scheme to solve the Euler-Lagrange equations [see Eq. (4.86)]. This can be accomplished by an approach detailed in Appendix D.2.1 that starts from the set of (exact) morphologies $\{M^\alpha\}$ compiled in Table 4.1 at $T = 0$ as starting solutions for a temperature $T \gtrsim 0$. Once convergence has been attained, the algorithm yields new morphologies for this sufficiently

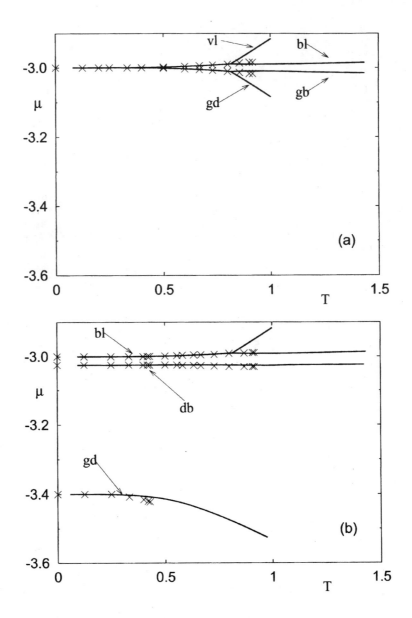

Figure 4.10: Phase diagrams $\mu_x(T)$ for a lattice fluid using a mean-field approximation (full lines) and from GCEMC ($\times$). Pairs of phases α and β indicated in the figure coexist along coexistence lines $\mu_x^{\alpha,\beta}(T)$ (see Table 4.1). For $T > 0$ a monolayer film (m) arises as a thermodynamically stable phase for $n_x = 20$, $n_s = n_s = 10$, $n_z = 10$. (a) $\varepsilon_{fs} = 1.0$, $\varepsilon_{fs} = 0.0$; (b) $\varepsilon_{fs} = 1.5$, $\varepsilon_{fs} = 0.0$; (c) $\varepsilon_{fs} = \varepsilon_{fw} = 1.5$.

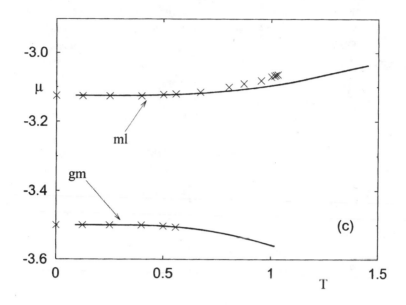

Figure 4.10: Continued.

small but nonvanishing temperature. These morphologies are then taken as
new starting solutions at a slightly higher temperature $T + \delta T$. Thus, by
varying μ at each fixed T, we obtain plots of $\Omega^\alpha(\mu, T)$ similar to the ones
plotted in Figs. 4.9(b)–4.9(d). The difference between these latter plots and
those presented in Fig. 4.9(a) is that they can no longer be represented by
straight lines but are increasingly convex the higher T becomes because of
the inequality Eq. (1.82).

According to the discussion in Section 1.7, we are now dealing with co-
existence *lines* rather than isolated points $\mu_x^{\alpha\beta}$ along which phases α and β
coexist. The corresponding phase diagram is defined in Eq. (1.88) as the
union of all coexistence lines. Figure 4.10 shows examples for $\mu_x(T)$ for a
number of different systems. The simplest situation is the one depicted in
Fig. 4.10(c) where $\varepsilon_{fs} = \varepsilon_{fw} = 1.5$, which is a relatively strongly attractive but
chemically *homogeneous* solid substrate. In this case we observe first-order
phase transitions along the coexistence curve $\mu_x^{gm}(T)$ at fairly low chemical
potentials. Along $\mu_x^{gm}(T)$ gaseous phases coexist with adsorbed monolayer
films. If one increases μ along lines of constant temperature, a second line of
first-order phase transitions is encountered. That is to say, we have capillary
condensation at $\mu_x^{ml}(T)$ for temperatures $T \leq T_c^{ml} \simeq 1.452$ which is smaller
than the bulk gas liquid critical temperature $T_{cb} = \frac{3}{2}$ on account of confine-

ment. Simultaneously, the chemical potential is depressed in the confined system relative to the bulk because of the presence of the substrates.

Plots in Figs. 4.10(a) and 4.10(b), on the other hand, pertain to cases in which the substrate is chemically *heterogeneous*. To illustrate the impact of this chemical heterogeneity, we fix $\varepsilon_{fs} = 0$ and vary ε_{fs}. In other words, the "weak" portion of the substrate is purely repulsive and supports drying rather than wetting. Generally speaking, when comparing Figs. 4.10(a) and 4.10(b) with Fig. 4.10(c), we notice that the chemical heterogeneity apparently gives rise to a more complex phase diagram. For example, plots in Fig. 4.10(a) reveal coexistence lines between various phases and the fluid bridge, as, for example, $\mu_x^{db}(T)$, $\mu_x^{gb}(T)$, and $\mu_x^{bl}(T)$. The existence of the fluid bridge is a direct consequence of the chemical structure of the underlying solid substrate that serves to "imprint" its own structure on the fluid adjacent to it. Thus, the fluid bridge, while being a generic thermodynamic phase, has no counterpart in the bulk. The uniqueness of special morphologies induced by (nano)patterned substrates has been the focus of several studies over the past few years [1, 61, 63–83]. It is now an accepted fact that these morphologies have the status of legitimate thermodynamic phases like the ordinary gaseous, liquid, or solid phases in the bulk.

Moreover, a comparison between Figs. 4.10(a) and 4.10(b) shows that, as the "strong" portion of the substrate becomes more attractive, layering transitions become more pronounced. For instance, the gas droplet coexistence line $\mu_x^{gd}(T)$ becomes detached from the remainder of $\mu_x(T)$, thereby causing the one-phase region of droplet phases to increase in size substantially.

4.5.3   The impact of shear strain

As we pointed out in Section 4.3.1, the confined fluid can be exposed to a nonvanishing shear strain by misaligning the two chemically striped surfaces. Misalignment is specified quantitatively in terms of the parameter α in Eq. (4.48a). On account of the discrete nature of our model, α can only be varied discretely in increments of $\Delta\alpha = 1/n_x$. This section is devoted to a discussion of both structure and phase behavior of a confined lattice fluid exposed to a shear strain.

4.5.3.1   Substrates in registry

We begin with the simplest situation in which the substrates are in registry, that is, $\alpha = 0$ in Eq. (4.48a). Applying the numerical procedure detailed in Appendix D.2.1 permits us to calculate the local density $\rho(x, z)$ as a solution of Eq. (4.86). Because of the discrete nature of the our model,

$\rho(x, z)$ is defined only at lattice sites. However, to visualize $\rho(x, z)$, it proves convenient to interpolate between neighboring sites. Figure 4.11(a) shows the typical structure of a bridge phase, namely a high(er) density over the strongly attractive portions of the substrate alternating in the x-direction with a low(er)-density regime over the weakly attractive ones. In the z-direction, high(er)- and low(er)-density portions of the fluid span the entire space between the substrates with comparably little variation of $\rho(x, z)$ along cuts $x = \text{const}$. Under suitable thermodynamic conditions, a bridge phase may condense and form a liquid-like phase [see Fig. 4.11(d) for a typical liquid-like phase]. Alternatively, a bridge morphology may evaporate leaving behind a gas-like phase [see Fig. 4.11(c) for a typical gas-like phase].

The bridge phase is unique in the sense that it has no counterpart in the bulk because its structure is sort of "imprinted" on the fluid by the chemical structure of the confining substrates. The importance of confinement for the existence of bridge phases is illustrated by plots of phase diagrams for various degrees of confinement in Fig. 4.12. The horizontal line in Fig. 4.12(a) represents the bulk phase diagram, which we include for comparison. Thermodynamic states $\mu < \mu_{xb}(T) = -3$ and $\mu > \mu_{xb} = -3$ pertain to the one-phase region of bulk liquid and gas, respectively ($T \leq T_{cb} = \frac{3}{2}$).

More subtle effects are observed if the lattice fluid is confined by solid substrates as plots in Fig. 4.12(a) show. For sufficiently large n_z, chemical decoration of the substrate does not matter but confinement effects prevail. For example, for $n_z = 15$, the critical point is shifted to lower T_c^{gl} and μ_c^{gl} compared with bulk $T_{cb} = \frac{3}{2}$ and $\mu_{cb} = -3$. Moreover, $\mu_x^{gl}(T)$ is no longer parallel with the temperature axis as in the bulk.

If n_z decreases, a bifurcation appears at $T = T_{tr}$. Only (inhomogeneous) liquid- and gas-like phases coexist along the line $\mu_x^{gl}(T)$ ($T < T_{tr}$). At $T = T_{tr}$ the latter two are in thermodynamic equilibrium with a bridge phase. For $T > T_{tr}$, the coexistence curve consists of two branches. The upper one, $\mu_x^{bl}(T)$, can be interpreted as a line of first-order phase transitions involving liquid-like and bridge phases whereas the lower one, $\mu_x^{gb}(T)$, corresponds to bridge and gas-like phases, respectively. Both branches terminate at their respective critical points $\{\mu_c^{bl}, T_c^{bl}\}$ and $\{\mu_c^{gb}, T_c^{gb}\}$. The entire coexistence curve $\mu_x(T)$ of the lattice fluid is formed by $\mu_x^{gl}(T)$, $\mu_x^{gb}(T)$, $\mu_x^{bl}(T)$, and the point $\{\mu_{tr}, T_{tr}\}$. Moreover, we verified numerically that

$$\lim_{T \to T_c^{ij}} \Delta \bar{\rho}_x^{ij} \propto \left(T_c^{ij} - T\right)^{\beta_{ij}} \geq 0 \tag{4.114}$$

where $\Delta \bar{\rho}_x^{ij}$ is the average-density difference between coexisting phases i and j. For the critical exponents we obtain $\beta_{gb} \simeq \beta_{bl} \simeq \frac{1}{2}$ within numerical accuracy for our three-dimensional lattice fluid model, indicating that the

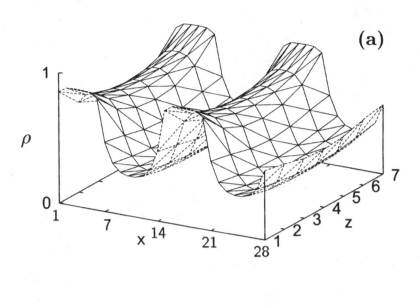

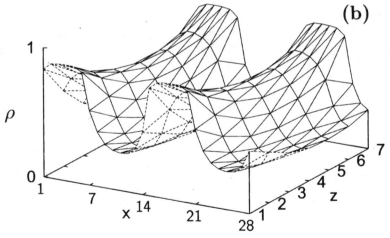

Figure 4.11: Local density $\rho(x, z)$ for confined lattice fluid at $T = 1.0$, $\mu = -3.03658$. Substrates are characterized by $n_x = 14$, $n_z = 7$, $n_s = 8$, $n_s = 6$, $\varepsilon_{fs} = 0.4$, and $\varepsilon_{fs} = 1.4$. (a) bridge morphology ($\alpha = 0$), (b) bridge phase ($\alpha = \frac{5}{14}$), (c) gaslike phase ($\alpha = \frac{1}{2}$), (d) liquidlike phase ($\alpha = \frac{1}{2}$). Plots in (c) and (d) correspond to coexisting phases. Two periods of $\rho(x, z)$ in the x–direction are shown because of lattice periodicity.

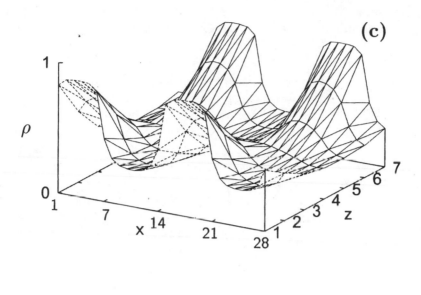

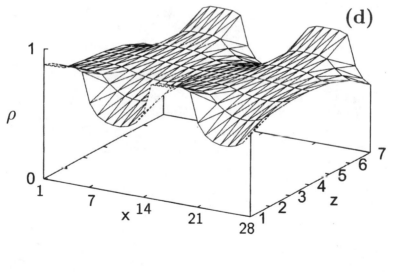

Figure 4.11: Continued.

mean-field character is preserved at both critical points (see Section 4.2.3). However, unlike for the van der Waals fluid discussed above, an analytic determination of the critical exponents is much more demanding here because a simple equation of state like the one given in Eq. (4.28) does not exist for the current model.

Comparing in Fig. 4.12(a) coexistence curves for $n_z = 8$ and 9, it is evident that the triple point is lowered the more severe the confinement becomes, that is, the smaller n_z is. Simultaneously, μ_c^{bl} increases, whereas μ_c^{gb} decreases such that the one-phase region for bridge phases widens. Because of these complex variations of $\mu_x(T)$ with n_z, it is conceivable that for a fixed thermodynamic state $\{\mu, T\}$ the confined phase is gas-like initially if n_z is sufficiently large. Upon lowering n_z, this gas-like phase may condense to a bridge and eventually to a liquidlike phase at even smaller m_z. This is illustrated in Fig. 4.12(b) for a specific thermodynamic state determined by $T = 1.325$ and $\mu = -3.0235$. From the plot it is clear that for $n_z \geq 10$ the confined fluid is gas-like because its thermodynamic state lies *below* all branches of $\mu_x(T)$. As the substrate separation decreases, however, one notices from the plot corresponding to $n_z = 9$ that the same thermodynamic state now pertains to the one-phase regime of liquid-like phases. That is to say, it falls *above* all branches of $\mu_x(T)$. Thus, in going from $n_z = 10$ to $n_z = 9$, the confined lattice fluid underwent a first-order phase transition from a gas- to a liquid-like phase. For an even smaller substrate separation $n_z = 8$, one sees from Fig. 4.12(a) that the triple point has shifted to rather small $\{\mu_{tr}, T_{tr}\}$ and that the one-phase region of bridge phases has widened considerably. Thus, as can be seen from the parallel Fig. 4.12(b), the thermodynamic state eventually belongs to the one-phase region of bridge phases where it remains for all smaller n_z. Hence, as one decreases the substrate separation from $n_z = 9$ to $n_z = 8$, an originally liquid-like phase is transformed into a bridge phase during a first-order phase transition.

4.5.3.2 Substrates out of registry

The preceding section clearly illustrates the complex phase behavior one can expect if fluids are confined between chemically decorated substrate surfaces. Three different length scales, which are present in our model, are primarily responsible for this complexity. In addition to the one corresponding to the range of interactions between lattice–fluid molecules (i.e., ℓ), another length scale refers to confinement (i.e., n_z) and is already present if the substrates are chemically homogeneous. It causes

1. A critical-point shift to lower $\{\mu_c^{gl}, T_c^{gl}\}$ ($\varepsilon_{fs} = \varepsilon_{fw} > \varepsilon_{ff}$) compared with the bulk $\{\mu_{cb}, T_{cb}\}$.

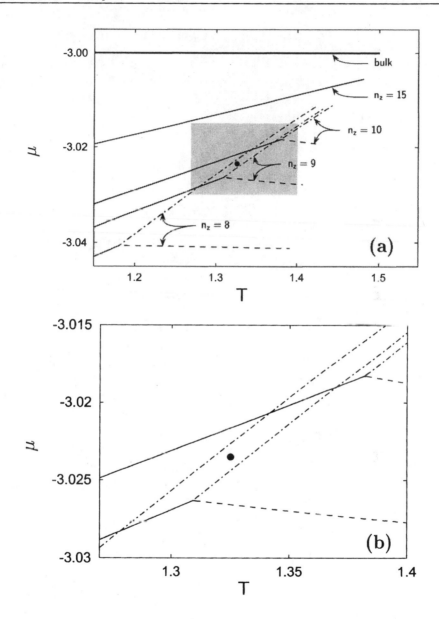

Figure 4.12: (a) Phase diagrams $\mu_x(T)$ for various confined lattice gases as functions of substrate separation n_z indicated in figure ($\alpha = 0$, $n_x = 14$, $n_s = 8$, $\varepsilon_{fs} = 0.3$, $\varepsilon_{fs} = 1.4$; (—) $\mu_x^{gl}(T)$, (— — —) $\mu_x^{gb}(T)$, and (—·—·—) $\mu_x^{bl}(T)$. (b) Enhancement of the shaded region in the plot of panel (a). In the plots of both panels (●) represents a (fixed) thermodynamic state of the confined fluid.

2. $\mu_x^{gl}(T)$ to form an angle larger than 0 with the temperature axis.

The third length scale, introduced by chemical decoration of the substrate, is set by n_s (or, equivalently, $n_x - n_s$), exceeding ℓ by almost an order of magnitude for the various coexistence curves plotted in Fig. 4.12. Consequences of this third length scale are

1. Existence of bridge phases as a new thermodynamic phase.

2. Two independent critical points $\{\mu_c^{gb}, T_c^{gb}\}$ and $\{\mu_c^{bl}, T_c^{bl}\}$.

Figure 4.12 already showed that the precise form of $\mu_x(T)$ is caused by an interplay of these different length scales.

To further elucidate this interplay, it seems interesting to expose the lattice fluid to a shear strain by varying α [see Eq. (4.48a)]. Comparing the plots in Fig. 4.11(a) and Fig. 4.11(b) illustrates the effect of a shear strain on the structure of a typical bridge phase. However, depending on the thermodynamic state, a bridge phase will sustain only a maximum shear strain and will then eventually be either "torn apart" and undergo a first-order phase transition to a gas-like phase [see Fig. 4.11(c)] or condense and form a liquid-like phase [see Fig. 4.11(d)]. Corresponding coexistence curves $\mu_x(T)$ plotted in Fig. 4.13 show that upon increasing α from its initial value of zero causes the triple point to shift to higher T_{tr} and μ_{tr}. Simultaneously, the one-phase region of bridge phases shrinks. The one-phase regime of bridge phases may, however, vanish completely for some $\alpha < \alpha_{max}$ depending on substrate separation (i.e., n_z), chemical corrugation (i.e., n_s/n_x), or strength of interaction with the chemically different parts of the substrate (i.e., ε_{fs}, ε_{fs}). Notice that for the special case $\alpha_{max} = \frac{1}{2}$ (i.e., n_x even) the one-phase region of bridge phases *must* vanish in the limit $\alpha = \alpha_{max}$ for symmetry reasons [see Eqs. (4.48)]. In addition, Fig. 4.13 shows that critical temperatures T_c^{bl} and T_c^{gb} depend only weakly on the shear strain unlike μ_c^{bl} and μ_c^{gb} such that the critical points are essentially shifted upward and downward, respectively, as α increases.

Consider now a specific isotherm $\mathbb{T} = 1.25$ in Fig. 4.13, intersecting with different branches of the (same) coexistence curve $\mu_x(T)$ at different chemical potentials. According to the definition of $\mu_x(T)$, each intersection corresponds to a pair of (separately) coexisting phases. For example, at $\mu_x^{gb}(\mathbb{T}) \simeq -3.053$ and $\alpha = 0$, a gas-like phase coexists with a (more dilute) bridge phase, whereas a (denser) bridge phase coexists with a liquid-like phase for $\mu_x^{bl}(\mathbb{T}) \simeq -3.029$. Because the one-phase region of bridge phases shrinks with α (see Fig. 4.13), the "distance" $\Delta\mu_x(\mathbb{T}) \equiv |\mu_x^{gb}(\mathbb{T}) - \mu_x^{bl}(\mathbb{T})| \to 0$ the larger α becomes, that is, with increasing shear strain. From the plot in

Fig. 4.13, it is clear that a shear strain exists such that $\Delta\mu_x = 0$, that is, $\mathbb{T} \leq T_{tr}(\alpha n_x)$. For this and larger shear strains only a single intersection remains, corresponding to coexisting gas- and liquid-like phases (see Fig. 4.13).

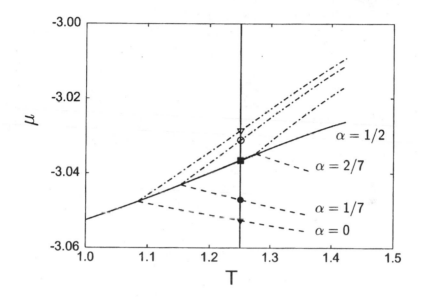

Figure 4.13: As Fig. 4.12, but for various shear strains α indicated in figure ($n_x = 14$, $n_s = 6$, $n_z = 7$, $\varepsilon_{fs} = 1.6$, $\varepsilon_{fs} = 0.4$). Intersections between isotherm $\mathbb{T}$ (vertical solid line, see text) and coexistence-curve branches represent coexisting phases. ($\blacktriangledown$) $\mu_x^{gb}(\mathbb{T})$, ($\triangledown$) $\mu_x^{bl}(\mathbb{T})$, $\alpha = 0$; ($\bullet$) $\mu_x^{gb}(\mathbb{T})$, ($\circ$) $\mu_x^{bl}(\mathbb{T})$, $\alpha = \frac{1}{7}$; ($\blacksquare$) $\mu_x^{gl}(\mathbb{T})$, $\alpha = \frac{2}{7}$.

Before returning to the issue of shear deformation of fluid bridges in a broader context in Section 5.6, we emphasize the mere fact that a fluid phase in confinement is capable of sustaining a nonvanishing shear strain. This is yet another feature of confined fluids that makes them distinct from other, more conventional, soft matter systems. As we will show below in Section 5.6, the fluid bridge "responds" to a shear strain in a fashion qualitatively similar to a bulk solid in terms of its rheological properties while maintaining a fluid-like structure. In retrospect it is this mixed fluid and solid-like nature of soft condensed matter in confinement which makes its thermodynamic treatment developed in Chapter 1 particularly insightful and appropriate.

4.6 Binary mixtures on a lattice

4.6.1 Model system

We now extend the previous discussion of pure confined lattice fluids to binary (A–B) mixtures on a simple cubic lattice of $\mathcal{N} = nz$ sites, whose lattice constant is again ℓ. We deviate from our previous notation (i.e., $\mathcal{N} = n_x n_y n_z$) because we concentrate on chemically homogeneous substrates where $n = n_x n_y$ located in a plane at some fixed distance from the substrate, which are energetically equivalent. Moreover, our subsequent development will benefit notationally by replacing henceforth n_z by just z.

The position of a fluid molecule on this lattice is then specified by a pair of integers (k, l), where $1 \leq k \leq n$ labels the position in an x–y plane and $1 \leq l \leq z$ determines the position of that plane along the z-axis. A specific site may be occupied either by a molecule of species A or B, or it may be altogether empty. Hence, this model accounts for mixed and demixed liquid phases as well as for gaseous ones. To describe individual configurations on the lattice, we introduce a matrix of occupation numbers $\mathbf{s}$ with elements

$$
s_{k,l} = \begin{cases} +1, & \text{site occupied by molecules of component A} \\ 0, & \text{empty site} \\ -1, & \text{site occupied by molecules of component B} \end{cases} \tag{4.115}
$$

For a given configuration $\mathbf{s}$, the total number of sites occupied by molecules of species A or B is given by

$$
N_A(\mathbf{s}) = \frac{1}{2} \sum_{k=1}^{n} \sum_{l=1}^{z} (s_{k,l} + 1) s_{k,l} = \frac{1}{2} \sum_{k=1}^{n} \sum_{l=1}^{z} (s_{k,l}^2 + s_{k,l}) \tag{4.116a}
$$

$$
N_B(\mathbf{s}) = \frac{1}{2} \sum_{k=1}^{n} \sum_{l=1}^{z} (s_{k,l} - 1) s_{k,l} = \frac{1}{2} \sum_{k=1}^{n} \sum_{l=1}^{z} (s_{k,l}^2 - s_{k,l}) \tag{4.116b}
$$

such that

$$
N(\mathbf{s}) = N_A(\mathbf{s}) + N_B(\mathbf{s}) = \sum_{k=1}^{n} \sum_{l=1}^{z} s_{k,l}^2 \tag{4.117}
$$

is the total number of *occupied* sites in a given configuration $\mathbf{s}$ (i.e., for a given occupation-number pattern). Equations (4.116) account for the fact that $(s_{k,l} \pm 1) s_{k,l}$ must not contribute to the sums if a site (k, l) is empty or occupied by a molecule of type B in Eq. (4.116a) nor must this term contribute to the sum in Eq. (4.116b) if the specific site is empty or occupied by a molecule of type A.

Moreover, it is straightforward to show that the total number of molecules of species A at either substrate is given by

$$N_{AW}\left(\mathbf{s}\right) = \frac{1}{2} \sum_{k=1}^{n} \left[\left(1 + s_{k,1}\right) s_{k,1} + \left(1 + s_{k,z}\right) s_{k,z}\right] \tag{4.118}$$

which follows from considerations similar to the ones leading to Eq. (4.116a). Thus, the total number of molecules of species B at the substrate is given by

$$N_{BW}\left(\mathbf{s}\right) = -\frac{1}{2} \sum_{k=1}^{n} \left[\left(1 - s_{k,1}\right) s_{k,1} + \left(1 - s_{k,z}\right) s_{k,z}\right] \tag{4.119}$$

Similarly, one can work out expressions for the number N_{AA} (N_{BB}) of A–A (B–B) pairs, which are directly connected sites, both of which are occupied by a molecule of species A (B). These somewhat more involved expressions are given by

$$
\begin{aligned}
N_{AA}\left(\mathbf{s}\right) &= \frac{1}{8} \sum_{k=1}^{n} \sum_{l=1}^{z} s_{k,l} \left(1 + s_{k,l}\right) \left[s_{k,l+1} \left(1 + s_{k,l+1}\right) + s_{k,l-1} \left(1 + s_{k,l-1}\right)\right. \\
&\quad \left. + \sum_{m=1}^{\tilde{\nu}(k)} s_{m,l} \left(1 + s_{m,l}\right)\right] \tag{4.120a}
\end{aligned}
$$

$$
\begin{aligned}
N_{BB}\left(\mathbf{s}\right) &= \frac{1}{8} \sum_{k=1}^{n} \sum_{l=1}^{z} s_{k,l} \left(1 - s_{k,l}\right) \left[s_{k,l+1} \left(1 - s_{k,l+1}\right) + s_{k,l-1} \left(1 - s_{k,l-1}\right)\right. \\
&\quad \left. + \sum_{m=1}^{\tilde{\nu}(k)} s_{m,l} \left(1 - s_{m,l}\right)\right] \tag{4.120b}
\end{aligned}
$$

where the summation over m extends over the 4 nearest neighbors $\tilde{\nu}\left(k\right)$ of lattice site k in the x–y plane. A slightly more complicated expression obtains for the number of A–B (nearest–neighbor) pairs, namely

$$
\begin{aligned}
N_{AB}\left(\mathbf{s}\right) &= -\frac{1}{8} \sum_{k=1}^{n} \sum_{l=1}^{z} \left\{ s_{k,l} \left(1 + s_{k,l}\right) \left[s_{k,l+1} \left(1 - s_{k,l+1}\right) + s_{k,l-1} \left(1 - s_{k,l-1}\right)\right. \right. \\
&\quad \left. + \sum_{m=1}^{\tilde{\nu}(k)} s_{m,l} \left(1 - s_{m,l}\right)\right] \\
&\quad + s_{k,l} \left(1 - s_{k,l}\right) \left[s_{k,l+1} \left(1 + s_{k,l+1}\right) + s_{k,l-1} \left(1 + s_{k,l-1}\right)\right. \\
&\quad \left. \left. + \sum_{m=1}^{\tilde{\nu}(k)} s_{m,l} \left(1 + s_{m,l}\right)\right] \right\} \tag{4.121}
\end{aligned}
$$

Because of the infinite repulsion "felt" by fluid molecules at vanishing distance from the substrate surface, we amend Eqs. (4.120) and (4.121) by the boundary conditions

$$s_{k,0} = s_{k,z+1} = 0, \qquad \forall k \tag{4.122}$$

The Hamiltonian function governing our system can then be cast as

$$
\begin{aligned}
H(\mathbf{s}) \;=\; & \varepsilon\left[N_{\mathrm{AA}}(\mathbf{s}) + \chi_B N_{\mathrm{BB}}(\mathbf{s})\right] + \varepsilon_{\mathrm{AB}} N_{\mathrm{AB}}(\mathbf{s}) \\
& + \varepsilon_{\mathrm{s}}\left[N_{\mathrm{AW}}(\mathbf{s}) + \chi_{\mathrm{s}} N_{\mathrm{BW}}(\mathbf{s})\right] \\
& - \mu\left[N_{\mathrm{A}}(\mathbf{s}) + N_{\mathrm{B}}(\mathbf{s})\right]
\end{aligned}
\tag{4.123}
$$

where for convenience

$$\mu = \mu_{\mathrm{A}} = \mu_{\mathrm{B}} \tag{4.124}$$

and

$$\varepsilon \equiv \varepsilon_{\mathrm{AA}} \tag{4.125a}$$

$$\varepsilon_{\mathrm{s}} \equiv \varepsilon_{\mathrm{AW}} \tag{4.125b}$$

$$\chi_{\mathrm{B}} \equiv \frac{\varepsilon_{\mathrm{BB}}}{\varepsilon_{\mathrm{AA}}} \tag{4.125c}$$

$$\chi_{\mathrm{s}} \equiv \frac{\varepsilon_{\mathrm{BW}}}{\varepsilon_{\mathrm{AW}}} \tag{4.125d}$$

In Eqs. (4.125), ε determines the depth of the attractive well (i.e., the attraction strength) of the A–A potential function. Likewise, ε_{s} describes the attraction of a molecule of species A by the solid substrate.

Parameter χ_{B} will henceforth be referred to as the "asymmetry" of the model mixture, where $\chi_{\mathrm{B}} > 1$ characterizes a binary mixture in which the formation of B–B pairs is energetically favored, whereas for $\chi_{\mathrm{B}} < 1$, this is the case for A–A pairs. For the special case $\chi_{\mathrm{B}} = 1$ the asymmetric mixture degenerates to the symmetric case previously studied in Refs. [84] and [85]. In addition, we define the "selectivity" of the solid surfaces by specifying χ_{s} in Eq. (4.125d) in a fashion similar to χ_{B} in Eq. (4.125c). Hence, the parameter space of our model is spanned by the set $\{\varepsilon, \varepsilon_{\mathrm{AB}}, \varepsilon_{\mathrm{s}}, \chi_{\mathrm{B}}, \chi_{\mathrm{s}}\}$.

4.6.2 Mean-field approximation

As we are again interested in determining the phase behavior of the binary mixture in confinement and near solid interfaces, we are essentially confronted with the same problem already discussed in Section 4.5, namely finding minima of the grand potential for a given set of thermodynamic (T, μ) and model parameters [see Eqs. (4.125)]. To obtain expressions for ω that are tractable, at least numerically, we resort again to a mean-field approximation. That is, we wish to replace H in Eq. (4.123) by its mean-field analog H_{mf}. Because we are dealing with a binary mixture, applying the approach taken in Section 4.4.2 is somewhat tedious. This is because in a binary mixture we do not only need to consider its density but also the composition to specify its physical nature without ambiguity. In the language of Section 4.4.2, we would thus have to involve a second field [besides $\{\Psi_i\}$, see Eq. (4.77)], which would render rather involved the approach taken in that section.

Alternatively, we *assume* that within each plane l parallel to the solid substrates the occupation number at each lattice site can be replaced by an *average* occupation number for the entire plane. On account of the symmetry-breaking nature of the solid substrate, these average occupation numbers will generally vary between planes; that is, they will change with l. Hence, we introduce the total *local* density

$$
\rho_l = \rho_l^A + \rho_l^B = \frac{1}{n} \sum_{k=1}^{n} s_{k,l}^2 \equiv \frac{n_l^A + n_l^B}{n} \equiv \frac{n_l}{n} \tag{4.126}
$$

and the *local* "miscibility" m_l

$$
m_l \rho_l = \rho_l^A - \rho_l^B = \frac{1}{n} \sum_{k=1}^{n} s_{k,l} \tag{4.127}
$$

as convenient alternative order parameters at the mean-field level. In the thermodynamic limit $n \to \infty$, ρ_l (in units of ℓ^3) is dimensionless and continuous on the interval $[0,1]$, which implies that m_l is continuous and dimensionless as well but on the interval $[-1,1]$.

Mathematically speaking, the mean-field assumption consists of mapping the $m \times z$ occupation-number matrix $\mathbf{s}$ onto the z-dimensional vectors $\mathbf{n}^A = \left(n_1^A, n_2^A, \ldots, n_z^A\right)$ and $\mathbf{n}^B = \left(n_1^B, n_2^B, \ldots, n_z^B\right)$ where n_l^i is the *total* number of molecules of species i on lattice plane l *regardless* of their specific arrangement. Hence, we replace $H(\mathbf{s})$ by its mean-field analog $H_{mf}\left(\mathbf{n}^A, \mathbf{n}^B\right)$ where we note in passing that the transformation $\mathbf{s} \to \mathbf{n}^A, \mathbf{n}^B$ is not bijective in general (see below).

To derive the mean-field analog of Eq. (4.53) for the current model we rewrite it more explicitly as

$$
\begin{aligned}
\Xi &= \sum_{s_{1,1}=-1}^{1} \sum_{s_{2,1}=-1}^{1} \cdots \sum_{s_{m,z}=-1}^{1} \exp\left[-\beta H(\mathbf{s})\right] \\
&= \prod_{l=1}^{z} \sum_{s_{1,l}=-1}^{1} \sum_{s_{2,l}=-1}^{1} \cdots \sum_{s_{m,l}=-1}^{1} \exp\left[-\beta H(\mathbf{s})\right]
\end{aligned} \tag{4.128}
$$

Hence, at the mean-field level, we may replace the $m \times z$ sums in parentheses above according to

$$\Xi \to \Xi_{\mathrm{mf}} = \prod_{l=1}^{z} \sum_{n_l^A=0}^{n} \sum_{n_l^B=0}^{n-n_l^A} \Theta\left(n^A, n^B\right) \exp\left[-\beta H_{\mathrm{mf}}\left(n^A, n^B\right)\right] \qquad (4.129)$$

where the combinatorial factor

$$\Theta\left(n^A, n^B\right) = \prod_{l=1}^{z} \binom{n}{n_l^A + n_l^B} \binom{n_l^A + n_l^B}{n_l^A} = \prod_{l=1}^{z} \binom{n}{n_l} \binom{n_l}{n_l^A} \qquad (4.130)$$

represents the a priori possible configurations corresponding to the same value of H_{mf}, that is, the degeneracy of a particular microstate characterized by vectors n^A and n^B.

In the thermodynamic limit (i.e., as $n \to \infty$) it is convenient to replace the discrete variables n_l^i by their (quasi-) continuous counterparts $\rho_l^i = n_l^i/n$ so that the double sums can be replaced by double integrals,

$$\sum_{n_l^A=0}^{n} \sum_{n_l^B=0}^{n-n_l^A} \ldots \xrightarrow{n \gg 1} n^2 \int_0^1 \mathrm{d}\rho_l^A \int_0^{1-\rho_l^A} \mathrm{d}\rho_l^B \ldots$$

where the z-dimensional vectors ρ^A and ρ^B are defined analogously to n^A and n^B, respectively. Changing variables $\rho_l^A, \rho_l^B \to \rho_l, m_l$ via Eqs. (4.126) and (4.127) in this last expression permits us to eventually cast Eq. (4.129) as

$$
\begin{aligned}
\Xi_{\mathrm{mf}} &= n^{2z} \prod_{l=1}^{z} \int_0^1 \mathrm{d}\rho_l^A \int_0^{1-\rho_l^A} \mathrm{d}\rho_l^B \Theta\left(\rho^A, \rho^B\right) \exp\left[-\beta H_{\mathrm{mf}}\left(\rho^A, \rho^B; \mu\right)\right] \\
&= \frac{n^{2z}}{2^z} \int \rho\, \mathrm{d}\rho \int \mathrm{d}m\, \Theta\left(\rho, m\right) \exp\left[-\beta H_{\mathrm{mf}}\left(\rho, m; \mu\right)\right] \\
&\equiv \frac{n^{2z}}{2^z} \int \rho\, \mathrm{d}\rho \int \mathrm{d}m\, \exp\left[-\beta \omega\left(\rho, m; T, \mu\right)\right] \qquad (4.131)
\end{aligned}
$$

where $\omega\left(\rho, m; T, \mu\right)$ defines an energy hyperplane in the multidimensional space spanned by the set of local order parameters $\{\rho, m\}$ for given values of T and μ.

The function $\omega\left(\rho, m; T, \mu\right)$ may have many extrema in ρ–m-space. The necessary conditions for these extrema to exist may be stated as

$$\frac{\partial \omega\left(\boldsymbol{\rho}, \boldsymbol{m}; T, \mu\right)}{\partial \rho_k} = h_1^k\left(\rho_{k-1}, m_{k-1}, \rho_k, m_k, \rho_{k+1}, m_{k+1}\right) = 0 \; (4.132\text{a})$$

$$\frac{\partial \omega\left(\boldsymbol{\rho}, \boldsymbol{m}; T, \mu\right)}{\partial m_k} = h_2^k\left(\rho_{k-1}, m_{k-1}, \rho_k, m_k, \rho_{k+1}, m_{k+1}\right) = 0 \; (4.132\text{b})$$

where explicit expressions for the functions h_1^k and h_2^k are given in Eqs. (D.32). Equations (4.132) may have several solutions $\alpha = 1, \ldots, i$. It is then sensible to introduce the notion of a phase M^α through the set of $2z$ elements

$$\mathsf{M}^\alpha = \{\boldsymbol{\rho}^\alpha, \boldsymbol{m}^\alpha\} \tag{4.133}$$

where $\boldsymbol{\rho}^\alpha$ and $\boldsymbol{m}^\alpha$ are not only simultaneous solutions of Eqs. (4.132) but also *minima* of $\omega\left(\boldsymbol{\rho}, \boldsymbol{m}; T, \mu\right)$. At this point, it is important to realize that in the thermodynamic limit (i.e. as $n \to \infty$) the global minimum $\boldsymbol{\rho}^*, \boldsymbol{m}^*$ of the function ω will completely determine the integral in Eq. (4.131). In the limit $m \to \infty$, this permits us to rewrite Eq. (4.131) as

$$\omega\left(T, \mu\right) = \frac{\Omega_{\mathrm{mf}}}{\mathcal{N}} = -\frac{\ln \Xi_{\mathrm{mf}}\left(\mathcal{N}, T, \mu\right)}{\beta \mathcal{N}} = -\frac{\ln \Theta\left(\boldsymbol{\rho}^*, \boldsymbol{m}^*\right)}{\beta \mathcal{N}} + \frac{H_{\mathrm{mf}}\left(\boldsymbol{\rho}^*, \boldsymbol{m}^*\right)}{\mathcal{N}}$$
$$(4.134)$$

where $\boldsymbol{\rho}^*$ and $\boldsymbol{m}^*$ represent the "configuration" at the absolute minimum of the grand-potential density $\omega\left(T, \mu\right)$, which is the thermodynamically stable phase (i.e., morphology) M^*, whereas all other $i-1$ phases are only metastable (except for points of phase coexistence, see Section 1.7).

4.6.3 Equilibrium states

4.6.3.1 The limit of vanishing temperature

Let us now briefly discuss the special case in which the transformation $s_{k,l} \longrightarrow \rho_l, m_l$ is bijective. From the definition of ρ_l and m_l in Eqs. (4.126) and (4.127), it is immediately clear that this can only be the case if all matrix elements in the mth row of $\mathbf{s}$ are equal assuming one of the three values given in Eq. (4.115). This then implies that $\rho_l = 0, 1$ is discrete and double-valued. In other words, across any given lattice plane l, *all* sites must be empty or occupied by molecules of one or the other species so that $\rho_l = \rho_l^A = 1$ or $\rho_l = \rho_l^B = 1$, respectively. To discriminate between these cases, Eq. (4.127) gives $m_l = 1$ if $\rho_l = \rho_l^A = 1$, whereas $m_l = -1$ if $\rho_l = \rho_l^B = 1$. Thus,

$$n_l = n \tag{4.135a}$$

$$n_l^A = \begin{cases} 0 \\ n_l \end{cases} \tag{4.135b}$$

implying $\Theta = 1$ from Eq. (4.130), which is mathematically equivalent to saying that the transformation $s_{k,l} \rightarrow \rho_l, m_l$ is bijective.

If this is so we conclude from Eq. (4.131) that

$$\omega^\alpha \left(\mathsf{M}^\alpha; T, \mu \right) = \frac{H_{\mathrm{mf}} \left(\mathsf{M}^\alpha \right)}{\mathcal{N}} \tag{4.136}$$

This latter expression is identical to Eq. (4.134) in the limit $T = 0$ replacing, however, in Eq. (4.136), M^α by M^*. Thus, in this sense

$$\omega_0 \left(\mu \right) = \omega_0 \left(\mathsf{M}^*; \mu \right) = \frac{H_{\mathrm{mf}} \left(\mathsf{M}^* \right)}{\mathcal{N}} \tag{4.137}$$

is a consequence of the fact that at $T = 0$ the mean-field treatment becomes exact (i.e., the transformation $s_{k,l} \rightarrow \rho_l, m_l$ becomes bijective) where the subscript "0" was introduced to emphasize the limit $T = 0$. Equation (4.136) is important because $H_{\mathrm{mf}} \left(\mathsf{M}^\alpha \right)$ can be calculated analytically for our current model.

4.6.3.2 Nonvanishing temperatures

For $T > 0$ we are concerned with solutions of Eqs. (4.132). To find these it is convenient to introduce the (transpose of the) $2z$-dimensional vector

$$\boldsymbol{x}^{\mathrm{T}} = \left(\rho_1, m_1, \ldots, \rho_l, m_l, \ldots, \rho_z, m_z \right) \tag{4.138}$$

which permits us to rewrite Eqs. (4.132) as

$$\boldsymbol{f} \left(\boldsymbol{x} \right) = \begin{pmatrix} h_1^1 \left(\rho_0, m_0, \rho_1, m_1, \rho_2, m_2 \right) \\ h_2^1 \left(\rho_0, m_0, \rho_1, m_1, \rho_2, m_2 \right) \\ \vdots \\ h_1^l \left(\rho_{l-1}, m_{l-1}, \rho_l, m_l, \rho_{l+1}, m_{l+1} \right) \\ h_2^l \left(\rho_{l-1}, m_{l-1}, \rho_l, m_l, \rho_{l+1}, m_{l+1} \right) \\ \vdots \\ h_1^z \left(\rho_{z-1}, m_{z-1}, \rho_z, m_z, \rho_{z+1}, m_{z+1} \right) \\ h_2^z \left(\rho_{z-1}, m_{z-1}, \rho_z, m_z, \rho_{z+1}, m_{z+1} \right) \end{pmatrix} \stackrel{!}{=} \boldsymbol{0} \tag{4.139}$$

Suppose a solution $\boldsymbol{x}_0$ of Eq. (4.139) exists for a given temperature T_0 and chemical potential μ_0. We are then seeking a solution $\boldsymbol{x}$ for slightly different thermodynamic conditions

$$T = T_0 + \delta T \tag{4.140a}$$

$$\mu = \mu_0 + \delta \mu \tag{4.140b}$$

where δT and $\delta\mu$ are sufficiently small so that we may expand Eq. (4.139) in a Taylor series around $\boldsymbol{x}_0$

$$\boldsymbol{f}(\boldsymbol{x}) = \boldsymbol{f}(\boldsymbol{x}_0) + \left(\boldsymbol{\nabla}\boldsymbol{f}^{\mathrm{T}}\right)\big|_{\boldsymbol{x}=\boldsymbol{x}_0} \cdot (\boldsymbol{x} - \boldsymbol{x}_0) + \mathcal{O}\left(|\boldsymbol{x} - \boldsymbol{x}_0|^2\right) \equiv 0 \qquad (4.141)$$

retaining only the linear term. In Eq. (4.141), the z-dimensional vector $\boldsymbol{\nabla} = (\partial/\partial\rho_1, \partial/\partial m_1, \ldots, \partial/\partial\rho_z, \partial/\partial m_z)$. Introducing the functional matrix $\mathbf{D}$ through the dyad $\boldsymbol{\nabla}\boldsymbol{f}^{\mathrm{T}}(\boldsymbol{x})$, that is

$$\mathbf{D} \equiv \boldsymbol{\nabla}\boldsymbol{f}(\boldsymbol{x}) = \begin{pmatrix} \partial h_1^1/\partial\rho_1 & \partial h_2^1/\partial\rho_1 & \cdots & \partial h_2^z/\partial\rho_1 \\ \partial h_1^1/\partial m_1 & \partial h_2^1/\partial m_1 & \cdots & \partial h_2^z/\partial m_1 \\ \vdots & \vdots & \ddots & \vdots \\ \partial h_1^1/\partial\rho_z & \partial h_2^1\partial\rho_z & \cdots & \partial h_2^z/\partial\rho_z \\ \partial h_1^1/\partial m_z & \partial h_2^1/\partial m_z & \cdots & \partial h_2^z/\partial m_z \end{pmatrix} \qquad (4.142)$$

we can solve Eq. (4.141) iteratively by rewriting it as

$$\boldsymbol{x}_{i+1} = -\mathbf{D}^{-1} \cdot \boldsymbol{f}(\boldsymbol{x}_i) + \boldsymbol{x}_i = \delta\boldsymbol{x}_i + \boldsymbol{x}_i \qquad (4.143)$$

where Eq. (4.139) has also been used and the elements of $\mathbf{D}$ can easily be computed with the aid of Eqs. (D.32). However, we may employ symmetry properties of $\mathbf{D}$ to simplify the numerical treatment. These symmetry properties are summarized in Appendix D.2.2.3.

4.7 Phase behavior of binary lattice mixtures

We begin with the simplest case of a confined binary mixture, which is a symmetric mixture confined between chemically homogeneous, nonselective planar substrates (slit-pore). The grand-potential density governing the equilibrium properties of such a mixture is given by Eq. (D.29) for the special case $\chi_{\mathrm{B}} = \chi_{\mathrm{s}} = 1$ and $\varepsilon_{\mathrm{AW}} = \varepsilon_{\mathrm{s}}$. These equilibrium states are obtained *in principle* by again solving Eqs. (4.132), where, however, h_1^k and h_2^k are now given by Eqs. (D.33) rather than by Eqs. (D.32). Except for this difference, we may, however, compute the phase diagram according to the algorithm detailed in Appendix D.2.2.3.

4.7.1 Symmetric binary bulk mixtures

We begin the discussion with bulk mixtures, which shall serve as a reference for confined binary mixtures to be discussed below in Section 4.7.2. For a more comprehensive discussion of the phase behavior of *general* bulk mixtures, the interested reader is referred to Ref. 86. Characteristic phase diagrams are displayed in Fig. 4.14 for selected values of ε_{AB}. To realize a binary bulk mixture, we choose $\varepsilon_W = 0$, $z = 1$ in Eqs. (D.33) and replace the hard-substrate boundary conditions $\rho_0 = \rho_{z+1} = 0$ by periodic boundary conditions $\rho_0 = \rho_{z+1} = \rho_1$ to account for the symmetry of the bulk mixture.

Results plotted in Fig. 4.14 for various values of ε_{AB} illustrate generic types of phase diagrams defined in Eq. (1.88). Bulk phase diagrams have also been discussed earlier by Wilding et al. [87]. These authors studied the phase behavior of a *continuous* square-well binary bulk mixture by means of Monte Carlo simulations and a mean-field approach. For $\varepsilon_{AB} = 0.40$, plots in Fig. 4.14(a) show that for temperatures $T \lesssim 1.32$ only gas and demixed liquid coexist along a line of first-order phase transitions. This line ends at a tricritical point located at $\mu_{tri} \simeq -1.75$ and $T_{tri} \simeq 1.32$. For temperatures exceeding T_{tri}, gas and demixed liquid coexist along the so-called λ-line [i.e., a line of critical points indicated by the thin solid line in Fig. 4.14(a)]. This type of phase diagram resembles the one shown by Wilding et al. in their Fig. 1(c) [87].[4]

For higher $\varepsilon_{AB} = 0.5$, the phase diagram differs qualitatively from the previous one. This can be seen from Fig. 4.14(b) where a bifurcation appears (i.e., at a triple point) for $\mu_{tr} \simeq -2.25$ and $T_{tr} \simeq 1.075$ at which a gas phase coexists simultaneously with both a mixed and a demixed fluid phase. Consequently a critical point exists ($\mu_{cb} \simeq -2.25$, $T_{cb} \simeq 1.15$) at which the line of first-order transitions between mixed liquid and gas states ends. The line of first-order transitions involving mixed and demixed liquid states ends at a higher temperature and chemical potential of $\mu_{tri} \simeq -2.00$ and $T_{tri} \simeq 1.18$, and the λ-line is shifted toward lower temperatures as one can see from the plot in Fig. 4.14(b). This type of phase diagram comports with the one shown in Fig. 1(b) of Wilding et al. [87].

A further slight increase of ε_{AB} to 0.56 does not cause the phase diagram to change *qualitatively* but *quantitatively* from the previously discussed case. This can be seen in Fig. 4.14(c) where for $\varepsilon_{AB} = 0.56$ the triple point is shifted to a lower temperature and chemical potential compared with $\varepsilon_{AB} = 0.50$. Likewise, the line of first-order transitions between gas and mixed liquid appears at lower chemical potential but is somewhat longer because the critical point is elevated to a higher $T_{cb} \simeq 1.18$. The opposite is true for the coexistence between mixed and demixed liquid phases as one can see from Figs. 4.14(b) and 4.14(c).

Eventually, as ε_{AB} becomes sufficiently large, first-order transitions be-

---

[4] As was shown recently in Ref. 86 the classification scheme proposed by Wilding et al. [87] is incomplete.

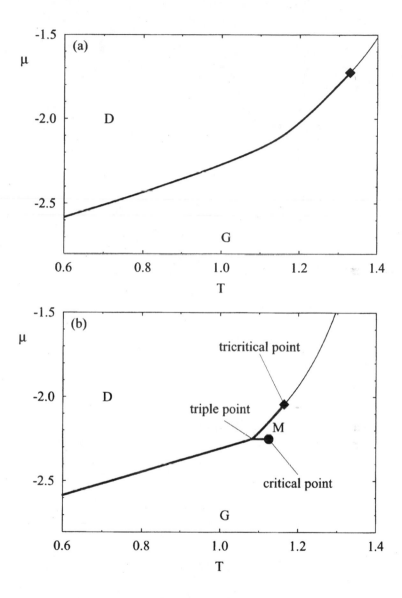

Figure 4.14: Bulk phase diagrams $\mu_x(T)$ [see Eq. (1.88)] where G, M, and D refer to one-phase regions of gaseous, mixed liquid, and demixed liquid phases, respectively. Pairs of neighboring phases coexist for state points represented by solid lines where thick and thin lines refer to first- and second-order phase transitions, respectively. (a) $\varepsilon_{AB} = 0.40$, (b) $\varepsilon_{AB} = 0.50$, (c) $\varepsilon_{AB} = 0.56$, (d) $\varepsilon_{AB} = 0.70$.

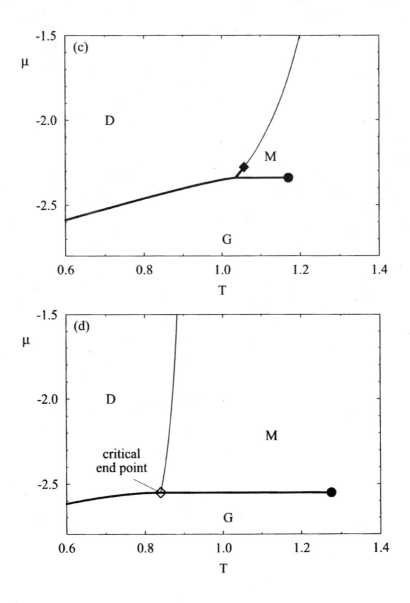

Figure 4.14: Continued.

tween mixed and demixed liquid phases disappear as the plot in Fig. 4.14(d) shows. For $\varepsilon = 0.70$ the λ-line intersects a line of first-order phase transitions at a critical end point $\mu_{\mathrm{cep}} \simeq -2.55$, $T_{\mathrm{cep}} \simeq 0.84$ because the nature of the participating phases along the λ-line differs from those involved in the first-order transitions for $T < T_{\mathrm{cep}}$ or $T > T_{\mathrm{cep}}$. This type of phase diagram resembles the one plotted in Fig. 1(a) in the paper of Wilding et al. [87].

In the limit $\varepsilon_{\mathrm{AB}} = 1.0$, the symmetric binary mixture degenerates to a pure fluid. In this case $T_{\mathrm{cep}} \to 0$ and the λ-line becomes formally indistinguishable from the μ-axis (and therefore physically meaningless). The remaining coexistence line $\mu_{\mathrm{xb}} = -3 = \mu_{\mathrm{cb}}$ (i.e., the phase diagram) involving gas (G) and liquid phases (L) becomes parallel with the T-axis and ends at the critical point where $T_{\mathrm{cb}} = \frac{3}{2}$ as expected for the bulk lattice gas [16] [see, for example, Fig. 4.12(a)].

4.7.2 Decomposition of symmetric binary mixtures

If we now confine the binary mixture to a slit-pore of nanoscopic dimension, we may, in fact, change the topolgy of the phase diagram. For example, by varying the degree of confinement (i.e., z in our current notation), it turns out to be possible to switch between various types of phase diagrams with profound consequences for liquid liquid and gas liquid phase equilibria. This phenomenon may have practical implications for the decomposition of mixtures of immiscible liquids in nanoporous matrices.

Consider as an example the case $\varepsilon_{\mathrm{AB}} = 0.5$ for which the bulk phase diagram is plotted in Fig. 4.15(a). It consists of a line of first-order phase transitions involving gaseous and *demixed* liquid states for $T \lesssim 1.08$. At $T_{\mathrm{tr}} \simeq 1.08$, the phase diagram bifurcates into a line of first-order phase transitions between gaseous and *mixed* liquid states ending at the critical point $\mu_{\mathrm{c}} \simeq -2.25$, $T_{\mathrm{c}} \simeq 1.13$, and a line of first-order transitions involving mixed and demixed liquid states. The latter ends at the tricritical point $\mu_{\mathrm{tri}} \simeq -2.04$ and $T_{\mathrm{tri}} \simeq 1.16$. If this binary mixture is now confined to a relatively wide slit-pore, the phase diagram remains of the same type, but the plot referring to $z = 12$ in Fig. 4.15(a) clearly shows the confinement-induced downward shift of coexistence lines and the displacement of characteristic (i.e., triple, critical, and tricritical) points discussed in the preceding section. However, if the degree of confinement becomes more severe [see plot for $z = 6$ in Fig. 4.15(a)], the topology of the phase diagram changes. In other words, by going from $z = 12$ to $z = 6$, the mixed liquid vanishes as a thermodynamically stable phase, whereas the entire phase diagram is further shifted to lower chemical potentials. This latter trend persists if the pore width is reduced even more with no further change in the topology of the phase diagram.

A classification of mixtures with respect to the topology of their phase diagram has been presented by van Konynenburg and Scott [88]. More recently, Woywod and Schoen have also studied the topography of phase diagrams of binary bulk mixtures [86]. This latter study was inspired by the geometrical approach to equilibrium thermodynamics discussed in the book by Wightman [89]. In their study, Woywod and Schoen present an argument which precludes the existence of tricritical points in binary mixtures in *general* [86]. This is a consequence of a purely geometrical argument based on an analysis of the number of ways in which coexistence surfaces can be joined in the (Euclidian) space of the three thermodynamic fields T, $\overline{\mu} \equiv (\mu_A + \mu_B)/2$, and $\Delta\mu \equiv (\mu_A - \mu_B)/2$ specifying the thermodynamic state of a binary fluid mixture. However, Woywod and Schoen show that, by the same token, tricritical points may exist in cases, where the mixture possesses some special symmetry.

If one then fixes the thermodynamic state such that the bulk mixture is a gas [represented by $*$ in the inset in Fig. 4.15(a)], confinement to a relatively wide pore (i.e., $z = 12$) may first cause capillary condensation to a mixed liquid mixture analogous to ordinary capillary condensation in pure fluids. If the fluid is confined to a narrower pore ($z = 6$), however, decomposition into A-rich and B-rich liquid phases is triggered by confinement upon condensation. Thus, by choosing an appropriate pore width, one can either promote condensation of a gas to a mixed liquid phase or, alternatively, initiate liquid liquid phase separation in the porous matrix where both processes are solely confinement-driven because the pore walls are nonselective for molecules of either species in our present model.

This process is further illustrated by the plots in Fig. 4.15(b) where the mean density $\overline{\rho}$ of thermodynamically stable confined phases is plotted as a function of z (i.e., the pore width). Three different branches are discernible. For small $z < 8$, $\overline{\rho}$ is relatively high indicating that the pore is filled with liquid. A corresponding plot of the local densities of a representative phase for $z = 5$ shows that this liquid consists locally of A- (or B-)rich, high-density fluid (because the two cannot be distinguished in a symmetric mixture). Hence, for $z < 8$, we observe (local) decomposition of liquid mixtures.

Along an intermediate branch of pore widths, that is, for $8 < z < 16$, $\overline{\rho}$ is somewhat smaller than for the tightest pores ($z < 8$). An inspection of a prototypical plot of the local densities for $z = 12$ reveals that the confined phase now consists of a locally equimolar mixture. Hence, for intermediate pore sizes, the confined phase is a mixed liquid.

Finally, for $z > 16$, $\overline{\rho}$ is still smaller than along the two previously discussed branches. The local density of a representative state for $z = 20$ now

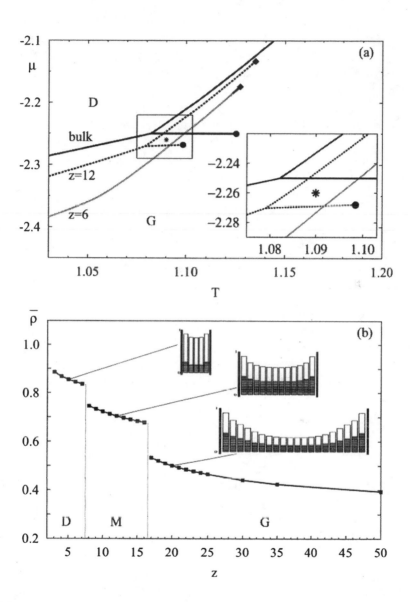

Figure 4.15: (a) As Fig. 4.14, where the inset is an enhanced representation of that part of the phase diagrams bounded by the box with a fixed thermodynamic state represented by ✱. (b) Mean pore density $\bar{\rho}$ as a function of pore width z, where stability limits between pairs of phases are demarcated by vertical lines; also shown are histograms of local density of representative phases where shading of the bars refers to ρ_l^A and ρ_l^B, respectively.

clearly shows that a comparatively low-density fluid exists at the center of the pore. As either substrate is approached, the density increases, indicating that this mixture wets the substrates. However, as expected for such a "gas" state, the fluid is composed locally of an equimolar mixture similar to states along the intermediate branch $8 < z < 16$.

The change of $\bar{\rho}$ between a pair of branches is discontinuous at characteristic pore widths where first-order transitions occur between these phases. The confinement-induced change in topology of the phase diagram may have important repercussions for the decomposition of binary mixtures in sorption experiments where one may envision pore condensation in nanoscopic solid matrices leading either to a mixed or demixed liquid such that the physical nature of the confined phase depends solely on the pore width.

4.8 Neutron scattering experiments

The discussion in the preceding section already showed that binary mixtures may decompose on account of the presence of a nanoscopic porous matrix. In general, binary liquid mixtures separate into two phases of different compositions below a critical solution point. If such mixtures are imbibed by a mesoporous matrix, phase separation cannot occur on a macroscopic length scale. Microphase separation of the system may lead to metastable local geometries of the two phases in the pore, depending on the relative amounts and strength of interaction of the two components with the surface [4]. The structure of microphase-separated mixtures in porous materials is of interest in a variety of different fields, ranging from liquid chromatography or microfiltration and related membrane processes to techniques by which liquids may be extracted from porous materials (for a review see Ref. 90).

A porous medium affects a liquid mixture not only by mere confinement to volumes of nanoscopic dimensions [91] but also by the energetic preference of the solid substrate for molecules of one of the components of the mixture [92, 93]. This selectivity causes an enrichment of the component in the proximity of the pore walls. For sufficiently wide pores, the decay length of the resulting concentration profile corresponds to the correlation length ξ_C of concentration fluctuations [94]. In narrow pores, on the other hand, when the mean pore width D is less than ξ_C, concentration profiles near the pore walls overlap, thereby causing enhanced adsorption.

Because the correlation length ξ_C increases as the critical solution point is approached, enhanced adsorption occurs in the pore space specifically in the near-critical region of the phase diagram. In the two-phase region the pore walls are wet, either partially or completely, by that phase in which the preferred component is the major component [95–97]. Incipient wetting of the pore walls manifested by multilayer adsorption is expected to occur in the one-phase region close to the liquid liquid coexistence curve in that region of thermodynamic state space where the energetically favored component is the minor component of the mixture [98–100].

To test some of the structural implications of these predictions, small-angle neutron scattering (SANS) experiments were carried out on the binary mixture iso-butyric acid (iBA, component A)+D_2O (component W) in a controlled-pore glass (CPG-10) of about 10 nm mean pore width (see Fig. 4.1). In this section, some results of the experimental work are compared with the predictions of a theoretical model. We employ the lattice model for an asymmetric binary mixture confined to a slit-pore with selective walls that we already introduced in Section 4.6.1. By adjusting model parameters properly according to criteria spelled out in Sections 4.8.2 and 4.8.3, we are able to represent a binary mixture of the amphiphile iBA and D_2O studied in the parallel SANS experiments to be described in some detail in the subsequent Section 4.8.1. Unlike in the preceding sections of this chapter dealing with mixtures, we are not able to focus on the *entire* phase diagram of the current mixture. The relative crudeness of our model, on the one hand, and the complexity of the current mixture, on the other hand, do not permit us to do so. Rather, we shall be focusing on liquid liquid phase equilibria henceforth. This narrower scope of our discussion below is motivated and justified by the conditions under which the parallel SANS experiments have been carried out.

4.8.1 Experimental details

Small-angle neutron scattering was used to study the temperature-dependent mesoscale structure of the confined liquid iBA+D_2O mixture. As the CPG-10 materials are not monoliths but consist of granules (mesh size ca. 100 nm), the sample was compressed to cylindrical pellets of about 1 mm thickness and 12 mm diameter. The pellets were soaked with a mixture at a temperature well above the phase separation temperature to prevent demixing during the filling process. A 10% excess of liquid relative to the pore space of the sample was added in order to fill completely the pore space and the interstitial space among the granules of the pellet. The SANS experiments were made in a temperature range from 10°C to 70°C after an equilibration time of at least 45 minutes.

The scattering data were analyzed by a function $I(q)$ similar to that proposed by Formisano and Teixeira [101, 102], which is composed of additive contributions accounting for the scattering of the silica matrix and an adsorbed film, as well as contributions due to concentration fluctuations in the confined liquid mixture and the microphase separation in the pores at low temperatures; that is

$$I(q) = \left[k_{\mathrm{G}} \sqrt{S_{\mathrm{G}}(q)} - k_{\mathrm{F}} \sqrt{S_{\mathrm{F}}(q)} \right]^2 + \frac{I_{\mathrm{P}}}{q^4} + \frac{I_{\mathrm{C}}}{1 + q^2 \xi_{\mathrm{C}}^2} + \frac{I_{\mathrm{S}}}{(1 + q^2 \xi_{\mathrm{S}}^2)^2} + I_{\mathrm{BG}}$$

(4.144)

The first term on the right-hand side of Eq. (4.144) describes the coherent superposition of the scattering by the silica matrix and an adsorbed film caused by preferential adsorption at the walls. Here, $S_{\mathrm{G}}(q)$ represents the structure factor arising from the quasi-periodic structure of the pore network in CPG-10, and $S_{\mathrm{F}}(q)$ is a structure factor representing the adsorbed layer at the pore walls. As in the work of Formisano and Teixeira [101, 102], these structure factors are represented by scaling functions proposed by Furukawa [103], which exhibit a correlation peak characterizing the periodicity of the matrix and adsorbed layer. The two structure factor terms are weighted by contrast factors, k_{G} and k_{F}, resulting from the difference in scattering-length density between the silica matrix and the adsorbed layer and between the layer and the core fluid, respectively (see Fig. 4.16). The temperature and composition dependence of the contrast parameters are of central importance to the theoretical analysis of the SANS experiments in the current section.

The term I_{P}/q^4 in Eq. (4.144) accounts for the Porod scattering [104], which is caused by the granular structure of the CPG-10 and the resulting contrast between the matrix and the liquid in the interstitial space. The term I_{BG} accounts for the noncoherent scattering background resulting from the nondeuterated organic liquid.

At high temperatures (i.e., in the one-phase region of the liquid mixture) the scattering of the pore liquid is described by an Ornstein-Zernike term, $I(q) \propto (1 + q^2 \xi_{\mathrm{C}}^2)^{-1}$, with a correlation length ξ_{C} characterizing the composition fluctuations. At temperatures well below the bulk-phase separation temperature of the pore liquid, the leading contribution to the scattering of the liquid is expected to arise from domains of the two liquid phases, which have sharp interfaces. This contribution, represented by the theory of Debye et al. [105], has the form $I(q) \propto (1 + q^2 \xi_{\mathrm{S}}^2)^{-2}$, where ξ_{S} is a correlation length representing the mean separation of the scattering objects.

Analysis of the SANS data [106] reveals that the Ornstein-Zernike term can be neglected at low temperatures, whereas the Debye term can be neglected at high temperatures. Details of the data analysis based on the expression given in Eq. (4.144) are presented in Ref. 106. The mean-field model described in Section 4.6.2 cannot account for fluctuation-induced effects. In the current work, we focus on the effects caused by confinement and

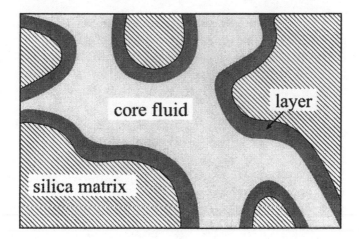

Figure 4.16: Sketch of the irregular silica network, an adsorbed layer at the pore wall, and the liquid in the core of the pore (cf. Fig. 4.1). The contrast parameters arise from differences in the scattering length density between the matrix and adsorbed layer (k_G) and between the core fluid and the adsorbed layer (k_F), respectively.

by the interaction of the components with the pore walls. In that context, fluctuation-induced effects of the sort just described are believed to be of minor importance.

4.8.2 Lattice model of water iBA mixtures

To gain an understanding of the experimental findings, we adopt a lattice model of a binary fluid mixture similar to the one introduced in Section 4.6.1. As in Section 4.6.1, we consider a simple cubic lattice with lattice constant ℓ. However, unlike in Section 4.6.1, we now assume molecules to occupy the cubic cells of volume ℓ^3 formed by the surrounding lattice sites rather than occupying the sites themselves. This approach allows us to account for the different sizes of water and iBA molecules (see below).

Each cell may be occupied by either D_2O (W) or iBA (A), or it may be empty. The attractive interactions between the molecules are accounted for by a square-well potential where the width of the attractive well is ℓ as before, which is to say that only molecules located in nearest-neighbor cells on the lattice interact directly with each other. The depths of the attractive wells are ε_{WW}, ε_{WA}, and ε_{AA} for water water, water iBA, and iBA iBA interactions, respectively. Because iBA molecules are about five times larger than water molecules, the cells occupied by water contain more parallel molecules than those occupied by iBA. Thus, the strength of the attraction between two water-occupied cells is greater than that between two iBA-occupied cells. By the same token we expect that the interaction parameters decrease in the order $\varepsilon_{WW} > \varepsilon_{WA} \gg \varepsilon_{AA}$. In this analysis, we deliberately set $\varepsilon_{AA} \equiv 0$ to reduce the number of model parameters.

In the model of the pore, the mixture is confined between two planar, homogenous substrates perpendicular to the z-axis. Thus the two substrates are at $\widetilde{z} = 0$ and $\widetilde{z} = z + 1$, where z is the number of lattice layers of the mixture parallel with the x–y plane. The width of the slit-pore is $z\ell$. Molecules do not occupy lattice cells at $\widetilde{z} = 0$ and $\widetilde{z} = z + 1$, which reflects the hard-core repulsion of the substrates. In the experimental system the water molecules are favored by the pore wall. This preferential interaction with the substrate is modeled by a potential [107]

$$V_S(\widetilde{z}) = \varepsilon_S \left[\frac{2}{15} \left(\frac{\ell}{\widetilde{z}} \right)^9 - \left(\frac{\ell}{\widetilde{z}} \right)^3 \right] \tag{4.145}$$

Equilibrium states of the water–iBA model mixture are characterized by minima of the grand-potential density

$$\omega(\boldsymbol{\rho}, \boldsymbol{m}; T, \mu, \Delta\mu) = f_{\text{int}}(\boldsymbol{\rho}, \boldsymbol{m}) - \frac{1}{z} \sum_{k=1}^{z} \left[\overline{\mu} \left(\rho_k^W + \rho_k^A \right) + \Delta\mu \left(\rho_k^W - \rho_k^A \right) \right.$$
$$\left. - \left\{ V_S(k\ell) + V_S([z + 1 - k]\ell) \right\} \rho_k^W \right] \tag{4.146}$$

where the mean-field intrinsic free energy is defined in Eqs. (D.30) and (D.31). Unlike in our previous discussion of the confined binary mixture, we now abandon the restriction stated in Eq. (4.124) as indicated in Eq. (4.146) where the mean chemical potential

$$\overline{\mu} \equiv \frac{1}{2} \left(\mu_s + \mu_A \right) \tag{4.147}$$

arises. According to its definition, $\overline{\mu}$ couples to the mean density of the water–iBA mixture $\overline{\rho}^W + \overline{\rho}^A$. where

$$\overline{\rho}^\alpha \equiv \frac{1}{z} \sum_{k=1}^{z} \rho_k^\alpha, \qquad \alpha = W, A \tag{4.148}$$

On the other hand, the incremental chemical potential in Eq. (4.146) defined as

$$\Delta\mu \equiv \frac{1}{2} \left(\mu_s - \mu_A \right) \tag{4.149}$$

controls the composition $\overline{\rho}^W + \overline{\rho}^A$ of the mixture. To obtain the density profiles of the confined binary mixture, which are of primary concern in this study, we minimize $\omega(\boldsymbol{\rho}, \boldsymbol{m}; T, \mu, \Delta\mu)$ according to the recipe described in Appendix D.2.2.3 for given values of $\overline{\mu}$, $\Delta\mu$, T, ε_{WW}, ε_{WA}, z, and ε_S. Only the latter two of these parameters describe the confinement effect due to the substrates.

However, in the experimental systems, the pore geometry of CPG is more or less that of cylinders rather than of slits as in the theoretical model. Thus, the surface-to-volume ratio a_S/v_p is larger for a cylindrical pore geometry. To take this into account, we introduce the concept of a hydraulic pore radius $r_h \equiv 2a_S/v_p$ following the suggestion of Rother et al. [108]. Whereas for cylindrical pores r_h is just the radius of the cylinders, for slit-pores r_h is taken to be twice the pore width; that is, $r_h = 2z\ell$. Setting the lattice constant equal to the mean distance of molecules of type W and A, and $\ell = 1$ nm, we find for the nominal pore size $r_h = 6.8$ nm of CPG-10-75 pores [108], and therefore, the overall number of layers corresponds to $z = r_h/\ell = 7$. Additionally, all densities ρ are measured in units of ℓ^{-3}, whereas $k_B T$, μ, $\Delta\mu$, c_{WA}, and ε_S are in units of ε_{WW}.

4.8.3 Phase diagram

4.8.3.1 Bulk Mixture

As explained in Section 4.8.2, our bulk model depends on four adjustable parameters, namely T, μ, $\Delta\mu$, and ε_{WA}. Among these the mean chemical potential μ determines mainly gas-liquid phase coexistence. Because the bulk experimental mixture is always in the liquid state, μ should be larger than -2.0 (in our dimensionless units) so that we are dealing with a (bulk) liquid state in the model calculations for all considered temperatures. The incremental chemical potential $\Delta\mu$, on the other hand, is the thermodynamic variable conjugate to the composition cast here in terms of $\overline{\rho}^W - \overline{\rho}^A$ for convenience [see Eq. 4.146]. In accord with the experimental conditions we adjust $\Delta\mu$ such that the model mixture is at the liquid-liquid coexistence for a given temperature below the critical point. Therefore, $\Delta\mu$ is fixed by this coexistence between water-rich and iBA-rich liquid phases.

To link the dimensionless model temperature T to that of the experimental study, we use the criterion that the critical solution temperature of the experimental system, T_c^{exp}, is to be matched by that of our model mixture, T_c. Moreover, we introduce a temperature offset, T_{off}, because this lattice fluid model is less suitable at low temperatures. Because this model is based on a mean-field treatment and does not account for the difference in molecular size of the two components, the calculated coexistence curve does not represent the phase behavior of the real system over an extended temperature range, especially at low temperatures where packing effects play an important role. This deficiency of the model is compensated by introducing a temperature shift as an empirical parameter by which the coexistence curve of the model can be tuned to that of the real system. Thus, the (dimensionless) model temperature T is transformed into the model temperature T^{mod} in K by

$$T^{\text{mod}} = \frac{T}{T_{\text{c}}} \left(T_{\text{c}}^{\text{exp}} - T_{\text{off}} \right) + T_{\text{off}} \tag{4.150}$$

In the experiment, mass fractions rather than volume fractions are used. The conversion from the volume densities ρ^{W} and ρ^{A} into the mass fractions w_{s} and w_{A} is effected by

$$w_{\text{A}} = 1 - w_{\text{s}} = \frac{1 + m}{1 + \eta + m\,(1 - \eta)} \tag{4.151}$$

with [cf. Eq. (4.127)]

$$m \equiv \frac{\overline{\rho}^{\text{W}} - \overline{\rho}^{\text{A}}}{\overline{\rho}^{\text{W}} + \overline{\rho}^{\text{A}}} \tag{4.152}$$

and $\eta = 0.850$ is the ratio of the mass densities of iBA and D_2O at 30°C.

The model bulk parameters T_{off}, μ, and ε_{WA} are adjusted to the experimental bulk phase diagram in the following way. Figure 4.17 shows the experimental and calculated coexistence curves for the adjusted parameters $T_{\text{off}} = 261$ K ($-12°$C), $\mu = -1.05$, and $\varepsilon_{\text{WA}} = 0.30$. The resulting model critical temperature, T_{c}, is 0.5 K higher than the experimental value reported by Gansen and Woermann [109] ($T_{\text{c}}^{\text{exp}} = 318.19$ K).[5] Close to the critical point the coexistence curve is expected to conform to a power law

$$\Delta w_{\text{A}} \propto (T_{\text{c}} - T)^{\beta} \tag{4.153}$$

where β is a critical exponent [cf. Eq. (1.86)]. Our calculation verifies the usual mean-field critical exponent $\beta = \frac{1}{2}$ (see Section 4.2.3 for comparison), whereas Gansen and Woermann [109] found $\beta \simeq 0.330$ as expected for a real mixture. The critical point of the system is located on the water-rich side, reflecting the fact that the strength of the attraction between water molecules exceeds that between iBA molecules.

We note that the influences of $\overline{\mu}$ and ε_{WA} on the bulk phase diagram in Fig. 4.17 are essentially independent. Changes in $\overline{\mu}$ lead to a horizontal

---

[5]The value of $T_{\text{c}}^{\text{exp}}$ reported by Gansen and Woermann [109] may be too high on account of impurities in the sample [110], which is, however, inconsequential for this analysis.

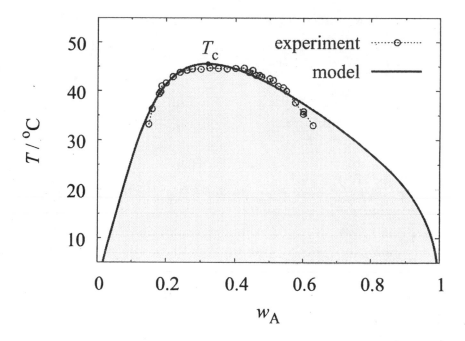

Figure 4.17: Liquid-liquid bulk coexistence curve of iBA+D$_2$O (A+W) in the w_A-T projection. The shaded region marks the region of liquid-liquid phase coexistence.

shift of the whole coexistence curve along the w_A-axis, whereas changes in ε_{WA} modify the curvature. The shaded area in Fig. 4.17 marks the liquid liquid coexistence at which the bulk system separates into a water-rich and an iBA-rich phase.

4.8.3.2 Confined system

Now we consider the water+iBA mixture confined in pores. As outlined in Section 4.8.2, we use the concept of the hydraulic radius to compare the results for the controlled-pore glass material (CPG-75, $r_h = 6.8$ nm) with those for a model slit-pore of width approximately 3.5 nm. Taking the adjusted parameter set from bulk, that is, $T_{off} = 261$ K, $\bar{\mu} = -1.05$, and $\varepsilon_{WA} = 0.30$, we have the only remaining model parameter ε_S, which measures the preferential strength of attraction of D$_2$O for the substrates [see Eq. (4.145)]. The parallel experiments of Rother et al. [108] suggest that ε_S should be larger than 1.5 but less than 5 (in our current units) to avoid unrealistic strong

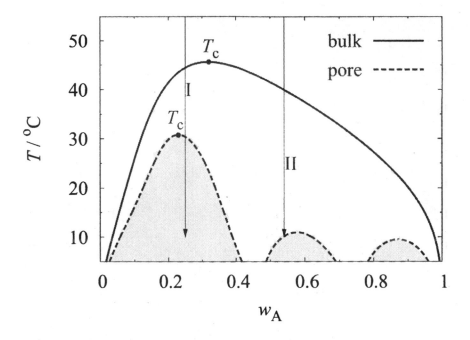

Figure 4.18: Phase diagram of the model mixture in the w_A-T projection for the bulk system (—) and confined in the slit-pore (- - -) with $\varepsilon_S = 3.0$. Shaded regions are phase coexistences of the confined system. Dots (•) indicate critical points. Paths I and II display temperature quenches at two fixed compositions (mean mass fractions $w_A = 0.25$ and $w_A = 0.54$, respectively).

adsorption effects.

The model coexistence curves for our choice of $\varepsilon_S = 3.0$ are shown in Fig. 4.18. The shaded areas indicate regions of phase coexistence of the confined system. The remarkable change of the phase diagram relative to that of the bulk system is caused by the strong confinement together with the strong selectivity of the pore for water. As expected, the critical temperature of the pore fluid is shifted downward. The critical composition has moved toward the water-rich side because of the selective character of the substrates.

In addition to the liquid liquid coexistence curve, the confined fluid exhibits two further, smaller phase coexistence regions at larger w_A and lower T. The coexisting phases represent water-rich films of a thickness corresponding to one or two layers, which are distinguishable only at lower temperatures. The existence of such first-order layering transitions may be overestimated by our lattice model on a homogeneous surface and enforced unrealistically

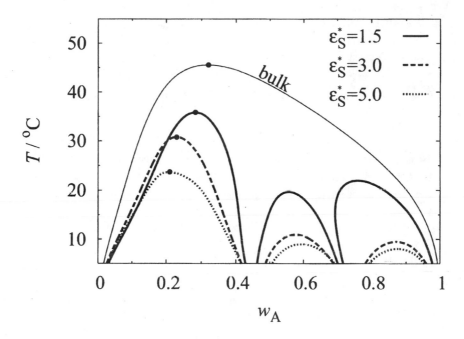

Figure 4.19: Phase diagrams of the mixture in the T-w_A projection for the model slit-pore for different water substrate interaction strengths ε_S. The bulk coexistence curve is represented by the thin solid line. Dots (•) indicate critical points.

by the geometry of the underlying (simple-cubic) lattice. Such transitions are, in fact, weakened or absent in continuous models or in model systems with rough or heterogenous walls [111, 112].

As shown in Fig. 4.18, confinement of the liquid mixture leads to a strong depression of the critical temperature T_c, in accordance with earlier studies [84, 91, 113, 114]. This critical point shift can be attributed to the cutoff of the range of concentration fluctuations when the correlation length ξ_c becomes equal to the pore width. Confinement also leads to a shift of the critical point to a more water-rich composition (i.e., in the direction of the component, which is preferred by the walls). This finding is consistent with results of earlier studies [93, 115] and can be rationalized by the fact that the pore liquid is highly inhomogeneous.

Figure 4.19 shows the influence of the substrate attraction strength ε_S. As can be seen, the larger the selectivity of the substrates, the lower the critical temperature and the smaller the critical iBA-concentration, w_A. Also, we observe a weakening of the film phase transitions for strongly adsorbing walls as almost all water molecules stick to the substrates. But for all ε_S values, these film phases are distinguishable only at temperatures well below the corresponding critical solution point.

4.8.4   Concentration profiles and contrast factors

We now focus on the behavior of the mixture as a function of T at constant overall composition in order to compare the predictions of the model with the results of the experimental study. Specifically, we consider the "trajectory" paths I and II indicated in Fig. 4.18, which correspond to the mass fractions w_A chosen in the experimental study. Schemmel et al. [106] showed that along these trajectories the Porod parameter I_P (see Section 4.8.1) does not change with temperature, which indicates that the mean composition of the mixture in the pore stays constant for all temperatures.

We adopt the following prescription to compare the measured contrast factors, k_G and k_F (see Section 4.8.1), with those predicted by our model. The densities, ρ_{lay}^W and ρ_{lay}^A, of the first layer at the pore wall and the mean densities, ρ_{core}^W and ρ_{core}^A, inside the remaining pore space are given by

$$\rho_{\text{lay}}^\alpha \equiv \frac{1}{2}\left(\rho_1^\alpha + \rho_z^\alpha\right) \tag{4.154a}$$

$$\rho_{\text{core}}^\alpha \equiv \frac{1}{z-2}\sum_{k=2}^{z-1}\rho_k^\alpha \tag{4.154b}$$

where $\alpha = A, W$ and we set again $z = 7$. Taking the values of scattering length densities d [116] in units of 10^{10} cm^{-2}, $d_s = 6.37$, $d_A = 0.537$, and $d_{\text{SiO}_2} = 3.64$, for D$_2$O, iBA, and SiO$_2$ (silica matrix), respectively, we may define the contrast parameters in units of 10^{10} cm^{-2} for a given model phase by

$$k_G \equiv \left|d_{\text{lay}} - d_{\text{SiO}_2}\right| = \left|6.37\rho_{\text{lay}}^W + 0.537\rho_{\text{lay}}^A - 3.64\right| \tag{4.155a}$$

$$k_F \equiv \left|d_{\text{core}} - d_{\text{SiO}_2}\right|$$
$$= \left|6.37\left(\rho_{\text{core}}^W - \rho_{\text{lay}}^W\right) + 0.537\left(\rho_{\text{core}}^A - \rho_{\text{lay}}^A\right)\right| \tag{4.155b}$$

To achieve the stable phase of constant mean concentration, w_A, for fixed temperature T, we have to distinguish two scenarios. If the state is in the one-phase region (outside the shaded regions in Fig. 4.18), then the differential chemical potential $\Delta\mu$ couples naturally to and behaves monotonically with the concentration w_A. Therefore, $\Delta\mu$ has to be varied until the chosen value of w_A is reached. On the other hand, for states in the phase coexistence region, the mixture separates into two stable phases [i.e., (1) and (2)] with mean composition such that $w_A^{(1)} < w_A < w_A^{(2)}$. Also, $\Delta\mu$ is fixed by this coexistence at the fixed T. The constant mean concentration, w_A, is then realized such that the condition

$$x w_A^{(1)} + (1 - x) w_A^{(2)} = w_A \qquad (4.156)$$

is met, where x is the mole fraction of phase (1) and $(1 - x)$ that of phase (2). The contrast parameters of the two coexistent phases are assumed to be additive such that the *mean* contrast parameters are given by

$$k_G = x k_G^{(1)} + (1 - x) k_G^{(2)} \qquad (4.157a)$$
$$k_F = x k_F^{(1)} + (1 - x) k_F^{(2)} \qquad (4.157b)$$

With the above prescriptions we can compute the density profiles of all stable phases along paths I and II in Fig. 4.18, respectively. From these density profiles, we calculate the contrast parameters using Eqs. (4.155) and, if necessary, Eqs. (4.156) and (4.157) to compare them with the values extracted from the experiments.

4.8.4.1 Path I ($w_A = 0.25$)

Here we consider the behavior of the confined water-rich liquid mixture (mean mass fraction of amphiphile $w_A = 0.25$) on the trajectory I in Fig. 4.18. SANS measurements of the intensity $I(q)$ [106] show that $I(q)$ is independent of temperature in the experimental range $(10 - 70°C)$. In terms of Eq. (4.144), this implies that within this temperature range the two contrast parameters, k_G and k_F, remain constant, which is surprising in view of the fact that the mixture undergoes a phase separation in this temperature range.

Consider now the results of the model calculations. As shown for path I in Fig. 4.18, the confined mixture separates into two phases (shaded region) at temperatures below $T \simeq 30°C$. Volume density profiles for a 3.5nm slit-pore are shown in Fig. 4.20 for four different temperatures. It is seen that water is preferentially adsorbed at the pore walls. Figure 4.20(a) and (b) show the situation in the single-phase region at $T = 70°C$ and $50°C$, whereas the graphs in Fig. 4.20(c1),(c2) and (d1),(d2) illustrate the density profiles for the two-phase region at $25°C$ and $10°C$. The mean mass fraction of phase (1) [i.e., $w_A^{(1)}$; see Fig. 4.20(d1)] is lower and that of phase (2) [i.e., $w_A^{(2)}$; see Fig. 4.20(d2)] is larger than $w_A = 0.25$, which gives the portions x of phase (1) and $(1 - x)$ of phase (2) across all pores according to Eq. (4.156).

If the system exhibits a phase separation, almost the entire amount of water in the iBA-rich phase is adsorbed by the walls and may not contribute

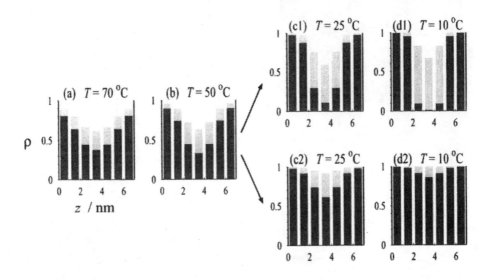

Figure 4.20: Volume density profiles for constant mean mass fraction of $w_A = 0.25$ for temperatures given in the graphs. (a) and (b) show stable single phases for $T = 70°C$ and $T = 50°C$, respectively; (c1),(c2) and (d1),(d2) show coexisting phases at the same temperature, respectively. Dark bars represent D_2O, and bright bars iBA amounts.

to the phase coexistence [see, e.g., Fig 4.20(d1)]. Thus, the iBA-enriched liquid inside the core region, which is confined by the water-rich layers, needs more water for phase separation. This explanation is consistent with the result that the shift of the critical composition becomes more pronounced for stronger matrix selectivity, as shown in Fig. 4.19. In addition, preference for the minor component by the walls leads to surface-induced phase transitions [94, 97]. Figure 4.18 displays film phases of one and two water-rich layers at low temperatures on the iBA-rich side in the phase diagram. Because the lattice fluid model may overemphasizes these layering transitions, we believe that they are irrelevant in this context.

In the model the two coexisting phases are well separated and the influence of the interface between them is not included. Therefore, questions of the morphology of the phase-separated liquid in the pore space are beyond the scope of this model. For the experimental system, it is believed that the two coexistent phases form a domain structure on a length scale of the pore size [106], with the domains of the iBA-rich phase [Figs. 4.20(c1) or 4.20(d1)] located mostly in rather wide pores or pore junctions of the network [96] and the water-rich domains [Figs. 4.20(c2) or 4.20(d2)] in narrow pore regions. Such a distribution of the domains is suggested by Kohonen and Christenson [117] who showed that the phase in which the preferred component is the majority component is found in regions of smaller pore widths. The sandwich-like structure of the iBA-rich phase as in Fig. 4.20(d1) for slits suggests a tube-like structure of that phase in cylindrical pores and in CPG-10. Such shapes have been found in earlier studies [95–97]. The SANS measurements [106] on the mixture of constant mean mass fraction of $w_A = 0.25$ showed that the contrast factors k_G and k_F remain constant in the temperature range of $10 - 70°C$. A constant k_G means that composition of the first layer adsorbed at the walls is unchanged in all phases in this temperature range, which can be seen in the model density profiles in Fig. 4.20. This also implies the constant contrast between this layer and the composition of the core region because the total mean mass fraction stays constant at $w_A = 0.25$.

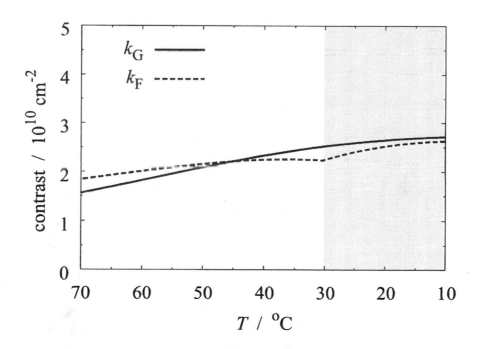

Figure 4.21: Model contrast parameters k_G and k_F calculated with Eqs. (4.155) and (4.157) as functions of temperature for $w_A = 0.25$. The shaded area indicates the region where phase separation occurs.

Contrast parameters k_G and k_F computed from Eqs. (4.155), (4.156), and (4.157) for various temperatures are displayed in Fig. 4.21. The shaded

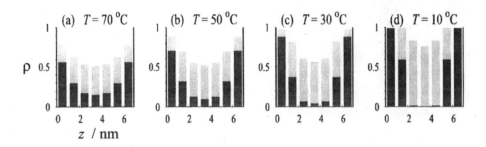

Figure 4.22: As Fig. 4.20 but for $w_A = 0.54$. Here, we have no phase transitions.

region again denotes the two-phase coexistence below $T \simeq 30°C$. It is seen
that neither of the two parameters exhibits a pronounced temperature dependence. This result can be rationalized by the fact that water is the major
component of the mixture ($w_s = 1 - w_A = 0.75$) and is preferred by the pore
wall. Thus the first layer at the substrate consists of almost pure water at all
temperatures, corresponding to a nearly temperature-independent value of its
mean density. Because the mean composition of the pore liquid is constant,
this leads to the fact that the mean density of the core region is also nearly
independent of temperature. Accordingly, on the basis of Eq. (4.155), k_G and
k_F exhibit just a weak dependence on temperature. Hence, we conclude that,
although the local composition of the mixture varies strongly within the considered temperature range (see Figs. 4.19 and 4.21), the contrast parameters
k_G and k_F do not reflect this. This weak temperature dependence supports
the experimental findings. However, it also implies that the contrast factors
may not be suitable for analyzing and characterizing the structure of phases
where water is the major component.

4.8.4.2   Path II ($w_A = 0.54$)

We now consider the behavior of the confined iBA-rich liquid mixture (mean
mass fraction of iBA $w_A = 0.54$) along path II in Fig. 4.18. Density profiles from the model calculation for four different temperature are shown in
Fig. 4.22. At this composition (mean mass fraction $w_A = 0.54$) the mixture
does not undergo phase separation (see path II in Fig. 4.19). As water is the
minor component of the mixture and is strongly preferred by the substrate,
almost all water accumulates in the first and second layer at low temperatures ($T \lesssim 30°C$), whereas the core liquid is almost depleted of water [see
Figs. 4.22(c) and 4.22(d)]. For this reason the contrast factor k_F attains
a high and nearly constant value at low temperatures, as can be seen in

Fig. 4.23(b). Furthermore, the contrast parameter k_G in Fig. 4.23(a) shows a similar behavior in this temperature range.

In this low temperature regime ($T \lesssim 30°C$) the layer next to the substrate is composed almost completely of water [see Figs. 4.23(c) and 4.23(d)] and there is a sharp interface from this water-rich layer to the iBA-enriched core region of the pore. Thus, the surface-directed structure of such a phase reveals a sharp layering reminiscent of the tube-like morphology in cylindrical pores [95–97]. We again emphasize that the thermodynamic state of the system here is away from two-phase coexistence. These surface layers have a thickness of about half of the pore width [see Figs. 4.23(c) and 4.23(d)]. In this temperature range the SANS measurements suggest that the Debye term corresponding to sharp structures of size ξ_S is now dominant [106]. These structures have a size of about $\xi_S = 3 - 4$ nm, which is the half of the nominal pore size of 6.8 nm [108]. Thus, our model calculations reproduce the experimental findings.

At higher temperatures, the density profiles exhibit more gradual concentration changes from the pore walls into the core region. As can be seen in Fig. 4.22(a) and 4.22(b), the layer core interface becomes more diffuse corresponding to water enrichment of the core region. Thus, the two contrast factors are decreasing with increasing temperature (see Fig. 4.23). Because the experimentally determined contrast factors are uncertain to within a temperature-independent scale factor, they are plotted in Fig. 4.23 in arbitrary units adjusted so that the values at $T = 10°C$ coincide with the corresponding values predicted by the model. Therefore, although we cannot compare the predicted and experimental *magnitudes* of k_G and k_F, we can see from Fig. 4.23 that the dependence of k_G and k_F on temperature for the model agrees qualitatively with that of the experiment.

Comparing Figs. 4.23(a) and 4.23(b), we find good qualitative agreement between the model and experimental contrast factors in the model and in the experimental study. The experimental curves of Fig. 4.23(b) also show that no phase transition occurs within this temperature range. As water is in the minor component ($w_A = 0.54$), a first-order transition would exhibit a discontinuous change in the composition of the adsorbed layer corresponding to a discontinuity in both, k_G and k_F, which is not apparent in Fig. 4.23(b). However, the graph of k_G shows a pronounced adsorption of water by the pore walls even for high temperatures. The plot of k_F [see Fig. 4.23(b)] indicates a decreasing contrast between the layer and the core region, because the sharp interfaces become diffuse.

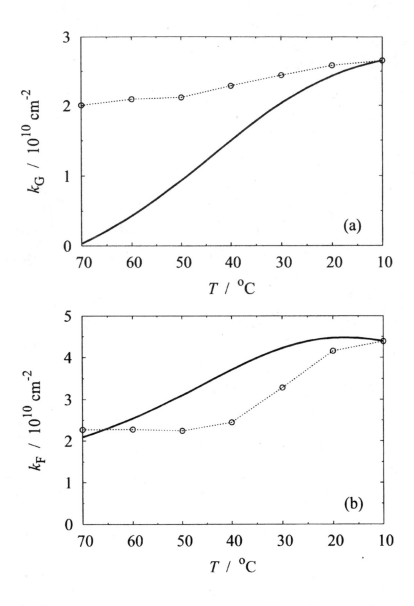

Figure 4.23: Contrast parameters (a) k_G and (b) k_F as functions of temperature obtained from the model calculations with Eq. (4.155) (—) and from the scattering measurements (○) at $w_A = 0.54$. The experimental data are scaled arbitrary as described in the text.

Chapter 5

Confined fluids with short-range interactions

5.1 Introductory remarks

As we mentioned at the beginning of Chapters 3 and 4, the key problem in statistical physics of equilibrium systems is to somehow get a handle on calculating the partition function Q [see Eqs. (2.38), (2.111)] or, equivalently, the configuration integral Z [see Eq. (2.112)]. This problem arises because in general the microscopic constituents forming the system of interest are not independent but rather correlated in their spatial arrangements because of intermolecular interactions. This, in turn, makes it impossible in general to factorize the multidimensional integral in Eq. (2.112) into a number of simpler ones that are tractable individually. Such a factorization is only possible if the system is simple enough. An example is the one-dimensional hard-rod fluid confined between hard substrates that we discussed in Chapter 3. Unfortunately, this model is virtually useless if one wishes to address one of the key issues of this text, namely that of phase transitions in confined fluids.

Therefore, we tackled the problem of evaluating the partition function of a three-dimensional many-particle system of interacting constituents by a different strategy in Chapter 4. In this chapter we introduced various versions of mean-field theory. In essence, and regardless of whether one is dealing with lattice or off-lattice models, all these approaches share the complete neglect of intermolecular correlations. Examples, are given in Eqs. (4.16)–(4.19) for the van der Waals treatment, Eqs. (4.77) and (4.84) for the one-component lattice fluid, or Eqs. (4.129), (4.130), and (4.131) for the binary lattice-fluid mixture. Despite the differences in the precise form of the mean-field approximation, it always reduces the original problem of solving the overwhelmingly complex integral Z to the extent that the remaining expressions can be handled analytically, which is a great achievement indeed.

Invoking additional approximations like a high-temperature expansion of the free energy [see Eqs. (4.3), (4.4), or (4.63)] or the Bogoliubov variational theorem [see Eq. (4.76)] combined with a restriction to nearest-neighbor interactions eventually leads to closed expressions for the relevant thermodynamic potential through which equilibrium properties of many-particle systems can be calculated [see Eqs. (4.26), (4.83), or (4.134]. Unfortunately, despite its surprising power in predicting properties of many-particle systems, the mean-field approximation is crude and breaks down in particular in the near-critical regime. However, it is noteworthy that this breakdown is quantitative rather than qualitative. This is reflected by the fact that in mean-field theories order parameters usually exhibit a power-law dependence on the relevant thermodynamic field where only the value of the (critical) exponent governing the power law is usually wrong compared with more sophisticated theories or experimental data (see Section 4.2.3).

If one wishes to abandon the mean-field approximation in dealing with complex and realistic systems (with respect to parallel experiments), there is essentially only one alternative to solve the key problem in statistical physics of equilibrium systems. This alternative route is offered by computer simulations where one aims at representing numerically the evolution of a many-particle system in phase (i.e., momentum and configuration) space by explicitly considering the interactions between the microscopic constituents of such a system. Because evaluating the interactions between the constituents is numerically demanding, simulation systems are usually microscopic in size, accommodating between, say, of the order of 10^2 to about 10^6 molecules given the current capacity of computers.

Perhaps the biggest advantage associated with the advent of computer simulations during the 1950s and 1960s is that they should be regarded as *first-principles* methods: Apart from adopting a specific interaction potential, no other approximations are involved. In particular, intermolecular correlations, which were completely neglected in the mean-field theories, are exactly accounted for, at least *in principle*, with the exception of the near-critical regime where special techniques need to be invoked to deal with the problem of (quasi-) macroscopic correlation lengths in microscopic simulation systems [118, 119]. However, the theory of critical phenomena is not of central interest here so that we defer the interested reader to the vast literature in this field [15, 120, 121].

Another great advantage of computer simulations is that they offer the possibility to directly visualize the spatio-temporal evolution of a system in addition to being amenable to a mathematical analysis in terms of rigorously defined statistical physical entities like local densities or various other correlation functions. In fact, the graphical analysis of either individual configurations generated in a computer simulations (i.e., "snapshots") or video sequences often gives one an idea of how to define a meaningful quantity in a mathematical rigorous way to analyze the results of such a simulation in terms of well-defined statistical physical quantities. Ordered structures with many defects that can be analyzed in terms of suitably defined bond-orientation correlation functions provide an example [122]. The choice of a proper bond-orientation correlation function is usually dictated by the symmetry inherent in the ordered structure, which can be determined best by directly looking at individual snapshots from the simulation. Another example illustrating the suitability of such a graphical analysis will be discussed below in Section 6.4.2.

From a more general perspective, computer simulations can be grouped into two different classes. Monte Carlo (MC) methods aim at generating a sequence of configurations in a specific statistical physical ensemble according to the relevant probability density $p\left(r^{N}; X\right)$ [see, for example, Eq. (2.117)] governing the distribution of these configurations in configuration space where X is a vector consisting of the implicitly fixed variables on which p depends [for example, N and s_{z} in Eq. (2.117)]; in molecular dynamics (MD) simulations, on the other hand, one solves numerically the equation of motion of a many-particle system and obtains the full information about the temporal evolution of the many-particle system in phase space. If the equilibrium system is ergodic [123–125], that is, if

$$\langle O \rangle = \int O\left(r^{N}\right) p\left(r^{N}; X\right) dr^{N} = \lim_{t \to \infty} \frac{1}{t} \int_{0}^{t} O\left[r^{N}(t)\right] dt \qquad (5.1)$$

both MC and MD can be expected to yield identically the same results for equilibrium properties $\langle O \rangle$.[1]

We briefly note that glasses are an important class of systems for which Eq. (5.1) frequently turns out to be invalid[2] even for quite long periods of observation t. Glasses are characterized by complex free-energy landscapes with rather deep "valleys" that cannot easily be surmounted given the low thermal energy typical of a glass. As a consequence the system becomes locked into a rather narrow region of (classic) phase space Γ (see Section 2.5)

---

[1]We explicitly disregard the presence of inaccuracies associated with, for example, inadequate sampling of configuration space.

[2]Of course, Eq. (5.1) would *always* be valid if one could let the observation time t become strictly infinite, which is generally precluded *in practice*.

for a long time until, eventually, a fluctuation arises permitting the system to escape and to explore a wider range of Γ.

In Eq. (5.1) the sequence of points $\left\{ r^N \left(t \right) \right\}$ represents a trajectory in configuration space obtained as a solution of the (classic) equation of motion and t denotes time. Moreover, we tacitly assumed that the microscopic quantity $O \left(r^N \right)$ whose average $\langle O \rangle$ we wish to calculate depends only on the positions r^N of the N molecules of the system but not on their momenta p^N.

In this work we shall exclusively concentrate on the MC method for two main reasons:

1. Compared with MD, MC is conceptually closer to the Gibbsian version of equilibrium statistical physics on which we essentially based the analysis in Chapter 2.

2. MC can be more easily adapted to specific statistical physical ensembles.

The second point is of particular importance if one wishes to study phase equilibria where one needs to have access to a thermodynamic potential. This is rather cumbersome in MD so that it is generally not advisable to employ MD to investigate phase equilibria. Consequently, MC has been much more widely used in numerical studies of equilibrium properties of confined systems. The advantage of MD is that it permits one to calculate transport properties in equilibrium systems such as diffusion constants. Investigation of transport phenomena is hampered in MC where one (at least in most cases) does not know on what time scale the system of interest evolves.

The only serious disadvantage of both simulation techniques (MC *and* MD) is that they will always be limited to systems that are extremely tiny on a macroscopic scale because of storage limitations and lack of computational speed. Because of the small sizes of simulation systems, special techniques have been devised to save computer time and to avoid domination of surface effects in the minute samples one is simulating. These techniques can be implemented in a more or less straightforward manner as long as the interaction potentials between the constituents of the system decay sufficiently rapidly.

Special precaution is, however, required, if this proviso is not met, that is, in systems where the interaction potentials decay rather slowly with intermolecular distance. Two prominent examples pertaining to this latter class are Coulomb interactions between charged molecules or interactions between molecules with a permanent dipole. In these cases, highly sophisticated techniques have been devised to obtain reliable results in computer simulations despite the smallness of the samples under study. Because of the great importance of confined fluids with such "long-range" interaction potentials, we defer a discussion of these techniques to the subsequent Chapter 6. Accordingly, this chapter will be devoted to confined fluids with *short-range* interaction potentials where the aforementioned limitation of system sizes poses a far less serious problem.

5.2 Monte Carlo simulations

5.2.1 Importance sampling

Because MC is a numerical technique to calculate multidimensional integrals like the one over configuration space in Eq. (5.1), we begin by discretizing configuration space and rewrite the integral as

$$\langle O \rangle = \lim_{M \to \infty} \sum_{m=1}^{M} O\left(r_m^N\right) p\left(r_m^N; \boldsymbol{X}\right) \tag{5.2}$$

where M denotes the number of "points" r_m^N in configuration space.[3] The basic idea then is to somehow generate a sequence of "points" in configuration space $\left\{r_m^N\right\}$. However, because of the high dimension of configuration space one should not generate the points $\left\{r_m^N\right\}$ according to some regular array because one can demonstrate that the overwhelming number of points on this array will lie on the surface of this multidimensional space. Hence, in this case one focuses on a narrow and perhaps largely irrelevant region of configuration space, which consequently might be sampled rather poorly.

To appreciate this latter point, consider the trajectory of a *single* molecule in space. Let us discretize this trajectory such that we represent the trajectory of the molecule by a succession of regularly spaced points. Thus, these points may be viewed as the p^3 nodes of a cubic lattice. Clearly, in each spatial dimension, 2 out of the total number of the p nodes in that direction lie on the surface of the cube. Extending these considerations to N instead of just a single molecule, it is immediately clear that we need to replace the original cube by an N-dimensional hypercube such that in each dimension the fraction $p - 2/p$ represents the ratio of nodes *not* on the surface of the hypercube relative to the total number of nodes. To estimate this fraction

---

[3]For a more comprehensive introduction to Monte Carlo simulations we refer the interested reader to the excellent text by Landau and Binder [126]. In Ref. 126 the authors discuss many applications of the Monte Carlo technique beyond the scope of the present book.

for the entire hypercube, we therefore need to consider

$$\left(\frac{p-2}{p}\right)^{3N} = \left(1 - \frac{2}{p}\right)^{3N} = \exp\left[3N\ln\left(1 - \frac{2}{p}\right)\right] \tag{5.3}$$

Moreover, to represent the continuous trajectory of each molecule by a succession of regularly spaced nodes, the lattice constant of the hypercube should be sufficiently small; that is, p should be large enough so that $x \equiv 2/p \ll 1$. In this case, we can approximate the logarithm in Eq. (5.3) through $\ln(1 - x) = -x + \mathcal{O}(x^2) \simeq -x$ such that

$$\left(\frac{p-2}{p}\right)^{3N} \simeq \exp\left(-\frac{6N}{p}\right) \tag{5.4}$$

which vanishes for each fixed and finite value of p in the limit $N \to \infty$. In other words, in the limit of a sufficiently large number of molecules, the sampling of configuration space is severely biased because almost all configurations $\{r_m^N\}$ turn out to be represented by nodes on the surface of the hypercube. As a result configuration space may be sampled highly inadequately, which may cause the sum in Eq. (5.2) to converge very slowly toward the correct $\langle O \rangle$; that is, M needs to be very large. This argument has been adopted from the book of Binder and Heermann [127].

The problem becomes even more severe in view of our discussion in Section 2.2.2.2, which showed that in a sufficiently large system the relevant region of (quantum mechanical) state space contributing to statistical thermodynamical averages is very small [see, for example, Eq. (2.30)]. Translating this observation into the language of classic statistical thermodynamics that we are using here, we conclude that as N becomes large the region in configuration space contributing to the sum in Eq. (5.2) declines as well. This, however, makes it even harder to sample configuration space adequately by using a regularly spaced hyperlattice in configuration space on account of the above considerations. Because of this additional effect, M in Eq. (5.2) must be made even larger than was already necessary due to the biased distribution of nodes on the hypercube discussed above.

It is therefore obvious that a numerical solution of Eq. (5.1) based on a regular distribution of nodes in configuration space is quite inefficient if not completely prohibitive. A much better idea would be to generate the "points" $\{r_m^N\}$ at *random*, which guarantees that configuration space will be sampled without any bias provided M in Eq. (5.2) can be made large enough. One may then calculate $O\left(r_m^N\right)$ at these points, multiply each value by its associated probability (density), and estimate the left side of Eq. (5.2) by summing up all these values.

However, there is a twofold problem with this naive approach. First, because the amount of computer time is inevitably finite, we need to limit the summation in Eq. (5.2) to some maximum value $M = M_{\mathrm{max}}$. It turns out that in most cases $M_{\mathrm{max}} = \mathcal{O}(10^5 - 10^7)$ is sufficient from a practical perspective so that this problem can be surmounted.

The second and by far more serious problem with an application of Eq. (5.2) involves the probability density $p\left(r_m^N; X\right)$ which is a priori unknown. An inspection of Eq. (2.117) reveals that $p\left(r_m^N; X\right)$ depends on the (classic) partition function, which involves the configuration integral as Eq. (2.118) shows. However, the partition function itself is unknown such that the probability with which $O\left(r_m^N\right)$ needs to be sampled at points $\left\{r_m^N\right\}$ remains undetermined.

Alternatively, one could envision generating points $\left\{r_m^N\right\}$ according to their *importance* determined by $p\left(r_m^N; X\right)$. In this case, it would be possible to replace Eq. (5.2) by a simpler one, namely

$$\langle O \rangle = \frac{1}{M_{\mathrm{max}}} \sum_{m=1}^{M_{\mathrm{max}}} O\left(r_m^{\prime N}\right) \tag{5.5}$$

where the prime has been attached to remind the reader that in the sequence $\left\{r_m^{\prime N}\right\}$ each configuration is realized according to its correct probability density. At this point one may wonder what one might have gained by Eq. (5.5). On the surface it seems much more straightforward to generate the random sequence of points rather than $\left\{r_m^{\prime N}\right\}$, which have to satisfy the additional constraint of compatibility with some a priori *unknown* probability density. The key to appreciating the great improvement represented by Eq. (5.5) is to realize that in order to generate numerically the sequence of configurations $\left\{r_m^{\prime N}\right\}$ it turns out that only the *relative* probability matters with which any two "neighboring" members $r_m^{\prime N}$ and $r_{m+1}^{\prime N}$ in the sequence $\left\{r_m^{\prime N}\right\}$ occur. This is a direct consequence of the *Principle of Detailed Balance*, which we introduce in Appendix E.1.2 as a stationary solution of the Chapman-Kolmogoroff equation.

Based on Eq. (E.20) we may design an algorithm compatible with the importance-sampling concept. We shall illustrate this below for two separate cases that are particularly relevant to confined fluids. The first of these concerns the grand canonical ensemble, which mimics situations encountered in sorption experiments [31, 128–132]. The second set of experiments concerns measurements employing the so-called *surface forces apparatus* (SFA), which permits one to deduce in an indirect way information about the local structure of confined fluids [133–137]. The algorithms we shall be introducing below are adapted versions of the classic Metropolis algorithm proposed in

1953 to simulate properties of a two-dimensional fluid of hard disks [138–140]. By this algorithm one can generate a numerical representation of a so-called Markov process. The sequence of configurations generated by the Metropolis algorithm is therefore frequently termed a "Markov chain" (see Appendix E.1).

5.2.2   The grand canonical ensemble

By analogy with Eqs. (2.117) and (2.118), it follows that the probability density in the grand canonical ensemble is given by

$$p\left(\widetilde{\boldsymbol{r}}^{N}; N\right) = \frac{1}{N! \Lambda^{3N} \Xi_{\mathrm{cl}}} \exp\left[\frac{\mu N}{k_{\mathrm{B}} T}\right] \exp\left[-\frac{U\left(\widetilde{\boldsymbol{r}}^{N}; N\right)}{k_{\mathrm{B}} T}\right] \qquad (5.6)$$

where the grand canonical ensemble partition function for a classic system is given in Eq. (2.120). For convenience, we introduce in Eq. (5.6) "reduced" coordinates through the transformation

$$\boldsymbol{r}_i = \begin{pmatrix} x_i \\ y_i \\ z_i \end{pmatrix} \rightarrow \widetilde{\boldsymbol{r}}_i = \begin{pmatrix} \widetilde{x}_i \\ \widetilde{y}_i \\ \widetilde{z}_i \end{pmatrix} = \begin{pmatrix} x_i/s_{\mathrm{x}0} \\ y_i/s_{\mathrm{y}0} \\ z_i/s_{\mathrm{z}0} \end{pmatrix} \qquad (5.7)$$

such that the simulation cell is a unit cube rather than the original parallelepiped with side length s_α, where $\alpha = $ x, y, or z. From the form of $p\left(\widetilde{\boldsymbol{r}}^{N}; N\right)$, one may guess intuitively that the generation of a Markov chain of configurations (see Appendix E.1.1) should involve two types of processes, namely

1. A random displacement of molecules, that is $\widetilde{\boldsymbol{r}}_{n-1}^{N} \rightarrow \widetilde{\boldsymbol{r}}_{n}^{N}$,

2. A random change of the number of molecules accommodated by the system, that is, $N_{n-1} \rightarrow N_n$

both of which we may associate separately with the abstract random process $y_{n-1} \rightarrow y_n$ introduced in Appendix E.1. It is then possible to identify with P_1 in Eq. (E.20), $p\left(\widetilde{\boldsymbol{r}}^{N}; N\right)$ from Eq. (5.6). As we realize from the discussion in Appendix E.1.2, the quantity that matters for the transition between $y_{n-1} \longleftrightarrow y_n$ is the transition probability Π introduced in Eq. (E.20). Thus, we realize that a calculation of Π does not require knowledge of the grand canonical partition function Ξ_{cl}, which turns out to cancel between numerator and denominator in Eq. (E.20).

As we also point out in Appendix E.1.2, the Chapman-Kolmogoroff equation is derived under the assumption of small changes in the random processes represented by y. Hence, the Metropolis algorithm proceeds in two consecutive steps, namely.

1. Pick molecule i and replace it according to

$$\widetilde{\boldsymbol{r}}_{i,n} = \widetilde{\boldsymbol{r}}_{i,n-1} + \delta_{\mathrm{r}} \left(\mathbf{1} - 2\boldsymbol{\xi}\right), \qquad i = 1, \ldots, N \qquad (5.8)$$

where $\mathbf{1}^{\mathrm{T}} = (1, 1, 1)$, δ_{r} is the side length of a small cube centered on $\widetilde{\boldsymbol{r}}_{i,n-1}$ (usually $1-10\%$ of the "diameter" of a molecule), and $\boldsymbol{\xi}$ is a vector whose three components are (pseudo-) random numbers uniformly distributed on the interval $[0, 1]$. There are two options to realize this. The first option is to pick molecules consecutively according to their storage location in the computer's memory. The second one consists of picking a molecule *at random*. In practice, however, it turns out that both approaches lead to the same results provided the Markov chain generated is sufficiently long.

2. Change the number of molecules according to

$$N_n = N_{n-1} \pm 1 \qquad (5.9)$$

where the "decision" to add or remove one molecule has to be drawn at random with equal probability for both options (addition and removal) to avoid biasing the creation or deletion frequency in the long run, that is, for a large number of creation/destruction attempts.

Henceforth, the sequence of N displacement attempts followed by N_{n-1} creation/destruction attempts in grand canonical ensemble MC (GCEMC) simulations will be referred to as a "GCEMC cycle."

For step 1 of this cycle, we notice that $N_{n-1} = N_n = N$ remains constant between members $n-1$ and n in the Markov chain. Hence, we compute [see Eq. (E.20)]

$$\frac{p\left(\widetilde{\boldsymbol{r}}_n^N; N\right)}{p\left(\widetilde{\boldsymbol{r}}_{n-1}^N, N\right)} = \exp\left\{-\frac{1}{k_{\mathrm{B}}T} \left[U\left(\widetilde{\boldsymbol{r}}_n^N; N\right) - U\left(\widetilde{\boldsymbol{r}}_{n-1}^N; N\right)\right]\right\}$$

$$\equiv \exp\left[-\frac{\Delta U_{n-1 \to n}}{k_{\mathrm{B}}T}\right] \qquad (5.10)$$

where $\widetilde{\boldsymbol{r}}_n^N$ and $\widetilde{\boldsymbol{r}}_{n-1}^N$ are consecutive configurations distinguished by the location of molecule i [see Eq. (5.8)]. The transition probability for step 1 of our GCEMC cycle is then given by

$$\Pi_1 = \min\left[1, \exp\left(-\Delta U_{n-1 \to n}/k_{\mathrm{B}}T\right)\right] \qquad (5.11)$$

Hence, if the energy decreases in the course of a displacement of a molecule, that is, if $\Delta U_{n-1 \to n} \leq 0$, the displacement attempt will immediately be accepted. If on the other hand, $\Delta U_{n-1 \to n} > 0$, the displacement attempt is not immediately rejected despite the *increase* in potential energy. Instead another random number $\xi \in [0, 1]$ will be picked. A decision about the outcome of the displacement process will then be made on the basis of

$$\exp\left[-\Delta U_{n-1 \to n}/k_B T\right] \quad > \quad \xi \longrightarrow \text{accept displacement} \qquad (5.12a)$$

$$\exp\left[-\Delta U_{n-1 \to n}/k_B T\right] \quad \leq \quad \xi \longrightarrow \text{reject displacement} \qquad (5.12b)$$

Equations (5.11) and (5.12) form the core of the Metropolis algorithm in its *classic* form [138], which is for the canonical ensemble where, by definition, $N = \text{constant}$.

At this point it may seem a bit difficult to immediately see why Eqs. (5.12) work in practice, that is, to see why Eqs. (5.12) will generate a distribution of configurations that comply with the partition function of a given statistical physical ensemble. Let us therefore elaborate on a simple intuitive argument from which the validity of Eqs. (5.12) emerges.

Consider a specific configuration r^N such that displacing a single molecule gives rise to an increase in total configurational potential energy corresponding to $\exp\left[-\Delta U_{n-1 \to n}/k_B T\right] = 0.1$, say. It is conceivable that there are numerous different configurations $\left\{r^N\right\}$ all being characterized by identically the same value of $\exp\left[-\Delta U_{n-1 \to n}/k_B T\right]$ upon particle displacement.[4]

Moreover, we may assume to have access to all these configurations in parallel, that is, at the same time. If we then pick a random number distributed uniformly on the interval $[0, 1]$ and compare this number with the quantity $\exp\left[-\Delta U_{n-1 \to n}/k_B T\right] = 0.1$ to reach a decision about whether to accept each individual displacement, it is clear that in 90% of the cases the decision will be to reject the displacement, whereas in 10% the displacement will be accepted according to Eqs. (5.12).

However, whether we have access to the configurations in parallel or sequentially is irrelevant, which permits us to conclude that, for a *sufficiently long* and *ergodic* Markov chain, displacements will be accepted *on average* with the correct probability dictated by the principles of statistical physics (i.e., the probability density of a given statistical physical ensemble).

Notice that, on account of displacing some of the N molecules during step 1, a molecule may eventually end up outside the unit cube (i.e., the simulation cell) in which it was placed originally. The most convenient way of preventing this from happening is to apply periodic boundary conditions at the faces of the unit cube. That is, one surrounds the simulation cell with other unit cubes accomodating precisely the same configuration as the simulation cell.

---

[4]In fact, an *infinite* number of such configurations is conceivable.

If a molecule originally belonging to the simulation cell crosses the boundary between it and a neighboring unit cell, another molecule from the opposite unit cell will simultaneously enter the simulation cell, thereby guaranteeing a constant number of molecules in the simulation cell. In practice, one needs to store only molecules in the simulation cell. The periodic boundary conditions are accounted for by replacing

$$\widetilde{\alpha}_i \rightarrow \widetilde{\alpha}'_i = \widetilde{\alpha}_i - \text{sign}\,(\widetilde{\alpha}_i) \left\{ \begin{array}{ll} 1, & \frac{1}{2} \leq |\widetilde{\alpha}_i| \leq 1 \\ 0, & 0 \leq |\widetilde{\alpha}_i| < \frac{1}{2} \end{array} \right., \qquad i = 1, \ldots, N \qquad (5.13)$$

where $\text{sign}\,(\widetilde{\alpha}_i)$ is a function returning the sign of its argument and $\widetilde{\alpha}_i = \widetilde{x}_i, \widetilde{y}_i, \widetilde{z}_i$ for a bulk system, whereas $\widetilde{\alpha}_i = \widetilde{x}_i, \widetilde{y}_i$ is for a slit-pore where the fluid substrate repulsion at short distance between a molecule and the substrate serves to constrain the z-coordinate of a fluid molecule to the simulation cell. In addition we assume in Eq. (5.13) that the origin of the coordinate system is located at the center of the simulation cell such that $\widetilde{x}_i, \widetilde{y}_i, \widetilde{z}_i \in \left[-\frac{1}{2}, +\frac{1}{2}\right]$.

However, because of periodic boundary conditions, one needs to make sure that as far as short-range interaction potentials are concerned, a molecule in the simulation cell interacts only with another particle in the simulation cell or one of its periodic images depending on which is closest. This so-called *minimum image convention* can easily be implemented through the equations

$$\Delta\widetilde{\alpha}_{ij} = \widetilde{\alpha}_i - \widetilde{\alpha}_j \qquad (5.14a)$$

$$\Delta\widetilde{\alpha}_{ij} \rightarrow \Delta\widetilde{\alpha}'_{ij} = \Delta\widetilde{\alpha}_{ij} - \text{sign}\,(\Delta\widetilde{\alpha}_{ij}) \left\{ \begin{array}{ll} 1, & \frac{1}{2} \leq |\widetilde{\alpha}_i| \leq 1 \\ 0, & 0 \leq |\widetilde{\alpha}_i| < \frac{1}{2} \end{array} \right. \qquad (5.14b)$$

where $\widetilde{\alpha}_i = \widetilde{x}_i, \widetilde{y}_i, \widetilde{z}_i$ for a bulk system, whereas $\widetilde{\alpha}_i = \widetilde{x}_i, \widetilde{y}_i$ is for a slit-pore as before [see Eq. (5.13)].

If one contemplates Eqs. (5.13) and (5.14), one realizes that Eq. (5.13) adds and subtracts a box length (or ± 1 in reduced units) depending on whether the molecule has left the simulation cell in the $-\alpha$- and $+\alpha$-direction, thereby restoring it by a periodic image. Similarly, Eq. (5.14) serves to ensure that a molecule interacts either with another one in the central cell or one of its images depending on which one is closer. In cases where molecules i and j are separated by more than half the relevant side length of the simulation cell, a periodic image of molecule j will be considered rather than molecule j itself. Thus, in effect the interaction between molecules is truncated at some cut-off distance, which is determined by the dimensions of the simulation cell and its geometry. Although this is not critical for systems governed by short-range interaction potentials, the truncation is the source of considerable difficulty in simulating systems with long-range interactions. We shall return to this issue in greater detail in Chapter 6.

In the second step of our adapted Metropolis algorithm, we change the number of molecules in the simulation cell by ± 1. More specifically, we either attempt to create a new molecule at a randomly chosen position in the simulation cell, that is

$$\widetilde{r}_{N+1,n} = \frac{1}{2}\left(1 - 2\xi\right) \tag{5.15}$$

or an existing molecule will be deleted. Both processes need to be carried out with equal probability to avoid biasing the generation of a (numerical representation of) Markov chain of configurations in favor of one or the other process. To determine the transition probability in this case, it turns out to be convenient to introduce the auxiliary quantity

$$B \equiv \frac{\mu}{k_{\mathrm{B}}T} - \ln\frac{\Lambda^3}{V} \tag{5.16}$$

following the original proposal by Adams [41] where Λ is the thermal de Broglie wavelength defined in Eq. (2.103). From Eqs. (2.79), (2.111), and (2.112) it follows that the chemical potential of the ideal gas is given by $[U\left(r^N\right) = 0]$

$$\mu^{\mathrm{id}} = \left(\frac{\partial \mathcal{F}^{\mathrm{id}}}{\partial N}\right)_{T,\sigma} = k_{\mathrm{B}}T\ln\frac{N\Lambda^3}{V} \tag{5.17}$$

where we used also the thermodynamic definition of the free energy [see Eq. (1.50)] as well as the Gibbs fundamental equation in its most general form [see Eq. (1.22)]. Introducing the excess chemical potential via

$$\mu^{\mathrm{ex}} = \mu - \mu^{\mathrm{id}} \tag{5.18}$$

and using Eq. (5.17), it turns out that Eq. (5.16) may be recast as

$$B = \frac{\mu^{\mathrm{ex}}}{k_{\mathrm{B}}T} - \ln N \tag{5.19}$$

which has a somewhat more transparent physical interpretation than the original (but numerically more useful) expression given in Eq. (5.16).

The decision of whether the attempt to create or destroy a molecule is accepted will again be based on a transition probability defined analogously to the one in Eq. (5.11). It depends on the ratio

$$\frac{p\left(\widetilde{r}_n^{N\pm 1}; N\pm 1\right)}{p\left(\widetilde{r}_{n-1}^{N}; N\right)} = \exp\left(r_{\pm}\right) \tag{5.20}$$

where the argument of the pseudo-Boltzmann factor in Eq. (5.20) is given by

$$r_\pm \equiv \pm B \mp \ln N \mp \frac{U_\pm}{k_\mathrm{B} T} \tag{5.21}$$

where the upper sign refers to addition and the lower one to removal of one fluid molecule, respectively. In Eq. (5.21) the meaning of N is that of either the number of molecules *after* adding a new one to the system or *prior* to the removal of an already existing one. Similarly, $U_\pm$ denotes the configurational energy of the molecule to be added or to be removed from the simulation cell. The creation/destruction attempt is then realized based on the transition probability

$$\Pi_2 = \min\left[1, \exp\left(r_\pm\right)\right] \tag{5.22}$$

In case creation or destruction is unfavorable, that is, if $r_\pm < 0$ the attempt will not be rejected immediately but realized according to a modified Metropolis criterion [see Eqs. (5.12)], that is

$$\exp\left(r_\pm\right) > \xi \longrightarrow \text{accept creation/destruction} \tag{5.23a}$$
$$\exp\left(r_\pm\right) \leq \xi \longrightarrow \text{reject creation/destruction} \tag{5.23b}$$

where again $\xi \in [0, 1]$ is a pseudo-random number.

Because step 2 of the adapted Metropolis algorithm for GCEMC simulations involves a change in density by $\pm 1/V$ between members $n-1$ and n of the Markov chain, some care has to be taken in computing $U_\pm$ if the interaction potential is short-range, that is, if it decays sufficiently rapidly but does not go to zero at any finite separation between molecules. An example is the Lennard-Jones (12,6) (LJ) potential defined by

$$u_\mathrm{LJ}\left(r_{ij}\right) = 4\varepsilon \left[\left(\frac{\sigma}{r_{ij}}\right)^{12} - \left(\frac{\sigma}{r_{ij}}\right)^6\right] \tag{5.24}$$

which is frequently employed to model the interactions between spherical molecules of "diameter" σ that are separated by a distance $r_{ij} = |\boldsymbol{r}_i - \boldsymbol{r}_j|$. The strength of repulsive (proportional to r_{ij}^{-12}) and attractive interactions (proportional to r_{ij}^{-6}) is scaled by $\varepsilon > 0$, which determines the depth of the attractive well. The attractive part of the LJ potential represents *dispersive* (or van der Waals) interactions arising from induced dipole moments generated by fluctuations in the electronic charge distributions of two interacting particles. Because $u\left(r_{ij}\right)$ decays rather quickly, it is convenient to employ some cut-off parameter and compute $U_\pm$ only for those molecules located inside some subdomain $\widetilde{V}$ of the entire system volume centered on the molecule to be created or removed. Assuming pairwise additivity of intermolecular interactions represented by $u\left(r_{ij}\right)$, we may write

$$U_{+} = \sum_{\substack{j=1 \in \widetilde{V}}}^{N} u\left(r_{ij}\right) + \Delta U_{c}, \qquad i = N+1 \qquad (5.25\text{a})$$

$$U_{-} = \sum_{\substack{j=1 \neq i \in \widetilde{V}}}^{N} u\left(r_{ij}\right) + \Delta U_{c} \qquad (5.25\text{b})$$

where ΔU_{c} is a correction due to longer-range attraction that is neglected by limiting the sums in Eqs. (5.25) to interactions within the cut-off solid of volume $\widetilde{V}$. Explicit analytic expressions for ΔU_{c} are derived below in Section 5.2.3 for a slit-pore, which is the most important confined geometry in the context of this book.

5.2.3   Corrections to the configurational energy

For a confined fluid

$$\Delta U_{c} = \Delta U_{c,\text{ff}} + \Delta U_{c,\text{fs}} \qquad (5.26)$$

where $\Delta U_{c,\text{ff}}$ and $\Delta U_{c,\text{fs}}$ represent cut-off corrections due to fluid-fluid and fluid substrate interactions. The latter arise in cases where the substrate itself is composed of individual atoms arranged according to some solid structure and interacting with a fluid molecule via a LJ (12,6) potential, say [see Eq. (5.24)]. Formal expressions for both corrections to the potential energy can be derived by noting that [cf., Eq. (4.16)]

$$U_{c,\text{ff}} = \frac{1}{2} \int_{\widetilde{V}} d\boldsymbol{r}_{1} \int_{V \backslash \widetilde{V}} d\boldsymbol{r}_{2} u_{\text{ff}}\left(r_{12}\right) \left[\rho_{n}^{(2)}\left(\boldsymbol{r}_{1}, \boldsymbol{r}_{2}\right) - \rho_{n-1}^{(2)}\left(\boldsymbol{r}_{1}, \boldsymbol{r}_{2}\right)\right] \quad (5.27\text{a})$$

$$U_{c,\text{fs}} = 2N_{s} \int_{\widetilde{V}} d\boldsymbol{r}_{2} u_{\text{fs}}\left(r_{12}\right) \left[\rho_{n}^{(1)}\left(\boldsymbol{r}_{2}\right) - \rho_{n-1}^{(1)}\left(\boldsymbol{r}_{2}\right)\right] \qquad (5.27\text{b})$$

where N_{s} denotes the number of solid atoms of which the confining solid substrate consists and $\rho^{(1)}\left(\boldsymbol{r}_{1}\right)$ and $\rho^{(2)}\left(\boldsymbol{r}_{1}, \boldsymbol{r}_{2}\right)$ are one- and two-particle densities in configurations $n-1$ and n, respectively whose standard definition can be found in textbooks on statistical mechanics (see, for example, Eqs. (4.18) and, for more details, Ref. 17). These corrections differ in step 2 of the Metropolis algorithm adapted for GCEMC because the density differs by $\pm 1/V$ on account of addition/deletion attempts of one molecule as pointed out in Section 5.2.2. In the first step of this algorithm, N is the same *before* and *after* displacement of one molecule so that both expressions in Eqs. (5.27) vanish identically. Notice that the integration over the coordinates of molecule 1 extends over the volume of the cut-off solid, whereas the integration over coordinates of molecule 2 is restricted to the surrounding volume $V \backslash \widetilde{V}$ in Eq. (5.27a).

It is customary to relate $\rho^{(1)} (r_1)$ and $\rho^{(2)} (r_1, r_2)$ to the pair correlation function $g (r_1, r_2)$ via Eq. (4.17). To proceed we introduce two key assumptions, namely

1. A mean-field approximation by assuming [cf. Eq. (4.19)]

$$g (r_1, r_2) = 1 \quad \forall r_2 \in V \backslash \widetilde{V} \tag{5.28}$$

2. We assume the confined fluid to be homogeneous represented by a constant density ρ [see Eq. (4.20)].

Under these conditions we can rewrite Eq. (5.27a) as

$$
\begin{aligned}
U_{\text{c,ff}} &= \frac{1}{2} \left(\frac{N}{V} \right)^2 \int\limits_{s_{x0}/2}^{s_{x0}/2} dx_1 \int\limits_{-s_{y0}/2}^{s_{y0}/2} dy_1 \int\limits_{-s_{z0}/2}^{s_{z0}/2} dz_1 \\
&\quad \times \int\limits_0^{2\pi} d\phi \int\limits_{-s_{z0}/2-z_1}^{s_{z0}/2-z_1} dz \int\limits_{r_c}^{\infty} d\rho \, \rho u_{\text{ff}} (\rho, z) \\
&= A_{z0} \pi \left(\frac{N}{V} \right)^2 \int\limits_{-s_z/2}^{s_z/2} dz_1 \int\limits_{s_z/2-z_1}^{s_z/2-z_1} dz \int\limits_{r_c}^{\infty} d\rho \, \rho u_{\text{ff}} (\rho, z) \tag{5.29}
\end{aligned}
$$

where we used $z = z_2 - z_1$ and associate with $\widetilde{V}$ a cut-off cylinder of radius r_c and height s_z in the z-direction. In addition we use cylindrical coordinates (ρ, z) for the integration over positions outside the cut-off cylinder. In Eq. (5.29), A_{z0} is the area of the z-directed face of the undeformed lamella introduced in Section 1.3.

Assuming now the radius r_c to be sufficiently large, we may approximate the relevant interaction potentials by their attractive contributions only [see Eq. (5.24)], that is

$$u_{\text{ff}} (\rho, z) \simeq \frac{-4\varepsilon_{\text{ff}} \sigma^6}{(\rho^2 + z^2)^3} \tag{5.30}$$

We insert Eq. (5.30) into Eq. (5.29), carry out the integration over ρ, and obtain

$$
U_{c,ff} = -\pi \varepsilon_{ff} \sigma^6 A_{z0} \left(\frac{N}{V}\right)^2 \int\limits_{-s_z/2}^{s_z/2} dz_1 \int\limits_{-s_z/2-z_1}^{s_z/2-z_1} dz \frac{1}{\left(r_c^2 + z^2\right)^2} \tag{5.31}
$$

The remaining two integrations can be carried out with the help of tabulated integrals [141]. One finally arrives at

$$
U_{c,ff} = -\frac{\pi \varepsilon_{ff} \sigma^6}{V r_c^3} N^2 \arctan\left(\frac{s_{z0}}{r_c}\right) \tag{5.32}
$$

Now, as configurations $n-1$ and n in the second step of the GCEMC–adapted Metropolis algorithm differ in $N_n = N_{n-1} \pm 1$, we obtain from the previous expression

$$
\Delta U_{c,ff} = -\frac{\pi \varepsilon_{ff} \sigma^6}{V r_c^3} (2N - 1) \arctan\left(\frac{s_{z0}}{r_c}\right) \tag{5.33}
$$

where N denotes either the number of molecules *before* removal or *after* addition.

In a bulk fluid, similar considerations may be used to derive an expression for cut-off corrections to the configurational energy. Using in this case a cut-off *sphere* rather than a *cylinder* gives rise to

$$
\Delta U_{c,ff} = -\frac{16\pi \varepsilon_{ff} \sigma^6}{3 V r_c^3} (N - 1) \tag{5.34}
$$

where N has the same meaning as above.

One may also derive a closed expression for the correction $U_{c,fs}$ using the homogeneity approximation $\rho^{(1)}\left(\boldsymbol{r}_1\right) \approx \rho$ together with Eqs. (5.27b) and (5.30) to obtain

$$
U_{c,fs} = -4\pi N_s \varepsilon_{fs} \sigma^6 \frac{N}{V} \int\limits_{-s_{z0}/2}^{s_{z0}/2} dz_2 \frac{1}{\left[r_c^2 + \left(z_2 - s_{z0}/2\right)^2\right]^2} \tag{5.35}
$$

from which

$$
U_{c,fs} = -\frac{2\pi \varepsilon_{fs} \sigma^6}{V r_c^3} N N_s \left[\arctan\left(\frac{s_{z0}}{r_c}\right) + \frac{s_{z0} r_c}{r_c^2 + s_{z0}^2}\right] \tag{5.36}
$$

follows after performing the remaining integration over z_2. Assuming as before that $N_n = N_{n-1} \pm 1$, we yield from the previous expression

$$
\Delta U_{c,fs} = -\frac{2\pi \varepsilon_{fs} \sigma^6 N_s}{V r_c^3} \left[\arctan\left(\frac{s_{z0}}{r_c}\right) + \frac{s_{z0} r_c}{r_c^2 + s_{z0}^2}\right] \tag{5.37}
$$

which permits us to estimate ΔU_c in Eq. (5.26) using Eqs. (5.33) and (5.37).

In Ref. 45 it was shown that the homogeneity assumption [see Eq. (4.20)] is very reliable in estimating ΔU_c. However, care has to be taken with respect to the mean-field approximation [see Eq. (5.28). As was shown by Wilding and Schoen, this assumption is prone to break down if the thermodynamic state of the fluid is in the near-critical regime because intermolecular correlations become long range and the assumption of $g(r_1, r_2)$ being unity outside the cut-off solid is invalid [142]. Hence, Eqs. (5.33) and (5.37) [and therefore ΔU_c in Eq. (5.26)] may lead to erroneous results if employed uncritically in GCEMC simulations.

To avoid these complications, it is advisable in most cases to replace the infinitely long-range LJ(12,6) potential in Eq. (5.24) by

$$u_{\text{ff}}(r_{ij}) = \begin{cases} u_{\text{SR}}(r_{ij}), & r_{ij} \leq r_c \\ 0, & r_{ij} > r_c \end{cases} \qquad (5.38)$$

where the explicitly short-range interaction potential $u_{\text{SR}}(r_{ij})$ is defined by

$$u_{\text{SR}}(r_{ij}) = u_{\text{LJ}}(r_{ij}) - u_{\text{LJ}}(r_c) + \left.\frac{\mathrm{d}u_{\text{LJ}}(r_{ij})}{\mathrm{d}r_{ij}}\right|_{r_{ij}=r_c} (r_c - r_{ij}) \qquad (5.39)$$

which vanishes continuously together with its first derivative at the cut-off radius r_c and is equal to zero for all larger intermolecular separations $r_{ij} > r_c$. The advantage is that in this case $\Delta U_c = 0$ because u_{SR} vanishes identically everywhere outside $\widetilde{V}$. Clearly, the same "trick" can also be applied to other continuous interaction potentials provided they decay with intermolecular distance r faster than r^{-3}. A decay *slower* than (and equal to) r^{-3} is typical for electrostatic potentials such as the dipole dipole potential. In this case longer range interactions are important and must not be eliminated in the spirit of Eq. (5.39). In these latter cases, special techniques have been devised to treat the longer range interactions properly. We defer a discussion of these methods to the subsequent Chapter 6.

5.2.4   A mixed isostress isostrain ensemble

GCEMC was introduced as a way to compute thermal properties of a system in contact with an infinitely large reservoir of heat and matter in Section 5.2.2. We shall now turn our attention to a situation where the thermodynamic system may exchange compressional (dilational) work with its surroundings. To simplify the treatment below, we shall assume there exists no longer any coupling to a reservoir of matter for the time being. However, at the end of this section, we shall describe how such an additional coupling can be reimplemented.

As we saw in Section 1.3, the prototypical lamella representing the confined fluid from a purely thermodynamic perspective may be deformed in a number of ways. For example, the most general expression for the exact differential of the internal energy in Eq. (1.43) shows that in the context of the current analysis there are three compressional strains (proportional to s_x, s_y, s_z) and one shear strain (proportional to αs_{x0}) acting on the lamella. To mimic a real experimental situation encountered, for instance, in the SFA, one may fix a subset of stresses and strains and study thermal properties of the lamella under these conditions. For one such example, we present the parallel statistical physical analysis culminating in an expression for the *grand* mixed isostress isostrain partition function χ in Eqs. (2.70a) [or the classic analog in Eq. (2.118)]. Here we shall consider a slightly simpler version of mixed isostress isostrain ensembles characterized by $N = \mathrm{const}.$

Suppose we set $N = N^\star$ so that the sum on N in Eq. (2.70b) can be represented by its maximum term $\exp\left(\mu N^\star / k_B T\right)$ (cf. Section 2.4), where

$$\chi = \exp\left(\frac{\mu N^\star}{k_B T}\right) \sum_{s_z} \exp\left(\frac{\tau_{zz} A_{z0} s_z}{k_B T}\right) \mathcal{Q} \equiv \exp\left(\frac{\mu N^\star}{k_B T}\right) \Upsilon_\perp \qquad (5.40)$$

Taking the logarithm of this expression and using Eq. (2.71), we realize that

$$\Phi = -k_B T \ln \chi = -k_B T \ln \Upsilon_\perp - \mu N^\star \qquad (5.41)$$

which permits us to introduce a generalized Gibbs potential

$$\mathcal{G}_\perp \equiv \Phi + \mu N^\star = \mathcal{U} - T\mathcal{S} - \tau_{zz} A_{z0} s_z \qquad (5.42)$$

as the thermodynamic potential of the current mixed isostress isostrain ensemble. Moreover, replacing on the right side of Eq. (2.118) the sum over N by its maximum term, one realizes that the factor $\exp\left(\mu N^\star / k_B T\right)$ cancels in Eq. (2.117) so that we may write[5]

$$p\left(\boldsymbol{r}^{N^\star}; s_z\right) = \frac{1}{N^\star! \Lambda^{3N^\star} \Upsilon_{\mathrm{cl},\perp}} \exp\left[\frac{\tau_{zz} A_{z0} s_z}{k_B T}\right] \exp\left[-\frac{U\left(\boldsymbol{r}^{N^\star}; s_z\right)}{k_B T}\right] \qquad (5.43)$$

for the probability density in the mixed isostress isostrain ensemble where

$$\Upsilon_{\mathrm{cl},\perp} = \frac{1}{N^\star! \Lambda^{3N^\star}} \sum_{s_z} \exp\left[\frac{\tau_{zz} A_{z0} s_z}{k_B T}\right] Z\left(N^\star, s_z\right) \qquad (5.44)$$

---

[5]Obviously, the factor $N^\star! \Lambda^{3N^\star}$ would also cancel in Eq. (5.43). However, we shall leave it alone for purely formal reasons to preserve $p\left(\boldsymbol{r}^{N^\star}; s_z\right)$ as a probability *density*.

is the associated partition function in the classic limit. To ease the notational burden henceforth, we shall drop the superscript "$\star$" as well as the subscript "cl".

Under these premises, the analog of Eq. (2.116) may now be cast as

$$
\begin{aligned}
\langle O \rangle &= \sum_{s_z} \int \mathrm{d}\boldsymbol{r}^N O\left(\boldsymbol{r}^N; s_z\right) p\left(\boldsymbol{r}^N; s_z\right) \\
&= \left(s_{x0} s_{y0}\right)^N \sum_{s_z} s_z^N \int \mathrm{d}\widetilde{\boldsymbol{r}}^N O\left(\widetilde{\boldsymbol{r}}^N; s_z\right) p\left(\widetilde{\boldsymbol{r}}^N; s_z\right) \\
&= \left(s_{x0} s_{y0}\right)^N \sum_{s_z} \int \mathrm{d}\widetilde{\boldsymbol{r}}^N O\left(\widetilde{\boldsymbol{r}}^N; s_z\right) p'\left(\widetilde{\boldsymbol{r}}^N; s_z\right)
\end{aligned}
\tag{5.45}
$$

where we introduced unit-cube coordinates [see Eq. (5.7)] replacing, however, s_{z0} by its (variable) analog s_z. From the last line of Eq. (5.45), we also realize that the weighting factor for the microscopic quantity $O\left(\widetilde{\boldsymbol{r}}^N; s_z\right)$ now becomes

$$
p'\left(\widetilde{\boldsymbol{r}}^N; s_z\right) = s_z^N p\left(\widetilde{\boldsymbol{r}}^N; s_z\right)
\tag{5.46}
$$

rather than $p\left(\widetilde{\boldsymbol{r}}^N; s_z\right)$ in the original coordinate system. Therefore, applying the Importance Sampling concept [see Eq. (5.5)] to estimate $\langle O \rangle$ in a mixed isostress isostrain MC (MIEMC) simulation, microstates need to be generated according to the scaled quantity $p'\left(\widetilde{\boldsymbol{r}}^N; s_z\right)$ rather than $p\left(\widetilde{\boldsymbol{r}}^N; s_z\right)$ itself. Hence, by analogy with Eqs. (5.10) and (5.21), the generation of the Markov chain will be determined by the ratio

$$
\frac{p'\left(\widetilde{\boldsymbol{r}}_n^N; s_{z,n}\right)}{p'\left(\widetilde{\boldsymbol{r}}_{n-1}^N; s_{z,n-1}\right)} = \exp\left(r_z\right)
\tag{5.47}
$$

where from Eqs. (5.43) and (5.46)

$$
r_z = -\frac{\Delta U - \tau_{zz} A_{z0} \Delta s_z}{k_\mathrm{B} T} + N \ln\left(\frac{s_{z,n}}{s_{z,n-1}}\right)
\tag{5.48}
$$

In Eq. (5.48) we also used

$$
\begin{aligned}
\Delta U &\equiv U\left(\widetilde{\boldsymbol{r}}_n^N; s_{z,n}\right) - U\left(\widetilde{\boldsymbol{r}}_{n-1}^N; s_{z,n-1}\right) \tag{5.49a} \\
\Delta s_z &\equiv s_{z,n} - s_{z,n-1} \tag{5.49b}
\end{aligned}
$$

In this case, we realize the random process $y_{n-1} \to y_n$ [see Appendix E.1] via

$$
s_{z,n} = s_{z,n-1} + \delta_z \left(1 - 2\xi\right)
\tag{5.50}
$$

by analogy with Eq. (5.8). With this process we associate a transition probability [see Eq. (E.20)]

$$\Pi_3 \equiv \min\left[1, \exp\left(r_z\right)\right] \tag{5.51}$$

As before, we realize this by immediately accepting any change in substrate separation if $r_z \geq 0$; if, on the other hand, $r_z < 0$,

$$\exp\left(r_z\right) \; > \; \xi \longrightarrow \text{accept change in substrate separation} \tag{5.52a}$$

$$\exp\left(r_z\right) \; \leq \; \xi \longrightarrow \text{reject change in substrate separation} \tag{5.52b}$$

according to a modified Metropolis criterion where again ξ denotes a pseudo-random number distributed uniformly on the interval $[0, 1]$.

In practice, we generate a new configuration in two steps. Step 1 is identical with step 1 of the Metropolis algorithm adapted for GCEMC, namely a random displacement of molecules governed by a transition probability Π_1 [see Section 5.2.2, Eq. (5.11)]. In step 2, the substrate separation is changed according to Eq. (5.50) so that

$$\widetilde{\boldsymbol{r}}_n^N = \begin{pmatrix} \widetilde{x}_n^N \\ \widetilde{y}_n^N \\ \widetilde{z}_n^N \end{pmatrix} = \begin{pmatrix} \widetilde{x}_{n-1}^N \\ \widetilde{y}_{n-1}^N \\ \widetilde{z}_{n-1}^N \end{pmatrix} \cdot \begin{pmatrix} 1 \\ 1 \\ s_{z,n}/s_{z,n-1} \end{pmatrix} \tag{5.53}$$

Because in the step 2 of the current MIEMC algorithm all N z-coordinates are changed at once, steps 1 and 2 are carried out with a frequency $N : 1$. Some care must also to be taken if a potential cut-off is employed. Then one has to make sure that after rescaling particle coordinates according to Eq. (5.53) the same subset of molecules employed in calculating $U\left(\widetilde{\boldsymbol{r}}_{n-1}^N; s_{z,n-1}\right)$ is also considered in calculating $U\left(\widetilde{\boldsymbol{r}}_n^N; s_{z,n}\right)$.

Last but not least we emphasize that we may amend both steps of the MIEMC algorithm by step 2 of the GCEMC algorithm; that is, the number of molecules accommodated by the slit-pore may fluctuate as well. In this case, all three steps are realized with a frequency of $N : 1 : N$ on account of the computational effort associated with generating a new configuration in all three steps. In this latter case, the distribution of microstates in configuration space complies with the probability density given in Eq. (2.117). Regardless of whether a fixed number of molecules is used in a MIEMC simulation, energy corrections need to be added to ΔU if a long-range interaction potential is employed such as the one in Eq. (5.24). These corrections can be worked out from Eqs. (5.33) and (5.37). However, we emphasize that for reasons pointed out in Section 5.2.2 it is advantageous to employ finite-range interaction potentials such as the one introduced in Eq. (5.39).

5.3 Chemically homogeneous substrates

Employing the MC simulation technique introduced in the previous section we now turn to a detailed discussion of thermophysical properties of confined fluids. In particular, we intend to illustrate the intimate relation between these properties and unique structural features caused by the competition between various length scales pertinent to specific confinement scenarios. These studies are largely motivated by parallel experimental work employing the SFA. Therefore, we begin with a concise description of some key aspects of SFA experiments.

5.3.1 Experiments with the surface forces apparatus

The main purpose of the SFA is to measure the forces exerted by a thin fluid film on a solid substrate with nearly molecular precision [143]. In the SFA, a thin film is confined between the surfaces of two macroscopic cylinders arranged such that their axes are at a right angle [143]. In an alternative setup, the fluid is confined between the surface of a macroscopic sphere and a planar substrate [144]. However, crossed-cylinder and sphere-plane configurations can be mapped onto one another by differential-geometrical arguments [145]. The surface of each macroscopic object is covered by a thin mica sheet with a silver backing, which permits one to measure the separation h between the surfaces by optical interferometry [143].

The radii of the curved surfaces in either setup are macroscopic as we mentioned so that they may be taken as approximately parallel on a *molecular* length scale around the point of minimum distance h between the opposite bodies (i.e., the two cylinders or the sphere and the plane). In addition, they are *locally* planar, because mica can be prepared with atomic smoothness over molecularly large areas.

This setup is then immersed in a bulk reservoir of the same fluid of which the confined film consists. Thus, at thermodynamic equilibrium, T and μ are the same in both subsystems (i.e., bulk reservoir and confined fluid). By applying an external force in the direction normal to both substrate surfaces, the thickness of the film can be altered by either expelling molecules from it or imbibing them from the reservoir until thermodynamic equilibrium is reestablished, that is, until the force exerted by the film on the surfaces equals the applied external normal force at the same T and μ. Plotting this force per radius R, F/R, as a function of h yields a damped oscillatory curve in most cases. This is illustrated by plots in Fig. 5.1 where typical curves are shown for several fluids consisting of branched and unbranched hydrocarbons [146]. As one can see, both the period and the amplitude of oscillations depend on the details of the molecular architecture of the fluid molecules.

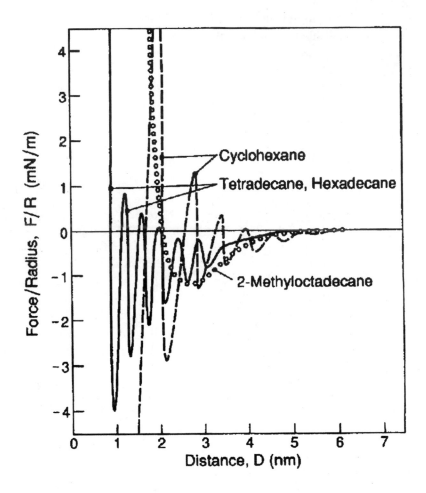

Figure 5.1: Force-distance F/R curves measured in the SFA for various hydro-carbon fluids (from Ref. 146). In this plot, D corresponds to h in Fig. 5.2.

In another mode of operation of the SFA, a confined fluid can be exposed to a shear strain by attaching a movable stage to the upper substrate via a spring characterized by its spring constant k [147–151] and moving this stage at some constant velocity in, say, the x-direction parallel to the film substrate interface. Experimentally it is observed that the upper wall first "sticks" to the film, as it were, because the upper substrate remains stationary. From the known spring constant and the measured elongation of the spring, the shear stress sustained by the film can be determined. Beyond some critical shear strain (i.e., at the so-called "yield point") corresponding to the maximum shear stress sustained by the film, the shear stress declines abruptly and the upper substrate "slips" across the surface of the confined film. If the stage moves at a sufficiently low velocity, the movable substrate eventually comes to rest again until the critical shear stress is once again attained so that the stick–slip cycle repeats itself periodically.

The stick-slip cycle, observed for all types of compounds ranging from long-chain (e.g., hexadecane) to spheroidal [e.g., octamethyltetracyclosiloxane (OMCTS)] hydrocarbons [136], has been attributed by Gee et al. [146] and later on by Klein and Kumacheva [150, 151] to solidification of the confined fluid. This suggests that the atomic structure of the walls induces the formation of a solid-like film when the substrates are properly registered and that this film "melts" when the substrates are moved out of the correct registry. As was first demonstrated in Ref. 45, such films may, in fact, form between commensurate substrate surfaces on account of a template effect imposed on the film. However, noting that the stick-slip phenomenon is quite general, in that it is observed in every liquid investigated regardless of whether its solid structure is commensurate with that of the confining substrates, Granick [136] has argued that mere confinement may so slow mechanical relaxation of the film that flow must be activated on a time scale comparable with that of the experiment. This more general mechanism does not necessarily involve solid films. In the discussion below, we shall therefore concentrate on this latter, more interesting and spectacular scenario in which confined fluids sustain a nonvanishing shear stress without attaining a highly ordered solid-like structure.

5.3.2 Derjaguin's approximation

To make contact with the SFA experiment, one has to realize that the confining surfaces are only locally parallel. Because of the macroscopic curvature of the substrate surfaces, the stress τ_{zz} exerted by the fluid on these curved substrates becomes a local quantity varying with the vertical distance $s_z(x, y)$ between the substrate surfaces (see Fig. 5.2). As the sphere-plane arrangement (see Section 5.3.1) is immersed in bulk fluid at some pressure $P_b(T, \mu)$, the total force exerted on the sphere by the film in the z-direction can be expressed as [152]

$$F(h; T, \mu) = - \int dx \int dy \left\{ \tau_{zz} [s_z(x, y); T, \mu] + P_b(T, \mu) \right\} \qquad (5.54)$$

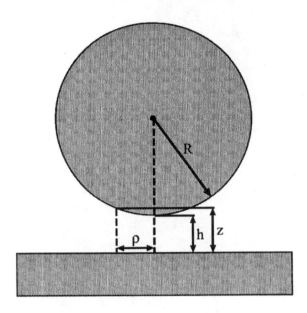

Figure 5.2: Side view of the geometry in which a fluid film (not shown) is confined between a sphere of macroscopic radius R and a planar substrate surface. The shortest distance between two points located on the surface of the sphere and the substrate is denoted by h.

which must be regarded as an *effective* rather than a typical intermolecular force because it depends on the thermodynamic state through T and μ. This solvation, or depletion, force plays a vital role in the context of binary mixtures of colloidal particles of different sizes [153, 154].

To evaluate the integral in Eq. (5.54), it is convenient to transform from Cartesian to cylindrical coordinates $\mathrm{d}x\,\mathrm{d}y \rightarrow \det\mathbf{J}\,\mathrm{d}\rho\,\mathrm{d}\phi$ to obtain

$$
\begin{aligned}
F\left(h\right) &= -\int_0^{2\pi}\mathrm{d}\phi\int_0^R \rho\,\mathrm{d}\rho\left[\tau_{zz}\left(s_z\right)+P_{\mathrm{b}}\right] = 2\pi\int_0^R \rho\,\mathrm{d}\rho f\left(s_z\right) \\
&= 2\pi\int_h^{h+R}\mathrm{d}s_z\left(R-s_z+h\right)f\left(s_z\right) \qquad (5.55)
\end{aligned}
$$

where the determinant of the Jacobian matrix is $\det\mathbf{J} = \rho$. For pedagogic reasons we restrict the current discussion to fluids interacting with chemically *homogeneous* substrates where the fluid substrate interaction is modeled according to Eq. (5.71), and dropped the arguments T and μ to simplify notation. In Eq. (5.55) we used (see Fig. 5.2)

$$s_z = h + R - \sqrt{R^2 - \rho^2} \tag{5.56a}$$

$$\rho\,d\rho = \sqrt{R^2 - \rho^2}\,ds_z = (R - s_z + h)\,ds_z \tag{5.56b}$$

which follows from elementary geometrical considerations.

In Eq. (5.55) we also introduced the disjoining pressure [cf., Eq. (3.71)]

$$f(s_z) \equiv -\tau_{zz} - P_b \tag{5.57}$$

which may be interpreted as the excess pressure exerted by the confined fluid on the substrate surfaces. This interpretation readily follows from Eqs. (1.60) and (1.63), which permit us to write

$$f(s_z) = -\frac{1}{A_0}\left(\frac{\partial\Omega}{\partial s_z}\right)_{T,\mu,A} + \left(\frac{\partial\Omega_b}{\partial V}\right)_{T,\mu} \equiv -\frac{1}{A_0}\left(\frac{\partial\Omega^{\mathrm{ex}}}{\partial s_z}\right)_{T,\mu,A} \tag{5.58}$$

where Ω_b is the grand potential of the bulk fluid and Eq. (1.31) has also been employed. In Eq. (5.58) we also used the fact that $V = A_0 s_z$, which follows from the isotropy of bulk phases (see Section 1.3.1). Equation (5.58) then permits us to define the excess grand potential $\Omega^{\mathrm{ex}} \equiv \Omega - \Omega_b$ of the confined fluid.

In Eq. (5.55), $F(h)$ still depends on the curvature of the substrate surfaces through R. Experimentally, one is typically concerned with measuring $F(h)/R$ rather than the solvation force itself [143] because, for macroscopically curved surfaces, this ratio is independent of R and therefore is independent of the specific experimental setup. This can be rationalized by realizing that $f(s_s)$ must vanish on a molecular length scale because this quantity is nonzero only over a range of substrate separations comparable with the range of the fluid substrate interaction potential, which is orders of magnitude smaller than R. We may therefore take the upper integration limit in Eq. (5.55) to infinity, which gives

$$\frac{F(h)}{2\pi R} = \int_h^\infty ds_z f(s_z) = -\frac{1}{A_0}\int_h^\infty ds_z \left(\frac{\partial\Omega^{\mathrm{ex}}}{\partial s_z}\right)_{T,\mu,A} = \frac{\Omega^{\mathrm{ex}}(h)}{A_0} \equiv \omega_A^{\mathrm{ex}}(h) \tag{5.59}$$

because Ω^{ex} vanishes in the limit $s_z \to \infty$ according to its definition in Eq. (5.58). In Eq. (5.59) we introduce the grand potential per unit area, $\omega_A^{\mathrm{ex}}(h)$ of a fluid confined between two *planar* substrate surfaces separated by a distance h. Equation (5.59) is the celebrated Derjaguin approximation [see Eq. (6) in Ref. 145].

It was pointed out by Götzelmann et al. that the Derjaguin approximation is exact in the limit of a macroscopic sphere, which is the only case of interest here [155]. A rigorous proof can be found in the Appendix of Ref. 156. A similar "Derjaguin approximation" for shear forces exerted on curved substrate surfaces has been proposed by Klein and Kumacheva [150].

Equation (5.59) is a key expression because it links the quantity $F(h)/R$ that can be measured in an SFA experiment directly to the local stress τ_{zz} available from MC simulations. Moreover, it is interesting to note that from Eq. (5.59) we obtain

$$\frac{1}{2\pi}\frac{\mathrm{d}}{\mathrm{d}h}\frac{F(h)}{R} = \frac{\mathrm{d}\omega^{\mathrm{ex}}(h)}{\mathrm{d}h} = \frac{\mathrm{d}}{\mathrm{d}h}\int\limits_{h}^{\infty}\mathrm{d}s_z f(s_z) = -f(h) \tag{5.60}$$

which shows that a derivative of the experimental data is directly related to the stress exerted locally on the macroscopically curved surfaces at the point $(0, 0, s_z = h)$.

5.3.3 Normal component of the stress tensor

To derive a molecular expression for the stress tensor component τ_{zz}, which is the basic quantity if one wishes to compute pseudo-experimental data $F(h)/R$, we start from the thermodynamic expression for the exact differential given in Eq. (1.59), where the strain tensor $\boldsymbol{\sigma}$ is given in Eq. (1.41). From these two expressions, it follows that

$$\tau_{zz} = \frac{1}{V_0}\left(\frac{\partial\Omega}{\partial\sigma_{zz}}\right)_{\{\cdot\}\backslash\sigma_{zz}} = \frac{1}{A_{z0}}\left(\frac{\partial\Omega}{\partial s_z}\right)_{\{\cdot\}\backslash\sigma_{zz}} = -\frac{k_{\mathrm{B}}T}{A_{z0}}\left(\frac{\partial\ln\Xi}{\partial s_z}\right)_{\{\cdot\}\backslash\sigma_{zz}} \tag{5.61}$$

where we used Eq. (2.81) and the fact that $\mathrm{d}\sigma_{zz} = \mathrm{d}s_z/s_{z0}$. The last expression in Eq. (5.61) may be recast as

$$\tau_{zz} = -\frac{k_{\mathrm{B}}T}{A_{z0}\Xi}\sum_{N=0}^{\infty}\frac{1}{N!\Lambda^{3N}}\exp\left(\frac{\mu N}{k_{\mathrm{B}}T}\right)\frac{\partial Z}{\partial s_z} \tag{5.62}$$

which follows from Eqs. (5.61), (2.112), and (2.120). As we demonstrate in Appendix E.3

$$\tau_{zz} = \tau_{zz}^{\mathrm{FF}} + \tau_{zz}^{\mathrm{FS}} \tag{5.63}$$

where [see Eqs. (E.33) and (E.40a)]

$$\tau_{zz}^{\mathrm{FF}} = \tau_{zz}^{\mathrm{id}} + \frac{\langle W_{zz}\rangle}{2V} \tag{5.64}$$

and τ_{zz}^{FS} is defined in Eq. (E.40b). As demonstrated in Appendix E.3.1.2, an alternative route to calculate τ_{zz} is provided by the so-called force expression given in Eq. (E.46). Together, virial and force routes provide a check on internal consistency of the simulations, an assessment that is highly recommended in practice. The consistency check is possible because of the different functional forms of the molecular expressions for τ_{zz} in Eqs. (E.40) and (E.46). Moreover, as the system is not supposed to move in space, the *total* force on the slit-pore must vanish on average. That is to say [see Eq. (E.47)]

$$\left\langle F_z^{[1]} \right\rangle = - \left\langle F_z^{[2]} \right\rangle \tag{5.65}$$

Thus, the symmetry of the force expression provides another useful check on the simulations. The accuracy to be expected is illustrated by entries in Table 5.1.

Table 5.1: Normal component of the (microscopic) stress tensor τ_{zz} from virial [see Eqs. (5.63), (5.64), and (E.40b)] and force [see Eq. (E.46)] expressions for $\mu = -11.50$ and $\varepsilon_{fs} = 1.00$.

s_z	τ_{zz} [Eq. (E.46)]	τ_{zz} [Eq. (5.63)]	τ_{zz}^{FF}	τ_{zz}^{FS}
1.90	-2.251	-2.261	-0.112	-2.149
2.10	0.020	0.021	-0.140	0.161
2.20	0.341	0.339	-0.138	0.477
2.30	0.379	0.385	-0.136	0.521
2.50	0.227	0.232	-0.140	0.372
2.70	-0.043	-0.056	-0.195	0.139
3.00	-0.183	-0.177	-0.271	0.093
3.80	-0.040	-0.037	-0.187	0.150
4.50	0.052	0.055	-0.111	0.170
5.00	0.015	0.020	-0.120	0.141
10.00	-0.026	-0.028	-0.048	0.020

5.3.4 Stratification of confined fluids

To illustrate the relation between microscopic structure and experimentally accessible information, we focus on the computation of pseudo–experimental solvation-force curves $F(h)/R$ [see Eqs. (5.57), (5.59), (5.63), and (E.46)] as they would be determined in SFA experiments. However, here these curves are computed from computer simulation data for τ_{zz} and P_b where P_b is

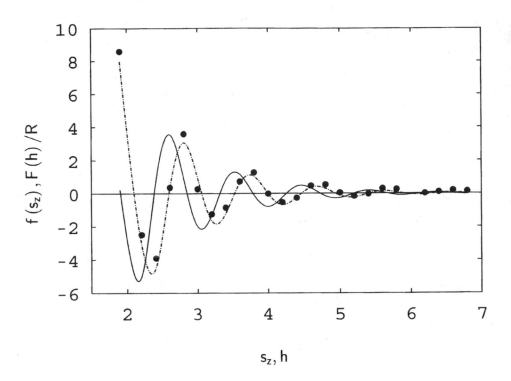

Figure 5.3: Excess pressure $f(s_z)$ [see Eq. (5.57)] ($\bullet$, $-\cdot-$) and the solvation force $F(h)/R$ (—) as a functions of s_z and h, respectively.

calculated from Eqs. (E.70), (E.73), and (E.74) in a separate simulation of a bulk fluid maintained at the same T and μ. Results are correlated with the microscopic structure of a thin film confined between plane parallel substrates separated by a distance $s_z = h$. Again we focus on "simple" fluids, which serve as a suitable model for the approximately spherical OMCTS molecules between mica surfaces, which is perhaps the most thoroughly investigated system in SFA experiments [143, 146]. Because OMCTS is chemically inert and electrically neutral, the influence of charges on the mica surfaces may safely be ignored (see Chapter 6 for a discussion of electrostatic interactions in confined fluids).

Plots of $f(s_z)$ and $F(h)/R$ versus s_z and h, respectively, are shown in Fig. 5.3. The oscillatory decay of both quantities is a direct consequence of the oscillatory dependence of τ_{zz} on s_z, which has also been investigated by integral equations of varying degree of sophistication [157–161]. As can be seen in Fig. 5.3, zeros of $f(s_z)$ correspond to successive extrema of $F(h)/R$

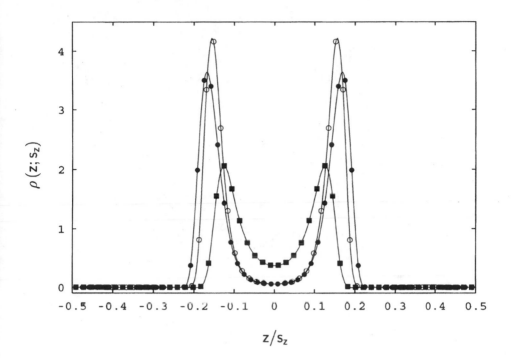

Figure 5.4: Local density $\rho(z)$ as a function of substrate separation s_z; $s_z = 2.60$ (■), $s_z = 2.80$ (○), and $s_z = 3.00$ (●).

because of Eq. (5.60). In actual SFA experiments, the only portions of the $F(h)/R$ curve generally accessible are those where the inequality

$$\frac{\mathrm{d}F(h)}{\mathrm{d}R} \leq 0 \qquad (5.66)$$

holds because ω^{ex} increases upon compression of the film [see Eq. (5.60)].

Alternatively, one may employ colloidal probe atomic force microscopy (AFM) to measure force distance curves such as the ones plotted in Fig. 5.1 [162]. The important difference between SFA and colloidal probe AFM experiments is that in the latter the *entire* force distance curve is accessible rather than only that portion satisfying Eq. (5.66) [163, 164]. In Ref. 164 a comparison is presented between theoretical and experimental data for confined poly-electrolyte systems.

In any case, structural changes accompanying the variation of $F(h)/R$ are rather obscure regardless of the experimental technique. These changes can be inferred more directly from Figs. 5.4–5.6 where plots of the local

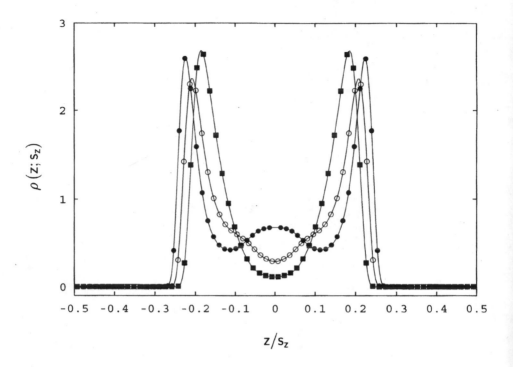

Figure 5.5: As Fig. 5.4, but for $s_z = 3.20$ (■), $s_z = 3.40$ (○), and $s_z = 3.55$ (●).

density

$$\rho(z) = \frac{\langle N(z) \rangle}{s_{x0} s_{y0} \delta_z} \tag{5.67}$$

are presented. In Eq. (5.67), $N(z)$ is the number of fluid molecules whose center of mass is located in a prism of dimensions $s_{x0} \times s_{y0} \times \delta_z$, where δ_z is typically of the order of $10^{-2} - 10^{-1} \sigma$ for a LJ(12,6) fluid [see Eq. (5.24)]. In general, $\rho(z) \to 0$ as $|z| \to s_z/2$ because of the increasing repulsion of fluid molecules by the substrates. Maxima in $\rho(z)$ reflect stratification, which is the arrangement of fluid molecules in individual layers parallel with the solid substrates. Because of the layered structure of the confined fluid, neighboring maxima in plots of $\rho(z)$ are separated by minima that reflect a reduced probability of finding the center of mass of fluid molecules in this region. Oscillations in $\rho(z)$ are damped as one moves away from the substrates because of diminishing fluid substrate interaction. In other words, if the slit-pore is sufficiently wide, stratification is pronounced only in the vicinity of the substrate surfaces such that the inhomogeneity of the fluid persists only over distances roughly comparable with the range of intermolecular forces.

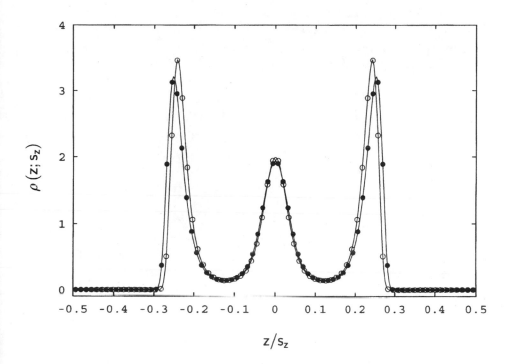

Figure 5.6: As Fig. 5.4, but for $s_z = 3.20$ (■), $s_z = 3.40$ (O), and $s_z = 3.55$ (●).

Because of Eq. (5.60), experimentally accessible portions of the pseudo-experimental data can be related to the local stress at the point $(0, 0, s_z = h)$ of minimum distance between the surfaces of the macroscopic sphere and the planar substrate (see Fig. 5.2). By correlating the local stress $\tau_{zz}(h)$ with the confined fluid's local structure at $(0, 0, h)$ via $\rho(z)$, one can establish a direct correspondence between pseudo-experimental data [i.e., $F(h)/R$] and the local microscopic structure of the confined fluid.

Plots of a sequence of local densities $\rho(z)$ in Figs. 5.4–5.6 over the range $2.60 \leq h \leq 4.00$ illustrate this correlation. In an actual SFA experiment $2.59 \leq h \leq 3.06$ and $3.53 \leq h \leq 4.00$ are accessible portions of the solvation-force curve, whereas $3.06 < h < 3.53$ demarcates the inaccessible range of distances because here the inequality stated in Eq. (5.66) is violated. Plots in Figs. 5.4 and 5.5 show that in the experimentally accessible regions the film consists locally of two and three molecular strata, respectively. For $h = 2.60$, the film is locally compressed because $F(h) > 0$ whereas it is stretched for $h = 3.00$ because here $F(h) < 0$. Under compression the film appears to be less stratified, as is reflected by smaller heights of less well-separated peaks of $\rho(z)$ compared with the two other curves in Fig. 5.4. For $h = 2.80$, $F(h) \approx 0$, and $\tau_{zz}(s_z = h)$ has almost assumed its minimum value, indicating that for this particular value of h film molecules are locally accommodated most satisfactorily between the surfaces of the macroscopic sphere and the planar substrate. It is therefore not surprising that peaks in $\rho(z)$ are taller for $h = 2.80$ compared with the two neighboring values of h (see Fig. 5.4).

In the next accessible region $3.53 \leq h \leq 4.00$, the film consists of three molecular strata for which the most pronounced structure is observed for $h \simeq 3.80$, corresponding to a point at which $F(h)/R$ nearly vanishes (see Fig. 5.6). As before in Fig. 5.4 this is reflected by the peak height of the contact strata (i.e., those layers being closest to the substrate surfaces), whereas inner portions of the film remain largely unaffected by the change in pore width.

Plots of $\rho(z)$ in the experimentally inaccessible regime of pore widths in Fig. 5.5 show that here the film undergoes a local reorganization characterized by the vanishing (appearance) of a whole layer of fluid molecules. The reorganization is gradual, as one can see in the plot of $\rho(z)$ for $h = 3.4$ where two shoulders appear at $z/s_z \simeq \pm 0.1$.

Stratification, as illustrated by the plots in Figs. 5.4–5.6, is due to constraints on the packing of molecules next to the substrate surface and is therefore largely determined by the repulsive part of the intermolecular potential [38]. Stratification is observed even in the complete absence of intermolecular attractions, such as in the case of a hard-sphere fluid confined between planar hard walls [165–167]. For this system Evans et al. [168] demonstrated that, as a consequence of the damped oscillatory character of the local density in the vicinity of the walls, τ_{zz} is itself a damped oscillatory function of s_z, if s_z is of the order of a few molecular diameters, which is confirmed by the plot in Fig. 5.3.

5.4   Chemically heterogeneous substrates

In the previous section we employed GCEMC simulations to illustrate the close relation between thermophysical properties [i.e., $F(h)/R$ or τ_{zz}] and the microscopic structure of the confined fluid [i.e., $\rho(z)$]. The characteristic damped oscillatory dependence of $F(h)/R$ on h observed in both SFA experiments [135, 169] and computer simulations [5, 39, 44, 170]) is a direct consequence of the interplay between two relevant length scales, namely the range of fluid fluid interactions and the degree of confinement represented by h or s_z.

GCEMC simulations at fixed T, μ, and s_z [39, 42–45] and in a grand mixed isostress isostrain ensemble [170–172] demonstrate that the fluid piles up in layers parallel with the walls and that, in coincidence with the oscillations in τ_{zz}, whole layers of fluid abruptly enter the pore. This stratification, due to constraints on the packing of molecules against the rigid planar walls, thus accounts for the oscillatory dependence of τ_{zz} on s_z [38]. GCEMC simulations [173, 174] of a monoatomic film between walls comprising like atoms fixed in the configuration of the face-centered cubic (fcc) (100) plane show that if the walls are in the right registry they can induce freezing of a molecularly thin film. The frozen film resists shearing (i.e., the walls stick) until a critical shear strain is surpassed, whereupon the film melts and the walls slip past one another. This effect has been invoked to explain stick-slip lateral movement observed by the SFA [146, 148].

In this section we shall focus on the behavior of confined phases exposed to a shear strain. However, unlike in these earlier studies, the confined phase will not be solid-like but will remain fluidic, which, in our opinion, makes the rheology of confined phases even more fascinating. It will turn out that the fluid's capability to resist a shear strain can be linked to a third length scale competing with the other two mentioned above. This third length scale can be identified with some inherent structure of the solid surfaces themselves. The inherent structure could be geometrical in nature such as a sequence of nanoscopic grooves rendering the confining substrates nonplanar. It could also be chemical in nature like some sort of imprinted structure by which the wettability of the solid surfaces varies locally. We shall focus on the second situation in this and Sections 5.5 and 5.6, where in the latter section we address the rheology of confined fluids. In all three sections, the third length scale entering our discussion can be associated with the dimensions of the chemical pattern with which the confining substrates are endowed.

5.4.1   The model

5.4.1.1   Continuum description of the substrate potential

For simplicity we employ a model system sketched in Fig. 5.7. It consists of a film composed of spherically symmetric molecules that is sandwiched between the surfaces of two solid substrates. The substrate surfaces are planar, parallel, and separated by a distance s_z along the z-axis of the coordinate system. They are semi-infinite in the z-direction, occupying the half spaces $s_z/2 \leq z \leq \infty$ and $-\infty \leq z \leq -s_z/2$, and infinite in the x- and y-directions. Each substrate comprises alternating slabs of two types: strongly adsorbing and weakly adsorbing. The "strong" and "weak" slabs have widths d_s and d_w, respectively, in the x-direction and are infinite in the y-direction. The

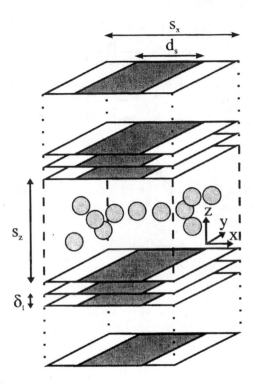

Figure 5.7: Scheme of a simple fluid confined by a chemically heterogeneous model pore. Fluid molecules (gray spheres) are spherically symmetric. Each substrate consists of a sequence of crystallographic planes separated by a distance δ_ℓ along the z-axis. The surface planes of the two opposite substrates are separated by a distance s_z. Periodic boundary conditions are imposed in the x- and y-directions (see Section 5.2.2).

system is thus periodic in the x-direction of period $d_s + d_w$ such that its properties are translationally invariant in the y-direction. In practice we take the system to be a finite piece of the film, imposing periodic boundary conditions [140] (see Section 5.2.2) on the planes $x = \pm s_x/2$ and $y = \pm s_y/2$.

The substrates are in registry meaning that slabs of the same type are exactly opposite each other. Substrate atoms are assumed to be of the same "diameter" (σ) and to occupy the sites of the fcc lattice [the substrate surfaces are taken to be (100) planes] having lattice constant ℓ, which is taken to be the same for both species. Thus, substrate species are distinguished only by the strength of their interaction with film molecules. We assume the total potential energy to be a sum of pair-wise additive LJ (12,6) potentials, all of which have the form given in Eq. (5.24). For the interaction between a pair of film molecules $\varepsilon = \varepsilon_{\text{ff}}$ [i.e., $u_{\text{ff}}(r)$]. The nanoscale heterogeneity of the substrate is characterized by $\varepsilon = \varepsilon_{\text{fs}}$ [i.e., $u_{\text{fs}}(r)$] for the interaction of a film molecule with a substrate atom in the strong (central) slab, and by $\varepsilon = \varepsilon_{\text{fw}}$ [i.e., $u_{\text{fw}}(r)$] for the interaction of a film molecule with a substrate atom in either of the two weak (outer) slabs (see Fig. 5.2). We take $\varepsilon_{\text{fs}} \geq \varepsilon_{\text{ff}}$ and $\varepsilon_{\text{fw}} \ll \varepsilon_{\text{ff}}$ (see Section 5.4.2 for specific values).

To avoid complications resulting from a detailed description of the atomic structure of the heterogeneous substrate surfaces, we employ a continuum representation of the fluid substrate potential. Details of its derivation can be found in Appendix E.2. Combining the expressions in Eqs. (E.21), (E.24), (E.26), (E.27), and (E.29), we obtain for the potential energy of a film molecule in the continuum representation of the substrate k $(= 1, 2)$

$$
\begin{aligned}
\Phi^{[k]} = \\
-\frac{3\pi}{2} n_A \sigma^2 \sum_{m=-\infty}^{\infty} \sum_{m'=0}^{\infty} \Bigg\{ (\varepsilon_{\text{fw}} - \varepsilon_{\text{fs}}) \, \Delta \left(x + \frac{d_s}{2} - m s_x, \frac{s_z}{2} + m' \delta_\ell \pm z \right) \\
- (\varepsilon_{\text{fw}} - \varepsilon_{\text{fs}}) \, \Delta \left(x - \frac{d_s}{2} - m s_x, \frac{s_z}{2} + m' \delta_\ell \pm z \right) \\
- \varepsilon_{\text{fw}} \, \Delta \left(x + \frac{s_x}{2} - m s_x, \frac{s_z}{2} + m' \delta_\ell \pm z \right) \\
+ \varepsilon_{\text{fw}} \, \Delta \left(x - \frac{s_x}{2} - m s_x, \frac{s_z}{2} + m' \delta_\ell \pm z \right) \Bigg\}
\end{aligned}
\tag{5.68}
$$

where the sign on z is chosen according to the convention $+ \leftrightarrow k = 1$ and $- \leftrightarrow k = 2$ (see Fig. 5.2).

Before discussing the implementation of Eq. (5.68) in the GCEMC simulation, we comment briefly on the properties of the *whole* fluid substrate potential $\Phi = \Phi^{[1]} + \Phi^{[2]}$ that follow strictly from considerations of symmetry. When the z-coordinate of the fluid molecule is reflected through the mirror plane $z = 0$, $-z$ in the arguments $z'' = s_z/2 + m' \delta_\ell \pm z$ of $\Phi^{[1]}$ changes to $+z$ in the arguments of $\Phi^{[2]}$ and *vice versa*. That is, $\Phi^{[1]}(x, -z) \rightarrow \Phi^{[2]}(x, z)$ and *vice versa*. The sum Φ is therefore invariant under reflection in the $z = 0$ plane. Likewise, Φ is invariant under reflection in the x-plane, although the proof involves more subtle interconversions. For example, under the transformation $x \rightarrow -x$, the first term in braces (for m) is converted into the second term in braces (for $-m$). Likewise, the third term in Eq. (5.68) is converted into the fourth term. Of course, as the potential is periodic in x, of period s_x, we need to represent the whole continuum fluid substrate potential in only one quadrant (say, $0 \leq x \leq s_x/2$, $0 \leq z \leq s_z/2$) of the x–z plane.

5.4.1.2 Computation of film-wall contribution to configurational energy

As we demonstrated in Section 5.2.2, the generation of a Markov chain of configurations in GCEMC simulations is governed by a change in configurational energy associated with both particle displacement and creation/destruction attempts. For this system the configurational energy U can be written as

$$U = \frac{1}{2} \sum_{i=1}^{N} \sum_{j \neq i=1}^{N} u_{ff}(r_{ij}) + \sum_{k=1}^{2} \sum_{i=1}^{N} \Phi^{[k]}(x_i, z_i; d_s, s_x, s_z) \equiv U_{FF} + U_{FS} \quad (5.69)$$

where $u_{ff} = u_{SR}$ is given in Eq. (5.39), $\Phi^{[k]}$ in Eq. (5.68), and $r_{ij} \equiv |\mathbf{r}_i - \mathbf{r}_j|$ is the distance between the centers of film molecules i and j located at $\mathbf{r}_i$ and $\mathbf{r}_j$, respectively. Equation (5.69) also defines the film-film and film-substrate contributions U_{FF} and U_{FS} to U. To implement the expression for $\Phi^{[k]}$ in Eq. (5.68), we truncate the infinite summations according to $\sum_{m=-\infty}^{\infty} \sum_{m'=0}^{\infty} \rightarrow \sum_{m=-M}^{M} \sum_{m'=0}^{M'}$, where integers M and M' are sufficiently large to yield $\Phi^{[k]}$ with a prescribed precision. For a system size of $s_x \geq 10$ and a lattice spacing of $\delta_\ell = 1.0$, we find that $M = 2$ and $M' = 50$ are large enough to give $\Phi^{[k]}$ to a precision of 0.3% regardless of the position of a film molecule with respect to the substrate.

However, M and M' are still too large to employ the truncated version of Eq. (5.68) directly in each GCEMC step. Tests show that for $M = 2$ and $M' = 50$ the evaluation of $\Phi^{[k]}$ for a *single* film molecule requires approximately the same amount of computer time as the computation of U_{FF} for $N = 100$, so that a GCEMC simulation of a typical length of 10^5 or 10^6 cycles (see Section 5.2.2) would be prohibitively expensive. Instead of computing $\Phi^{[k]}$ during each step of the GCEMC simulation, we adopted the following procedure. Prior to the simulation we computed $\Phi^{[k]}$ by the truncated version of Eq. (5.68) and stored it at the nodes of a square grid $\{x_k, z_k\}_{k=1,...,K}$ in the quadrant $0 \leq x \leq s_x/2$, $0 \leq z \leq s_z/2$. During the simulation the value of $\Phi^{[k]}$ at the the actual (instantaneous) position (x_i, z_i) of film molecule i (which does not necessarily coincide with any node) is obtained by bilinear interpolation [175] among the values of $\Phi^{[k]}$ at the four nearest-neighbor nodes of (x_i, z_i).

We tested the interpolation scheme for a special case in which $\Phi^{[k]}$ can be readily evaluated during each GCEMC step. The substrate consists of a single, chemically homogeneous plane for which we set $\varepsilon_{fs} = \varepsilon_{fw} = \varepsilon_{ff} = \varepsilon$. Thus, the discrete sum on m and piece-wise integrations over the strips are replaced by a single integration on x' from $-\infty$ to ∞. The summation on m' also reduces to a single term $m' = 0$. Under these conditions Eq. (E.21) can be rewritten as

$$\Phi^{[k]}\left(z; s_z\right) = 4\varepsilon n_A \int\limits_0^{2\pi} \mathrm{d}\phi \int\limits_0^\infty \mathrm{d}\rho\rho \left[\left(\frac{\sigma^2}{\rho^2 + z''^2}\right)^6 - \left(\frac{\sigma^2}{\rho^2 + z''^2}\right)^3\right] \qquad (5.70)$$

in cylindrical coordinates, so that the integrations on ϕ and ρ can be carried out in closed form to yield

$$\Phi^{[k]}\left(z; s_z\right) = 2\pi\varepsilon n_A \sigma^2 \left[\frac{2}{5}\left(\frac{\sigma}{z \pm s_z/2}\right)^{10} - \left(\frac{\sigma}{z \pm s_z/2}\right)^4\right] \qquad (5.71)$$

where $(+ \leftrightarrow k = 1, - \leftrightarrow k = 2)$. Employing a mesh of $\delta_x = \delta_z = 0.025$, corresponding to $K = 7.6 \times 10^3$ ($s_x = 10$, $s_z = 1.9$) for the smallest substrate separation and $K = 5.0 \times 10^4$ ($s_x = 10$, $s_z = 12.50$) for the largest, we compare (see Table 5.2) results obtained using the interpolation scheme [Eq (5.68)] with those based on direct evaluation of the potential given by Eq. (5.71). It is noteworthy that the agreement is good, even for fluid fluid (FF) and fluid substrate (FS) contributions to the normal component of the stress tensor τ_{zz}, which are particularly sensitive to numerical inaccuracies in case the magnitude of these contributions is small.

Table 5.2: Comparison of interpolation [truncated version of Eq. (5.68)] and direct evaluation [Eq. (5.71)] of fluid substrate potential energy for various properties of a fluid confined by substrates consisting of single, chemically homogeneous planes. Entries, given in dimensionless units defined in Table 1, refer to simulations based on either direct evaluation (D) or interpolation (I).

s_z	$-\mu$	$-\langle U_{FF}/N\rangle$		$-\langle U_{FS}/N\rangle$		$-\tau_{zz}^{FF}$		$-\tau_{zz}^{FS}$	
		D	I	D	I	D	I	D	I
3.00	9.56	3.676	3.668	2.920	2.912	1.09	1.10	0.21	0.20
2.70	9.46	3.086	3.088	3.021	3.013	2.45	2.44	3.16	3.20
2.20	9.26	2.103	2.113	4.860	4.858	0.42	0.42	-1.53	-1.54

5.4.2   Structure and phase transitions

In Section 5.3.4, we demonstrated that fluids confined to nanoscopic volumes are highly inhomogeneous in that they form molecular strata. The most direct way of realizing this was through plots of the local density (see

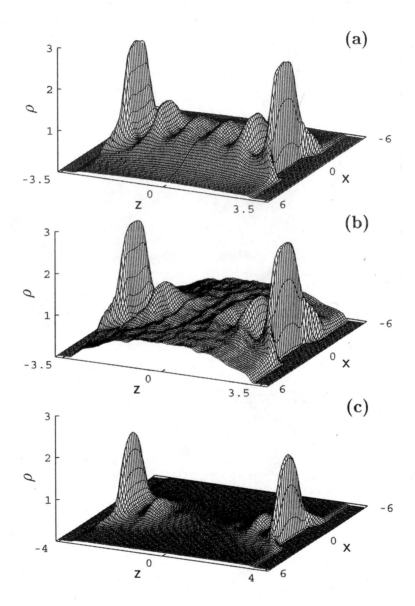

Figure 5.8: Local density $\rho(x, z)$ as a function of position in the x–z plane for $s_x = 12$, $\varepsilon_{\mathrm{fw}} = 10^{-3}$, $\varepsilon_{\mathrm{fs}} = 1.25$, $d_{\mathrm{s}} = 4.0$ [see Eq. (5.68)]; (a) $s_z = 7.2$, (b) $s_z = 7.5$, (c) $s_z = 8.2$.

Figs. 5.4–5.6) based on the definition of $\rho(z)$ defined in Eq. (5.67). However, because in the current model, the fluid substrate interaction potential is a function of x and z [see Eq. (5.68)], the local density must also depend on both (Cartesian) coordinates, which we redefine by writing

$$\rho(x,z) \equiv \frac{\langle N(x,z) \rangle}{\delta_x s_{y0} \delta_z} \qquad (5.72)$$

In Eq. (5.67), $N(x,z)$ is the number of fluid molecules in a given configuration that are located in a square prism of dimensions $\delta_x \times s_{y0} \times \delta_z$ centered on a point (x,z). Three characteristic examples for the local density are plotted in Fig. 5.8 for different substrate separations. Because of the symmetry of $\Phi^{[k]}$, $\rho(x,z)$ must be symmetric about the $x = 0$ and $z = 0$ planes (see Fig. 5.8). Similar to the plot of $\rho(z)$ in Figs. 5.4–5.6, peaks in its two–dimensional counterpart represent positions of molecular strata. For $s_z = 7.2$ a stratified fluid bridges the gap between the strongly attractive central portions of the opposite substrates [i.e., for $|x| \lesssim 2.0$, see Fig. 5.8(a)]. Because of the decay of the fluid substrate interaction potential, stratitification diminishes as z increases along lines of constant x. Stratification is absent over the weakly attractive portion of the substrate [see Fig. 4.13)]. Here an inhomogeneous low-density fluid exists, as indicated by the relatively low value of $\rho(x,z)$ and its weak dependence on x and z for $|x| \gtrsim 4.0$. By analogy with the mean-field data plotted in Figs. 4.11(a) and 4.11(b), we refer to situations akin to the one depicted in Fig. 5.8(a) as "bridge" phase: A stratified high-density fluid stabilized by the attractive part of the substrate *plus* a surrounding gas over the outer two, weakly attractive portions of the substrate material.

For larger $s_z = 7.5$ [see Fig. 5.8(b)], the structure of the fluid changes significantly. Over the strongly attractive portion of the substrate, the fluid remains stratified. However, the low-density portion has given way to an inhomogeneous high-density fluid over the weakly attractive part of the substrate. Consequently, the interface between higher- and lower-density portions of the confined fluid visible in the plot of $\rho(x,z)$ in Fig. 5.8(a) has disappeared, and can no longer be seen in Fig. 5.8(b). As the weak portions of the substrate are essentially repulsive, $\rho(x,z)$ decreases for $|x| \gtrsim 4.0$ from the center of the fluid ($z = 0$) toward the substrate ($|z| \to s_z/2$).

If the distance between the substrates is increased even further, another structural change occurs in the fluid. It is illustrated by the plot of $\rho(x,z)$ for $s_z = 8.2$ in Fig. 5.8(c), where the fluid bridge disappeared and only two strata of fluid molecules "cling" to the strongly attractive portion of the substrate. For example, for $|z| \lesssim 3.0$ and $x = 0$, the density is rather low and decreases monotonically toward the center of the confined fluid located at $z = 0$. The height of the two maxima of $\rho\,(x, z)$ appears to be substantially reduced compared with the plots in Figs. 5.8(a) and 5.8(b). Thus, by increasing s_z the high-density morphology of the fluid illustrated by the plot in Fig. 5.8(b) eventually evaporates, leaving behind two inhomogeneous fluid columns (because of the translational invariance of the density in the y-direction) that are stabilized by the strongly attractive portions of the opposite substrates. These columns are surrounded by a portion of fluid of relatively low density, as revealed by the plot in Fig. 5.8(c).

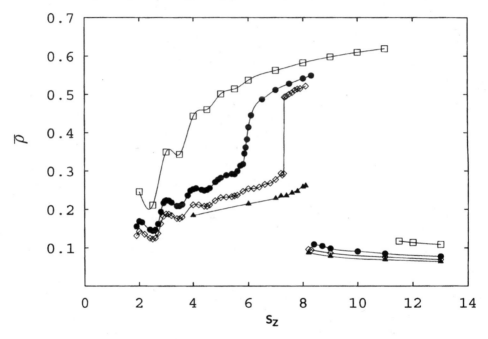

Figure 5.9: Mean density $\bar{\rho}$ [see Eq. 5.74)] as a function of substrate separation s_z. Data are plotted for various degrees of chemical corrugation of the substrate $c_r = \frac{4}{7}$ ($\square$), $c_r = \frac{4}{10}$ ($\bullet$), $c_r = \frac{4}{12}$ ($\diamond$), and $c_r = \frac{4}{14}$ ($\blacktriangle$) [see Eq. (5.73)].

The sequence of plots in Figs. 5.8(a)–5.8(c) illustrates the peculiar phase transition from a fluid bridge to a liquidlike phase and eventually to a thin adsorbed film with increasing s_z. That the morphologies plotted in those figures do, in fact, have the status of legitimate thermodynamic phases is borne out by the corresponding mean-field lattice density functional calculations discussed in detail in Section 4.5.2. In particular the plots presented in Figs. 4.12 illustrate the sequence of phase transitions discussed in the preceding section. As can be seen from Figs. 4.12, the change from an initial bridge phase to a liquid-like and eventually gas-like confined fluid is caused by a shift of the gas bridge liquid triple point to higher temperatures and an associated narrowing of the one-phase region of the bridge phases.

At this point it seems worthwhile to investigate in some depth the impact of chemical corrugation on this phase behavior by varying

$$c_r \equiv \frac{d_s}{d_s + d_w} = \frac{d_s}{s_x} \qquad (5.73)$$

which is the relative width of the strongly adsorbing stripe on the substrate surfaces. Instead of visualizing the associated phase changes by plots of the *local* density as before, we shall focus below on the *mean* density related to the local density via

$$\bar{\rho} \equiv \frac{1}{s_{x0} s_{z0}} \int\limits_{-s_{x0}/2}^{s_{x0}/2} dx \int\limits_{-s_{z0}/2}^{s_{z0}/2} dz \rho\left(x, z\right) \qquad (5.74)$$

Plots of $\bar{\rho}$ versus s_z are shown in Fig. 4.15 for various degrees of chemical corrugation of the substrates. Starting with the largest degree of corrugation $c_r = \frac{4}{7}$, we notice from Fig. 5.9 that $\bar{\rho}$ oscillates for $s_z \lesssim 6.0$ with a period of approximately one molecular "diameter." A similar behavior of $\bar{\rho}$ is found for $c_r = \frac{4}{10}$ and $\frac{4}{12}$, which can be interpreted as a fingerprint of stratification [see, for example, Fig. 5.8(a)], that is, the change in the number of molecular strata accommodated between the substrates with vanishing s_z [38, 172].

However, in the limit $s_z \to \infty$ the confined fluid becomes increasingly bulk-like on account of the vanishing influence of fluid substrate interactions. Because the bulk phase at the current values of $T = 1.0$ and $\mu = -11.5$ turns out to be a gas, one intuitively expects at least one phase transition from a denser confined fluid at small s_z to a lower-density fluid at some characteristic larger substrate separation. This transition, known as capillary condensation/evaporation (see also Section 4.2), is, in fact, observed for $c_r = \frac{4}{7}$ around $s_z \simeq 11.0$.

For $c_r = \frac{4}{10}$, capillary condensation shifts to a smaller value $s_z \simeq 11.0$, which is reasonable in view of the reduced *net* strength of attractive fluid substrate interactions compared with the previously discussed case. The shift of capillary condensation to lower substrate separation also persists for $c_r = \frac{4}{12}$ and $c_r = \frac{4}{14}$, but it is much less pronounced, as Fig. 5.9 reveals. However, comparing only the latter two values of c_r, chemical corrugation of the substrate seems to be only of marginal importance for the location of the discontinuity (that is, the location of the phase transition) in the plot of $\bar{\rho}$.

For $c_r = \frac{4}{10}$ this second discontinuity vanishes in favor of a rather steep increase of $\bar{\rho}$ over the range $5.5 \lesssim s_z \lesssim 6.5$. A corresponding plot of the isothermal compressibility κ_{yy} in Fig. 5.10 has a tall, cusp-like peak in the same range of substrate separations. The derivation of κ_{yy} parallels precisely the one for $\kappa_{\parallel}$ in Section 1.6.2 where, however, here

$$\kappa_{yy} \equiv \frac{1}{s_y} \left(\frac{\partial s_y}{\partial \tau_{yy}} \right)_{\{\cdot\}\backslash s_y} \tag{5.75}$$

From Eqs. (1.66) and the associated Gibbs Duhem Eq. (1.68) it follows that

$$\left(\frac{\partial^2 \omega}{\partial \mu^2} \right)_{\{\cdot\}\backslash \mu} = - \left(\frac{\partial \rho}{\partial \mu} \right)_{\{\cdot\}\backslash \mu} = -\rho \kappa_{yy} \tag{5.76}$$

where the analysis follows precisely the one between Eqs. (1.79) and (1.82) and we used $V = A_{y0} s_y$. However, here the grand-potential density is defined in a slightly different fashion as [see Eq. (1.77)]

$$\omega \left(T, \mu, s_x, s_y, s_z \right) \equiv \frac{\Omega}{A_{y0} s_y} = \tau_{yy} \tag{5.77}$$

Using the statistical expression for Ω given in Eq. (2.81) [using the (classic) expression for Ξ given in Eq. (2.120)], we can apply precisely the same derivation as the one presented in Section 2.3.4 to obtain

$$\kappa_{yy} = \frac{V}{k_B T} \frac{\langle N^2 \rangle - \langle N \rangle^2}{\langle N \rangle^2} \tag{5.78}$$

In the vicinity of the cusp-like peak in the plot of κ_{yy} for $c_r = \frac{4}{10}$ in Fig. 5.10, the isothermal compressibility depends on the size of the simulation system. This is illustrated by the plot of the density distribution $P(\rho)$ plotted in Fig. 5.11 as a function of s_y. To analyze the data plotted in Fig. 5.11, we have normalized $P(\rho)$ numerically such that

$$\int_{-\infty}^{\infty} d\Delta\rho P(\rho) = 1 \tag{5.79}$$

where $\Delta\rho \equiv \rho - \overline{\rho}$ in which ρ and $\overline{\rho}$ are the instantaneous density (of a given configuration of the Markov chain) and the mean density of the confined fluid, respectively. For a sufficiently small density fluctuations, $P(\rho)$ should be approximately Gaussian as demonstrated in Appendix C.3. We can therefore use Eq. (C.36) as a basis for a finite-size scaling analysis of κ_{yy} obtained from GCEMC simulations in which the computational cell is inevitably finite in

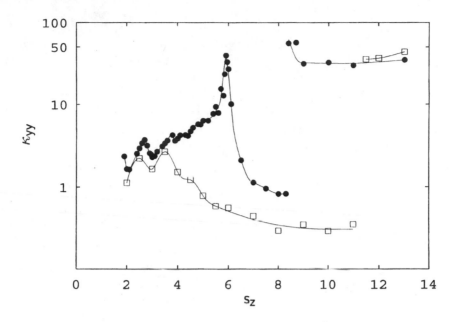

Figure 5.10: As in Fig. 5.9, but for the isothermal compressibility κ_{yy}: $c_r = \frac{4}{7}$ ($\square$) and $c_r = \frac{4}{10}$ ($\bullet$).

size. To this end we notice from Eq. (5.78) that the variance of $P(\rho)$ is given by

$$\sigma_\rho = \bar{\rho} \sqrt{\frac{\kappa_{yy} k_B T}{A_{y0}}} \frac{1}{\sqrt{s_y}} \tag{5.80}$$

Thus, we may construct $P(\rho)$ as a histogram during a GCEMC simulation and calculate σ_ρ by fitting Eq. (C.36) to the discrete data points that constitute this histogram. We may then plot σ_ρ as a function of $1/\sqrt{s_y}$, which should give us a straight line if the simulation cell is large enough in the y-direction. From the slope of this linear relationship, we can then also estimate κ_{yy} via Eq. (5.80). An additional scaling relationship is obtained for the height of the density histogram. Because of Eq. (C.36) we also realize that the height of the peak in the plot of $P(\rho)$ should scale with $\sqrt{s_y}$ because the normalization constant in Eq. (C.36) is inversely proportional to σ_ρ in the Gaussian limit (i.e., in the limit $s_y \to \infty$).

The system-size dependence of density fluctuations is illustrated by the plot of $P(\rho)$ in Fig. 5.11. If $s_y = 10$, for example, the density distribution is by no means Gaussian but rather bimodal, indicating that the system "oscillates" between higher- and lower-density states. As s_y increases, this bimodal nature of $P(\rho)$ gradually declines such that for $s_y \geq 50$, $P(\rho)$ turns out to be Gaussian with the required dependence of peak height and standard deviation on s_y. Data for κ_{yy} obtained from the finite-size analysis just described are plotted in Fig. 5.10. However, as we pointed out in Ref. 176, a similar finite-size dependence was not observed for any other quantity computed in the present context. For the system sizes employed here, these findings are completely in accord with a more recent study of system-size effects in computer simulations [177].

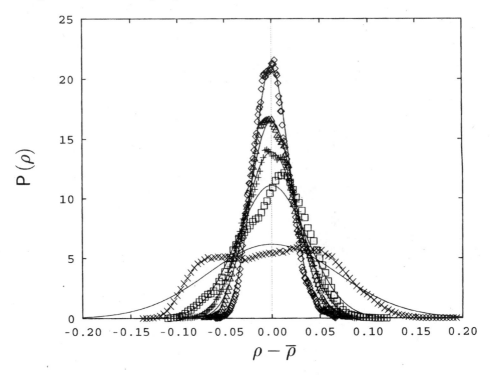

Figure 5.11: Normalized density distribution $P(\rho)$ as a function of the deviation of the instanteneous density $\rho = N/V$ from the average density $\overline{\rho} = \langle N \rangle /V$ for various values of s_y: ($\times$) $s_y = 10$, ($\square$) $s_y = 30$, ($+$) $s_y = 50$, ($\triangle$) $s_y = 65$, and ($\diamond$) $s_y = 100$. Solid lines represent the fit of a Gaussian to $P(\rho)$.

The observed system-size dependence clearly indicates that the correlation length associated with density fluctuations in the confined fluid exceeds the dimensions of the simulation cell [178]. This is indicative of a near-critical thermodynamic state of the confined fluid. Because of the density of the participating phases in this near-critical region we conclude that the critical point is the one at which fluid bridge and liquid-like phases become indistinguishable. As the variation of $\overline{\rho}$ with s_z is continuous (see Fig. 5.9), the thermodynamic state of the confined fluid is still slightly supercritical with respect to the critical point where the phase diagram is qualitatively similar to the mean-field phase diagram plotted in Fig. 4.12 for $n_z = 8 - 10$. As can be seen from that figure, a triangular-shaped region exists corresponding to the one-phase region of fluid bridges.

On either side of the cusp-like peak, κ_{yy} decays rapidly for $c_r = \frac{4}{10}$ to rather small values typical of dense LJ (12,6) fluids (see Fig. 5.10). For $s_z \lesssim$ 5.5, κ_{yy} oscillates with a period of about one molecular diameter reflecting stratification in the sense of the discussion in Section 5.3.4. In addition to its cusp-like maximum κ_{yy} also changes discontinuously during a first-order phase transition at $s_z \simeq 8.3$. ($c_r = \frac{4}{10}$). For all $s_z \gtrsim 8.3$ the magnitude of κ_{yy} corresponds to that of a typical LJ (12,6) gas in accord with the corresponding plot of $\overline{\rho}$ in Fig. 4.15. Stratification-induced oscillations of κ_{yy} can also be seen for $c_r = \frac{4}{7}$ and small s_z. However, in this case, κ_{yy} remains rather small up to the substrate separation where its discontinuous change again signals a first-order phase transition to a low-density phase ($s_z \simeq 11.0$). The smaller value of κ_{yy} ($c_r = \frac{4}{7}$) compared with $c_r = \frac{4}{10}$ indicates the presence of a denser fluid. This seems sensible because the net attraction of fluid molecules by the substrate is larger for $c_r = \frac{4}{7}$ than for $c_r = \frac{4}{10}$.

5.5 Chemical patterns of low symmetry

5.5.1 Nanopatterned model substrate

The situation studied in the previous section becomes more complicated if the chemical pattern with which the substrate surface is decorated is of finite extent. For example, the model substrates employed in Section 5.4 were endowed with chemically distinct stripes that are infinitely long in one spatial direction. As a result, the potential function, describing the fluid substrate interaction turned out to be independent of the y-coordinates of fluid molecules [see Eq. (5.68)]. As a consequence, properties of the confined fluid are translationally invariant in this direction, which has important repercussions for the investigation of the phase behavior of fluids confined between such (chemically) heterogeneous surfaces. For example, if translational invariance of fluid properties is not preserved on account of the symmetry of the heterogeneity on the substrate, we demonstrated in Section 1.6 that a Gibbs Duhem equation may not exist, which, in turn, precludes the existence of a "mechanical" expression for the grand potential, which is the key quantity on which an investigation of phase behavior is based.

Precisely this latter situation arises if the confining solid surface is endowed with a chemical pattern that is both nanoscopic in size and finite
in extent. Such chemical patterns may be created by lithographic methods
[179]. Atomic beams have been employed to produce hexagonal nanostruc-
tures [180]. Other methods capable of creating chemically nanostructured
substrate surfaces involve microphase separation in diblock copolymer films
[181] or the use of force microscopy to locally oxidize silicon surfaces [182].

Another example is the imprinting of ring patterns over the surface [53].
In this situation, the pattern is characterized by an inner radius R_{in} and
an outer radius R_{out} forming an annulus of constant width in which the
surface displays preferential interaction for the fluid. Those surfaces may be
prepared by microcontact printing using alkanethiols on gold [53, 54]. In
this method an elastomer stamp is used to deposit molecules on surfaces.
The stamp is first "inked" with a solution of alkanethiol molecules and then
pressed onto a gold surface, which results in well-defined hydrophilic sites.
The remaining bare gold surface is then made hydrophobic by dipping it into
another alkanethiol. A liquid is then subsequently adsorbed onto the surface
in a closed cell. The experiments are conducted by cooling down or heating
up the system, thereby changing the volume of liquid formed on the surface
[54].

Experiments performed on these ring-patterned substrates show that,
when the volume increases, the liquid geometry formed over the pattern
undergoes a transition from a ring morphology (where the liquid forms a homogeneous covering over the pattern) to a bulge morphology where a droplet
of liquid is formed over one side of the pattern. This bulge geometry "breaks"
the symmetry imposed by the pattern. As the volume of liquid increases, the
bulge progressively spreads over the nonwetting disk inside the pattern and
finally a spherical cap is formed over the whole disk of radius R_{out}. This
bulge geometry can also be observed on surfaces patterned with micrometric
stripes [73].

If one aims at understanding this rather peculiar phase behavior from
a microscopic perspective, one again needs to know the relevant thermodynamic potential. However, one is immediately confronted with a complication
because a "mechanical" expression for thermodynamic potentials cannot be
derived due to the low symmetry of the confined fluid (see also discussion in
Section 1.6.2). Therefore, a different means of calculating these potentials
must be devised. This alternative computational technique will be based on
a perturbational approach to which the current section is devoted.

Specifically, we consider the situation depicted schematically in Fig. 5.12:
a "simple" fluid confined between two planar substrate surfaces composed
of like atoms where the fluid fluid interaction is described by Eq. (5.39) for

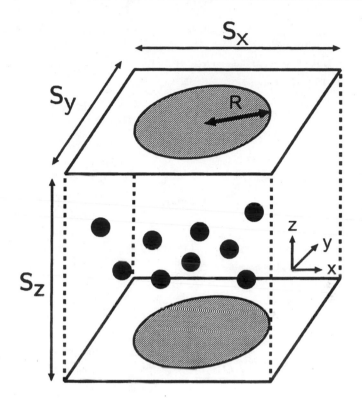

Figure 5.12: Schematic representation of a fluid (black spheres) confined between two planar substrates decorated with a circular region of radius R that attracts the fluid molecules. Outside the circular region fluid–substrate interaction is purely repulsive.

the LJ (12,6) potential. If the substrates are free of any nanoscopic chemical patterns the fluid substrate interaction is described by Eq. (5.71), which we split into attractive and repulsive contributions according to

$$\varphi_{\text{rep}}^{[k]}\left(z\right) \equiv \frac{4\pi\varepsilon_{\text{fs}}n_{\text{A}}\sigma^2}{5}\left(\frac{\sigma}{z \pm s_{\text{z}}/2}\right)^{10} \tag{5.81a}$$

$$\varphi_{\text{att}}^{[k]}\left(z\right) \equiv 2\pi\varepsilon_{\text{fs}}n_{\text{A}}\sigma^2\left(\frac{\sigma}{z \pm s_{\text{z}}/2}\right)^{4} \tag{5.81b}$$

such that

$$\Phi^{[k]}\left(z; s_{\text{z}}\right) = \varphi_{\text{rep}}^{[k]}\left(z\right) - \varphi_{\text{att}}^{[k]}\left(z\right) \tag{5.82}$$

employing again the convention $+ \leftrightarrow k = 1$ and $- \leftrightarrow k = 2$. In Eqs. (5.81), ε_{fs} sets the energy scale of fluid substrate interactions. For simplicity we take

$n_A \sigma^2 = 1$, $\varepsilon_{fs} = \varepsilon_{ff}$, and treat the substrates as semi-infinite solids. The fluid substrate attraction is long range, that is $\varphi_{att}^{[k]}(z) \propto z^{-3}$ [see Eq. (5.81b)].

To model substrate surfaces with imprinted chemical nanopatterns (see Fig. 5.12), we modify Eq. (5.82) according to

$$
\begin{aligned}
\Phi^{[k]}(r) &= \varphi_{rep}^{[k]}(z) - s(x, y; R, \kappa) \varphi_{att}^{[k]}(z) \\
&= \varphi_{rep}^{[k]}(z) - \phi_{att}^{[k]}(r)
\end{aligned}
\tag{5.83}
$$

where r denotes the (vector) position of a fluid molecule and the "switching" function

$$
s(x, y; R, \kappa) = \frac{1}{1 + \exp\left[\kappa\left(x^2 + y^2 - R^2\right)\right]}
\tag{5.84}
$$

is introduced as a continuous representation of the Heaviside function (i.e., the Fermi function [17]; see Fig. 5.13) such that $\Phi^{[k]}(r)$ describes the interaction between a fluid molecule and an infinitesimally smooth, repulsive solid surface endowed with an attractive circular area of (fixed) radius R (and infinite height in the $\pm z$-directions) centered at $(0, 0, \pm s_z/2)$. In Eq. (5.84), $\kappa \geq 0$ is a measure of "softness" with which the attractive part of $\Phi^{[k]}(r)$ is turned off as a fluid molecule moves away from the center of the circular area, that is, from the point $(0, 0)$ in the x–y plane. In other words, the range of in-plane distances $r' \equiv \sqrt{x^2 + y^2}$ over which $s(x, y; R)$ varies between 0 and 1 is determined by κ.

5.5.2   Thermodynamic perturbation theory

To proceed we employ a perturbational approach that enables us to calculate (the absolute value of) the grand-potential density. The key idea in any perturbation theory is to somehow link properties of the system of interest to those of a reference system whose properties can readily be calculated. One then needs to "perturb" the reference system in a controllable manner so that one switches continuously from the known reference system to the system of interest assuming that one can obtain system properties at any instant of the applied perturbation.

Let us apply this general philosophy to the grand potential for the system depicted schematically in Fig. 5.12. Clearly, Eq. (1.66) represents the (exact differential of the) grand potential for the situation shown in Fig. 5.12. Because the fluid substrate potential in Eq. (5.83) depends on the (vector) position r of a fluid molecule rather than a subset of Cartesian coordinates, it is immediately clear that we are dealing with a fluid that is inhomogeneous in all three spatial directions. According to our discussion in Section 1.6.1,

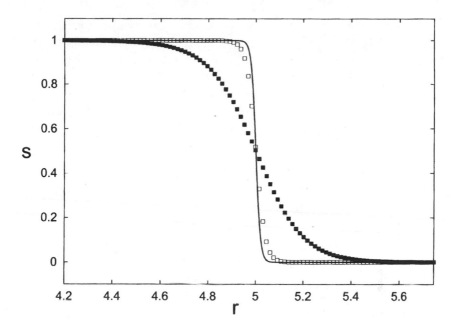

Figure 5.13: Switching function $s(x, y; R, \kappa)$ [see Eq. (5.84)] as a function of distance $r = \sqrt{x^2 + y^2}$ from the center of the coordinate system: ($\blacksquare$) $\kappa = 1$, ($\square$) $\kappa = 5$, (—) $\kappa = 10$.

this low symmetry of the fluid precludes translational invariance of its properties in any direction. Consequently a closed "mechanical" expression for Ω is not obtained because there is no Gibbs Duhem equation pertinent to the system depicted in Fig. 5.2.

However, we also notice from Eq. (5.83) that in the absence of the switching function, the remaining fluid substrate potential would depend only on the z-coordinate of a fluid molecule. In this case, fluid properties would be translationally invariant in the x- and y-directions. Hence, in this case, the exact differential of the grand potential would be given by Eq. (1.63), where this higher symmetry of the confined fluid has already been exploited. As a consequence we obtain a closed expression for the grand potential [see Eq. (1.65)] in terms of the transverse stress $\tau_\parallel$ as we show in Section 1.6.1.

The discussion at the beginning of this section therefore suggests taking the fluid confined between undecorated surfaces as a reference system. For the reference system, we may derive a molecular expression for $\tau_{yy} = \tau_{xx} = \tau_\parallel$, where

$$\tau_{\alpha\alpha} = -\frac{\langle N \rangle k_B T}{A_{\alpha 0} s_\alpha} + \frac{\langle W_{\alpha\alpha} \rangle}{2 A_{\alpha 0} s_\alpha}, \qquad \alpha = x, y \qquad (5.85)$$

following the derivation for these stress tensor elements as outlined in Appendix E.3.3 because in the reference system the fluid substrate interaction potential does not depend on either the x- or the z-coordinate of a fluid molecule. Equation (5.85) can easily be evaluated in a GCEMC simulation so that $\omega = \tau_\parallel$ can also be obtained for the reference system.

Following the general philosophy of perturbation theory, we replace the fluid substrate potential for the system of interest by

$$\Phi^{[k]}(\boldsymbol{r}; \lambda) = \varphi_{\text{rep}}^{[k]}(z) - \lambda \phi_{\text{att}}^{[k]}(\boldsymbol{r}) \tag{5.86}$$

where $\lambda \in [0, 1]$ is a perturbation parameter and the contributions $\varphi_{\text{rep}}^{[k]}(z)$ and $\phi_{\text{att}}^{[k]}(\boldsymbol{r})$ to the fluid substrate potential are introduced in Eq. (5.83). Clearly, if $\lambda = 0$, we are dealing with the unperturbed reference system in which the solid substrates are chemically homogeneous and nonwettable due to the absence of attractive contributions to the fluid substrate potential. For $\lambda = 1$, on the other hand, we are concerned with the situation of interest shown in Fig. 5.12. Because λ is assumed to be continuous, we can continuously switch between the reference system and the system of interest by increasing λ from 0 to its maximum value of 1.

What is the effect of introducing the perturbation parameter λ from a molecular perspective? The reader will immediately realize that the configurational energy

$$U\left(\boldsymbol{r}^N; \lambda\right) = \frac{1}{2} \sum_{i=1}^{N} \sum_{j \neq i=1}^{N} u_{\text{ff}}\left(r_{ij}\right) + \sum_{k=1}^{2} \sum_{i=1}^{N} \left[\varphi_{\text{rep}}^{[k]}(z_i) - \lambda \phi_{\text{att}}^{[k]}(\boldsymbol{r}_i)\right] \tag{5.87}$$

becomes a function of λ as well as

$$\Omega(\lambda) = -k_{\text{B}}T \ln \Xi_{\text{cl}}(\lambda) = -k_{\text{B}}T \ln \sum_{N=0}^{\infty} \frac{1}{N! \Lambda^{3N}} \exp\left(\frac{\mu N}{k_{\text{B}}T}\right) Z(\lambda) \tag{5.88}$$

This is apparent from Eq. (2.120), the definition of $U\left(\boldsymbol{r}^N; \lambda\right)$ in Eq. (5.87), and that of the configuration integral in Eq. (2.112). From Eqs. (2.112), (5.88), and (5.87) it is also evident that, for fixed values of the set of natural variables $\{T, \mu, s_x, s_y, s_z, \alpha s_{x0}\}$, Ω is solely a function of the perturbation parameter so that we may write

$$\frac{\mathrm{d}\Omega(\lambda)}{\mathrm{d}\lambda} = -\frac{1}{\Xi_{\text{cl}}(\lambda)} \sum_{N=0}^{\infty} \frac{1}{N! \Lambda^{3N}} \exp\left(\frac{\mu N}{k_{\text{B}}T}\right) \int \mathrm{d}\boldsymbol{r}^N \left(\sum_{k=1}^{2} \sum_{i=1}^{N} \phi_{\text{att}}^{[k]}(\boldsymbol{r}_i)\right)$$

$$\times \exp\left[-\frac{U\left(\boldsymbol{r}^N; \lambda\right)}{k_{\text{B}}T}\right] \tag{5.89}$$

Introducing

$$\phi_{\text{att}}^{\text{tot}} \equiv \sum_{k=1}^{2} \sum_{i=1}^{N} \phi_{\text{att}}^{[k]} \left(\boldsymbol{r}_i \right) \tag{5.90}$$

we may rewrite the previous expression as

$$\frac{\mathrm{d}\Omega \left(\lambda \right)}{\mathrm{d}\lambda} = - \left\langle \phi_{\text{att}}^{\text{tot}} \right\rangle_\lambda \tag{5.91}$$

where the right side can be obtained as an ensemble average in a GCEMC simulation but depends on λ because $\Xi \left(\lambda \right)$ as well as $U \left(\boldsymbol{r}^N ; \lambda \right)$ both are functions of that parameter. If we integrate this latter expression, we finally arrive at

$$\omega \left(\lambda \right) = \tau_\parallel - \frac{1}{V_0} \int_0^\lambda \mathrm{d}\lambda' \left\langle \phi_{\text{att}}^{\text{tot}} \right\rangle_{\lambda'} , \qquad T, \mu, s_{\text{x}}, s_{\text{y}}, s_{\text{z}}, \alpha s_{\text{x}0} = \text{const} \tag{5.92}$$

according to the foregoing discussion of the limit $\lambda = 0$. This approach, usually referred to as "λ-expansion," has been employed frequently as a suitable starting point in thermodynamic perturbation theories [183] (see also Ref. 30). In Eq. (5.92) we assume that the confined fluid lamella is not strained during the process of "switching on" the circular pattern on the substrate so that $V_0 = s_{\text{x}0} s_{\text{y}0} s_{\text{z}0}$.

From the above derivation it is also clear that ω is still a function of the set $\{T, \mu, s_{\text{x}}, s_{\text{y}}, s_{\text{z}}, \alpha s_{\text{x}0}\}$ regardless of the value of λ. This fact may be exploited to devise two additional routes on which ω can be calculated from GCEMC data. For example, differentiating the expression for $\Omega \left(\lambda \right)$ with respect to μ one finds [cf. Eq. (2.71)]

$$\omega \left(\mu_2 \right) - \omega \left(\mu_2 \right) - \int_{\mu_1}^{\mu_2} \mathrm{d}\mu \, \overline{\rho} \left(\mu \right) , \qquad T, \lambda, s_{\text{x}}, s_{\text{y}}, s_{\text{z}}, \alpha s_{\text{x}0} = \text{const} \tag{5.93}$$

where $\overline{\rho} \left(\mu \right) = \left\langle N \left(\mu \right) \right\rangle / V_0$. In addition we find from Eq. (1.66) that

$$\frac{1}{A_{\text{z}0}} \left(\frac{\partial \Omega}{\partial s_{\text{z}}} \right)_{\{\cdot\} \backslash s_{\text{z}}} = \tau_{\text{zz}} \tag{5.94}$$

Defining $\omega = \Omega / A_{\text{z}0} s_{\text{z}} = \Omega / V$, we may formally integrate the last expression to obtain

$$s_{\text{z}2} \omega \left(s_{\text{z}2} \right) = s_{\text{z}1} \omega \left(s_{\text{z}1} \right) + \int_{s_{\text{z}1}}^{s_{\text{z}2}} \mathrm{d}s_{\text{z}} \, \tau_{\text{zz}} \left(s_{\text{z}} \right) , \qquad T, \mu, s_{\text{x}}, s_{\text{y}}, \lambda, \alpha s_{\text{x}0} = \text{const}$$

$$\tag{5.95}$$

which can be evaluated using molecular expressions for the stress tensor derived in Section 5.3.3. Because of the specific form of the fluid substrate potential, we have

$$F_{z,i}^{[k]}(\lambda) = -\frac{\partial \Phi^{[k]}}{\partial z_i} = \left[10\varphi_{\mathrm{rep}}^{[k]}(z_i) - 4\lambda\phi_{\mathrm{att}}^{[k]}(\boldsymbol{r}_i) \right] \frac{1}{z_i \pm s_z/2} \qquad (5.96)$$

so that the stress tensor element

$$\tau_{zz}(\lambda) = \tau_{zz}^{\mathrm{id}} + \tau_{zz}^{\mathrm{FF}} + \tau_{zz}^{\mathrm{FS}}(\lambda) \qquad (5.97)$$

where τ_{zz}^{id} and τ_{zz}^{FF} are given in Eqs. (E.33) and (E.40a), respectively, and

$$\tau_{zz}^{\mathrm{FS}}(\lambda) = \sum_{k=1}^{2} \sum_{i=1}^{N} \left[10\varphi_{\mathrm{rep}}^{[k]}(z_i) - 4\lambda\phi_{\mathrm{att}}^{[k]}(\boldsymbol{r}_i) \right] \qquad (5.98)$$

and an alternative "force" expression follows from Eqs. (5.96) and Eq. (E.46).

5.5.3 Computation of the grand potential

5.5.3.1 Chemically homogeneous substrate

As a test of Eq. (5.92) we apply the procedure outlined in Section 5.5.2 to a fluid confined to a slit-pore with infinitesimally smooth, chemically homogeneous substrate surfaces. These can be realized by replacing in Eq. (5.83) $s(x, y; R, \kappa) \equiv 1$ so that $\phi_{\mathrm{att}}^{[k]}(z_i)$ in Eq. (5.87) is replaced by $\varphi_{\mathrm{att}}^{[k]}(z_i)$. According to the discussion in Section 5.5.2 we may then also replace Eq. (5.92) by

$$\tau_{\|}(\lambda) = \tau_{\|}(0) - \frac{1}{V_0} \int_0^{\lambda} d\lambda' \left\langle \phi_{\mathrm{att}}^{\mathrm{tot}} \right\rangle_{\lambda'} \qquad (5.99)$$

such that $\Delta\tau_{\|}(\lambda) \equiv \tau_{\|}(\lambda) - \tau_{\|}(0)$ can be calculated either directly from Eq. (5.85) or by numerically integrating $\langle \phi_{\mathrm{att}}^{\mathrm{tot}} \rangle_{\lambda}$.

Plots of $\Delta\tau_{\|}(\lambda)$ are shown in Fig. 5.14 for representative gas ($\mu = -7.8$) and liquid phases ($\mu = -7.0$) at a temperature $T = 0.65$ and $s_{z0} = 12$. Although low, this temperature is still expected to exceed the triple-point temperature in the bulk according to Fan and Monson [99]. The plots show that $\Delta\tau_{\|}(\lambda)$ is a monotonically decreasing function of λ regardless of the thermodynamic state considered. For the gas, the dependence on λ is largest in the vicinity of $\lambda \simeq 1$ [see Fig. 5.14(a)]. One also notices that the statistical uncertainty is somewhat larger for liquid than for gas states. This is not surprising in view of typical liquid and gas densities listed in Table 5.3

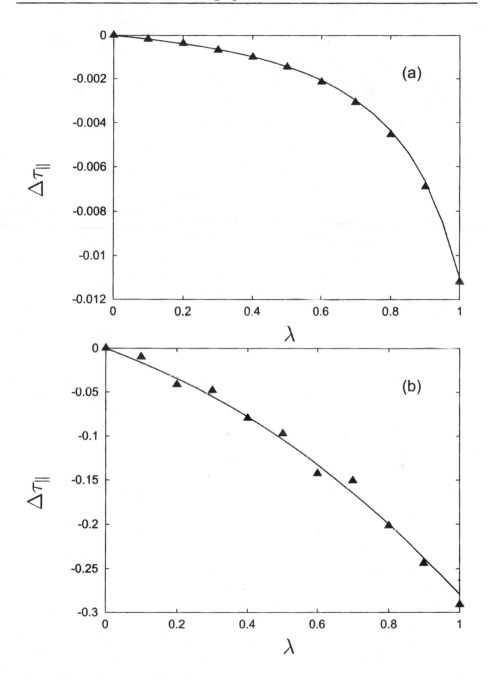

Figure 5.14: The incremental shear stress $\Delta\tau_{\parallel}$ as a function of the perturbation parameter λ. (▲) from Eq. (5.85), (—) from Eq. (5.99); (a) gas, (b) liquid (see text).

Table 5.3: Acceptance ratio and mean density for selected gas and liquid states of a fluid confined between homogeneous substrates (see Fig. 5.14).

State	λ	Acceptance ratio	$\bar{\rho}$
liquid	0.0	$3.790 \cdot 10^{-3}$	$6.667 \cdot 10^{-1}$
liquid	1.0	$2.010 \cdot 10^{-4}$	$7.567 \cdot 10^{-1}$
gas	0.0	$3.836 \cdot 10^{-1}$	$7.000 \cdot 10^{-3}$
gas	1.0	$1.690 \cdot 10^{-1}$	$3.400 \cdot 10^{-2}$

and the associated acceptance probabilities. Acceptance probabilities for gaseous states exceed those for liquid-like states by approximately two orders of magnitude. For the same number of GCEMC cycles one therefore expects configuration space to be sampled more efficiently for gas than for liquid states. However, it is noteworthy that even for the highest-density state listed in Table 5.3 the acceptance probability exceeds, by more than a factor of two, the threshold value of approximately 10^{-4} at which point the current GCEMC algorithm becomes prohibitively inefficient.

5.5.3.2   Chemically heterogeneous substrate

Consider now the chemically decorated substrates where the fluid substrate interaction potential is given by Eqs. (5.81)–(5.84). Over the temperature range $0.60 \leq T \leq 0.75$ to which this study is restricted, the confined fluid may form three distinctly different morphologies characterized by the local density defined here as

$$\rho(r, z) = \frac{\langle N(r, z) \rangle}{2\pi r \delta r \delta z} \tag{5.100}$$

where $N(r, z)$ is the number of fluid molecules in an annulus of width δr and thickness δz located at a distance $r = \sqrt{x^2 + y^2}$ and z from the center of the coordinate system and the substrate surface, respectively. Because the system is symmetric with respect to the plane $z = 0$, we enhance the statistical accuracy of the histogram representing $\rho(r, z)$ [see Eq. (5.100)] by averaging over spatially equivalent points in the upper ($z > 0$) and lower ($z < 0$) half of our system. From the plots in Fig. 5.15(a), it is evident that for $T = 0.75$ and $\mu = -8.36$ a small portion of fluid is adsorbed by the circular area on each substrate; the remainder of the system volume is occupied by low-density gas. The portion of fluid in the immediate vicinity of the substrates is stratified as indicated by the non-monotonic decay of $\rho(r, z)$ with decreasing $|z|$ along lines of constant r. Morphologies similar to

the one depicted in Fig. 5.15(a) are therefore referred to as "gas." Notice
also that the statistical error increases substantially as $r \to 0$ because the
area of the ring segment [and therefore $N(r, z)$] goes to zero in that limit.

If, on the other hand, $\mu = -8.30$, the entire volume is occupied by fluid at
a much higher density compared with the gas morphology [see Fig. 5.15(c)].
The plot in Fig. 5.15(c) shows that $\rho(r, z)$ decays more or less monotonically
as $|z| \to s_z/2$ and $r > R$ (i.e., outside the circular attractive region). This
is characteristic of repulsive substrates, which are not wet by the confined
fluid. The morphology illustrated by the plot in Fig. 5.15(c) is representative
of what we shall call "liquid." From the plot in Fig. 5.15(c), it is furthermore
evident that the liquid is not only stratified in the direction *perpendicular*
to the substrate (i.e., along the z-axis and lines $r = $ const), but also in the
contact layer (i.e., the one closest to either substrate) as one moves out of the
attractive circular region in radial direction, which is with increasing distance
r from the center of that region. The separation between successive maxima
in $\rho(r, z)$ as r increases in the contact layer is approximately 1.

It also seems worthwhile pointing out that these radial oscillations of the
local density in the contact layer become more pronounced with increasing
r as one approaches $r = R$, which is the boundary separating the circular
attractive from the purely repulsive part of the substrate. Thus, the fluid is
more ordered along the circumference compared with portions controlled by
"inner" parts of the circular area. Near the center of the circular attractive
region (i.e., for $r \simeq 0$), fluid order has nearly vanished as reflected by $\rho(r, z)$
that is nearly independent of r in this regime of the contact layer. However,
in the z-direction (along lines of constant $r \leq R$), the separation between
successive maxima of $\rho(r, z)$ is also approximately 1, indicating stratification as it would be expected in fluids confined among planar, chemically
homogeneous, and attractive solid substrates (see, for example, Ref. 5).

This packing effect is due solely to the geometry of the chemical decoration of the substrate and has not been observed for other geometries, which is
for, say, alternating strip-like domains composed of different solid materials
(see Figs. 5.8).

For intermediate chemical potentials, the confined fluid may condense
only partly in a subvolume $\widetilde{V} \simeq \{(r, z) \,|\, 0 \leq r \lesssim R, -s_z/2 < z < s_z/2\}$ "con-
trolled" approximately by the circular attractive area with which the sub-
strates are decorated. This is illustrated by the plot of $\rho(r, z)$ in Fig. 5.15(b)
where high(er)-density fluid is spanning the gap between the circular attractive regions on the opposite substrates. Therefore, this morphology will be
referred to as "bridge." Notice that in the contact layers the fluid is stratified
in radial directions similar to the liquid [see Fig. 5.15(c)]. However, here the
radial ordering of fluid molecules is less pronounced compared with the liquid

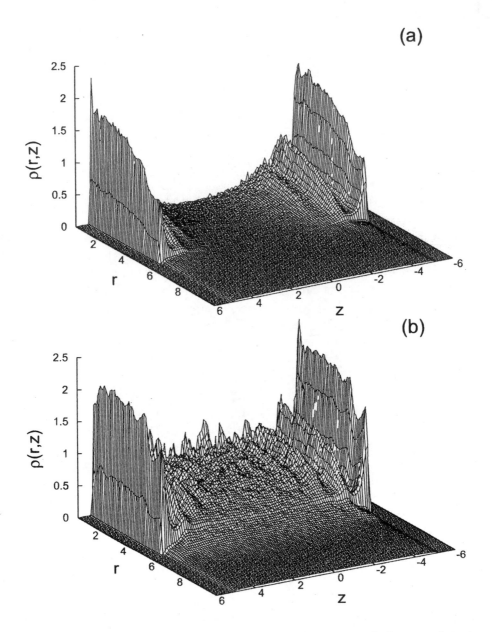

Figure 5.15: Local density $\rho(r, z)$ as function of position relative to substrate plane (z) and distance r from center of attractive circular nanopattern (see Fig. 5.12): (a) gas ($\mu = -8.36$), (b) bridge ($\mu = -8.33$), (c) liquid ($\mu = -8.30$). In all cases $T = 0.75$, $R = 5$, $s_z = 12$, and $s_x = 20$.

(c)

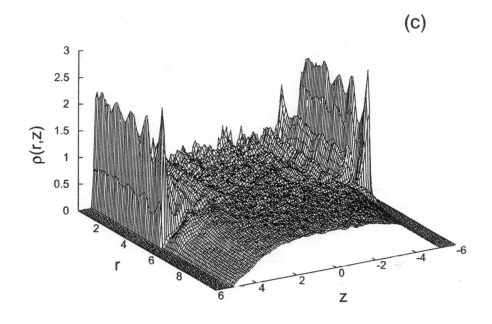

Figure 5.15: Continued.

state.

For gas and liquid states, $\omega(\mu)$ can be calculated in GCEMC simulations by thermodynamic integration employing Eqs. (5.92) and (5.93) for fixed T and $s_z = 12$. In general, ω is a monotonically decreasing function of μ [see Eq. (1.78)]. Because of Eq. (1.78) and Fig. 5.15, we expect different slopes for $\omega(\mu)$ depending on the morphology in question (e.g., gas, bridge, or liquid), that is, fluids characterized by distinctly different local densities (and therefore manifestly different $\bar{\rho}$'s). For sufficiently low $\mu_1 = -7.60$ ($T = 0.63$) and $\lambda = 0$, a gas forms between the purely repulsive substrate surfaces. This chemical potential is sufficiently low to guarantee that as $\lambda \to 1$ the original gas is not subject to any discontinuous phase transition [see Fig. 5.16(a)]. For $\lambda = 1$ the interaction between fluid molecules and the decorated substrate has been fully "switched on." Thermodynamic integration then proceeds by raising the chemical potential and employing Eq. (5.93) along the remainder of the integration path as before. Only points along this latter path are shown in Fig. 5.16(a).

For a corresponding liquid state one begins with $\lambda = 0$ and a sufficiently high $\mu_1 = -7.15$ such that a liquid morphology is stable even though the substrates are not wet (because they are purely repulsive) [see Eq. (5.87)]. Once $\lambda = 1$ the chemical potential is now *lowered* and ω is again calculated by

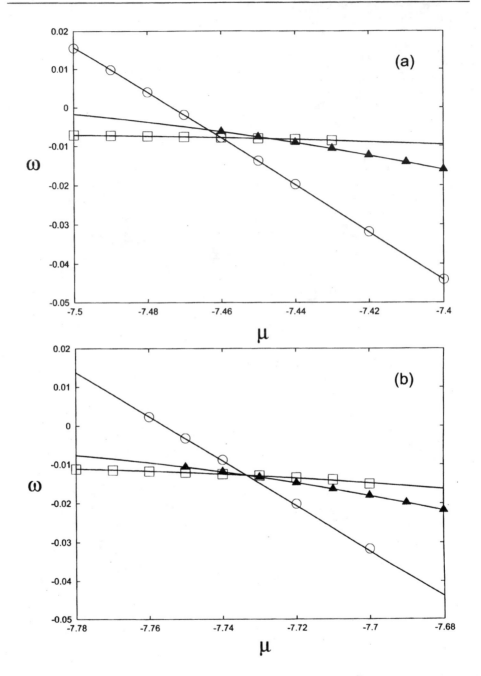

Figure 5.16: Grand-potential density ω as function of chemical potential μ for gas ($\square$), bridge ($\blacktriangle$), and liquid morphologies ($\bigcirc$): (a) $T = 0.63$, (b) $T = 0.67$, (c) $T = 0.75$. Solid lines are fits to simulation data intended to guide the eye.

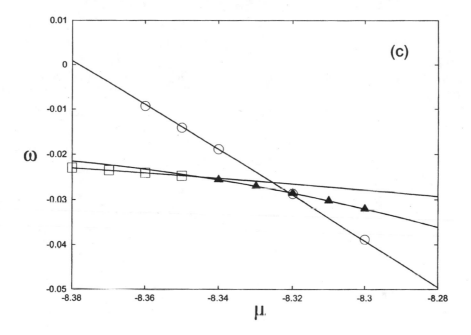

Figure 5.16: Continued.

thermodynamic integration employing Eq. (5.93). Again only points along the second integration path are shown in Fig. 5.16(a).

Plots in Fig. 5.16 also indicate that, over certain ranges of μ, $\omega(\mu)$ is a multivalued function where the lowest value of ω obviously corresponds to the thermodynamically stable morphology (i.e., phase); the others are only metastable. Metastability ends (i.e., the confined fluid becomes unstable) if the inequality in Eq. (1.82) can no longer be satisfied. The reader should realize that in general metastability in MC simulations is an artifact caused by the limited system size and insufficient length of the Markov chain (i.e., the finite computer time available) [184]. Metastability would not be observed in an infinite system where the evolution of the system could be followed indefinitely. In other words, metastability vanishes in the thermodynamic limit.

Here the situation is slightly more delicate. Because the nanopatterns are finite in extent by definition, there is no way of increasing the system size without altering the physical conditions of the confined fluid. Hence, in a sense, metastability here is "real" and associated with the (physically meaningful) small size of the fluid bridge. However, this also raises the question of whether the morphologies triggered by finite-size chemical patterns should be regarded as thermodynamic phases in the strict sense. Nevertheless, these morphologies are characterized by distinctly different grand-potential densities as plots in Fig. 5.16 clearly show. Therefore, the notion of a "thermodynamic phase" does not seem to be totally nonsensical even in light of the above remarks.

Suppose now two morphologies α and β exist with associated grand-potential densities ω^α and ω^β. Under isothermal conditions (i.e., with T, s_x, s_y, and s_z fixed) and according to the above rationale Eq. (1.76a) may have a solution $\mu_x^{\alpha\beta} \equiv \mu^{\alpha\beta}$ at which the *morphologies* α and β correspond to coexisting phases. If a third morphology γ exists, one may have three solutions $\mu^{\alpha\beta}$, $\mu^{\beta\gamma}$, and $\mu^{\gamma\alpha}$ from equations analogous to Eq. (1.76a) involving pairs of these morphologies. Suppose the mean densities associated with the morphologies satisfy the inequality

$$\overline{\rho}^\alpha < \overline{\rho}^\gamma < \overline{\rho}^\beta \tag{5.101}$$

irrespective of μ. Because of Eq. (1.82) this implies

$$\left(\frac{\partial \omega^\alpha}{\partial \mu}\right)_{\{\cdot\}\backslash\mu} > \left(\frac{\partial \omega^\gamma}{\partial \mu}\right)_{\{\cdot\}\backslash\mu} > \left(\frac{\partial \omega^\beta}{\partial \mu}\right)_{\{\cdot\}\backslash\mu} \tag{5.102}$$

Denoting by $\omega^{\alpha\beta}$ the value of the grand-potential density at $\mu^{\alpha\beta}$, three different scenarios are discernible as one can verify geometrically:

1. $\omega^{\gamma\beta} > \omega^{\alpha\beta} > \omega^{\alpha\gamma}$
   $\mu^{\gamma\beta} < \mu^{\alpha\beta} < \mu^{\alpha\gamma}$
   In this case only morphologies α and β comport with thermodynamically coexisting phases at $\mu^{\alpha\beta} \equiv \mu_x^{\alpha\beta}$; at $\mu^{\gamma\beta}$ and $\mu^{\alpha\gamma}$ morphologies γ, β and α, γ are only metastable.

2. $\omega^{\gamma\beta} = \omega^{\alpha\beta} = \omega^{\alpha\gamma}$
   $\mu^{\gamma\beta} = \mu^{\alpha\beta} = \mu^{\alpha\gamma}$
   The three intersections coincide at the given temperature thereby defining a triple point $\{T_{tr}, \mu_{tr}\}$ at which all three morphologies are thermodynamically stable.

3. $\omega^{\gamma\beta} < \omega^{\alpha\beta} < \omega^{\alpha\gamma}$
   $\mu^{\gamma\beta} > \mu^{\alpha\beta} > \mu^{\alpha\gamma}$
   This describes a situation in which two pairs of separately coexisting morphologies are thermodynamically stable phases, namely γ and β at $\mu^{\gamma\beta} \equiv \mu_x^{\gamma\beta}$ and α and γ at $\mu^{\alpha\gamma} \equiv \mu_x^{\alpha\gamma}$; at $\mu^{\alpha\beta}$ morphologies α and β are only metastable.

According to this logic, plots of ω *versus* μ in Fig. 5.16(a) indicate that for $T = 0.63$ gas morphologies are thermodynamically stable over the range $-\infty < \mu \lesssim -7.46$, whereas liquid morphologies are thermodynamically stable over the range $-7.46 \lesssim \mu \leq \mu_x^{ls}$, keeping in mind the general possibility of solidification of the confined phase at a sufficiently high chemical potential μ_x^{sl} where liquid and solid phases may coexist. At the intersection $\mu_x^{gl} \simeq -7.46$ gas and liquid morphologies coexist. Figure 5.16(a), amended by a parallel calculation of $\rho(r, z)$, also shows that bridge morphologies form as metastable phases over the range of chemical potentials where this morphology may exist subject to thermodynamic consistency as spelled out in Eq. (1.82). Apparently, this situation resembles scenario 1 above.

Calculating $\omega(\mu)$ for bridge morphologies is significantly more demanding in terms of the thermodynamic integration procedure. To avoid a discontinuous phase transition during the initial stage where the substrate attraction is "turned on" [i.e., as $\lambda \to 1$, see Eq. (5.92)], one needs to start from small values $s_z = 3 - 3.5$, which are too small for any discontinuous transition to occur [185]. Thus, as one increases λ from 0 to 1, more and more molecules *gradually* assemble in the vicinity of the increasingly attractive circular regions between the substrates. Once $\lambda = 1$, an additional integration must be done to carry the substrate separation to the desired value $s_z = 12$. Along this path ω is calculated via Eq. (5.95). Because τ_{zz} depends non-monotonically on s_z, as the plot in Fig. 5.3 shows, this curve needs to be known with high resolution for Eq. (5.95) to provide sufficiently accurate results. Here we calculate $\tau_{zz}(s_z)$ in steps of $\Delta s_z = 0.1$. Once the substrate separation has reached $s_z = 12$, the remaining integration proceeds as discussed above for liquid and gas.

5.6 Rheological properties of confined fluids

In the preceding two sections we demonstrated that confined fluids are highly inhomogeneous on account of the external field represented by the confining substrates. This is because the external field adds a new relevant length scale to the system competing with the characteristic length of fluid fluid intermolecular interactions. As a result confined fluids appear generally to be stratified, at least to some extent, which manifests itself in a characteristic oscillatory dependence of the solvation force with respect to a variation of substrate separation. If the substrates themselves are structured either chemically, as in the example discussed in some detail in Section 5.4, or geometrically the external field may depend on more than just one (Cartesian) coordinate. In these cases, confinement may give rise to new thermodynamic phases that have no counterpart in the bulk. An example is the fluid bridge to which we devoted considerable attention in Section 5.4.2.

An equally remarkable feature to which we shall turn now is the fact that confined fluids may sustain a certain shear stress without exhibiting structural features normally pertaining to solid-like phases; that is, they do not necessarily assume any long-range periodic order. We tacitly assumed this from the very beginning of this book in our development of a thermodynamic description of confined fluids, which closely resembles that appropriate for solid-like bulk phases (see Section 1) [12].

In addition we pointed out in Section 5.3.1 that the shear deformation can be measured experimentally in one mode of operation of the SFA. Hence, this section will be devoted to an analysis of these experiments in the framework of various computer simulation approaches.

5.6.1 The quasistatic approach

Many attempts have been made to elucidate details of the behavior of confined fluids under shear using theory. The approaches can be grouped into two different categories, which may be labeled "dynamical" [186–192] and "quasistatic" [122, 171–173, 193–195]. In the dynamical approaches a stationary nonequilibrium state is created either by applying an external driving force [186] or by explicitly moving a substrate [187, 189–192] in nonequilibrium molecular dynamics (NEMD) simulations in order to mimic dynamical aspects of a corresponding SFA experiment directly on a molecular scale. However, the relationship between NEMD simulations [187, 189–192] and SFA experiments remains elusive for a number of reasons.

First, to describe the motion of the substrate on a physical time scale, an equation of motion needs to be solved that inevitably involves the substrate mass. However, there are no physical criteria on which the choice of a specific value for this mass could be based. Second, even though the substrate is a macroscopic object in the SFA experiment, its mass cannot be too much larger than the mass of a film molecule in the NEMD simulations because otherwise the wall would remain at rest on the time scale on which film molecules move. In fact, the ratio of the mass of a single film molecule to that of the entire wall is sometimes as small as $1/8$ [191, 192] so that one can expect relaxation phenomena in the film to depend sensibly (and therefore unphysically from an experimental perspective) on this arbitrarily selected wall mass [170]. Third, the speed at which the walls are slid in the SFA experiment is typically of the order of $10^{-9} - 10^{-7} \text{Åps}^{-1}$ [136] so that under realistic conditions the walls remain practically stationary on a typical length and time scale of molecular relaxation processes.

To avoid these problems and in view of the characteristic low shear rates in the actual SFA experiments, we employ a "quasistatic" or reversible approach in which the thermodynamic state of the film passes through a succession of equilibrium states (see Section 3.3 in Ref. 196), each being distinguished by a different (average) lateral alignment of the walls [122, 170–173, 193]. Equilibrium properties of the film can be computed within the framework of MC simulations designed to capture key characteristics of a corresponding SFA experiment to a maximum degree.

5.6.2   Molecular expression for the shear stress

Because of the chemical decoration of each substrate, a confined fluid can be exposed to a shear strain by misaligning the substrates in the $+x$-direction according to

$$\overline{x} \equiv \begin{cases} x + \alpha s_{x0}/2, & k = 1 \\ x - \alpha s_{x0}/2, & k = 2 \end{cases} \tag{5.103}$$

where $\alpha \equiv \delta_\alpha/s_x$ is a dimensionless number and δ_α is the magnitude of the relative displacement of the substrates with respect to each other where $\{\alpha \,|\, 0 \leq \alpha \leq \frac{1}{2}\}$ may vary continuously between its limits. In this range $\alpha = 0$ refers to substrates in registry, whereas $\alpha = \frac{1}{2}$ if the substrates are out of registry. A key quantitative measure of the resistance of any confined fluid to an external shear strain is the shear stress τ_{xz}. As before for the compressional stress τ_{zz} it follows from Eqs. (E.54) and (E.57) that a "virial" expression for the shear stress can be derived here, too (see Appendix E.3.2.1). It may be cast as

$$\tau_{xz} = \tau_{xz}^{FF} + \tau_{xz}^{FF} \tag{5.104}$$

Alternatively, we may derive a "force" expression for τ_{xz} following the derivation presented in Appendix E.3.2.2 for the stress tensor component τ_{zz}. It follows if we combine Eqs. (E.62) and (E.63) with Eq. (E.49) from which we obtain

$$\tau_{xz} = -\frac{\langle \overline{F}_x \rangle}{2A_{x0}} \tag{5.105}$$

As before [see Eq. (5.65)] mechanical stability of the entire system requires

$$\langle F_x^{[1]} \rangle = -\langle F_x^{[2]} \rangle \tag{5.106}$$

which serves as a useful internal check on consistency of the GCEMC simulation together with Eqs. (5.104) and (5.105).

Another interesting quantity is the shear modulus c_{44} where we use Voigt's notation ([12], see p. 14 in Ref. 196). We reemphasize the appropriateness of this notation, originally devised for a thermodynamic description of solids, for the current case in which one is dealing with fluid phases that are nevertheless distinct from solids by their lack of any long-range order. The shear modulus is defined by

$$
c_{44} \equiv \frac{1}{A_{x0}} \left(\frac{\partial^2 \Omega}{\partial (\alpha s_{x0})^2} \right)_{\{\cdot\}\backslash \alpha s_{x0}} = \frac{1}{A_{x0}} \left(\frac{\partial \tau_{xz}}{\partial (\alpha s_{x0})} \right)_{\{\cdot\}\backslash \alpha s_{x0}}
\tag{5.107}
$$

A microscopic definition of c_{44} can be derived directly from Eq. (E.49); that is,

$$
\begin{aligned}
\left(\frac{\partial \tau_{xz}}{\partial (\alpha s_{x0})} \right)_{\{\cdot\}\backslash \alpha s_{x0}} &= -\frac{k_B T}{A_{z0}} \sum_{N=0}^{\infty} \exp \left(\frac{\mu N}{k_B T} \right) \frac{\partial}{\partial (\alpha s_{x0})} \left(\frac{1}{\Xi} \frac{\partial Z}{\partial (\alpha s_{x0})} \right) \\
&= \frac{k_B T}{A_{z0}} \left[\frac{1}{\Xi} \sum_{N=0}^{\infty} \exp \left(\frac{\mu N}{k_B T} \right) \frac{\partial Z}{\partial (\alpha s_{x0})} \right]^2 \\
&\quad - \frac{k_B T}{A_{z0} \Xi} \sum_{N=0}^{\infty} \exp \left(\frac{\mu N}{k_B T} \right) \frac{\partial^2 Z}{\partial (\alpha s_{x0})^2}
\end{aligned}
\tag{5.108}
$$

Focusing for convenience on the "force" treatment it is a simple matter to verify from Eq. (E.62) that

$$
\begin{aligned}
\frac{\partial^2 Z}{\partial (\alpha s_{x0})^2} &= -\frac{1}{k_B T} \prod_{i=1}^{N} \int \overset{\cdots}{\underset{\cdots}{\mathrm{d}y_i}} \int \overset{\cdots}{\underset{\cdots}{\mathrm{d}z_i}} \int \overset{\cdots}{\underset{\cdots}{\mathrm{d}x_i}} \frac{\partial^2 U_{FS}}{\partial (\alpha s_{x0})^2} \exp \left(-\frac{U}{k_B T} \right) \\
&\quad + \frac{1}{(k_B T)^2} \prod_{i=1}^{N} \int \overset{\cdots}{\underset{\cdots}{\mathrm{d}y_i}} \int \overset{\cdots}{\underset{\cdots}{\mathrm{d}z_i}} \int \overset{\cdots}{\underset{\cdots}{\mathrm{d}x_i}} \left(\frac{\partial U_{FS}}{\partial (\alpha s_{x0})} \right)^2 \\
&\quad \times \exp \left(-\frac{U}{k_B T} \right)
\end{aligned}
\tag{5.109}
$$

where the integration limits (represented symbolically by "...") are the same as in Eq. (E.62) and we employ the argument for differentiating Z presented above [see Eqs. (E.58)–(E.62)]. Thus, with the aid of Eqs. (E.62), (E.63), (5.107), and (5.109) it follows from Eq. (5.108) that

$$
c_{44} = \frac{1}{k_B T A_{x0}^2} \left[\langle \overline{F}_x \rangle^2 - \langle \overline{F}_x^2 \rangle \right] + \frac{1}{A_{x0}^2} \left\langle \frac{\partial^2 U_{FS}}{\partial (\alpha s_{x0})^2} \right\rangle
\tag{5.110}
$$

which shows that the shear modulus depends on fluctuations of the fluid substrate force and on the curvature of the fluid substrate contribution to the configurational potential energy with respect to the shear strain.

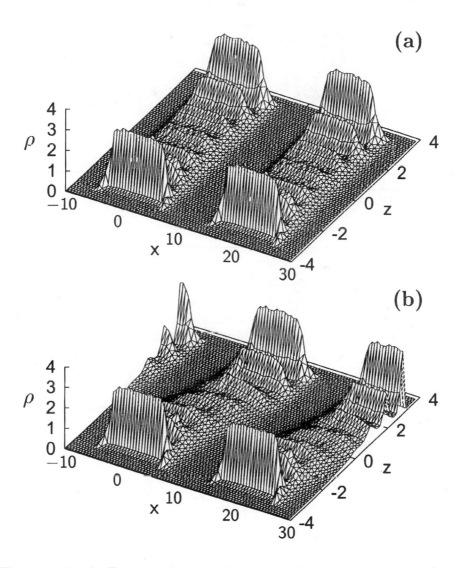

Figure 5.17: As Fig. 5.8, where the fluid bridge is unsheared in part (a) but exposed to a shear strain in part (b). To enhance the clarity of the presentation two periods in the x-direction are shown.

5.6.3 Fluid bridges exposed to a shear strain

As a first illustration we consider the model discussed in Section 1.3.3, namely a fluid of "simple" molecules confined between chemically striped solid surfaces (see Fig. 5.2). As before in Section 5.4 we treat the confined fluid as a thermodynamically open system. Hence, equilibrium states correspond to minima of the grand potential Ω introduced in Eqs. (1.66) and (1.67). The fluid fluid interaction is described by the intermolecular potential $u_{\text{ff}}(r)$ introduced in Eq. (5.38) where the associated shifted-force potential is introduced in Eq. (5.39). The fluid substrate interaction is described by $\Phi^{[k]}(\overline{x}, z)$ in the continuum representation [see Eq. (5.68)], where $\overline{x}$ replaces x because of the misalignment of the substrates relative to each other [see Eq. (5.103)].

From a morphological perspective, the confined fluid can exist either as a thin inhomogeneous film [see Fig. 5.8(a)], a high-density inhomogeneous liquid phase [see Fig. 5.8(b)], or a fluid bridge [see Fig. 5.8(c)]. As was already evident from the mean-field calculations described in Section 4.5.3, the bridge morphology is distinct from the other two in that it can be deformed in a direction parallel to the solid substrates without breaking apart instantaneously [see Figs. 4.11(a) and 4.11(b)]. Likewise, in the current GCEMC simulations the fluid bridge may be deformed by applying a shear strain in the x-direction [see Eq. 5.103)] as plots of the local density $\rho(x, z)$ in Fig. 5.17 clearly show.

As a quantitative measure of the extent to which a confined phase is capable of resisting a shear deformation, we introduce in Section 5.6.2 the shear stress τ_{xz}. For a fluid bridge a typical shear-stress curve $\tau_{\text{xz}}(\alpha s_{\text{x0}})$ is plotted in Fig. 5.18. Regardless of the thermodynamic state and the thickness (i.e., s_{z}) of a bridge phase, a typical stress curve exhibits the following features:

1. For vanishing shear strain (i.e., $\alpha s_{\text{x0}} = 0$), $\tau_{\text{xz}}(0) \equiv 0$ for symmetry reasons.

2. $\tau_{\text{xz}}(\alpha s_{\text{x0}})$ depends linearly on the shear strain αs_{x0} in the limit $\alpha \to 0$. That is to say, the response of the bridge phase to small shear strains follows Hooke's law.

3. For larger shear strains, negative deviations from Hooke's law are observed, eventually leading to a yield point $\left(\alpha^{\text{yd}}, \tau_{\text{xz}}^{\text{yd}}\right)$ defined by the constitutive equation

$$\left(\frac{\partial \tau_{\text{xz}}}{\partial (\alpha s_{\text{x0}})}\right)_{\{\cdot\}\backslash \alpha s_{\text{x0}}}\Bigg|_{\alpha = \alpha^{\text{yd}}} = 0 \qquad (5.111)$$

or, alternatively [see Eq. (5.110)],

$$c_{44}\left(\alpha^{yd}s_{x0}\right) = 0 \qquad (5.112)$$

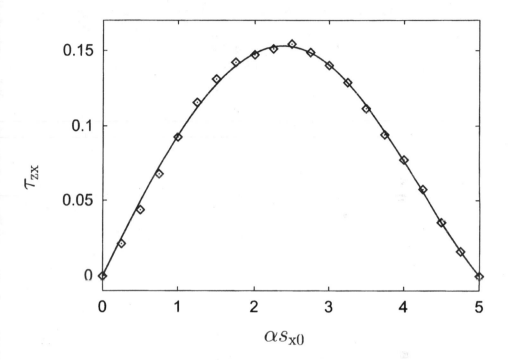

Figure 5.18: Typical stress curve $\tau_{xz}\left(\alpha s_{x0}\right)$ for a monolayer bridge phase and $c_r = \frac{5}{10}$. Solid line is a least-squares fit of a polynomial to the (discrete) MC data points ($\diamond$) intended to guide the eye.

As far as the current model is concerned, the degree of chemical corrugation of the substrate has significant consequences for the yield-point location $\left(\alpha^{yd}, \tau_{xz}^{yd}\right)$. Plots of stress curves for various values of c_r are shown in Fig. 5.19(a). For monolayer bridge phases and fixed $s_x = 10$ one can see from Fig. 5.19(a) that both τ_{xz}^{yd} and α^{yd} are smallest for the smallest $c_r = \frac{2}{10}$. For $c_r < \frac{2}{10}$ only gas phases are thermodynamically stable because the strongly attractive portion of the substrate is too small to support formation of denser (bridge) phases. As c_r increases both τ_{xz}^{yd} and α^{yd} increase until they reach their maximum values $\left(\alpha^{yd}s_x, \tau_{xz}^{yd}\right) \approx (2.740, 0.169)$ for $c_r = \frac{5}{10}$. For larger $c_r > \frac{5}{10}$ the plots in Fig. 5.19(a) show that both τ_{xz}^{yd} and α^{yd} decrease again until $\left(\alpha^{yd}s_x, \tau_{xz}^{yd}\right) \approx (1.550, 0.069)$ for $c_r = \frac{8}{10}$, which is the largest substrate corrugation for which bridge phases were observed. For $c_r > \frac{8}{10}$ only thermodynamically stable liquid phases formed in the simulations, incapable of sustaining a shear strain.

One also notices from Fig. 5.19(a) that stress curves for $c_r = \frac{2}{10}, \frac{3}{10}$, and $\frac{4}{10}$ apparently do not cover the entire range of shear strains. In these cases the strongly attractive portion of the substrates is too narrow to stabilize the denser portion of a bridge phase regardless of the applied shear strain. At some strain threshold $\alpha_c s_{x0}$, the bridge phase is simply "torn apart" and undergoes a first-order phase transition to an inhomogeneous film. This film, by virtue of its microscopic structure [see Fig. 5.8(c)], is incapable of sustaining a shear stress. Thus, at $\alpha_c s_{x0}$, τ_{xz} drops to zero discontinuously such that $\tau_{xz} = 0$ for all $\left\{\alpha \,\middle|\, \alpha_c \leq \alpha \leq \frac{1}{2}\right\}$. For the sake of clarity we do not plot this part of the stress curves in Fig. 5.19(a).

Despite this non-monotonic variation of the yield-point location with c_r it turns out that within the theory of corresponding states [197] it is feasible to renormalize stress curves such that all data points fall onto a unique master curve. Renormalization is effected by introducing dimensionless variables $\widetilde{\tau}_{zx} \equiv \tau_{zx} \left(\widetilde{\alpha} s_{x0}; c_r\right) / \tau_{xz}^{yd} \left(c_r\right)$ and $\widetilde{\alpha} \equiv \alpha / \alpha^{yd} \left(c_r\right)$. Normalization by α^{yd} and τ_{xz}^{yd} is consistent with the theory of corresponding states because it was pointed out in [198] that the yield point may be perceived as a shear critical point analogous to the liquid gas critical point in pure homogeneous fluids. If the simulation data plotted in Fig. 5.19(a) are renormalized according to this recipe, they can indeed be represented by a master curve as the plot in Fig. 5.19(b) shows.

The remarkable insensitivity of $\widetilde{\tau}_{zx} \left(\widetilde{\alpha} s_{x0}\right)$ to variations of c_r can be rationalized as follows. Because of the Hookean regime in the limit $\alpha s_x \to 0$, c_{44} should be approximately constant and positive in this limit. A typical plot in Fig. 5.20 confirms this notion. However, because of Eq. (5.112) one expects c_{44} to decline from its Hookean value as $\alpha s_x \to \alpha^{yd} s_{x0}$ also in agreement with Fig. 5.20. Furthermore, as Fig. 5.20 shows that the variation of c_{44} with αs_x is not too strong over the range $\left\{\alpha \,\middle|\, 0 \leq \alpha \leq \alpha^{yd}\right\}$, it seems sensible to expand c_{44} in a power series according to

$$c_{44} \left(\alpha s_{x0}\right) = \sum_{k=0}^{\infty} \frac{1}{k!} \left.\frac{d^{(k)} c_{44}}{d \left(\alpha s_{x0}\right)^k}\right|_{\alpha=0} \left(\alpha s_{x0}\right)^k = \sum_{k=0}^{\infty} a_k \left(\alpha s_{x0}\right)^k \simeq a_0 + a_2 \left(\alpha s_{x0}\right)^2$$

$$(5.113)$$

where we refer to the far right side as the small-strain approximation in the current context (cf., Section 1.2.1). Notice that the set of coefficients $\{a_k\}$ refer to the *unstrained* bridge phase (i.e., $\alpha = 0$). A molecular expression for $a_0 \equiv c_{44} \left(0\right)$ is given in Eq. (5.110). In the small-strain approximation a_2 accounts for deviations from Hookean behavior and may therefore be

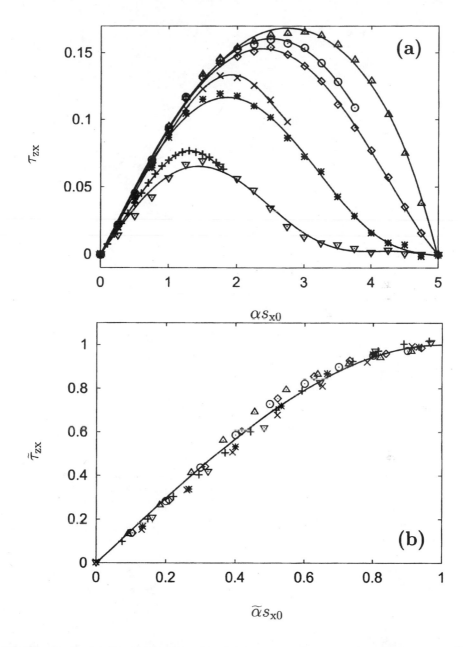

Figure 5.19: (a) Stress curve $\tau_{xz}(\alpha s_x)$ for various chemical corrugations $c_r = \frac{2}{10}$ (+), $\frac{3}{10}$ (×), $\frac{4}{10}$ (○), $\frac{5}{10}$ (△), $\frac{6}{10}$ (◇), $\frac{7}{10}$ (∗), $\frac{8}{10}$ (▽). Solid lines are intended to guide the eye. (b) Reduced stress curve $\widetilde{\tau}_{zx}(\widetilde{\alpha}s_{x0})$ [see Eq. (5.116)] where symbols are referring to data plotted in (a). The solid line is a representation of Eq. (5.117).

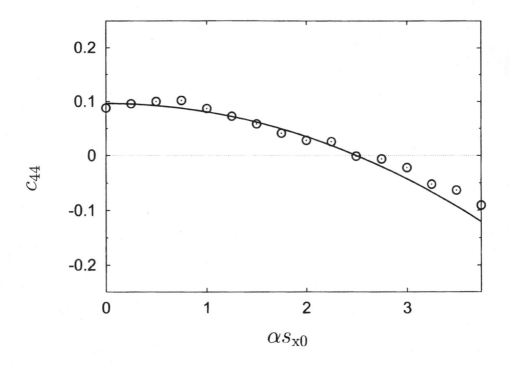

Figure 5.20: Shear modulus c_{44} as a function of shear strain αs_{x0}. ($\bigcirc$): MC simulations in grand mixed isostress isostrain ensemble; (—): representation of small-strain approximation $c_{44}(\alpha s_{x0}) = a_0 + a_2(\alpha s_{x0})^2$ [see Eqs. (5.113) and (5.114)].

interpreted as a measure of plasticity of the unsheared confined film. Moreover for $\alpha = 0$ symmetry requires $a_{2k-1} \equiv 0$ ($k = 1, \ldots, \infty$). However, we note in passing that these coefficients do not vanish a priori for $\alpha \neq 0$. From Eqs. (5.107) and (5.113) we obtain the (shear stress) equation of state

$$\tau_{xz}(\alpha s_{x0}) = \int\limits_0^{\alpha s_{x0}} d(\alpha s_x)' c_{44}\left[(\alpha s_{x0})'\right] \simeq a_0 \alpha s_{x0} + \frac{1}{3} a_2 (\alpha s_{x0})^3 \qquad (5.114)$$

based on the small-strain approximation. In principle, a_0 and a_2 are determined by ordinate and initial curvature of the function $c_{44}(\alpha s_{x0})$ ($\alpha \to 0$) (see Fig. 5.20). The latter is extremely difficult to extract given the typical accuracy with which the shear modulus can be calculated in our MC simulations (see Fig. 5.20). However, an accurate estimate is possible based on

Table 5.4: Comparison of shear modulus c_{44} from molecular expression and yield-point location (see text).

				Eq. (5.115a)	Eq. (5.110)
c_r	$\langle s_z \rangle$	$\alpha^{yd} s_{x0}$	τ^{yd}_{xz}	$c_{44}(0)$	$c_{44}(0)$
$\frac{2}{10}$	2.113	1.350	0.075	0.084	0.079
$\frac{4}{10}$	2.075	2.499	0.161	0.096	0.088
$\frac{4}{10}$	3.057	2.588	0.101	0.058	0.060
$\frac{5}{10}$	2.069	2.743	0.169	0.092	0.101
$\frac{6}{10}$	3.044	2.412	0.095	0.059	0.066

Eq. (5.111), which, together with Eq. (5.114), leads to

$$u_0 \equiv c_{44}(0) = \frac{3}{2} \frac{\tau^{yd}_{xz}}{\alpha^{yd} s_{x0}} \tag{5.115a}$$

$$a_2 \equiv \frac{1}{2} \frac{d^2 c_{44}(\alpha s_{x0})}{d(\alpha s_{x0})^2}\bigg|_{\alpha \to 0} = -\frac{3}{2} \frac{\tau^{yd}_{xz}}{(\alpha^{yd} s_{x0})^3} \tag{5.115b}$$

in terms of yield stress and strain. These latter quantities can be determined with high precision from Eqs. (5.105), (5.104), and plots similar to the ones shown in Figs. 5.18, 5.19(a). The validity of Eq. (5.115a) is illustrated by Table 5.4 where we compare it with the shear modulus obtained directly from the molecular expression Eq. (5.110) for a selection of unsheared bridge phases. Inserting now Eqs. (5.115a) and (5.115b) into the equation of state Eq. (5.114) (in the small-strain approximation) together with the transformations

$$\alpha \to \widetilde{\alpha} \equiv \frac{\alpha}{\alpha^{yd}} \tag{5.116a}$$

$$\tau_{xz} \to \widetilde{\tau}_{zx} \equiv \frac{\tau_{xz}}{\tau^{yd}_{xz}} \tag{5.116b}$$

permits us to recast Eq. (5.114) as

$$\widetilde{\tau}_{zx} = \frac{1}{2}[\widetilde{\alpha} s_{x0}(3 - \widetilde{\alpha} s_{x0})] \tag{5.117}$$

It is furthermore noteworthy that universality of stress curves, as defined here, is not restricted to monolayer fluids. Plots of $\widetilde{\tau}_{zx}$ *versus* $\widetilde{\alpha} s_x$ in

Fig. 5.21(b) show that simulation data for mono-, bi-, and trilayer bridge phases can also be mapped onto the master curve Eq. (5.117) according to the treatment detailed in this section. Again, the stress curves in Fig. 5.21(a) end at some $\alpha_c s_x$ because the bridge phases evaporate if they are strained beyond this limit.

5.6.4   Thermodynamic stability

From a fundamental point of view, bridge phases comprising different numbers of molecular strata may be viewed as different thermodynamic phases. This interpretation is evident if one considers the thermodynamic potential Φ defined in Eq. (2.71). Together with Eq. (1.43), we obtain

$$
d\Phi = -\mathcal{S}dT - Nd\mu + A_{x0}\tau_{xx}ds_x + A_{y0}\tau_{yy}ds_y - A_{x0}s_zd\tau_{zz} + A_{x0}\tau_{xz}d\left(\alpha s_{x0}\right)
$$
$$(5.118)$$

Applying the arguments detailed in Section (1.6) and in view of the fact that the fluid substrate potential does not depend on y, we immediately conclude from Eq. (5.118) that

$$
\phi \equiv \frac{\Phi}{A_{y0}s_y} = \tau_{yy} \tag{5.119}
$$

Because fluid bridges of different length in the z-direction will generally be characterized by different τ_{yy}, it is clear that these bridges will have different values of ϕ and must therefore be considered as legitimate thermodynamic phases. This, however, causes a complication because the thermodynamic state is uniquely specified by the set $\{T, \mu, s_x, s_y, \tau_{zz}, \alpha s_{x0}\}$. Because bridge phases of variable length in the z-direction may generally be compatible with the same fixed set of thermodynamic state variables, we apparently have a multiplicity of phases despite the fact that the thermodynamic state is uniquely specified. However, from an equilibrium perspective, only the morphology corresponding to the *global* minimum of ϕ is a thermodynamically stable phase; the others must be metastable.

Fortunately, only a small, finite number of possible morphologies can exist under the current thermodynamic constraints. This can be understood by considering the (normal) compressional stress τ_{zz} plotted as a function of substrate separation s_z in Fig. 5.22(a). Data plotted in Fig. 5.22 were obtained in GCEMC simulations in which a thermodynamic state is specified by a choice of natural variables similar to the ones determining Φ, replacing, however, τ_{zz} by its conjugate variable s_z [cf. Eqs. (1.66) and (5.118)]. Again, the plot in Fig. 5.22(a) shows that τ_{zz} is a damped oscillatory function of s_z. As we saw in Section 5.3.4, these oscillations are fingerprints of stratification, which corresponds to the formation of new fluid layers as

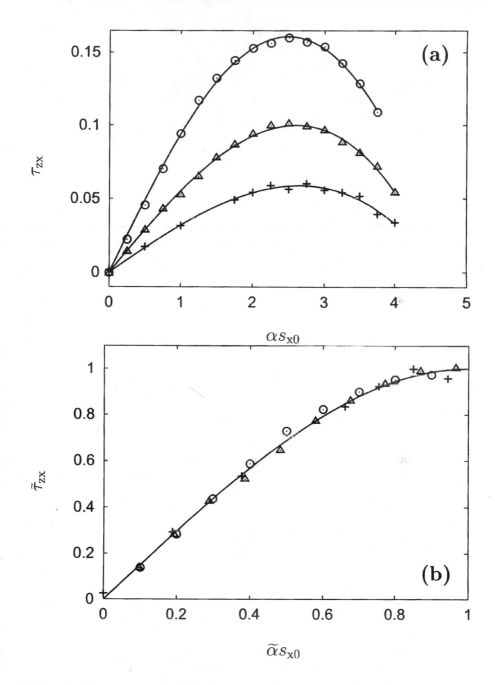

Figure 5.21: (a) As Fig. 5.19(a), but for mono- (○), bi- (△, and trilayer (+) morphologies and $c_r = \frac{4}{10}$. (b) As Fig. 5.19(b) but for data points plotted in (a).

the substrate separation increases at constant T and μ. Damping can be ascribed to the decreasing influence of the fluid substrate potential, which becomes negligible if s_z exceeds some critical value $s_{z,c}$. For $s_z \geq s_{z,c}$ one expects a homogeneous region to exist in the confined fluid. The homogeneous region is centered halfway between both substrates, increases in size with s_z, and its local density (which is independent of position) equals that of a corresponding bulk phase for the same T and μ. As a result

$$\lim_{s_z \to \infty} \tau_{zz}(s_z) = -P_b \qquad (5.120)$$

where $P_b(\mu, T) \simeq 0.03$ is the bulk pressure. In other words, because stratification diminishes with increasing s_z, oscillations in $\tau_{zz}(s_z)$ also vanish eventually [168]. Therefore, the plot in Fig. 5.22(a) shows that, under the current conditions, and for $s_z \geq 6.0$, stratification becomes subdominant.

In the grand mixed isostress isostrain ensemble, morphologies consistent with the set $\{T, \mu, s_x, s_y, \tau_{zz}, \alpha s_{x0}\}$ of state variables can now be identified with intersections between the oscillatory curve $\tau_{zz}(s_z)$ and the isobar $\tau_{zz} = \mathrm{const} \leq 0$. However, only intersections for which $\mathrm{d}\tau_{zz}/\mathrm{d}s_z \geq 0$ correspond to (thermodynamically or meta-) stable states as pointed out in Section 5.3.4 [see Eq. (5.66)]: intersections for which $\mathrm{d}\tau_{zz}/\mathrm{d}s_z < 0$ pertain to unstable states, which cannot be realized in MC simulations in the grand mixed isostress isostrain ensemble. The thermodynamically stable morphology corresponds to the intersection having the smallest ϕ ($\tau_{zz} = 0$) according to Eq. (5.119). Based on this rationale, an inspection of Fig. 5.22 shows that the thermodynamically stable, unstrained morphology ($\alpha = 0.0$) is a monolayer film with $s_z \simeq 2.1$ ($\tau_{zz} = 0.0$). If confined films are progressively sheared, a parallel analysis of plots in Fig. 5.23 and 5.24 shows that the minimum of ϕ for $s_z \simeq 2.1$ becomes shallower, whereas another minimum around $s_z \simeq 3.1$, corresponding to a bilayer film, becomes deeper with increasing shear strain. Eventually the depth of the latter minimum exceeds that associated with the monolayer film so that a bilayer film becomes the thermodynamically stable morphology. Thus, a shear strain exists such that ϕ is the same for mono- and bilayer films. At this shear strain both morphologies may therefore be viewed as coexisting phases in the classic sense (see Section 1.7).

To obtain a more concise picture of thermodynamic stability of different film morphologies, we plot ϕ as a function of αs_{x0} in Fig. 5.25 for the same system analyzed in Figs. 5.22–5.24. In a sequence of MC simulations in the grand mixed isostress isostrain ensemble, we calculate ϕ directly from Eq. (5.119) using the molecular expression for τ_{yy} [see Eq. (5.85)], which does not contain any fluid substrate contribution between the fluid substrate

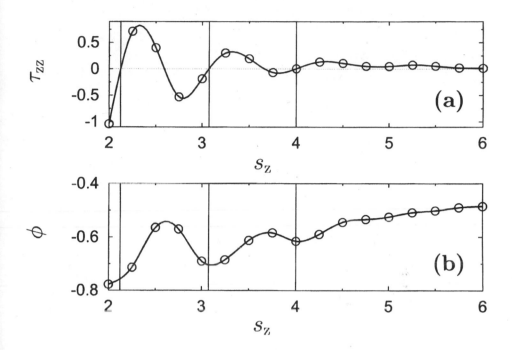

Figure 5.22: (a) Normal compressional stress τ_{zz} (see Appendix E.3 for molecular expressions) as a function of substrate separation from GCEMC simulations (○) ($\alpha s_{x0} = 0.0$). Solid lines are intended to guide the eye. (b) As (a) but for ϕ [see Eq. 5.119]. Intersections between the latter and the vertical lines demarcate (meta- or thermodynamically) stable states in the grand mixed isostress isostrain ensemble for $\tau_{zz} = 0.0$ (see text).

interaction potential as it does not depend on the y-position of fluid molecules. An alternative expression for $\phi(\alpha s_{x0})$ can be obtained by integrating Eq. (5.118)

$$\phi(\alpha s_{x0}) = \phi(0) + \int_0^{\alpha s_{x0}} d(\alpha s_{x0})' \tau_{xz}\left[(\alpha s_{x0})'\right], \quad \text{fixed } T, \mu, s_x, s_y, \tau_{zz}$$

$$\simeq \phi(0) + \frac{a_0}{2}(\alpha s_{x0})^2 + \frac{a_2}{12}(\alpha s_{x0})^4 \tag{5.121}$$

where the second line is based on the small strain approximation introduced in Eq. (5.113). Full lines in Fig. 5.25 are representations of Eq. (5.121). Solid lines plotted in Fig. 5.25(a) are therefore obtained without further adjusting a_0 and a_2; $\phi(0)$ is taken from MC simulations for unstrained bridge

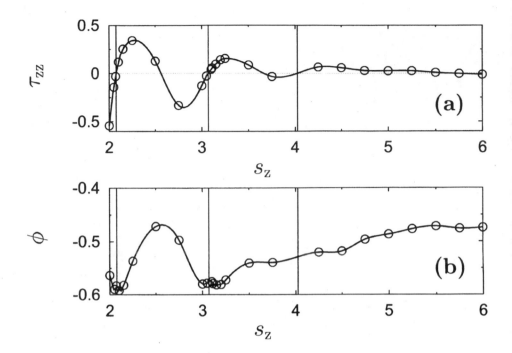

Figure 5.23: As Fig. 5.22, but for $\alpha s_{x0} = 2.25$.

phases. The excellent agreement between $\phi\,(\alpha s_{x0})$ from the MC simulations in the grand mixed isostress isostrain ensemble and the small strain approximation in Eq. (5.121) highlights once more the validity of the latter for all $\left\{\alpha\,|\alpha \leq \alpha^{yd}\right\}$. However, the plot in Fig. 5.25(a) also shows that the small strain approximation is doomed to fail for sufficiently large shear strains in accord with one's expectation.

From the plots in Fig. 5.25(a) one notices that ϕ is lowest for a monolayer bridge phase over the range $0.0 \leq \alpha s_{x0} \lesssim 2.2$, indicating that the monolayer is the thermodynamically stable morphology in this regime. Figure 5.25(a) also shows that intersections between the curves exist at which ϕ for a pair of different bridge morphologies assumes the same value. Thus, at the corresponding values αs_{x0}, these different morphologies coexist so that the intersections can be ascribed to first-order phase transitions between bridge phases comprising different numbers of molecular strata. Although there is no obvious relationship linking αs_{x0} at coexistence between mono- and bilayer morphologies $\alpha^{yd} s_{x0}$, we notice that for all cases investigated a monolayer film is the thermodynamically stable morphology for all $\left\{\alpha\,|\alpha \leq \alpha^{yd}\right\}$

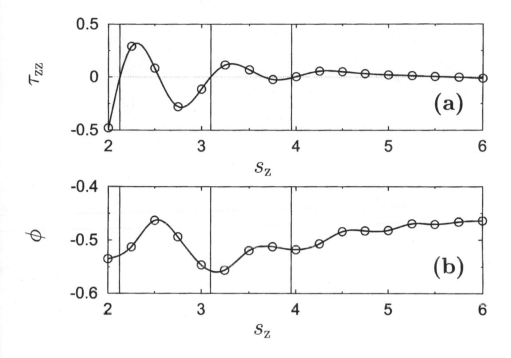

Figure 5.24: As Fig. 5.22, but for $\alpha s_{x0} = 2.50$.

so that, up to the yield point, plots in Fig. 5.19 apparently pertain to thermodynamically stable phases.

Thicker films are therfore thermodynamically stable only if the shear strain exceeds the yield strain. For example, plots in Fig. 5.25(a) for $c_r = \frac{6}{10}$ show that ϕ for a bilayer bridge phase is lower than for the corresponding monolayer bridge phase over the range $2.3 \lesssim \alpha s_{x0} \leq 5.0$ where the bilayer bridge phase is the thermodynamically stable one according to the above discussion. An additional trilayer bridge phase was observed for $c_r = \frac{4}{10}$ as plots in Fig. 5.25(b) show. For $c_r = \frac{4}{10}$ the bilayer is thermodynamically stable over the range $2.4 \lesssim \alpha s_{x0} \lesssim 3.3$, whereas the trilayer film seems to be thermodynamically stable over the range $3.3 \lesssim \alpha s_{x0} \lesssim 4.0$ where all three curves end. However, for the trilayer morphology the statistical error of $\phi\,(\alpha s_{x0})$ is already quite large because τ_{yy} is small (see Figs. 5.22–5.24). For $\alpha s_{x0} \simeq 4.0$ bridge phases become unstable and the system undergoes a first-order phase transition and evaporates.

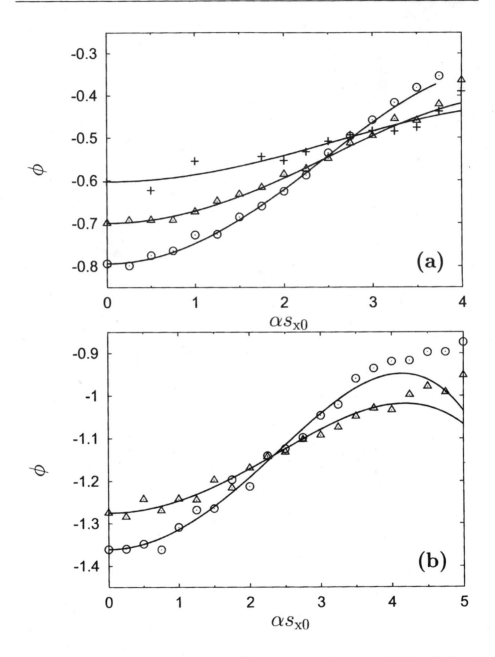

Figure 5.25: (a) ϕ as a function of shear strain αs_{x0} for mono- ($\bigcirc$), bi- ($\triangle$), and trilayer ($+$) morphologies calculated in grand mixed isostress isostrain ensemble MC simulations [see Eqs. (5.119) and (5.85)] for $c_r = \frac{6}{10}$. Solid lines are calculated from Eq. (5.121). (b) As (a), but for $c_r = \frac{4}{10}$.

5.6.5   Phase behavior of shear-deformed confined fluids

A confined fluid may undergo phase transitions among thin gaseous films, liquid-like, and bridge phases similar to those observed for the confined lattice fluid in Section 4.5.3. To demonstrate the close correspondence between the two models as far as the phase behavior is concerned, we calculate the average overall density defined in Eq. (5.74) for various substrate separations s_z. A plot of $\overline{\rho}$ in Fig. 5.9 for $c_r = \frac{4}{12}$ exhibits two discontinuities. By a parallel analysis of $\rho(x, z)$ in Fig. 5.8, the one around $s_z \simeq 8.2$ turns out to correspond to a first-order phase transition involving gas- and liquid-like phases, whereas the one at $s_z \simeq 7.5$ refers to a transition between a liquid-like phase and a bridge phase (upon reducing s_z). Therefore, the sequence of phase transitions in Fig. 5.9 resembles precisely the scenario observed for the lattice fluid in Fig. 4.12(b). However, depending on the precise chemical structure of these surfaces, different phase transitions are possible (see Fig. 5.9), which can also be explained qualitatively within the framework of the mean-field lattice fluid. Oscillations of $\overline{\rho}$ in Fig. 5.9 over the range $2 \lesssim s_z \lesssim 6$ reflect stratification of the confined fluid as decribed above.

To make direct contact with the mean-field calculations of the related discrete model in Section 4.5.3, we employ the grand canonical rather than the mixed grand isostress isostrain ensemble used in the preceding section. However, investigations of phase transitions by GCEMC simulations are frequently plagued by metastability, that is, the existence of a sequence of configurations $\left\{ \boldsymbol{r}_k^N \right\}_{k=1,\dots,M}$ corresponding only to a *local* minimum of the grand-potential density ω where M can be quite substantial. In other words, the "lifetime" of a metastable thermodynamic state can be large compared with the time over which the microscopic evolution of the system can be pursued on account of limited computational speed. The origin of metastability is lack of ergodicity in the immediate vicinity of a first-order phase transition that arises on account of the microscopically small systems employed in computer simulations [173].

As we pointed out above, metastability is manifest as hysteresis in a sorption isotherm (like the one plotted in Fig. 5.9). Metastability involves a range of finite width Δs_z around the true transition point over which for the same T and μ, $\overline{\rho}(s_z)$ is a double-valued function. To distinguish the metastable from the thermodynamically (i.e., *globally*) stable phase one needs to compare ω for the two states pertaining to different branches of the sorption isotherm at the same μ and s_z [see Eq. (1.68)]. The branch having lowest ω is the globally stable phase; the other one is only metastable. In Fig. 5.9 we plot only data for thermodynamically stable phases identified according to this rationale.

Because of the similarity between the lattice fluid calculations and the MC simulations for the continuous model, it seems instructive to study the phase behavior in the latter if the confined fluid is exposed to a shear strain. This may be done quantitatively by calculating $\overline{\rho}$ as a function of μ and αs_{x}. For sufficiently low μ, one expects a gas-like phase to exist along a subcritical isotherm (see Fig. 4.13) defined as the set of points $(T = \mathrm{const})$

$$\mathbb{T} = \left\{ (\mu, T) \left| \mu_{\mathrm{tr}} < \mu < \min\left(\mu_{\mathrm{c}}^{\mathrm{gb}}, \mu_{\mathrm{c}}^{\mathrm{bl}}\right), T_{\mathrm{tr}} < T < \min\left(T_{\mathrm{c}}^{\mathrm{gb}}, T_{\mathrm{c}}^{\mathrm{bl}}\right) \right. \right\} \quad (5.122)$$

At an intersection between $\mathbb{T}$ and $\mu_{x}^{\mathrm{gb}}(T)$, the gas-like phase will undergo a spontaneous transformation to a bridge phase. In a corresponding plot of $\overline{\rho}(\mu)$, one should see a discontinuous jump to a higher density. Eventually, another intersection between $\mathbb{T}$ and $\mu_{x}^{\mathrm{bl}}(T)$ exists and a second discontinuous jump to an even higher value of $\overline{\rho}(\mu)$ should be visible. Both of these transitions are indeed observed in Fig. 5.26 for $\alpha s_{x} = 0$, $\mu \simeq -8.40$, and $\mu \simeq -7.98$, respectively. Notice that, in Fig. 5.26, μ_{x}^{bl} for $\alpha s_{x0} = 0.0$ exceeds its bulk counterpart μ_{xb}; that is, for μ_{x}^{bl}, the corresponding bulk phase is liquid. This can be rationalized by noting that the low(er)-density part of a bridge phase is predominantly involved in this second transition. Recall also that this part of a bridge phase is stabilized by the *weak* portions of both (perfectly aligned) substrates characterized by $\varepsilon_{\mathrm{fw}} \ll \varepsilon_{\mathrm{ff}}$. Hence, the second first-order transition is inhibited rather than induced by the substrates (with respect to the bulk) because of the dominating repulsive interaction of a fluid molecule with the weak part of the substrate.

If a shear strain is applied, the region of overlap of the weak substrate parts in the x-direction shrinks [see Eq. 5.103] such that a fluid molecule located at $\{x \,|d_{\mathrm{s}}/2 \leq |x| \leq s_{x}/2, \alpha s_{x} = 0.0\}$ is exposed to a stronger net attractive fluid substrate interaction. Consequently, one expects an associated shift of μ_{x}^{bl} to lower values. The plot in Fig. 5.26 confirms the expectation. In addition, Fig. 4.13 shows that the one phase region shrinks because T_{tr} shifts to higher temperatures and because the slope of the coexistence lines does not change much. The plot in Fig. 4.13 therefore suggests that for $\alpha > 0$ the two discontinuities in $\overline{\rho}(\mu)$ approach each other so that the branch of $\overline{\rho}(\mu)$ belonging to bridge phases becomes narrower with increasing αs_{x0}. This effect is indeed visible in Fig. 5.26 where the width of the intermediate-density branch of $\overline{\rho}(\mu)$ (corresponding to thermodynamically stable bridge phases) diminishes from $|\Delta\mu| \simeq 0.42$ $(\alpha s_{x0} = 0.0)$ to $|\Delta\mu| \simeq 0.14$ $(\alpha s_{x0} = 7.5)$. Finally, if the shear strain is large enough, the lattice fluid results in Fig. 4.13 suggest that for a given temperature T^{*}, $T_{\mathrm{tr}}(\alpha s_{x}) > T^{*}$ for sufficiently large shear strains (see the curve for $\alpha = \frac{2}{7}$ in Fig. 4.13). Hence, under these

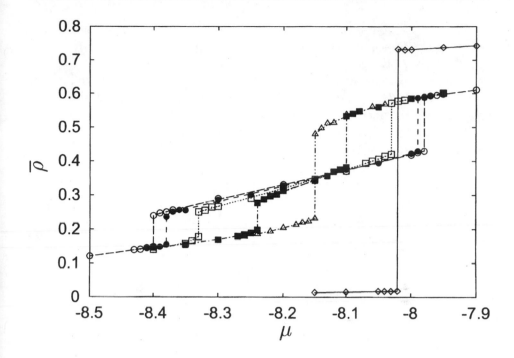

Figure 5.26: Sorption isotherms $\bar{\rho}(\mu)$ from GCEMC simulations: ($\circ$): $\alpha s_x = 0.0$; ($\bullet$): $\alpha s_x = 2.5$; ($\square$): $\alpha s_x = 5.0$; ($\blacksquare$): $\alpha s_x = 7.5$; ($\triangle$): $\alpha s_x = 10.0$. Also shown are corresponding bulk data ($\diamond$). Results were obtained for $T = 0.7$, $s_x = 20.0$, $d_s = 10.0$, and $s_z = 8.0$.

circumstances, one would expect $\bar{\rho}(\mu)$ to exhibit just a single discontinuity referring to a phase transition between gaseous film- and liquid-like phases. The plot in Fig. 5.26 for $\alpha s_{x0} = 10$ confirms this notion.

5.7   The Joule-Thomson effect

5.7.1   Experimental background and applications

After illustrating the rather fascinating structural and rheological properties of confined fluids we conclude our discussion of MC simulations of continuous model systems (i.e., models in which fluid molecules move along continuous trajectories in space) with yet another example of the unique behavior of confined fluids. For pedagogic reasons we selected a topic that is standard in physical chemistry textbooks [26, 199–203] as far as bulk fluids are concerned, namely the Joule-Thomson effect.

The Joule-Thomson effect refers to a phenomenon observed if a gas in a vessel 1 at temperature T_1 and pressure P_1 expands slowly through a valve or porous plug into another vessel 2 where its pressure $P_2 < P_1$. During this expansion a temperature change $\Delta T \equiv T_2 - T_1$ is observed, which can be positive, negative, or vanish altogether depending on the precise experimental conditions. This phenomenon is referred to as the Joule-Thomson effect and was originally reported by Joule and Thomson (later titled Lord Kelvin) [204].

During the expansion the gas does not exchange heat with its environment. However, it exchanges work because of the expansion against the nonzero pressure P_2. It is then a simple matter to demonstrate that the gas expands isenthalpically [26, 199–203]. This makes it convenient to discuss the Joule-Thomson process quantitatively in terms of a Joule-Thomson coefficient

$$\delta \equiv \left(\frac{\partial T}{\partial P}\right)_{\mathcal{H}} \gtreqless 0 \qquad (5.123)$$

where $\mathcal{H}$ is the enthalpy. From Eq. (5.123) it is clear that during an isenthalpic expansion $(\mathrm{d}P < 0)$ $\delta > 0$ if the gas is cooled and $\delta < 0$ if it is heated instead. The fact that the gas can be cooled during a Joule-Thomson expansion is of great technological relevance in applied fields like cryogenics and in particular in the liquefaction of gases [200].

From a fundamental perspective the Joule-Thomson effect is important because it can be linked directly to the nature of intermolecular forces between gas molecules [205]. Consider, for example, a classic ideal gas as the simplest case in which molecules do not interact by definition. For this model it is simple to show that as a consequence of the absence of any intermolecular interactions a Joule-Thomson effect does not exist, that is, $\delta \equiv 0$ [200, 201]. If, on the other hand, the ideal gas is treated quantum mechanically, it can be demonstrated [206] that a Joule-Thomson effect exists $(\delta \neq 0)$ despite the lack of intermolecular interactions.

The origin of the nonvanishing Joule-Thomson effect is the effective repulsive (Fermions) and attractive (Bosons) potential exerted on the gas molecules, which arises from the different ways in which quantum states can be occupied in systems obeying Fermi–Dirac and Bose–Einstein statistics, respectively [17]. In other words, the effective fields are a consequence of whether Pauli's antisymmetry principle, which is relativistic in nature [207], is applicable. Thus, a weakly degenerate Fermi gas will always heat up $(\delta < 0)$, whereas a weakly degenerate Bose gas will cool down $(\delta > 0)$ during a Joule-Thomson expansion. These conclusions remain valid even if the ideal quantum gas is treated relativistically, which is required to understand certain aspects of stellar matter [208, 209]. We shall return to these issues in Section 5.7.4.3 where we consider the Joule-Thomson effect in confined ideal quantum gases as a first application.

Beyond the (classic or quantum) ideal-gas level, molecules in a gas are subject to attractive and repulsive intermolecular interactions. Thus, intuitively, one expects a real gas to show both a positive and a negative Joule-Thomson effect depending on the thermodynamic conditions. In other words, the sign of δ depends essentially on the degree with which molecules probe attractive and repulsive portions of the intermolecular potential. From this line of arguments, one then expects an inversion temperature $T_{\text{inv}}(\rho)$ (ρ is the density) to exist along which $\delta = 0$, thus separating regions in thermodynamic state space that are characterized by a positive or negative Joule-Thomson effect, respectively. These notions are readily confirmed by treating a bulk van der Waals fluid [200]. We will extend these considerations to a confined van der Waals fluid below in Section 5.7.5.

Unfortunately, previous work is almost exclusively concerned with the inversion temperature in the limit of vanishing gas density, $T_{\text{inv}}(0)$. The inversion temperature can be linked to the second virial coefficient, which can be measured [210] or computed from rigorous statistical physical expressions [211] with moderate effort. Currently, only the fairly recent study of Heyes and Llaguno is concerned with the density dependence of the inversion temperature from a molecular (i.e., statistical physical) perspective [212]. These authors compute the inversion temperature from isothermal isobaric molecular dynamics simulations of the LJ (12,6) fluid over a wide range of densities and analyze their results through various equations of state.

All these considerations apply strictly to homogeneous bulk gases, that is, for gases in containers of macroscopic dimensions. Under this proviso only a vanishingly small portion of the gas will be perturbed by the interaction of its molecules with the container walls making this interaction inconsequential for gas properties. However, Rybolt [213] and Pierotti and Rybolt [214] have studied the Joule-Thomson effect in aerosols consisting of finely dispersed carbon powder in argon gas. In such an aerosol the ratio of solid–surface area to volume becomes large so that gas solid interactions can no longer be ignored (see, for example, Fig. 7.1). By applying the concepts of statistical physics, Pierotti and Rybolt [214] derived an expression for the Joule-Thomson coefficient in terms of gas gas and gas solid virial coefficients. An analysis of the adsorption data of Cole et al. [215] shows that the Joule-Thomson effect can be enhanced by up to an order of magnitude over that observed in pure bulk gases depending on powder concentration (i.e., the relative contribution of gas solid interactions), and this enhancement may have practical implications for refrigeration devices. Unfortunately, all this work is again restricted to the limit of low gas density.

From a more general perspective the interaction of soft condensed matter with solid substrates is of great importance whenever it is desirable to, say, miniaturize mechanical machines. As we demonstrated in Section 5.6, the presence of such substrates has profound consequences for thermophysical properties of soft condensed matter and especially so if the confinement is to spaces of nanoscopic dimensions. The availability of a variety of techniques to design and to construct devices on a nano- to micrometer lengthscale in a controlled manner has also given birth to a flourishing new field in applied science referred to as "microfabrication technology" or "microengineering" [49]. A particularly interesting example in the current context are micro-miniature refrigerators [216, 217]. By means of photolithographic techniques fine nozzles and channels can be designed in a controlled fashion on a micrometer lengthscale through which a gas can flow such that the Joule-Thomson effect can be employed for cooling purposes. Through microminiature refrigeration, superconducting electronic devices including fast A/D converters, precision voltage standards, and single chip, high-speed logical Josephson devices can be cooled efficiently [217]. Thus, all these examples illustrate that the consideration of the Joule-Thomson effect under nanoconfinement conditions is of broader than just academic interest and may very well have practical applications in the future.

5.7.2   Model system

The model we shall be employing below to investigate various aspects of the Joule-Thomson effect consists of a "simple" fluid confined between the chemically homogeneous and planar substrates of a slit-pore separated by a *fixed* distance s_z. For this system we may cast the configurational energy as

$$U = \frac{1}{2} \sum_{i=1}^{N} \sum_{j \neq i=1}^{N} u_{\mathrm{ff}}\left(r_{ij}\right) + \sum_{k=1}^{2} \sum_{i=1}^{N} \Phi^{[k]}\left(z_i; s_z\right) \qquad (5.124)$$

where u_{ff} is given in Eq. (5.38) and $\Phi^{[k]}$ in Eq. (5.71) where we are again using the LJ (12,6) potential [see Eq. (5.24)] for the intermolecular interaction potential u in Eq. (5.39).

To investigate the impact of the chemical nature of the (homogeneous) substrate, two different cases are studied. In the "strong" model A, fluid substrate interactions are described by $\Phi_A^{[k]}\left(z; s_z\right)$ as introduced in Eq. (5.71). In addition, the "weak" model B is considered in which fluid substrate interactions are purely repulsive; that is [cf. Eq. (5.71)]

$$\Phi_{\mathrm{B}}^{[k]}\left(z; s_z\right) = \frac{4\pi\varepsilon n_{\mathrm{A}}\sigma^2}{5}\left(\frac{\sigma}{z \pm s_z}\right)^{10} \tag{5.125}$$

Because of the absence of any fluid substrate attraction, the fluid in model B cannot wet the confining substrates.

In addition to classic fluids with interacting molecules, we shall also consider below the ideal quantum gas of Bosons and Fermions. The ideal quantum gases are confined by plane parallel, structureless, and chemically homogeneous substrates represented by

$$\Phi_{\mathrm{sw}}^{[k]}\left(z; s_z\right) = \begin{cases} \infty, & |z| \geq s_{z0}/2 - \sigma \\ -\varepsilon_{\mathrm{fs}}, & s_{z0}/2 - \lambda\sigma \leq |z| \leq s_{z0}/2 - \sigma \\ 0, & |z| \leq s_{z0}/2 - \lambda\sigma \end{cases} \tag{5.126}$$

This fluid substrate potential is chosen because it accounts for attractive as well as for repulsive interactions, it is short-range, and it permits an analytical treatment of confinement effects as we shall demonstrate below in Section 5.7.4.3.

5.7.3 Thermodynamic considerations

5.7.3.1 Joule-Thomson coefficient and inversion temperature

The central quantity in the context of this chapter is the Joule-Thomson coefficient, which we define by analogy with its bulk counterpart [see Eq. (5.123)] as

$$\delta_{\|} \equiv -\left(\frac{\partial T}{\partial \tau_{\|}}\right)_{\mathcal{H}} \tag{5.127}$$

It is positive if the confined fluid is cooled ($dT < 0$) upon transverse compression ($d\tau_{\|} > 0$) and negative if the fluid is heated instead. According to the assertions at the beginning of Section 5.7.1, the key thermodynamic potential in the current context is the enthalpy $\mathcal{H}$, which we obtain as a Legendre transform (see Section 1.5) of the internal energy via

$$d\mathcal{H} \equiv d\left(\mathcal{U} - \tau_{\|}As_{z0}\right) = TdS + \mu dN - As_{z0}d\tau_{\|} + \tau_{\perp}A_0 ds_z \tag{5.128}$$

where for the current model Eq. (1.22) for $d\mathcal{U}$ applies. From the exact differential for the enthalpy in Eq. (5.128) we readily conclude that the set $\left\{S, N, \tau_{\|}, s_z\right\}$ specifies the natural variables of $\mathcal{H}$. To proceed we immediately restrict the discussion to a situation in which the fluid lamella is confined to a slit-pore of fixed pore width $s_z = $ const and contains a fixed number of molecules $N = $ const. Moreover, from the definition of the Joule-Thomson coefficient in Eq. (5.127) it is clear that we need to establish a relation between $\mathcal{H}$ on the one hand and the variables T and $\tau_{\parallel}$ on the other hand.

We accomplish this via

$$\mathrm{d}\mathcal{S} = \left(\frac{\partial \mathcal{S}}{\partial T}\right)_{N,\tau_{\parallel},s_z} \mathrm{d}T + \left(\frac{\partial \mathcal{S}}{\partial \tau_{\parallel}}\right)_{T,N,s_z} \mathrm{d}\tau_{\parallel}, \qquad N, s_z = \text{const} \qquad (5.129)$$

At this point it is convenient to define a specialized isostress ($\tau_{\parallel} = $ const) heat capacity

$$c_{\parallel} \equiv T \left(\frac{\partial \mathcal{S}}{\partial T}\right)_{N,\tau_{\parallel},s_z} \qquad (5.130)$$

to eliminate the first term on the right side of Eq. (5.129) by some *in principle* measurable quantity. The second term in Eq. (5.129) can be replaced through a Maxwell relation [see Eq. (A.7)]. Therefore, we need to introduce yet another Legendre transform of the internal energy (see Section 1.5)

$$\begin{aligned}
\mathrm{d}\mathcal{G}_{\parallel} &\equiv \mathrm{d}\left(\mathcal{U} - T\mathcal{S} - \tau_{\parallel}As_{z0}\right)\\
&= -\mathcal{S}\mathrm{d}T + \mu\mathrm{d}N - As_{z0}\mathrm{d}\tau_{\parallel} + \tau_{\perp}A_0\mathrm{d}s_z\\
&= \left(\frac{\partial \mathcal{G}_{\parallel}}{\partial T}\right)_{\{\cdot\}\backslash T} \mathrm{d}T + \left(\frac{\partial \mathcal{G}_{\parallel}}{\partial N}\right)_{\{\cdot\}\backslash N} \mathrm{d}N\\
&\quad + \left(\frac{\partial \mathcal{G}_{\parallel}}{\partial \tau_{\parallel}}\right)_{\{\cdot\}\backslash \tau_{\parallel}} \mathrm{d}\tau_{\parallel} + \left(\frac{\partial \mathcal{G}_{\parallel}}{\partial s_z}\right)_{\{\cdot\}\backslash s_z} \mathrm{d}s_z
\end{aligned}$$

$$(5.131)$$

which can be interpreted as a specialized Gibbs potential depending on $\{T, N, \tau_{\parallel}, s_z\}$ as its set of natural variables. Applying Eq. (A.7) to Eq. (5.131) we realize that

$$\begin{aligned}
\left[\frac{\partial}{\partial \tau_{\parallel}}\left(\frac{\partial \mathcal{G}_{\parallel}}{\partial T}\right)_{\{\cdot\}\backslash T}\right]_{\{\cdot\}\backslash \tau_{\parallel}} &= -\left(\frac{\partial \mathcal{S}}{\partial \tau_{\parallel}}\right)_{\{\cdot\}\backslash \tau_{\parallel}} = \left[\frac{\partial}{\partial T}\left(\frac{\partial \mathcal{G}_{\parallel}}{\partial \tau_{\parallel}}\right)_{\{\cdot\}\backslash \tau_{\parallel}}\right]_{\{\cdot\}\backslash T}\\
&= -s_{z0}\left(\frac{\partial A}{\partial T}\right)_{\{\cdot\}\backslash T} \equiv -s_{z0}A\alpha_{\parallel} \qquad (5.132)
\end{aligned}$$

where $\alpha_{\parallel}$ is the (transverse) expansivity of the confined lamella. Replacing the partial derivatives on the right side of Eq. (5.129) by Eqs. (5.130) and

(5.132) and realizing that the Joule-Thomson process is carried out under isenthalpic conditions (i.e., $d\mathcal{H} = 0$) we can rearrange Eq. (5.128) such that

$$dT = \left(\frac{\partial T}{\partial \tau_{\|}}\right)_{\mathcal{H}} d\tau_{\|} = \frac{As_{z0}}{c_{\|}} \left(1 - T\alpha_{\|}\right) d\tau_{\|}, \qquad N, s_z = \text{const} \qquad (5.133)$$

which shows that under the current conditions T is solely a function of $\tau_{\|}$. With the definitions $\bar{\rho} \equiv N/(As_{z0})$ and $c_{\|}^N \equiv c_{\|}/N$ we can differentiate the previous expression with respect to $\tau_{\|}$ to obtain [see Eq. (5.127)]

$$\delta_{\|} = \frac{1}{\bar{\rho}\, c_{\|}^N} \left(T\alpha_{\|} - 1\right) \qquad (5.134)$$

Because all coefficients in Eq. (5.134) are positive definite we obtain as a thermodynamic expression for the inversion temperature ($\delta_{\|} = 0$)

$$T_{\text{inv}} = \frac{1}{\alpha_{\|}} \qquad (5.135)$$

5.7.3.2 Consistency relation

For subsequent checks on the MC simulations, from which we seek to determine T_{inv}, it will prove convenient to derive a consistency relation that must hold regardless of molecular details of the specific model under consideration. The derivation starts by assuming that an equation of state $\tau_{\|}(T, A)$ (fixed N, s_z) exists such that

$$d\tau_{\|} = \left(\frac{\partial \tau_{\|}}{\partial T}\right)_{N,A,s_z} dT + \left(\frac{\partial \tau_{\|}}{\partial A}\right)_{T,N,s_z} dA \qquad (5.136)$$

Focusing on thermodynamic transformations such that $\tau_{\|} = \text{const}$ (i.e., $d\tau_{\|} = 0$) we can rearrange this expression to give

$$\left(\frac{\partial \tau_{\|}}{\partial T}\right)_{N,A,s_z} = -\frac{\alpha_{\|}}{\kappa_{\|}} \qquad (5.137)$$

where we employed the definitions of $\kappa_{\|}$ and $\alpha_{\|}$ in Eqs. (1.81) and (5.132), respectively.

Inserting the equation of state Eq. (5.136) into Eq. (5.129) and using also Eqs. (5.130) and (5.132), we obtain

$$d\mathcal{S} = \left(\frac{c_{\|}}{T} - As_{z0}\frac{\alpha_{\|}^2}{\kappa_{\|}}\right) dT + s_{z0}\frac{\alpha_{\|}}{\kappa_{\|}} dA, \qquad N, s_z = \text{const} \qquad (5.138)$$

where Eqs. (1.81) and (5.137) have also been employed. Defining the isostrain heat capacity by analogy with Eq. (5.130) as

$$c_\sigma \equiv T \left(\frac{\partial S}{\partial T} \right)_{N,A,s_z} \tag{5.139}$$

we obtain from Eq. (5.138) the desired consistency relation

$$c_\parallel^N - c_\sigma^N = \frac{T}{\bar{\rho}} \frac{\alpha_\parallel}{\kappa_\parallel} \tag{5.140}$$

where $c_\sigma^N \equiv c_\sigma/N$ and $\bar{\rho}$ is defined as above.

5.7.4 The limit of low densities

5.7.4.1 Virial expansion

We now turn to a microscopic treatment of the Joule-Thomson effect and begin with the limit of vanishing density. The treatment below is very similar to the one presented in Section 3.2.2 where we derived molecular expressions for the first few virial coefficients of the one-dimensional hard-rod fluid. Here it is important to realize that a mechanical expression for the grand potential exists for a fluid confined to a slit-pore with chemically structured substrate surfaces as we demonstrated in Section 1.6.1 [see Eq. (1.65)]. Combining this expression with the molecular expression given in Eq. (2.81) we may write

$$\Xi_{cl} = \exp\left(-\frac{\tau_\parallel A s_{z0}}{k_B T} \right) = \sum_{k=0}^{\infty} \frac{(\pm 1)^k}{k!} \left(\frac{\tau_\parallel A s_{z0}}{k_B T} \right)^k \tag{5.141}$$

where we have expanded $\exp(-x)$ into a power series. Moreover, we write Ξ as a power series

$$\Xi_{cl} = \sum_{N=0}^{\infty} \frac{Z_N}{N!} z^N \tag{5.142}$$

in terms of the activity $z = \exp(\mu/k_B T) \Lambda^{-3N}$ [see Eq. (2.120)]. In the previous expression we denote the configuration integral [see Eq. (2.112)] by Z_N to emphasize its implicit dependence on the number of molecules. Equation (5.141) suggests that it should be possible to expand $\tau_\parallel$ in a power series in the activity as well. Thus, we employ again Eq. (3.22), which we insert into Eq. (5.141) to obtain

$$\begin{aligned}
\Xi_{cl} &= \sum_{k=0}^{\infty} \frac{(\pm A s_{z0})^k}{k!} \left(\sum_{j=1}^{\infty} b_j z^j \right)^k \\
&= 1 - A s_{z0} \left(b_1 z + b_2 z^2 \right) + \frac{(A s_{z0})^2}{2} b_1^2 z^2 + \mathcal{O}\left(z^3 \right) \tag{5.143}
\end{aligned}$$

where we replaced in the original Eq. (3.22), τ_b by $\tau_\parallel$ and retain terms only up to second order in z.

Comparison with Eq. (5.142) immediately gives

$$b_1 = -\frac{Z_1}{As_{z0}} \tag{5.144a}$$

$$b_2 = -\frac{Z_2 - Z_1^2}{2As_{z0}} \tag{5.144b}$$

Unfortunately, the original expansion of $\tau_\parallel$ in terms of the activity z is somewhat awkward in practice. Instead we would prefer an expansion of $\tau_\parallel$ in terms of the mean density $\bar\rho$ of the confined fluid. This can be accomplished by first noting from Eq. (5.142) that we may write

$$\bar\rho = \frac{1}{As_{z0}} \underbrace{z\frac{\partial \ln \Xi}{\partial z}}_{\langle N \rangle} = -\frac{z}{k_{\mathrm B}T}\frac{\partial \tau_\parallel}{\partial z} = -\sum_{j=1}^{\infty} jb_j z^j \tag{5.145}$$

where we also used Eqs. (5.141) and (3.22). Expressing now the activity in terms of a power series in $\bar\rho$ by the *ansatz* [cf. Eq. (3.26)]

$$z = a_1\bar\rho + a_2\bar\rho^2 + \mathcal{O}\left(\bar\rho^3\right) \tag{5.146}$$

we obtain from Eq. (5.145) the expression [cf. Eq. (3.27)]

$$\bar\rho = -a_1 b_1 \bar\rho - b_1\left(2a_1^2 + a_2\right)\bar\rho^2 + \mathcal{O}\left(\bar\rho^3\right) \tag{5.147}$$

where we retain terms only up to second order in $\bar\rho^2$. Equating in this expression terms of equal power in $\bar\rho$ on both sides of the equation, we can express the unknown coefficients a_1 and a_2 in terms of the known constants b_1 and b_2 as

$$a_1 = -\frac{1}{b_1} \tag{5.148a}$$

$$a_2 = -\frac{2b_2 a_1^2}{b_1} = -\frac{b_2}{b_1^3} \tag{5.148b}$$

Thus, replacing the expansion coefficients in Eq. (5.146) via Eqs. (5.148) and inserting the resulting expression into Eq. (3.22) we obtain [cf. Eq. (3.29)]

$$\tau_\parallel = -k_{\mathrm B}T\bar\rho - k_{\mathrm B}T\frac{b_2}{b_1^2}\bar\rho^2 + \mathcal{O}\left(\bar\rho^3\right) \equiv \tau_\parallel^{\mathrm{id}} - k_{\mathrm B}TB_2\left(T\right)\bar\rho^2 + \mathcal{O}\left(\bar\rho^3\right) \tag{5.149}$$

where $\tau_\parallel^{\mathrm{id}}$ is the ideal-gas contribution to the transverse stress and

$$B_2 = \frac{b_2}{b_1^2} = -\frac{As_{z0}}{2} \frac{Z_2 - Z_1^2}{Z_1^2} \qquad (5.150)$$

is the second virial coefficient of the confined fluid, which appears to be solely a function of temperature via Eqs. (5.158) and (5.163). Notice the similarity between Eq. (5.150) and its counterpart Eq. (3.30a) for the one-dimensional hard-rod fluid.

5.7.4.2 Inversion temperature

We can now derive an expression for the inversion temperature that is valid in the limit of sufficiently low densities. Therefore we differentiate [see Eq. (5.149)]

$$\left(\frac{\partial}{\partial T}\frac{\tau_\parallel}{k_{\mathrm{B}}T}\right)_{N,\tau_\parallel,s_z} = -\frac{\tau_\parallel}{k_{\mathrm{B}}T}\frac{1}{T} = [1 + 2\bar{\rho}B_2]\left(\frac{\partial\bar{\rho}}{\partial T}\right)_{N,\tau_\parallel,s_z} + \bar{\rho}^2\frac{\mathrm{d}B_2}{\mathrm{d}T} \qquad (5.151)$$

where we may replace $\tau_\parallel/k_{\mathrm{B}}T$ via Eq. (5.149). After multiplying both sides of the resulting expression by T, dividing them by $\bar{\rho}$, and rearranging terms we arrive at

$$-1 - \frac{T}{\bar{\rho}}\left(\frac{\partial\bar{\rho}}{\partial T}\right)_{N,\tau_\parallel,s_z} = \frac{\bar{\rho}}{1 + 2\bar{\rho}B_2}\left(T\frac{\mathrm{d}B_2}{\mathrm{d}T} - B_2\right) \qquad (5.152)$$

This expression may be simplified even further by noting that

$$\left(\frac{\partial\bar{\rho}}{\partial T}\right)_{N,\tau_\parallel,s_z} = \left(\frac{\partial\left(N/As_{z0}\right)}{\partial T}\right)_{N,\tau_\parallel,s_z} = \frac{N}{s_{z0}}\left(\frac{\partial A^{-1}}{\partial T}\right)_{N,\tau_\parallel,s_z}$$

$$= -\frac{N}{As_{z0}}\frac{1}{A}\left(\frac{\partial A}{\partial T}\right)_{N,\tau_\parallel,s_z} = -\bar{\rho}\alpha_\parallel \qquad (5.153)$$

where the definition of the expansivity in Eq. (5.132) has also been used. Using this expression we finally obtain

$$T\alpha_\parallel - 1 = -\frac{\bar{\rho}}{1 + 2\bar{\rho}B_2}\frac{\mathrm{d}}{\mathrm{d}T}\frac{B_2\left(T\right)}{T} \qquad (5.154)$$

Comparing this expression with the thermodynamic one for $\delta_\parallel$ defined in Eq. (5.134) it is clear that the inversion temperature can be obtained here from the zero of

$$\frac{\mathrm{d}}{\mathrm{d}T}\frac{B_2\left(T\right)}{T}\bigg|_{T=T_{\mathrm{inv}}} = \frac{1}{T_{\mathrm{inv}}^2}\Bigg[\underbrace{B_2\left(T_{\mathrm{inv}}\right)}_{\mathrm{I}} - T_{\mathrm{inv}}\underbrace{\frac{\mathrm{d}B_2\left(T\right)}{\mathrm{d}T}\bigg|_{T=T_{\mathrm{inv}}}}_{}\Bigg] = 0 \qquad (5.155)$$

which has a lucid geometrical interpretation in that it defines T_{inv} at that point at which a tangent to $B_2(T)$ through the origin (term II) touches that curve (term I).

Finally, in closing this section we notice that the inversion temperature defined by Eq. (5.155) is expected to depend on the presence and chemical nature of the solid substrate even in the limit of vanishing density at least in principle (see Section 5.7.5). This is because Z_1 and Z_2 depend on the fluid substrate potential [see Eqs. (5.158) and (5.161)], which is, in turn, expected to affect $B_2(T)$ through Eq. (5.150). Moreover, we note that, because the above treatment is valid only in the limit $\overline{\rho} \to 0$ [and because $B_2(T) \neq f(\overline{\rho})$], the inversion temperature does not depend on the mean density of the confined fluid.

5.7.4.3   Confined ideal quantum gas

The simplest system one might consider in the context of the Joule-Thomson effect is the ideal gas. As we showed in Eq. (5.149) the equation of state of the ideal gas in the *classic limit* is given by

$$\tau_{\parallel}^{id} = -\overline{\rho} k_B T \tag{5.156}$$

Using this expression it is easy to verify from Eq. (5.153) that

$$\alpha_{\parallel}^{id} = \frac{1}{T} \tag{5.157}$$

Hence, it follows from Eq. (5.134) that $\delta_{\parallel} = 0$ regardless of the thermodynamic conditions considered, which is in accord with standard textbook knowledge.

However, at the molecular level, symmetry properties of the quantum mechanical wave function give rise to deviations from the classic behavior as we showed in Section 2.5. These deviations may be interpreted as a net repulsion (Fermi–Dirac gas) or attraction (Bose–Einstein gas) between the molecules. As we emphasized in Section 2.5.3, quantum effects are maximized in semiclassic ideal gases. From this point of view, it then seems sensible to address the following questions:

1. Does a Joule-Thomson effect exist in ideal quantum gases?

2. What is the role of confinement to nanoscopic volumes?

In this section we shall answer both questions by considering an ideal quantum gas (of Fermions or Bosons) confined to a slit-pore with chemically homogeneous solid surfaces represented by the potential $\Phi_{\mathrm{sw}}^{[k]}(z; s_{z0})$ defined in Eq. (5.126).

For the ideal quantum gas

$$
Z_1 \equiv \int d\mathbf{r}_1 \exp\left[-\frac{U(\mathbf{r}_1)}{k_{\mathrm{B}}T}\right] = A \int\limits_{-s_{z0}/2}^{s_{z0}/2} dz_1 \exp\left[-\frac{1}{k_{\mathrm{B}}T}\sum_{k=1}^{2}\Phi_{\mathrm{sw}}^{[k]}(z_1; s_{z0})\right]
$$

(5.158)

where we shall use the subscript to indicate the number of molecules in the system; that is, Z_1 is the single-particle configuration integral. Note that for a bulk system $Z_1 = As_{z0}$ because in this case $\Phi_{\mathrm{sw}}^{[k]}(z; s_{z0})$ vanishes by definition. For the potential introduced in Eq. (5.126), Eq. (5.158) can be rewritten more explicitly as

$$
\begin{aligned}
\frac{Z_1}{A} &= \int\limits_{-s_{z0}/2}^{s_{z0}/2} dz_1 \exp\left[-\frac{1}{k_{\mathrm{B}}T}\sum_{k=1}^{2}\Phi_{\mathrm{sw}}^{[k]}(z_1; s_{z0})\right] \\
&= \exp\left(\frac{\varepsilon_{\mathrm{fs}}}{k_{\mathrm{B}}T}\right)\int\limits_{-s_{z0}/2+\sigma}^{-s_{z0}/2+\lambda\sigma} dz_1 + \int\limits_{-s_{z0}/2+\lambda\sigma}^{s_{z0}/2-\lambda\sigma} dz_1 + \exp\left(\frac{\varepsilon_{\mathrm{fs}}}{k_{\mathrm{B}}T}\right)\int\limits_{s_{z0}/2-\lambda\sigma}^{s_{z0}/2-\sigma} dz_1
\end{aligned}
$$

(5.159)

The evaluation of the remaining integrals then becomes trivial, and we obtain

$$
Z_1 = As_{z0}\left\{\frac{2\sigma}{s_{z0}}\left[(\lambda-1)\exp\left(\frac{\varepsilon_{\mathrm{fs}}}{k_{\mathrm{B}}T}\right)-\lambda\right]+1\right\}
$$

(5.160)

as a closed expression for the single-particle configuration integral. As expected, the bulk expression $Z_1 = As_{z0}$ is recovered from Eq. (5.160) in the limit $s_{z0} \to \infty$.

The semiclassic expression for the two-body configurational integral follows from Eqs. (2.110) – (2.112) as

$$
Z_2 = \int d\mathbf{r}_1 \int d\mathbf{r}_2 \left[1 \pm \exp\left(-\frac{2\pi r_{12}}{\Lambda^2}\right)\right]\exp\left[-\frac{U(\mathbf{r}_1, \mathbf{r}_2)}{k_{\mathrm{B}}T}\right]
$$

(5.161)

where for a confined *ideal* (quantum) gas

$$
U(\mathbf{r}_1, \mathbf{r}_2) = \sum_{k=1}^{2}\left[\Phi_{\mathrm{sw}}^{[k]}(z_1; s_{z0}) + \Phi_{\mathrm{sw}}^{[k]}(z_2; s_{z0})\right]
$$

(5.162)

and $\Phi_{sw}^{[k]}(z_i; s_{z0})$ is again given by Eq. (5.126). In Eq. (5.161) and below, the upper symbol of the shorthand notation, "$\pm$" always refers to a Boson gas, whereas the lower symbol refers to a gas of Fermions instead. Note also, that in the classic limit, $\exp(-2\pi r_{12}/\Lambda^2) \approx 0$ such that $Z_1^2 = Z_2$ regardless of whether the ideal gas is confined by solid substrates.

To evaluate the double integral in Eq. (5.161) it is advantageous to change variables according to $r_1, r_2 \to r_1, r \equiv r_1 - r_2$ and to employ cylindrical coordinates such that $dr = dx\, dy\, dz = \det \mathbf{J}\, d\phi\, d\rho\, dz$ where the determinant of the Jacobian for this transformation $\det \mathbf{J} = \rho$. Moreover, we realize from Eq. (5.150) that we need to compute the difference $Z_2 - Z_1^2$ to calculate the second virial coefficient of the confined quantum gas. Hence, by immediately carrying out the trivial integrations over x_1, y_1, and ϕ we obtain

$$
Z_2 - Z_1^2 = \pm 2\pi A \int_{-s_{z0}/2}^{s_{z0}/2} dz_1 \exp\left[-\frac{1}{k_B T} \sum_{k=1}^{2} \Phi_{sw}^{[k]}(z_1; s_{z0}) \right]
$$

$$
\times \int_{-s_{z0}/2-z_1}^{s_{z0}/2-z_1} dz \exp\left[-\frac{1}{k_B T} \sum_{k=1}^{2} \Phi_{sw}^{[k]}(z + z_1; s_{z0}) \right]
$$

$$
\times \int_{0}^{\infty} \rho d\rho \exp\left[-\frac{2\pi(\rho^2 + z^2)}{\Lambda^2} \right] \tag{5.163}
$$

by noting that the summand Z_1^2 can be expressed in terms of the first two integrals times the prefactor if we use the same set of coordinates.

If we then pull out the factor $\exp(-2\pi z^2/\Lambda^2)$ from the last integral in Eq. (5.163), the remaining integral over ρ can immediately be solved. With the aid of the transformation $\rho \to u \equiv \rho^2$ the remaining integral becomes

$$
\int_{0}^{\infty} \rho d\rho \exp\left(-\frac{2\pi \rho^2}{\Lambda^2} \right) = -\frac{1}{2}\frac{\Lambda^2}{2\pi} \exp\left(-\frac{2\pi u}{\Lambda^2} \right) \Bigg|_{0}^{\infty} = \frac{\Lambda^2}{4\pi} \tag{5.164}
$$

Next we consider

$$
\int_{-s_{z0}/2-z_1}^{s_{z0}/2-z_1} dz \exp\left[-\frac{1}{k_B T} \sum_{k=1}^{2} \Phi_{sw}^{[k]}(z + z_1; s_{z0}) \right] \exp\left(-\frac{2\pi z^2}{\Lambda^2} \right)
$$

$$
= \int_{-s_{z0}/2}^{s_{z0}/2} dz' \exp\left[-\frac{1}{k_B T} \sum_{k=1}^{2} \Phi_{sw}^{[k]}(z'; s_{z0}) \right] \exp\left[-\frac{2\pi(z' - z_1)^2}{\Lambda^2} \right]
$$

$$
\tag{5.165}
$$

where we have used the transformation $z \rightarrow z' \equiv z + z_1$. Focusing on a physical situation in which T and m are not too small we may assume Λ to be sufficiently small such that $\exp\left[-2\pi \left(z' - z_1\right)^2 / \Lambda^2\right]$ differs appreciably from 0 only if $|z' - z_1| \simeq 0$, that is, if particles 1 and 2 are very close to each other as far as their z-coordinates are concerned. Notice that this approximation is consistent with our semiclassic treatment in Section 2.5.3. We may then approximate the Gaussian function in the previous expression by the Dirac δ-function [see Eq. (B.75)] and write

$$\int\limits_{-s_{z0}/2}^{s_{z0}/2} \mathrm{d}z' \exp\left[-\frac{1}{k_B T} \sum_{k=1}^{2} \Phi_{\mathrm{sw}}^{[k]}\left(z'; s_{z0}\right)\right] \exp\left[-\frac{2\pi \left(z' - z_1\right)^2}{\Lambda^2}\right]$$

$$\overset{\Lambda \to 0}{\simeq} \frac{\Lambda}{\sqrt{2}} \int\limits_{-s_{z0}/2}^{s_{z0}/2} \mathrm{d}z' \exp\left[-\frac{1}{k_B T} \sum_{k=1}^{2} \Phi_{\mathrm{sw}}^{[k]}\left(z'; s_{z0}\right)\right] \delta\left(z' - z_1\right)$$

$$\simeq \frac{\Lambda}{\sqrt{2}} \int\limits_{-\infty}^{\infty} \mathrm{d}z' \exp\left[-\frac{1}{k_B T} \sum_{k=1}^{2} \Phi_{\mathrm{sw}}^{[k]}\left(z'; s_{z0}\right)\right] \delta\left(z' - z_1\right)$$

$$= \frac{\Lambda}{\sqrt{2}} \exp\left[-\frac{1}{k_B T} \sum_{k=1}^{2} \Phi_{\mathrm{sw}}^{[k]}\left(z_1; s_{z0}\right)\right] \tag{5.166}$$

where we replaced the integration limits $\pm s_{z0}$ by $\pm\infty$ on account of the sharpness of the Dirac δ-function and the fact that $\Phi_{\mathrm{sw}}^{[k]}\left(z'; s_{z0}\right)$ diverges to infinity as $|z'| \rightarrow s_{z0}/2 - \sigma$ [see Eq. (5.126)]. Putting all this together we finally realize that

$$Z_2 - Z_1^2 = \pm\frac{A\Lambda^3}{2^{3/2}} \int\limits_{-s_{z0}/2}^{s_{z0}/2} \mathrm{d}z_1 \exp\left[-\frac{2}{k_B T} \sum_{k=1}^{2} \Phi_{\mathrm{sw}}^{[k]}\left(z_1; s_{z0}\right)\right] \tag{5.167}$$

This turns out to be very similar to the expression for Z_1 given in Eq. (5.159). Hence, we can immediately carry out the remaining integration to obtain

$$Z_2 - Z_1^2 = \pm\frac{A s_{z0} \Lambda^3}{2^{3/2}} \left\{\frac{2\sigma}{s_{z0}}\left[(\lambda - 1)\exp\left(\frac{2\varepsilon_{\mathrm{fs}}}{k_B T}\right) - \lambda\right] + 1\right\} \tag{5.168}$$

Inserting this expression together with Eq. (5.160) into Eq. (5.150) we yield[6]

---

[6]The expression in Eq. (5.169) is correct except for a factor of $2S + 1$ due the total spin S of the ideal quantum gas, which we have ignored from the very beginning for simplicity (see Section 2.5.1 and Ref. 19).

$$B_2 = \mp \frac{\Lambda^3}{2^{5/2}} \frac{(2\sigma/s_{z0})\left[(\lambda-1)\exp(2x)-\lambda\right]+1}{\{(2\sigma/s_{z0})\left[(\lambda-1)\exp(x)-\lambda\right]+1\}^2} \equiv \mp \frac{\Lambda^3}{2^{5/2}} f(x;\lambda,s_{z,0})$$

$$(5.169)$$

for the second virial coefficient of a confined ideal gas of Bosons $(-)$ and Fermions $(+)$, respectively, where $1/x \equiv k_{\mathrm{B}}T/\varepsilon_{\mathrm{fs}}$ is a dimensionless (i.e., "reduced") temperature.

For Eq. (5.169) to be physically meaningful, the function $f(x;\lambda,s_{z,0})$ must not have any poles. Hence, the parameter λ must be in a range such that the denominator of $f(x;\lambda,s_{z,0})$ has no zeros. These zeros are obtained as a solution of the expression

$$\exp(x) = \frac{\lambda - s_{z0}/2\sigma}{\lambda - 1} \tag{5.170}$$

Obviously, for $x \gtrsim 0$ this relation is meaningful only if $\lambda \gtrsim 1$. Moreover, because the left side of the previous expression cannot become negative, the denominator of $f(x;\lambda,s_{z0})$ cannot have any zeros if $\lambda \lesssim s_{z0}/2\sigma$. Therefore, the range of physically meaningful values of λ is bounded from above and below according to the inequality

$$1 \leq \lambda \leq \frac{s_{z0}}{2\sigma} \tag{5.171}$$

which is consistent with the definition of $\Phi_{\mathrm{sw}}^{[k]}(z;s_{z0})$ in Eq. (5.126).

Recalling from Eq. (5.126) that $\varepsilon_{\mathrm{fs}}$ determines the strength of fluid substrate attraction, we first focus on the case $\varepsilon_{\mathrm{fs}} = 0$ (i.e., $x = 0$), that is a slit-pore with "hard," repulsive solid surfaces for which

$$f(0;\lambda,s_{z0}) = \frac{1}{1 - 2\lambda\sigma/s_{z0}} \geq 1 \tag{5.172}$$

on account of the inequality stated in Eq. (5.171). In Eq. (5.172) the equality holds in the bulk, that is, in the limit $s_{z0} \to \infty$. In this latter case we obtain from Eq. (5.169) the well-known result (see footnote 6 above)

$$B_2 = \mp \frac{\Lambda^3}{2^{5/2}} \tag{5.173}$$

for the second virial coefficient of a bulk gas composed of either Bosons $(+)$ or Fermions $(-)$. In other words, the function $f(x;\lambda,s_{z,0})$ in Eq. (5.169) is a quantitative measure of confinement effects on the second virial coefficient.

Let us now turn to cases where we have attractive fluid substrate interactions in addition to repulsion; that is, we now focus on cases where $\varepsilon_{\mathrm{fs}} > 0$

[see Eq. (5.126)]. In this case $x > 0$ and Eq. (5.171) implies the inequalities

$$(\lambda - 1)\exp(2x) > 0 \tag{5.174a}$$

$$1 - \lambda\frac{2\sigma}{s_{z0}} > 0 \tag{5.174b}$$

so that

$$f(x; \lambda, s_{z0}) > 0 \tag{5.175}$$

because the denominator of $f(x; \lambda, s_{z0})$ is always positive.

To gain some more insight into the effect of variations of temperature and/or fluid substrate attraction it is necessary to investigate the dependence of $f(x; \lambda, s_{z,0})$ on x. This becomes possible by considering small variations of x around some reference value x_0 by expanding $f(x; \lambda, s_{z0})$ in a Taylor series around this reference value $x_0 > 0$. Because of the definition of the variable x (see above) this may be considered either as an expansion in terms of ε_{fs} or, alternatively, $1/T$. More specifically, we write

$$
\begin{aligned}
f(x; \lambda, s_{z0}) &\simeq f(x_0; \lambda, s_{z0}) + \delta x \left.\frac{df(x; \lambda, s_{z0})}{dx}\right|_{x=x_0} \\
&= f(x_0; \lambda, s_{z0}) + \delta x f'(x_0; \lambda, s_{z0})
\end{aligned}
\tag{5.176}
$$

where we have truncated the Taylor series after the linear term, which is always possible because the small quantity $\delta x \equiv x - x_0 \ll 1$ is at our disposal. From Eq. (5.169), it is somewhat tedious but straightforward to verify that

$$
\begin{aligned}
f'(x_0; \lambda, s_{z0}) &= -2\left(\frac{2\sigma}{s_{z0}}\right)^2 \frac{\exp(2x_0) - \exp(x_0)}{\{(2\sigma/s_{z0})[(\lambda - 1)\exp(x_0) - \lambda] + 1\}^3}(\lambda - 1)^2 \\
&\leq 0
\end{aligned}
\tag{5.177}
$$

which follows because $\exp(2x) - \exp(x) \geq 0$ for all $x \geq 0$ and because the denominator is positive for all values of λ satifying the inequality given in Eq. (5.171). Hence, it follows that

$$f(x_{i+1}; \lambda, s_{z0}) \lesssim f(x_i; \lambda, s_{z0}), \qquad \delta x = x_{i+1} - x_i > 0 \tag{5.178}$$

The interpretation of this inequality is quite straightforward. First, because of the definition of the variable x, $\delta x > 0$ corresponds either to an increase of fluid substrate attractivity (i.e., an increase of ε_{fs}) or, alternatively, to a decrease in temperature T. Second, because x cannot become negative by definition, $f(0; \lambda, s_{z0})$ given in Eq. (5.172) is an upper limit for the confinement-induced shift of the second virial coefficient of confined ideal quantum gases relative to its bulk value. The change in $f(x; \lambda, s_{z0})$

can, however, be realized in different ways. On account of the definition of x the inequality in Eq. (5.178) permits us to conclude that *increasing* the attractivity of the substrate at any given fixed temperature T reduces the confinement-induced enhancement of the (magnitude of) the second virial coefficient as predicted by Eq. (5.172). Likewise, for a given attractivity of the substrate [i.e., for fixed ε_{fs}, see Eq. (5.126)], reducing the temperature (i.e., $T_{i+1} < T_i$) also reduces the shift of B_2 relative to its bulk value caused by the presence of "hard" repulsive substrates. Because of this analysis it is conceivable that the function $f(x; \lambda, s_{z0})$ will generally satisfy

$$0 < f(x; \lambda, s_{z0}) \leq f(0; \lambda, s_{z0}) \tag{5.179}$$

This implies that a temperature T exists at which B_2 for the confined ideal quantum gas intersects B_2 for its bulk counterpart. However, confinement by attractive (or repulsive) solid surfaces cannot change the sign of B_2. In other words, B_2 for a confined gas of Bosons will always be negative, whereas it will always be positive for a gas of Fermions. Moreover, the monotonicity of $f(0; \lambda, s_{z0})$ as well as that of Λ cause B_2 to decrease monotonically toward zero with increasing temperature.

This observation is important because it also permits us to conclude that it will not be possible to construct a tangent through the origin at any point of the curve $B_2(T)$ no matter whether we consider a bulk or confined ideal quantum gas and irrespective of whether the quantum particles are Fermions or Bosons. In other words, for the ideal (bulk and confined) quantum gases, an inversion temperature T_{inv} does not exist because Eq. (5.155) does not have a solution. However, the reader should note that a Joule–Thomson effect does exist as pointed out in Section 5.7.1, namely a dilute gas of Bosons is always cooled upon an isenthalpic expansion ($B_2(T) < 0$), whereas a gas of Fermions is always heated during this process ($B_2(T) > 0$). The extent to which this happens is modified in a nontrivial way by confinement according to the above discussion.

5.7.4.4   Nonideal classic fluids

Despite the insight gained by considering the confined ideal quantum gas as a model system, the model itself is rather special in that it ignores fluid fluid interactions altogether. Hence, we now turn to nonideal, classic fluids in which the total configurational potential energy is given by Eq. (5.124) with the fluid substrate interaction as represented by models A and B according to the description in Section 5.7.2. In addition, we assume that for nonvanishing fluid fluid interactions the factor $r/\Lambda \gg 1$ such that quantum effects can be ignored. Under these conditions

$$
Z_1 = A \int_{-s_{z0}/2}^{s_{z0}/2} dz_1 \exp\left[-\frac{1}{k_{\mathrm{B}}T} \sum_{k=1}^{2} \Phi^{[k]}\left(z_1; s_{z0}\right) \right] \tag{5.180}
$$

and by similar reasoning as before [see Eq. (5.163)]

$$
\begin{aligned}
Z_2 - Z_1^2 &= 2\pi A \int_{-s_{z0}/2}^{s_{z0}/2} dz_1 \exp\left[-\frac{1}{k_{\mathrm{B}}T} \sum_{k=1}^{2} \Phi^{[k]}\left(z_1; s_{z0}\right) \right] \\
&\quad \times \int_{-s_{z0}/2-z_1}^{s_{z0}/2-z_1} dz \exp\left[-\frac{1}{k_{\mathrm{B}}T} \sum_{k=1}^{2} \Phi^{[k]}\left(z + z_1; s_{z0}\right) \right] \\
&\quad \times \int_{0}^{\infty} \rho d\rho \left\{ \exp\left[-\frac{u_{\mathrm{ff}}\left(\rho, z\right)}{k_{\mathrm{B}}T} \right] - 1 \right\}
\end{aligned} \tag{5.181}
$$

similar to the expressions given in Eqs. (5.158) and (5.163) above. However, because of the form of u_{ff} and $\Phi^{[k]}\left(z; s_{z0}\right)$ for models A and B (see Section 5.7.2) these integrals cannot be evaluated analytically, but they are amenable to a numerical evaluation using standard quadrature techniques. This finally permits a numerical calculation of the second virial coefficient B_2 from Eq. (5.150) on which the subsequent results for the inversion temperature in the limit of vanishingly small gas densities will be based.

5.7.5   Confined fluids at moderate densities

The above considerations are only valid in the limit of very small gas densities. However, in general the inversion temperature can be expected to depend on density as well. To incorporate the density dependence we have to go beyond the second virial coefficient in our expansion of $\tau_{\parallel}$ in Eq. (5.149). Considering larger densities of the confined gas, the virial expansion of $\tau_{\parallel}$ would need to involve many more terms if such a power series in $\overline{\rho}$ at all converges. Hence, to calculate the inversion temperature at higher densities up to the critical density of the confined gas, an alternative approach is required. It becomes possible by employing a mean-field description of the confined fluid, which we discussed in Section 4.2.2. Differentiating $\tau_{\parallel}/k_{\mathrm{B}}T$ given in Eq. (4.28) we obtain [cf. Eq. (5.151)]

$$
\frac{\tau_{\parallel}}{k_{\mathrm{B}}T}\frac{1}{T} = \left[\frac{1}{\left(1 - b\overline{\rho}\right)^2} - \frac{2a_{\mathrm{p}}\left(\xi\right)\overline{\rho}}{k_{\mathrm{B}}T} \right] \left(\frac{\partial\overline{\rho}}{\partial T} \right)_{N,\tau_{\parallel},s_z} + \frac{a_{\mathrm{p}}\left(\xi\right)\overline{\rho}^2}{k_{\mathrm{B}}T}\frac{1}{T} \tag{5.182}
$$

Replacing on the left side of this expression $\tau_\parallel / k_B T$ as before by employing again the equation of state in Eq. (4.28), it is a simple matter to show that

$$\alpha_\parallel = \frac{k_B \left(1 - b\overline{\rho}\right)}{k_B T - 2a_p \left(\xi\right) \left(1 - b\overline{\rho}\right)^2 \overline{\rho}} \tag{5.183}$$

where we also used Eq. (5.153) for the expansivity. Using the mean-field expression for $\alpha_\parallel$ and the thermodynamic definition for the inversion temperature it requires nothing but straightforward algebra to demonstrate that at the mean-field level

$$k_B T_{\text{inv}} \left(\overline{\rho}\right) = \frac{2a_p \left(\xi\right)}{b} \left(1 - b\overline{\rho}\right)^2 \tag{5.184}$$

which shows that the inversion temperature depends on the density of the confined gas as anticipated.

However, in the limit $\overline{\rho} \to 0$, the mean-field treatment must be consistent with the one developed in Section 5.7.4. From Eq. (5.184) we see that in this limit

$$\lim_{\overline{\rho} \to 0} k_B T_{\text{inv}} \left(\overline{\rho}\right) = k_B T_{\text{inv}} \left(0\right) = \frac{2a_p \left(\xi\right)}{b} \tag{5.185}$$

This latter expression can be derived independently by expanding in the mean-field equation of state [see Eq. (4.28)] the term $1 / \left(1 - b\overline{\rho}\right)$ $(b\overline{\rho} \gg 1)$ in a MacLaurin series according to

$$\frac{1}{1 - b\overline{\rho}} = 1 + \sum_{k=1}^{\infty} \left(b\overline{\rho}\right)^k \tag{5.186}$$

Inserting this expansion into the mean-field equation of state and considering only terms up to second order in density, one can show that

$$\tau_\parallel = \tau_\parallel^{\text{id}} - k_B T \left[b - \frac{a_p \left(\xi\right)}{k_B T}\right] \overline{\rho}^2 + \mathcal{O} \left(\overline{\rho}\right) \tag{5.187}$$

which may be compared with Eq. (5.149) to conclude that the second virial coefficient at the mean-field level is given by

$$B_2^{\text{mf}} \left(T\right) = b - \frac{a_p \left(\xi\right)}{k_B T} \tag{5.188}$$

Inserting this expression into Eq. (5.155), we find

$$\frac{\mathrm{d}}{\mathrm{d}T} \frac{B_2^{\text{mf}} \left(T\right)}{T} \bigg|_{T=T_{\text{inv}}} = \frac{\mathrm{d}}{\mathrm{d}T} \left[\frac{b}{T} - \frac{a_p \left(\xi\right)}{k_B T^2}\right]\bigg|_{T=T_{\text{inv}}} = -\frac{b}{T_{\text{inv}}^2} + \frac{2a_p \left(\xi\right)}{k_B T_{\text{inv}}^3} = 0 \tag{5.189}$$

from which Eq. (5.185) follows immediately, thereby proving the consistency between the current mean-field theoretical treatment and the virial expansion in the limit of vanishing density.

As we already demonstrated that the mean-field treatment developed in Section 4.2.2 is capable of describing, for instance, capillary condensation in nanoscopic porous media in a qualitatively correct fashion (see Section 4.2.4), the above discussion permits us to draw some important preliminary conclusions concerning the Joule-Thomson effect in confined fluids. These conclusions, bolstered further by corresponding MC data to be presented below in Sections 5.7.8 and 5.7.9, can be summarized as follows:

1. The inversion temperature decreases with increasing density. This follows from Eqs. (5.184) and (5.185) from which the inequality

$$T_{\mathrm{inv}}\left(\overline{\rho}\right) = T_{\mathrm{inv}}\left(0\right) \underbrace{\left(1 - b\overline{\rho}\right)^2}_{\leq 1} \leq 0 \qquad (5.190)$$

is readily deduced. The equal sign holds in the limit of vanishing density.

2. The inversion temperature of a confined gas becomes lower the more severely confined is the gas. This follows from Eq. (5.184) and the definition of $a_{\mathrm{p}}\left(\xi\right)$ in Eq. (4.24) which turns out to become smaller the smaller is s_z (i.e., the smaller is ξ). This implies that a gas that might be *cooled* during an isenthalpic expansion in a wider porous medium $(\delta_{\|} < 0)$ may get *heated* in a narrower porous medium instead $(\delta_{\|} > 0)$. At the mean-field level, the magnitude of the associated confinement-induced shift of the inversion temperature is given quantitatively by the term in brackets in Eq. (4.24).

3. The inversion temperature does not depend on the chemical nature of the substrate because neither $a_{\mathrm{p}}\left(\xi\right)$ nor b in Eqs. (5.184) or (5.185) depend on any parameter describing a specific substrate. Similar conclusions could not be drawn on the basis of the much more involved expression for $B_2\left(T\right)$ in Eq. (5.150).

4. Existence of an inversion temperature is solely linked to *attractive* fluid fluid interactions modulated by confinement. In the absence of these attractions, $a_{\mathrm{p}}\left(\xi\right) = 0$ and therefore $B_2^{\mathrm{inf}}\left(T\right) = b$. If this expression is inserted into Eq. (5.155), one realizes that $T_{\mathrm{inv}} = 0$ is the only possible solution.

5.7.6 Exact treatment of the Joule-Thomson coefficient

Even though the mean-field treatment in the preceding section led to some detailed insight concerning the impact of confinement on the inversion temperature, the analysis in both Sections 5.7.4 and 5.7.5 appears to be somewhat hampered in the sense that it was either limited to very low gas densities or that it was based on a mean-field assumption explicitly stated in Eqs. (4.19) and (4.20). To test the predictions of the two previous approaches, we need to tackle the Joule-Thomson effect by an approach that is free of any additional assumptions. In this regard the MC technique is again ideally suited because MC simulations should be regarded as a first-principles method according to the opening discussion of this chapter in Section 5.1. As we showed in Eq. (5.135), a determination of the inversion temperature T_{inv} essentially requires computation of the expansivity $\alpha_\parallel$. Therefore, we begin by deriving an exact expression suitable for an evaluation in the subsequent MC simulations.

5.7.6.1 Partition function

As we showed in Section 2.5.4, thermal averages in isostress isostrain ensembles can be related through a Laplace transformation. Hence, for the conjugate stress $\tau_\parallel$ and strain A, we may employ Eq. (2.121) and change variables according to $\tau_{zz} \to \tau_\parallel$ and $s_z \to A$ giving

$$\langle O(\tau_\parallel) \rangle = \frac{Q_{cl}}{\Upsilon_{\parallel,cl}} \int_0^\infty dA \exp\left[\frac{\tau_\parallel A s_{z0}}{k_B T}\right] \langle O(A) \rangle$$

$$= \frac{1}{N!\Lambda^{3N}\Upsilon_{\parallel,cl}} \int_0^\infty dA \exp\left[\frac{\tau_\parallel A s_{z0}}{k_B T}\right]$$

$$\times \int dr^N O(r^N; A) \exp\left[-\frac{U(r^N; A)}{k_B T}\right] \qquad (5.191)$$

where we also replaced the grand canonical partition function Ξ_{cl} by the partition function of the canonical ensemble Q_{cl} because we are concerned with systems accommodating a fixed number of fluid molecules. As, on the other hand,

$$\langle O(\tau_\parallel) \rangle = \sum_A \int dr^N O(r^N; A) p(r^N; A) \qquad (5.192)$$

must also hold, where $p(r^N; A)$ is the probability density for a specific configuration r^N of fluid molecules in a lamella exposed to a compressional (dilational) strain proportional to Λ, a comparison with Eq. (5.191) suggests that

$$p\left(\boldsymbol{r}^{N};A\right)=\frac{1}{N!\Lambda^{3N}\Upsilon_{\parallel,\mathrm{cl}}}\exp\left[-\frac{U\left(\boldsymbol{r}^{N};A\right)-\tau_{\parallel}As_{z0}}{k_{\mathrm{B}}T}\right] \tag{5.193}$$

In Eq. (5.193) the normalization condition

$$\sum_{A}\int\mathrm{d}\boldsymbol{r}^{N}p\left(\boldsymbol{r}^{N};A\right)=1 \tag{5.194}$$

suggesting that the partition function is given by

$$\Upsilon_{\parallel,\mathrm{cl}}=\frac{1}{N!\Lambda^{3N}}\sum_{A}\int\mathrm{d}\boldsymbol{r}^{N}\exp\left[-\frac{U\left(\boldsymbol{r}^{N};A\right)-\tau_{\parallel}As_{z0}}{k_{\mathrm{B}}T}\right] \tag{5.195}$$

where we tacitly assumed that summation on and integration over A are equivalent operations.

In the thermodynamic limit we may apply the maximum term method (see Appendix B.4) to write

$$\begin{aligned}\Upsilon_{\parallel,\mathrm{cl}} &\simeq \frac{1}{N!\Lambda^{3N}}\exp\left[\frac{\tau_{\parallel}A^{\star}s_{z0}}{k_{\mathrm{B}}T}\right]\int\mathrm{d}\boldsymbol{r}^{N}\exp\left[-\frac{U\left(\boldsymbol{r}^{N};A^{\star}\right)}{k_{\mathrm{B}}T}\right] \\ &= \exp\left[\frac{\tau_{\parallel}A^{\star}s_{z0}}{k_{\mathrm{B}}T}\right]\mathcal{Q}_{\mathrm{cl}}\left(T,N,A^{\star},s_{z}\right)\end{aligned} \tag{5.196}$$

Taking the logarithm of this expression we have from Eqs. (2.79), (5.131), and the Legendre transform $\mathcal{F}\equiv\mathcal{U}-T\mathcal{S}$ that

$$\mathcal{G}_{\parallel}=-k_{\mathrm{B}}T\ln\Upsilon_{\parallel}\left(T,N,\tau_{\parallel},s_{z}\right) \tag{5.197}$$

is the relevant thermodynamic potential where here and below we shall drop the subscript "cl" of $\Upsilon_{\parallel}$ as well as the asterisk as a superscript of A to ease the notational burden.

5.7.6.2 Thermal averages and their fluctuations

Differentiating Eq. (5.197) with respect to T, we obtain

$$\left(\frac{\partial\mathcal{G}_{\parallel}}{\partial T}\right)_{\{\cdot\}\backslash T}=\frac{\mathcal{G}_{\parallel}}{T}-k_{\mathrm{B}}T\left(\frac{\partial\ln\Upsilon_{\parallel}}{\partial T}\right)_{\{\cdot\}\backslash T}=-\mathcal{S} \tag{5.198}$$

where the far right side follows from Eq. (5.131). The second term on the right side of this expression may be further evaluated by realizing that

$$
k_\mathrm{B}T\left(\frac{\partial \ln \Upsilon_\parallel}{\partial T}\right)_{\{\cdot\}\backslash T} = \frac{k_\mathrm{B}T}{N!\Lambda^{3N}\Upsilon_\parallel}\frac{\partial}{\partial T}\sum_A \int d\boldsymbol{r}^N \exp\left[-\frac{\mathsf{H}\left(\boldsymbol{r}^N;A\right)}{k_\mathrm{B}T}\right]
$$
$$
-\frac{3Nk_\mathrm{B}T}{\Lambda}\frac{\partial \Lambda}{\partial T}
$$
$$
= \frac{\langle \mathsf{H}\left(\boldsymbol{r}^N;A\right)\rangle}{T} + \frac{3Nk_\mathrm{B}}{2} = \frac{\mathcal{H}}{T} \qquad (5.199)
$$

where we introduce for convenience

$$
\mathsf{H}\left(\boldsymbol{r}^N;A\right) \equiv U\left(\boldsymbol{r}^N;A\right) - \tau_\parallel A s_{z0} \qquad (5.200)
$$

and with it a statistical expression for the enthalpy $\mathcal{H}$ that was obtained in Eq. (5.128) as a Legendre transform of the internal energy

$$
\mathcal{U} \equiv \frac{3Nk_\mathrm{B}T}{2} + \langle U\left(\boldsymbol{r}^N;A\right)\rangle \qquad (5.201)
$$

From Eq. (5.198) and the thermodynamic definition of $c_\parallel$ in Eq. (5.130), it is straightforward to realize that

$$
c_\parallel = \frac{\partial}{\partial T}\left(k_\mathrm{B}T^2\frac{\partial \ln \Upsilon_\parallel}{\partial T}\right)_{\{\cdot\}\backslash T} = \frac{\partial}{\partial T}\langle \mathsf{H}\left(\boldsymbol{r}^N;A\right)\rangle + \frac{3Nk_\mathrm{B}}{2} \qquad (5.202)
$$

To evaluate the partial derivative on the far right side of this expression we rewrite it more explicitly as

$$
\frac{\partial}{\partial T}\langle \mathsf{H}\left(\boldsymbol{r}^N;A\right)\rangle = \frac{1}{N!}\frac{\partial}{\partial T}\frac{1}{\Lambda^{3N}\Upsilon_\parallel}\sum_A \int d\boldsymbol{r}^N \mathsf{H}\left(\boldsymbol{r}^N;A\right)\exp\left[-\frac{\mathsf{H}\left(\boldsymbol{r}^N;A\right)}{k_\mathrm{B}T}\right]
$$
$$
= \frac{\partial}{\partial T}\frac{\sum_A \int d\boldsymbol{r}^N \mathsf{H}\exp\left[-\mathsf{H}/k_\mathrm{B}T\right]}{\sum_A \int d\boldsymbol{r}^N \exp\left[-\mathsf{H}/k_\mathrm{B}T\right]}
$$
$$
= \frac{1}{k_\mathrm{B}T^2}\left[\langle \mathsf{H}^2\rangle - \langle \mathsf{H}\rangle^2\right] \qquad (5.203)
$$

Thus, putting together these last two expressions, we obtain

$$
c_\parallel = \frac{1}{k_\mathrm{B}T^2}\left[\langle \mathsf{H}^2\rangle - \langle \mathsf{H}\rangle^2\right] + \frac{3Nk_\mathrm{B}}{2} \qquad (5.204)
$$

as an exact statistical physical expression for the isostress heat capacity where H is defined in Eq. (5.200).

An analogous expression may be derived by using $\Upsilon_{\parallel}$ [see Eq. (5.196)] in Eq. (5.202), which gives

$$\frac{\partial}{\partial T}\left(k_B T^2 \frac{\partial \ln \Upsilon_{\parallel}}{\partial T}\right) \simeq \frac{\partial}{\partial T}\left(k_B T^2 \frac{\partial \ln \mathcal{Q}}{\partial T} - \tau_{\parallel} A s_z\right) = \frac{\partial}{\partial T}\left(k_B T^2 \frac{\partial \ln \mathcal{Q}}{\partial T}\right)$$
(5.205)

Using the expression for $\mathcal{Q}$ we can apply the above considerations to obtain an expression parallel to that given in Eq. (5.204) replacing, however, $\mathsf{H}\left(r^N; A\right)$ by $U\left(r^N; \right)$. This way

$$c_{\sigma} = \frac{1}{k_B T^2}\left[\langle U^2\rangle - \langle U\rangle^2\right] + \frac{3N k_B}{2}$$
(5.206)

which we identify as the isostrain heat capacity defined thermodynamically in Eq. (5.139). This would become immediately apparent if we replace in Eq. (5.197), $\mathcal{G}_{\parallel} \to \mathcal{F}$ and $\Upsilon_{\parallel} \to \mathcal{Q}$ [see Eq. (2.79)] and repeat the above analysis step by step.

Besides the heat capacities $c_{\parallel}$ and c_{σ}, the isothermal compressibility $\kappa_{\parallel}$ is given by

$$\kappa_{\parallel} \equiv \frac{1}{A}\left(\frac{\partial A}{\partial \tau_{\parallel}}\right)_{\{\cdot\}\backslash\tau_{\parallel}} = -\frac{1}{As_{z0}}\left(\frac{\partial^2 \mathcal{G}_{\parallel}}{\partial \tau_{\parallel}^2}\right)_{\{\cdot\}\backslash\tau_{\parallel}}$$
(5.207)

From these thermodynamic definitions and the statistical expression for the generalized Gibbs potential in Eq. (5.197), it is straightforward to show that

$$\begin{aligned}
\kappa_{\parallel} &= \frac{s_{z0}}{k_B T}\frac{1}{\langle A\rangle}\left\{\frac{1}{N!\Lambda^{3N}\Upsilon_{\parallel}}\sum_A \int dr^N A^2 \exp\left[-\frac{\mathsf{H}\left(r^N; A\right)}{k_B T}\right]\right.\\
&\quad \left. - \left(\frac{1}{N!\Lambda^{3N}\Upsilon_{\parallel}}\sum_A \int dr^N A \exp\left[-\frac{\mathsf{H}\left(r^N; A\right)}{k_B T}\right]\right)^2\right\}\\
&= \frac{s_{z0}}{k_B T}\frac{1}{\langle A\rangle}\left[\langle A^2\rangle - \langle A\rangle^2\right]
\end{aligned}$$
(5.208)

which differs in an interesting way from the parallel expression that would be obtained in the grand canonical ensemble from Eqs. (1.81) and (2.75) where the transverse compressibility is expressed in terms of fluctuations in the number of molecules rather than the shape of the lamella.

However, the key quantity in the context of the Joule-Thomson effect is the expansivity defined in Eq. (5.132). From the thermodynamic definition and the statistical physical Eqs. (5.197) and (5.198), it is immediately clear that

$$\left[\frac{\partial}{\partial \tau_{\parallel}}\left(\frac{\partial \mathcal{G}_{\parallel}}{\partial T}\right)_{\{\cdot\}\backslash T}\right]_{\{\cdot\}\backslash\tau_{\parallel}} = -k_B\left(\frac{\partial \ln \Upsilon_{\parallel}}{\partial \tau_{\parallel}}\right)_{\{\cdot\}\backslash\tau_{\parallel}} - \frac{\partial}{\partial \tau_{\parallel}}\frac{\langle \mathsf{H}\left(r^N; A\right)\rangle}{T}$$
(5.209)

where we also used Eq. (5.199). It can then be shown that

$$
-k_{\mathrm{B}}\left(\frac{\partial \ln \Upsilon_{\|}}{\partial \tau_{\|}}\right)_{\{\cdot\}\backslash\tau_{\|}} = \frac{1}{T}\frac{1}{N!\Lambda^{3N}\Upsilon_{\|}}\sum_A \int dr^N \frac{\partial \mathsf{H}}{\partial \tau_{\|}}\exp\left[-\frac{\mathsf{H}\left(r^N; A\right)}{k_{\mathrm{B}}T}\right]
$$
$$
= -\frac{s_{z0}}{T}\langle A\rangle \tag{5.210}
$$

which follows directly from the definition of $\mathsf{H}\left(r^N; A\right)$ in Eq. (5.200) and the definition of the probability density $p\left(r^N; A\right)$ in Eq. (5.193). For the second term in Eq. (5.209), we obtain by straightforward differentiation

$$
-\frac{\partial}{\partial \tau_{\|}}\frac{\langle \mathsf{H}\left(r^N; A\right)\rangle}{T} = -\frac{1}{T}\frac{1}{N!\Lambda^{3N}\Upsilon_{\|}}\sum_A \int dr^N\left(\frac{\partial \mathsf{H}}{\partial \tau_{\|}} - \frac{\mathsf{H}}{k_{\mathrm{B}}T}\frac{\partial \mathsf{H}}{\partial \tau_{\|}}\right)
$$
$$
\times \exp\left[-\frac{\mathsf{H}\left(r^N; A\right)}{k_{\mathrm{B}}T}\right]
$$
$$
-\frac{1}{k_{\mathrm{B}}T^2}\frac{1}{\left(N!\Lambda^{3N}\Upsilon_{\|}\right)^2}\sum_A \int dr^N \mathsf{H}\exp\left[-\frac{\mathsf{H}\left(r^N; A\right)}{k_{\mathrm{B}}T}\right]
$$
$$
\times \sum_A \int dr^N \frac{\partial \mathsf{H}}{\partial \tau_{\|}}\exp\left[-\frac{\mathsf{H}\left(r^N; A\right)}{k_{\mathrm{B}}T}\right]
$$
$$
= \frac{s_{z0}}{T}\langle A\rangle - \frac{s_{z0}}{k_{\mathrm{B}}T^2}\left[\langle \mathsf{H}A\rangle - \langle \mathsf{H}\rangle\langle A\rangle\right] \tag{5.211}
$$

Combining these two expressions with the thermodynamic definition of $\alpha_{\|}$, we see that

$$
\alpha_{\|} = \frac{1}{k_{\mathrm{B}}T^2}\frac{\langle \mathsf{H}A\rangle - \langle \mathsf{H}\rangle\langle A\rangle}{\langle A\rangle} \tag{5.212}
$$

obtains without further ado. In Table 5.5 we summarize the results for the various response coefficients $\alpha_{\|}$, $\kappa_{\|}$, $c_{\|}$, and c_σ.

5.7.7   Isostress isostrain ensemble MC simulations

According to the above discussion the distribution of microstates in the current mixed isostress isostrain ensemble is governed by the probability density given in Eq. (5.193). The similarity between the present probability density and the one relevant in the closely related ensemble discussed in Section 5.2.4 suggests we should design an adapted Metropolis algorithm closely related to the one described in that section. In fact, from the detailed discussion in Section 5.2.4, it turns out that we just need to replace the substrate separation s_z by the area A. More specifically, we need to replace Eqs. (5.46)–(5.48)

Table 5.5: Overview of fluctuation-related response coefficients in the mixed isostress isostrain ensemble at constant T, N, $\tau_{\parallel}$, and s_{z0} (see text for details of the derivations).

Coefficient	Expression	See Equation
$c_{\parallel}$	$3Nk_{\mathrm{B}}/2 + \left[\langle \mathsf{H}^2 \rangle - \langle \mathsf{H} \rangle^2\right]/k_{\mathrm{B}}T^2$	(5.204)
c_{σ}	$3Nk_{\mathrm{B}}/2 + \left[\langle U^2 \rangle - \langle U \rangle^2\right]/k_{\mathrm{B}}T^2$	(5.206)
$\kappa_{\parallel}$	$s_{z0}\left[\langle A^2 \rangle - \langle A \rangle^2\right]/\left(\langle A \rangle k_{\mathrm{B}}T\right)$	(5.208)
$\alpha_{\parallel}$	$\left[\langle \mathsf{H}A \rangle - \langle \mathsf{H} \rangle \langle A \rangle\right]/\left(\langle A \rangle k_{\mathrm{B}}T^2\right)$	(5.212)

by

$$p'\left(\widetilde{\boldsymbol{r}}^N; A\right) = A^N p\left(\widetilde{\boldsymbol{r}}^N; A\right) \tag{5.213a}$$

$$\frac{p'\left(\widetilde{\boldsymbol{r}}^N; A_n\right)}{p'\left(\widetilde{\boldsymbol{r}}^N; A_{n-1}\right)} = \exp\left(r_{\mathrm{A}}\right) \tag{5.213b}$$

$$r_{\mathrm{A}} = -\frac{\Delta U - \tau_{\parallel}s_{z0}\Delta A}{k_{\mathrm{B}}T} + N\ln\left(\frac{A_n}{A_{n-1}}\right) \tag{5.213c}$$

where, by analogy with Eqs. (5.49),

$$\Delta U \equiv U\left(\widetilde{\boldsymbol{r}}_n^N; A_n\right) - U\left(\widetilde{\boldsymbol{r}}_{n-1}^N; A_{n-1}\right) \tag{5.214a}$$

$$\Delta A \equiv A_n - A_{n-1} \tag{5.214b}$$

Again we associate the random process $y_{n-1} \to y_n$ (see Appendix E.1) with probability

$$\Pi_3 \equiv \min\left[1, \exp\left(r_{\mathrm{A}}\right)\right] \tag{5.215}$$

where we realize a change in A according to

$$\exp\left(r_{\mathrm{A}}\right) > \xi \longrightarrow \text{accept change in area} \tag{5.216a}$$

$$\exp\left(r_{\mathrm{A}}\right) \leq \xi \longrightarrow \text{reject change in area} \tag{5.216b}$$

according to a modified Metropolis criterion if $r_{\mathrm{A}} < 0$, where ξ is a pseudo-random number as before; if, on the other hand, $r_{\mathrm{A}} \geq 0$ the process $A_{n-1} \to A_n$ is accepted without further ado.

As before in Section 5.2.4 generation of a sequence of configurations proceeds in two steps. Step 1 is again identical with step 1 of the corresponding adapted Metropolis algorithm for GCEMC simulations (see Section 5.2.2): a random displacement of a single molecule governed by a transition probability Π_1 [see Eq. (5.11)]. In step 2 the area $A = s_x s_y$ is changed according to

$$s_{x,n} = s_{x,n-1} + \delta_\parallel (1 - 2\xi) \qquad (5.217a)$$
$$s_{y,n} = s_{y,n-1} + \delta_\parallel (1 - 2\xi) \qquad (5.217b)$$

such that [cf. Eq. (5.53)]

$$\tilde{\boldsymbol{r}}_n^N = \begin{pmatrix} \tilde{x}_n^N \\ \tilde{y}_n^N \\ \tilde{z}_n^N \end{pmatrix} = \begin{pmatrix} \tilde{x}_{n-1}^N \\ \tilde{y}_{n-1}^N \\ \tilde{z}_{n-1}^N \end{pmatrix} \cdot \begin{pmatrix} s_{x,n}/s_{x,n-1} \\ s_{y,n}/s_{y,n-1} \\ 1 \end{pmatrix} \qquad (5.218)$$

Because in step 2 of the present algorithm the entire set of N x and y-coordinates are changed at once, steps 1 and 2 are carried out with a frequency $N:1$.

To demonstrate the validity of these operations in generating properly a numerical representation of a Markov chain of configurations, the thermodynamic consistency relation in Eq. (5.140) turns out to be particularly useful. It can be employed to calculate c_σ from the various response coefficients listed in Table 5.5 that we calculate as ensemble averages in the current isostress isostrain ensemble simulations. We also note from Eq. (5.206) that c_σ might as well be calculated in corresponding MC simulations in the canonical ensemble by fixing the dimensions of the simulation cell to $\langle s_x \rangle$ and $\langle s_y \rangle$ determined as ensemble averages in the isostress isostrain ensemble simulations. Entries in Table 5.6 show that the agreement between Eqs. (5.140) and (5.206) is always better than 5%, which seems remarkable in view of the relatively wide range of thermodynamic states considered.

5.7.8 Inversion temperature at low density

Attending now to T_{inv} as the key quantity of this study, it seems sensible to begin discussing the limit of vanishing fluid density. In this limit $T_{\text{inv}}(0)$ is obtained as a (numerical) solution of Eq. (5.155), where $B_2(T)$ is obtained from Eqs. (5.158), (5.161), and (5.150) by numerical integration. Results for $B_2(T)$ are plotted in Figs. 5.27 for various cases studied. Generally speaking, over the temperature range plotted, $B_2(T)$ is a monotonically increasing function of T, where $B_2(T) < 0$ below the Boyle temperature T_{Boyle} and $B_2(T) > 0$ otherwise; at $T = T_{\text{Boyle}}$, $B_2(T) = 0$ and the fluid behaves like an ideal gas [see Eq. (5.149)] (disregarding, of course, higher-order virial coefficients).

Confinement causes $B_2(T)$ to be shifted with respect to the bulk curve. If the substrate potential is wettable [model A, see Fig. 5.27(a)], $B_2(T)$ is

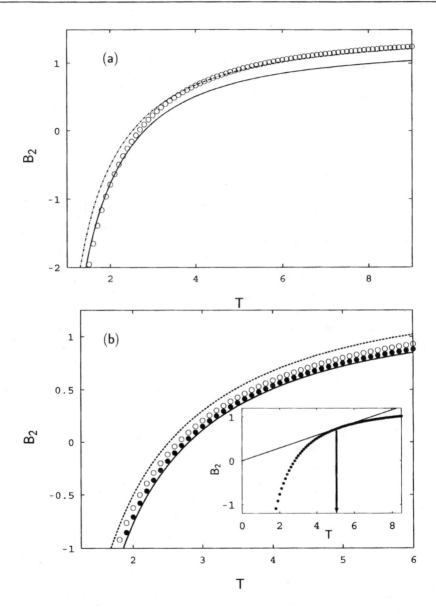

Figure 5.27: Second virial coefficient $B_2(T)$ as a function of temperature T for bulk (—), model A ($- \cdot -$), and model B (○) ($s_{z0} = 10$ for confined fluids). Curves are obtained by numerical integration (see text). Intersections between $B_2(T)$ and the solid horizontal line define the Boyle temperature. (b) As (a) but for bulk (—) and model A for $s_{z0} = 10$ (- - -), $s_{z0} = 20$ (○), and $s_{z0} = 50$ (●). Inset shows $B_2(T)$ for bulk (●) and the tangent through the origin ($- - -$) defining the inversion temperature demarcated by the vertical arrow [see Eq. (5.155)].

Table 5.6: Comparison of results for isostrain heat capacity from consistency relation Eq. (5.140) with directly computed values from canonical ensemble (CE) [see eq. (5.206)] for model A (see text).

T	$\bar{\rho}$	$-\tau_\parallel$	$\alpha_\parallel$	$\kappa_\parallel$	$c_\parallel^N$	c_σ^N	c_σ^N (CE)
1.50	0.577	1.50	0.384	0.173	4.243	2.027	2.040
1.50	0.514	1.00	0.532	0.327	4.488	1.962	1.943
1.50	0.461	0.75	0.676	0.522	4.765	1.917	1.933
1.50	0.378	0.50	1.028	1.224	5.345	1.920	1.921
1.50	0.203	0.35	1.256	2.581	5.279	2.085	2.016
1.50	0.287	0.25	1.305	4.526	4.970	2.190	2.111
2.00	0.532	2.00	0.292	0.173	3.784	1.931	1.933
2.00	0.478	1.50	0.375	0.282	3.977	1.886	1.861
3.00	0.409	2.00	0.233	0.255	3.320	1.758	1.763
3.00	0.267	1.00	0.336	0.748	3.369	1.660	1.684

shifted to more positive values irrespective of T. If the substrate is nonwettable [model B, see Fig. 5.27(a)], $B_2(T)$ turns out to be smaller than for model A for $T \lesssim 4$; it is even lower than the second virial coefficient in the bulk for $T \lesssim 2$. For high temperatures, however, the plots in Fig. 5.27(a) show that, for model B, $B_2(T)$ exceeds all other curves ($T \gtrsim 6$). The different temperature dependence of $B_2(T)$ between various models illustrates the impact of wettability of the substrate on thermophysical quantities of confined fluids in the limit of low densities.

If, on the other hand, the degree of confinement decreases (i.e., with increasing s_{z0}), $B_2(T)$ for a confined fluid is expected to approach its bulk counterpart because of the diminishing influence of fluid substrate interactions. This notion is confirmed by the plots in Fig. 5.27(b). However, it seems worthwhile emphasizing that, even for the largest substrate separation studied ($s_{z0} = 50$), $B_2(T)$ for confined and bulk fluids differ slightly but significantly even though the range of the fluid substrate interaction potential does not exceed a distance of a few molecular diameters from either substrate so that the dominant portion of the confined phase is not subjected to interactions with that substrate. The remarkably large range of distances over which substrate-induced effects prevail was also noted with respect to adsorption phenomena in the subcritical regime [218].

For the curves plotted in Figs. 5.27(a), $T_{inv}(0)$ is calculated from differential equation Eq. (5.155). Results are compiled in Table 5.7 for models A and B and $s_{z0} = 10$. They show that $T_{inv}(0)$ is higher for a hydrophobic

Table 5.7: Inversion temperature in the vanishing-density limit [see Eq. (5.155), Fig. 5.27(b)].

Model	s_{z0}	$T_{inv}(0)$	T_{Boyle}	$T_{Boyle}/T_{inv}(0)$
B	10.0	4.985	2.681	1.860
A	5.0	4.464	2.300	1.941
A	10.0	4.841	2.521	1.920
A	20.0	5.051	2.660	1.899
A	50.0	5.177	2.747	1.885
A	100.0	5.217	2.776	1.879
Bulk	∞	5.259	2.805	1.875

substrate (i.e., if fluid substrate interactions are purely repulsive). Table 5.7 also indicates that in the limit $s_{z0} \rightarrow \infty$ the bulk inversion temperature is approached in accord with the plots in Fig. 5.27(b).

From the mean-field expressions Eqs. (4.24) and (5.185) one expects the difference

$$\Delta T_{inv}(0; s_{z0}) \equiv T_{inv}(0; \infty) - T_{inv}(0; s_z) \propto s_{z0}^{-1} \qquad (5.219)$$

The plot in Fig. 5.28 shows that data compiled in Table 5.7 are consistent with this scaling relation except for $s_{z0} = 5.0$ where the assumption of homogeneity of the confined fluid, on which the mean-field theory is based (see Section 4.2.2), can hardly be expected to be valid. The results in this section therefore confirm our expectation that the inversion temperature should depend on the substrate separation and that it becomes higher the more severely confined is the fluid (i.e., the smaller s_{z0} becomes).

However, it seems worthwhile stressing that the relation between Boyle and inversion temperatures is only approximately described by the mean-field theory. For example, the mean-field Eqs. (5.185) and (5.188) predict $T_{Boyle}/T_{inv}(0) = 2$ irrespective of s_{z0}, but entries in Table 5.7 show that this ratio is lower and depends on s_{z0} as well as on the chemical nature of the substrate. This clearly indicates that the mean-field treatment developed in Sections 4.2.2 and 5.7.5 is not fully adequate as one would have expected. However, the deviation from the limiting value $T_{Boyle}/T_{inv}(0) = 2$ does not exceed 6.5% for $s_{z0} = 100$, where the mean-field treatment is expected to work best. This, on the other hand, shows that mean-field theory is quite useful to understand the behavior of confined fluids at least from a qualitative point of view.

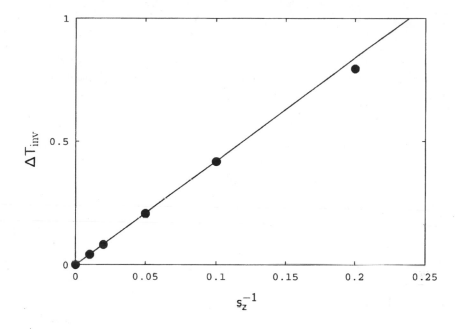

Figure 5.28: $\Delta T_{\mathrm{inv}}(0; s_{z0})$ as a function of inverse substrate separation $1/s_{z0}$ in the limit of vanishing fluid density [see Eq. (5.219)]. According to mean-field theory, data points should fall on a straight line through the origin (see text). The straight solid line is a fit to the data points using only entries for $1/s_{z0} = 0, 0.01$.

5.7.9   Density dependence of the inversion temperature

In accord with the mean-field theory developed in Section 5.7.5, the inversion temperature, however, does depend on the density of the fluid. This can be seen from plots of $T\alpha_\| - 1$ in Fig. 5.29 based on isostress isostrain ensemble simulations. Regardless of T, $T\alpha_\|/k_B - 1$ turns out to be a nonmonotonic function of density. It has a maximum that increases and shifts to lower densities with decreasing temperature. In the limit $\overline{\rho} \to 0$, one expects all curves to approach zero according to [see Eq. (5.155)]

$$
\begin{aligned}
T\alpha_\| - 1 \;&=\; -\frac{\overline{\rho}}{1 + 2\overline{\rho}B_2^{\mathrm{mf}}(T)} \frac{\mathrm{d}}{\mathrm{d}T} \frac{B_2^{\mathrm{mf}}(T)}{T} \\[2mm]
&=\; \overline{\rho}\left[-b + \frac{2a_{\mathrm{p}}(\xi)}{k_{\mathrm{B}}T}\right] + 2\overline{\rho}^2\left[b^2 - \frac{3ba_{\mathrm{p}}(\xi)}{k_{\mathrm{B}}T} + 2\left(\frac{a_{\mathrm{p}}^2(\xi)}{k_{\mathrm{B}}T}\right)^2\right] \\[2mm]
&\overset{\overline{\rho}\to 0}{\simeq}\; b\overline{\rho}\left[\frac{T_{\mathrm{inv}}(0)}{T} - 1\right]
\end{aligned}
\qquad (5.220)
$$

where the second line follows with the help of the mean-field Eq. (5.188) after expanding $\left[1 + 2\bar{\rho}B_2^{\mathrm{mf}}(T)\right]^{-1}$ $\left(\left|2\bar{\rho}B_2^{\mathrm{mf}}(T)\right| \ll 1\right)$. The last line of Eq. (5.220) is obtained by considering only the leading term of the expansion where Eq. (5.155) has also been used. Thus, in the limit $\bar{\rho} \to 0$, the curves in Fig. 5.29 become straight lines whose slope is determined by the hard core of the fluid molecules and their inversion temperature in the limit of vanishing density. For the cases plotted $T_{\mathrm{inv}}(0)/T > 1$ (see Table 5.7) so that the slope of $T\alpha_{\|} - 1$ should be positive for low densities, which is confirmed by the plots in Fig. 5.29. One also expects from Eq. (5.220) the slope of $T\alpha_{\|} - 1$ to be smaller for higher temperatures, which is also confirmed by Fig. 5.29.

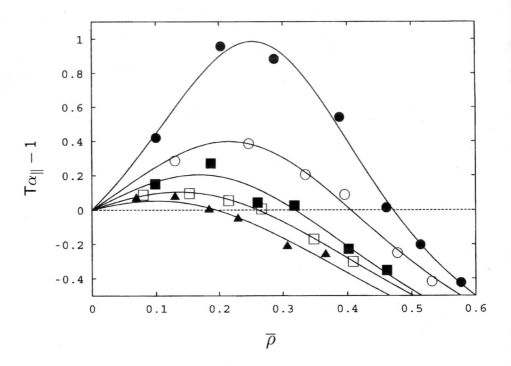

Figure 5.29: Plots of $T\alpha_{\|} - 1$ as functions of average fluid density $\bar{\rho} = N/\langle A\rangle\, s_{z0}$ for $T = 1.50$ (●), $T = 2.00$ (○), $T = 2.50$ (■), $T = 3.00$ (□), and $T = 3.50$ (▲) from MC simulations in the isostress isostrain ensemble for model A ($s_z = 10$). Solid lines are fits of Eq. (5.183) to simulation data. Intersections with dashed horizontal line define inversion temperature T_{inv} [see Eq. (5.135)].

Solid lines in Fig. 5.29 represent fits of the mean-field Eq. (5.183) to the simulation data taking $a_{\mathrm{p}}(\xi)$ and b as fitting parameters. Although these fits represent the simulation data remarkably well, both parameters turn out to be temperature dependent unlike the mean-field treatment in Section 4.2.2 suggests. Therefore it is not possible to extract any information about the location of the critical point of the confined (or bulk) fluid from the relations between critical temperature and density on the one hand and $a_p(\xi)$ and b on the other hand.

However, in view of the rather complex variation of $\alpha_{\|}$ with temperature and density, the mean-field approach is still very useful because it permits an estimate of the inversion temperature from an analytic expression [see Eq. (5.184)] at moderate computational expense. The computed inversion temperatures are plotted in Figs. 5.30 for various situations. In accord with the plots in Fig. 5.29, the inversion temperature depends strongly on the density. For example, plots in Fig. 5.30(a) show that, over a density range of $0.1 < \bar{\rho} < 0.5$, T_{inv} changes by about a factor of 3. Over wide ranges of temperature and density the data are again well represented by the mean-field expression in Eq. (5.184).

The curves in Fig. 5.30(a) for models A and B appear to be shifted downward compared with the bulk because of confinement. However, a more subtle phenomenon can be seen by comparing the plots referring to wettable and nonwettable substrates. Here one notices that for high densities the inversion temperature is generally lower for the hydrophobic substrate (model B) compared with the hydrophilic one (model A). However, in the limit of vanishing density, the inversion temperature is higher for the nonwettable compared with the wettable substrate (see Table 5.7). Therefore, the curves $T_{inv}(\bar{\rho})$ for models A and B must intersected at a sufficiently low density. From the fit of Eq. (5.184) to the simulation data, the intersection is located at $\bar{\rho} \simeq 0.08$ and $T \simeq 4.2$. This is roughly also the temperature at which $B_2(T)$ for the two models intersect [see Fig. 5.28(a)].

If the substrate separation increases one expects the inversion temperature of the confined fluid to eventually coincide with the bulk inversion temperature irrespective of the density. This notion is supported by plots of $T\alpha_{\|} - 1$ for $s_{z0} = 10$, 20, and bulk in Fig. 5.31(a). The maximum of this curve and its location are shifted toward the bulk curve with increasing substrate separation and so does T_{inv}. Similar plots are obtained for other temperatures, thus permitting one to construct the plot in Fig. 5.30(b) parallel to the one in Fig. 5.30(a). As before for $T_{inv}(0)$ one expects $\Delta T_{inv}(\bar{\rho}, s_{z0}) = T_{inv}(\bar{\rho}, \infty) - T_{inv}(\bar{\rho}, s_{z0}) \propto s_z^{-1}$ from the plot in Fig. 5.30(b) and the mean-field expressions in Eqs. (4.24) and (5.185). For $s_{z0} = 20$, for instance, a depression of T_{inv} of about 4% compared with the bulk value is deduced from Fig. 5.30(b). Thus, even if fluids are confined to spaces of mesoscopic dimension, the confinement-induced depression of the inversion temperature should *in principle* be accessible experimentally given the accuracy with which the

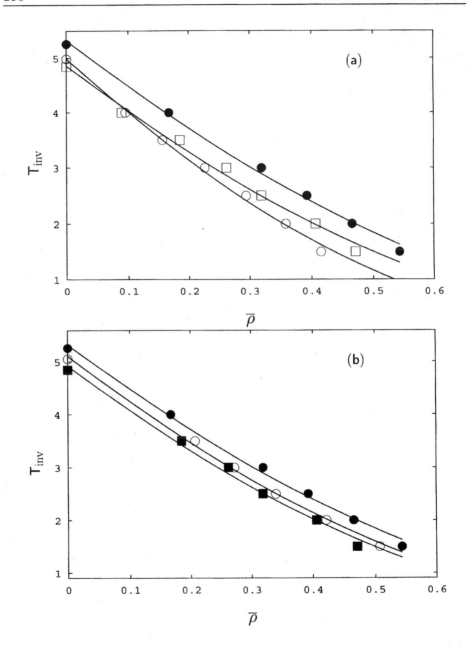

Figure 5.30: (a) Density dependence of inversion temperature from MIEMC simulations: (●) bulk, (□) model A, and (○) model B ($s_{z0} = 10$). Solid lines represent fits of Eq. (5.184) to simulation data. Data points for $\overline{\rho} = 0$ are obtained from Eq. (5.155) and were not included in the fit. (b) As (a) but for bulk (●) and model A [$s_{z0} = 10$ (■), $s_{z0} = 20$ (○)].

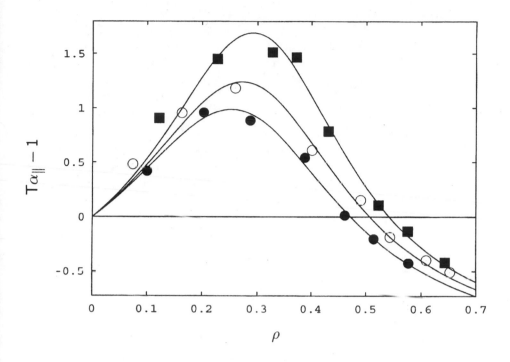

Figure 5.31: As Fig. 5.29 but for model A, $T = 1.50$, and $s_{z0} = 10$ (●), $s_{z0} = 20$ (○), and bulk (■).

phase behavior of confined fluids can nowadays be determined [31].

5.8   Lattice Monte Carlo simulations

5.8.1   Advantages and disadvantages of lattice models

The discussion of the Joule-Thomson effect in the previous section clearly showed that it is advantageous in theoretical treatments of confined fluids to tackle a given physical problem by a combination of different methods. This was illustrated in Section 5.7 where we employed a virial expansion of the equation of state, a van der Waals type of equation of state, and MC simulations in the specialized mixed isostress isostrain ensemble to investigate various aspects of the impact of confinement on the Joule-Thomson effect. The mean-field approach was particularly useful because it could predict certain trends on the basis of analytic equations. However, the mean-field treatment developed in Sections 4.2.2 and 5.7.5 is hampered by the assumption of homogeneity of the confined fluid [see Eq. (4.20) (which is not really justified as the discussion in Section 5.3.4 shows).

The assumption of homogeneity can be abandoned if the continuous mean-field treatment is replaced by a discrete treatment where the positions of fluid molecules are restricted to nodes on a lattice. The discussion in Section 5.4.2 and 5.6.5 showed that the mean-field lattice density functional theory developed in Section 4.3 was crucial in unraveling the complex phase behavior of fluids confined by chemically decorated substrate surfaces. A similar deep understanding of the phase behavior would not have been possible on the basis of simulation results alone. Nevertheless, the relation between these MC data and the lattice density functional results remained only qualitative on account of the continuous models employed in the computer simulations. Thus, we aim at a more *quantitative* comparison between MC simulations and mean-field lattice density functional theory in the closing section of this chapter.

This being the primary goal of the subsequent discussion we would also like to emphasize two other, perhaps more practical, aspects. On account of the rigidity of the underlying lattice it seems inconceivable to develop mixed isostress isostrain ensembles suitable for lattice MC simulations. On the other hand, lattice simulations are computationally much less demanding because molecules can occupy only discrete positions in space. Hence, the number of configurations possible on a lattice is greatly reduced compared with simulations of continuous model systems.

Another aspect of lattice models concerns the determination of phase behavior. As far as continuous models were concerned we emphasized already that an investigation of phase transitions in such models usually requires a "mechanical" representation of the relevant thermodynamic potential in terms of one or more elements of the microscopic stress tensor. The existence of such a mechanical representation was linked inevitably to symmetry considerations in Section 1.6, where it was also pointed out that such a mechanical expression may not exist at all. In this case a determination of the thermodynamic potential requires thermodynamic integration along some suitable path in thermodynamic state space, which may turn out to be computationally demanding.

For lattice models mechanical expressions for thermodynamic potentials are out of the question regardless of whether the lattice fluid possesses a sufficiently high symmetry in the sense of our discussion in Section 1.6. The reason is again the incompressibility of the lattice so that one always has to resort to thermodynamic integration techniques in MC simulations of lattice fluids.

A clear advantage of a lattice model, on the other hand, lies in the fact that one may expand the partition function in the limit $T \to \infty$ as we showed in Section 4.3.2 to obtain a closed analytic expression for thermodynamic state functions, which may in turn be used as a suitable starting point in a thermodynamic integration procedure with moderate computational effort. In the limit $T \to 0$, on the other hand, the mean-field treatment becomes exact and we already showed in Section 4.5 that closed analytic expressions for the grand potential may be derived [see Eq. (4.94)] so that a second starting point for a thermodynamic integration scheme exists. These expressions have no counterpart as far as continuous models are concerned.

5.8.2 Grand canonical ensemble Monte Carlo simulations

For pedagogic reasons it seems sensible to consider a fluid confined to a slit-pore with chemically heterogeneous substrates to make contact with the parallel mean-field calculations described in Section 4.3. As in that section we employ a simple cubic lattice of $\mathcal{N}$ sites. In accord with our previous notation, $s^{\mathcal{N}}$ represents a configuration of fluid molecules where the (double-valued, discrete) elements of the $\mathcal{N}$-dimensional vector $s^{\mathcal{N}}$ are represented by Eq. (4.51). Molecules of the (pure) lattice fluid interact with each other via a square-well potential where the width of the attractive well is equal to the lattice constant ℓ.

The fluid substrate interaction is modelled according to Eqs. (4.48). To minimize finite size effects, periodic boundary conditions are applied at the planes $x = 1, n_x$ and $y = 1, n_y$ such that a molecule located at the plane $\alpha = 1$ interacts with its nearest neighbors on lattice sites characterized by $\alpha = n_\alpha$ and vice versa where $\alpha = x, y$.

We treat the lattice fluid as an open thermodynamic system represented microscopically by the grand canonical ensemble. In this ensemble the probability density for the occupation of a given site i on the lattice is given by

$$p(s_i) = \frac{1}{\Xi} \exp\left[-\frac{\widetilde{h}(s_i)}{k_B T} \right] \tag{5.221}$$

where Ξ is the partition function of the grand canonical ensemble introduced in Eq. (4.54a). It seems worthwhile to emphasize the difference between this expression and the corresponding one for continuous model systems given in Eq. (5.6). A comparison reveals that in Eq. (5.221) the factor $1/N!$ is absent because fluid molecules on the lattice are distinguishable on account of the specific site to which they are restricted. Second, the factor of $1/\Lambda^{3N}$ in Eq. (5.6) is also missing. This is because molecules on the lattice have no kinetic energy, which is evident from the Hamiltonian function defined in Eq. (4.51).

Moreover, note that Eq. (5.221) depends on the single-particle Hamiltonian function unlike its counterpart in Eq. (5.6), which is governed by the configurational energy $U\left(\widetilde{r}^{N};N\right)$. Depending on the range of the intermolecular interaction potentials on which $U\left(\widetilde{r}^{N};N\right)$ depends, many intermolecular interactions have to be considered. For the current simple cubic lattice, on the other hand, only six nearest neighbors contribute to the fluid fluid part of $\widetilde{h}$ because of the short-range square-well potentials governing the intermolecular interactions. Hence, the computational effort in generating new configurations on the lattice is marginal compared with that required by off-lattice simulations. This is true even if similarly short-range potentials would be employed in off-lattice simulations because molecules can move continuously in space.

Therefore, molecules are restricted to lattice sites, and consequently, step 1 of the adapted Metropolis algorithm in the grand canonical ensemble for continuous model systems (i.e., random displacement of molecules) can be abandoned altogether (see Section 5.2.2). Instead the adapted Metropolis algorithm for lattice fluids proceeds as follows. The $\mathcal{N}$ sites of the lattice are visited consecutively. If a specific site is occupied (i.e., $s_i = 1$), an attempt is made to change the value of the occupation number to $s_i = 0$ (empty site). The associated change in the single-particle Hamiltonian function is calculated from

$$\Delta\widetilde{h} \equiv \widetilde{h}\left(s_{i,n}\right) - \widetilde{h}\left(s_{i,n-1}\right) \tag{5.222}$$

where the single-particle Hamiltonian function is defined in Eqs. (4.51) and (4.54b). Employing the importance sampling concept (see Section 5.2.1), we accept a local change in the occupation number with a probability

$$\Pi \equiv \min\left[1, \exp\left(-\Delta\widetilde{h}/k_{\mathrm{B}}T\right)\right] \tag{5.223}$$

based on the principle of detailed balance as we discussed in detail in Section 5.2.2. As before, an attempt to switch the occupation number between its two values at lattice site i is not immediately rejected if the associated $\Delta\widetilde{h} > 0$. In this case we draw a random number ξ distributed uniformly on the interval $[0,1]$ and apply the adapted Metropolis criterion,

$$\exp\left(-\Delta\widetilde{h}/k_{\mathrm{B}}T\right) \; > \; \xi \longrightarrow \text{accept change } s_{i,n-1} \rightarrow s_{i,n} \tag{5.224a}$$

$$\exp\left(-\Delta\widetilde{h}/k_{\mathrm{B}}T\right) \; \leq \; \xi \longrightarrow \text{reject change } s_{i,n-1} \rightarrow s_{i,n} \tag{5.224b}$$

The sequence of $\mathcal{N}$ attempts to change the occupation numbers constitutes a MC cycle.

5.8.3   Thermodynamic integration for lattice fluids

Based on the above numerical procedure, we shall eventually generate a distribution of microstates (i.e., configurations $s^{\mathcal{N}}$ on the lattice) corresponding to a minimum of the grand potential Ω defined in Eq. (4.55) (or its associated density $\omega \equiv \Omega/\mathcal{N}$). For the subsequent discussion of phase behavior of the lattice fluid, the absolute value of ω would be required according to the discussion in Section 1.7. However, as we pointed out in Section 5.8.1, such a calculation is not straightforward because we cannot directly evaluate the sum over configurations in Eq. (4.55) (i.e., the grand canonical partition function) nor does a mechanical expression for ω exist on account of the rigidity of the underlying lattice. Hence, we must resort to thermodynamic integration following ideas originally proposed by Binder [219].

We begin by realizing that [see Eq. (1.59)]

$$\mathrm{d}\Omega = -\mathcal{S}\mathrm{d}T - N\mathrm{d}\mu \qquad (5.225)$$

because the lattice is rigid and therefore $V_0 \mathrm{Tr}\,(\boldsymbol{\tau}\mathrm{d}\boldsymbol{\sigma}) = 0$. Thus, for $\mu = \mathrm{const}$, $\mathrm{d}\mathcal{S} = (\partial\mathcal{S}/\partial T)_\mu \, \mathrm{d}T$ so that we may write

$$\mathcal{S}\,(T,\mu) = \underbrace{\mathcal{S}\,(0,\mu)}_{=0} + \int_0^T \left(\frac{\partial\mathcal{S}}{\partial T'}\right)_\mu \mathrm{d}T' = \int_0^T \left(\frac{\partial\mathcal{U}_\mu}{\partial T'}\right)_\mu \frac{\mathrm{d}T'}{T'} \qquad (5.226)$$

where $\mathcal{S}\,(0,\mu)$ vanishes according to the third law of thermodynamics. The far right side of the previous expression obtains because [219]

$$\left(\frac{\partial\mathcal{S}}{\partial T}\right)_\mu = \frac{1}{T}\left(\frac{\partial\mathcal{U}_\mu}{\partial T}\right)_\mu \qquad (5.227)$$

where

$$\mathcal{U}_\mu \equiv \mathcal{U} - \mu N \qquad (5.228)$$

Inserting Eqs. (5.226) and (5.227) into Eq. (5.225) permits us to calculate

$$\beta\Omega = \beta\mathcal{U}_\mu - \int_0^T \left(\frac{\partial\mathcal{U}_\mu}{\partial T'}\right)_\mu \frac{\mathrm{d}T'}{k_\mathrm{B}T'} \qquad (5.229)$$

An alternative expression is obtained if we integrate $\mathrm{d}\mathcal{S}$ from $T' = \infty$ down to the desired temperature $T' = T$; that is

$$\frac{\mathcal{S}\,(T,\mu)}{k_\mathrm{B}} - \frac{\mathcal{S}\,(\infty,\mu)}{k_\mathrm{B}} = \int_\infty^T \left(\frac{\partial\mathcal{U}_\mu}{\partial T'}\right)_\mu \frac{\mathrm{d}T'}{k_\mathrm{B}T'} = \frac{\mathcal{U}_\mu}{k_\mathrm{B}T} - \int_0^{1/T} \mathcal{U}_\mu \mathrm{d}\left(\frac{1}{k_\mathrm{B}T'}\right) \qquad (5.230)$$

which follows after partial integration and by changing variables according to $T' \to 1/T'$. Hence, inserting Eq. (5.230) into Eq. (5.225), we obtain

$$\beta\Omega = -\mathcal{S}\left(\infty, \mu\right) + \int\limits_{0}^{1/T} \mathcal{U}_{\mu} \mathrm{d}\left(\frac{1}{k_{\mathrm{B}}T'}\right) \tag{5.231}$$

where for our lattice model $\mathcal{S}\left(\infty, \mu\right) = \mathcal{N}k_{\mathrm{B}}\ln 2$ because at infinite temperature intermolecular interactions become irrelevant and each lattice site can either be occupied or empty [219].

Last but not least we realize that because of [see Eq. (5.225)]

$$\frac{1}{\mathcal{N}}\left(\frac{\partial\Omega}{\partial\mu}\right)_T = \left(\frac{\partial\omega}{\partial\mu}\right)_T = -\frac{N}{\mathcal{N}} = -\overline{\rho} \tag{5.232}$$

a third route for thermodynamic integration exists. Integrating Eq. (5.232) we may write

$$\omega\left(T, \mu_2\right) = \omega\left(T, \mu_1\right) - \int\limits_{\mu_1}^{\mu_2} \overline{\rho}\left(T, \mu\right)\mathrm{d}\mu \tag{5.233}$$

and calculate the grand-potential density $\omega\left(T, \mu_2\right)$ provided we know its value for T and μ_1 and know the equation of state, that is, $\overline{\rho}\left(T, \mu\right)$.

Equations (5.229)–(5.233) provide the basis for the thermodynamic integration scheme employed in Section 5.8.4. However, it is worthwhile noting at this point that key quantities such as $\overline{\rho}$ and $\mathcal{U}$ in Eqs. (5.229)–(5.233) can be calculated readily as thermal averages of s_i and $\mathcal{H}\left(s^{\mathcal{N}}\right)$, respectively. These grand canonical ensemble averages will be calculated via GCEMC simulations described in Section 5.8.2.

5.8.4   Comparison with mean-field density functional theory

To apply the thermodynamic integration procedure introduced above in Section 5.8.3, we need to know roughly where to expect coexistence lines because one must not integrate across these lines where quantities like $\mathcal{U}$ and $\overline{\rho}$ change discontinuously. The mean-field results presented in Section 4.5 are taken as a guideline in this sense. In most cases we base our integration on isothermal paths using Eq. (5.233) starting at high (path I) or low values of μ_1 (path II). We realize that if μ_1 is sufficiently large, we can expect the lattice fluid to be in its liquid state with all lattice sites completely filled; that is, $\{\rho_l\} = 1$ regardless of T. In this case, $\Omega\left(\mu_1, T\right)$ is given by $\Omega^{\mathrm{l}}\left(\mu\right)$ at $T = 0$

(see Table 4.1). If, on the other hand, μ_1 is sufficiently small, we will always end up in the thermodynamically stable gas phase. Hence, in this case we approximate $\Omega(\mu_1, T)$ by $\Omega^g(\mu) = 0$ at $T = 0$ because in the gas phase all lattice sites are empty. We now have analytic expressions for $\Omega(\mu_1, T)$, we can calculate $\overline{\rho} = \langle N \rangle / \mathcal{N}$ from GCEMC, and we can perform the integration in Eq. (5.233) numerically. We carry out this procedure until we reach a discontinuous change in $\overline{\rho}$ (i.e., $\langle N \rangle$), indicating that the lattice fluid becomes unstable. Typical data are presented in Fig. 5.32.

From Fig. 5.32(b) one notices that, over a certain range of temperatures, $\overline{\rho}$ turns out to be a double-valued function of μ; that is, one may observe hysteresis. Hysteresis refers to the fact that somewhere in the temperature range where high(er)- or low(er)-density branches of $\overline{\rho}$ overlap one of the two phases will be metastable, the other one being thermodynamically (i.e., globally) stable. On the basis of the associated $\omega(\mu, T)$ in Fig. 5.32(a), we can readily discriminate between the two, thereby identifying the thermodynamically stable phase at this given temperature. Moreover, one may locate $\mu^{\alpha\beta}$ defined in Eq. (1.76a) [see Fig. 5.32(a)]. Repeating this procedure for all temperatures $0 \lesssim T \lesssim T_c^{\alpha\beta}$, we may eventually construct the coexistence line between phases α and β.

Some attention has to be paid to the vicinity of the critical points. Not only are density fluctuations increasing enormously as one approaches $T_c^{\alpha\beta}$, but the densities of the two branches of the adsorption isotherm become more and more alike, and therefore, it is much more difficult to distinguish between coexisting phases. If one were to locate the critical points with high precision, one would eventually have to employ more sophisticated simulation techniques [220, 221] as, for example, Wolff's algorithm, which applies in particular to the current (Ising-type of) system [118, 119]. However, as we are not interested in locating critical points with high precision, it is sufficient to define the critical point as that temperature at which plots of $\overline{\rho}$ versus μ do not exhibit any hysteresis.

For the bridge phase, where the one-phase region is triangular in shape bounded by, say, the coexistence lines $\mu^{gb}(T)$ and $\mu^{bl}(T)$ (see Fig. 4.10, for example), it turns out to be useful to first integrate $\mathcal{U}$ according to Eq. (5.231) to some temperature $T \leq \min\left(T_c^{gb}, T_c^{bl}\right)$ and then complete the integration employing Eq. (5.233) to higher and lower values of μ isothermally until one hits the chemical potentials at which $\overline{\rho}$ changes to a higher and lower value, respectively.

Based on these considerations we are eventually in a position to construct $\mu_x(T)$ for the case previously investigated on the basis of the mean-field approach. Comparing $\mu_x(T)$ from the mean-field calculations with GCEMC data in Figs. 4.10, we see that the overall topology of the phase diagrams

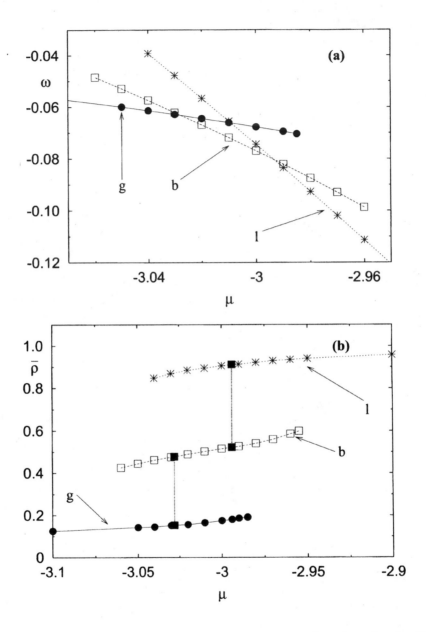

Figure 5.32: (a) Plots of grand-potential density ω as function of μ for $T = 0.73$, $n_x = 20$, $n_z = 10$, $n_s = n_w = 10$, $\varepsilon_{fw} = 0$, and $\varepsilon_{fs} = 1.5$; (∗) liquid, (□) bridge, and (●) gas phase. (b) As (a) but for mean density $\bar{\rho}$; vertical lines represent thermodynamic phase transitions between states represented by (■). Dotted, dashed, and full lines are intended to guide the eye.

is quite well represented by the mean-field treatment of $H\left(s^N\right)$. However, there are differences. In general, one notices that all GCEMC coexistence lines terminate at lower (critical) temperatures compared with their mean-field analogs, which is a well-known phenomenon in the bulk [15].

Comparison of the mean-field data with their analogs from GCEMC (see Fig. 4.10) also reveals somewhat more subtle differences. These differences concern the absence of the rather short coexistence lines $\mu^{\mathrm{vl}}\left(T\right)$ and $\mu^{\mathrm{gd}}\left(T\right)$. These somewhat more arcane phase coexistences turn out to be stabilized by the mean-field approximation relative to the GCEMC data because the former underestimates density fluctuations by disregarding them altogether. However, apart from these differences, it is noteworthy that the mean-field approximation yields $\mu_{\mathrm{x}}\left(T\right)$ in excellent agreement with the GCEMC data for those ranges of thermodynamic states where both approaches exhibit first-order phase transitions.

Chapter 6

Confined fluids with long-range interactions

6.1 Introductory remarks

In all fluid model systems discussed in Chapters 3–5 the molecules interact with each other via *short-range* potentials which decay with the intermolecular distance r as r^{-6} characteristic of dispersive interactions between polarizable molecules or even faster as in the nearest-neighbor lattice models discussed in Sections 4.3 or 5.8. There is, however, an increasing interest in complex systems governed by long-range interactions[1] such as the electrostatic (Coulomb) interactions between charged particles (r^{-1}) or the dipole dipole interactions (r^{-3}) between particles with permanent electric or magnetic dipole moments.

Indeed, Coulombic interactions are important or even dominant in almost all biological systems, such as proteins, DNA, or charged membranes. Dipolar interactions, on the other hand, play a prominent role in phospholipid bilayers [222, 223], and they are always important because of the omnipresence of dipolar water molecules in biological tissues and electrolyte solutions.

Apart from biological and electrochemical systems, there is a wealth of technologically important substances where long-range interactions play a central role. An example are polyelectrolytes, where the charged nature is the key ingredient for their functionality [224]. Dipole dipole interactions, on the other hand, are of fundamental importance in colloidal systems [225] such as ferrofluids [226–228], which are dispersions of ferromagnetic nanoparticles. Other examples are magnetic colloids consisting of superparamagnetic

---

[1]Strictly speaking, the term "long range" refers to interaction potentials decaying slower than r^{-4}.

particles and electrorheological fluids [229] represented by colloidal dispersions of polarizable particles.

In many applications involving such systems, one is faced with some sort of spatial confinement. Examples are cell membranes, soap bubbles, electrolyte solutions near charged surfaces, proteins near (charged) membranes, polyelectrolyte films [230], and thin films of magnetic colloids or ferrofluids [231]. Although this admittedly incomplete list illustrates the great importance of long-range electrostatic interactions, their correct treatment in computer simulations poses a highly nontrivial problem. This is because in any computer simulation one is inevitably restricted to microscopically small systems where the number of molecules is always many orders of magnitude smaller than in a macroscopic sample where typically $N = \mathcal{O}\left(10^{23}\right)$.

Indeed, long-range interactions still represent a computational challenge even for (large) *bulk* fluids, which are periodic in all three spatial dimensions. The reason is that the conventional strategy established for fluids with short-range interactions, namely to *truncate* the potentials at some cut-off distance r_c, leads to serious inaccuracies and artifacts when applied to Coulombic or dipolar systems. In view of this dilemma, several techniques to treat the long-range interactions in a more reliable way have been proposed. An early example is the reaction-field method [232, 233] for dipolar systems. Within this method, the interactions are truncated at a distance r_c from each particle, but (contrary to a simple truncation) the medium beyond r_c is taken into account as a dielectric continuum characterized by a dielectric constant ϵ_{RF} [140]. Apart from the approximation made via the replacement of microscopic interactions by a macroscopic "reaction field," a further drawback of this method is that ϵ_{RF} has to be chosen close to the dielectric constant of the fluid itself. The latter, however, is a priori unknown. Other techniques proposed more recently are the fast multipole method, particle-mesh methods [234–236], and the Lekner method [237], which is primarily intended for simulations of thin films.[2]

For two reasons we focus in this chapter on yet a different technique, the so-called Ewald summation method. First, Ewald sums are nowadays the most widely used and accepted method to handle long-range interactions [140, 238] at least as far as bulk systems are concerned. Second, the formulation of Ewald sums for confined systems is straightforward as we shall demonstrate below and in the accompanying Appendix F to which we refer for a detailed discussion of the derivation of the relevant equations. Moreover, during the last few years, there have been substantial improvements that led to an optimization of these methods such that nowadays they are

---

[2]For an overview of other techniques to treat long-range interactions, see, e.g., Ref. 238.

not only accurate but also computationally efficient.

Our goal in this chapter is to present a simple and physically meaningful derivation of various Ewald summation techniques. For a mathematically more rigorous presentation, we refer the reader to the original papers by de Leeuw et al. [239–241].

As an introduction, and for pedagogical reasons, we start in Section 6.2.1 with the energy of Coulombic systems in three spatial dimensions. Based on the resulting Ewald sum we then derive step-by-step corresponding formulas for bulk dipolar systems (see Section 6.2.2) and for systems in slab geometry where we consider point charges and dipoles confined by either insulating (see Section 6.4) or conducting walls (see Section 6.5). Mathematical details of the derivations and explicit Ewald expressions for forces, torques, and stress tensors can be found in Appendix F. Illustrating applications are presented in Sections 6.4 and 6.5.

6.2 Three-dimensional Ewald summation

6.2.1 Ionic systems

We start by considering an ionic system consisting of N charged particles forming a (rectangular) cell of volume $V = s_x s_y s_z$. We assume that the system as a whole is electrically neutral, that is, $\sum_{i=1}^{N} q_i = 0$, where q_i are the individual charges. We are interested in the Coulomb contribution to the potential energy of this N-particle system, which can be written as

$$U_C = \frac{1}{2} \sum_{i=1}^{N} q_i \Phi(\boldsymbol{r}_i) \tag{6.1}$$

where $\Phi(\boldsymbol{r}_i)$ is the electrostatic potential at the position of particle i.

We immediately specialize to a bulk-like situation where the central cell is surrounded by periodic replicas in all three spatial dimensions. In this case, the electrostatic potential is given by[3]

$$\Phi(\boldsymbol{r}_i) = \sum_{\{\boldsymbol{n}\}}{}' \sum_{j=1}^{N} \frac{q_j}{|\boldsymbol{r}_{ij} + \boldsymbol{n}|} \tag{6.2}$$

where $\boldsymbol{r}_{ij} = \boldsymbol{r}_i - \boldsymbol{r}_j$ is the connecting vector between particles i and j and $\{\boldsymbol{n}\}$ is a set of lattice vectors of the rectangular superlattice generated by the

---

[3]Throughout this chapter we employ Gaussian units to keep notation as compact as possible.

periodic replication of the original cell, that is to say, $\boldsymbol{n} = (n_x s_x, n_y s_y, n_z s_z)$ $(n_x, n_y, n_z \in \mathbb{Z})$. The prime attached to the first summation sign in Eq. (6.2) indicates that the term $j = i$ is omitted for $\boldsymbol{n} = \boldsymbol{0}$, i.e., within the central cell.

For the derivation of the Ewald method it is important to consider the charge density $\rho_i(\boldsymbol{r})$ corresponding to the electrostatic potential $\Phi(\boldsymbol{r}_i)$ given in Eq. (6.2). The link between these quantities is provided by Poisson's equation [242]

$$\Phi(\boldsymbol{r}_i) = \int d\boldsymbol{r}' \frac{\rho_i(\boldsymbol{r}')}{|\boldsymbol{r}_i - \boldsymbol{r}'|} \qquad (6.3)$$

Combining Eqs. (6.2) and (6.3) we see that the charge distribution generating the potential in Eq. (6.2) may be perceived as a sum of Dirac δ-functions (see Appendix B.6.2)

$$\rho_i(\boldsymbol{r}') = \sum_{\{\boldsymbol{n}\}}' \sum_{j=1}^{N} q_j \delta(\boldsymbol{r}' - \boldsymbol{r}_j + \boldsymbol{n}) \qquad (6.4)$$

where, as before, the sum over lattice vectors excludes the term corresponding to $j = i$ if $\boldsymbol{n} = \boldsymbol{0}$. We note in passing that it is this restriction of the lattice sum, which causes the charge distribution to depend on (particle) index i.

For each point charge q_i involved in Eq. (6.4), the resulting electrostatic potential decays in proportion to the inverse distance [see Eq. (6.2)], such that the lattice sum buried in the expression for the total energy U_C [see Eq. (6.1)] converges rather slowly. In view of this dilemma, the central idea of the Ewald summation techniques is to rewrite the δ-like charge density in Eq. (6.4) as a sum of three contributions, $\rho_i^{(1)}(\boldsymbol{r}')$, $\rho_i^{(2)}(\boldsymbol{r}')$, and $\rho_i^{(3)}(\boldsymbol{r}')$. Each one of these yields an electrostatic potential whose convergence is controllable by a few parameters.

In a first step we associate with each δ-function a diffuse cloud of point charges q_j of opposite sign located at $\boldsymbol{r}' = \boldsymbol{r}_j - \boldsymbol{n}$. It is convenient (yet not crucial) to represent these clouds by Gaussians; that is

$$\rho_{j,\boldsymbol{n}}(\boldsymbol{r}') = -q_j \left(\frac{\alpha}{\sqrt{\pi}}\right)^3 \exp\left[-\alpha^2(\boldsymbol{r}' - \boldsymbol{r}_j + \boldsymbol{n})^2\right] \qquad (6.5)$$

where α controls the width of the Gaussian, which is normalized such that

$$\int d\boldsymbol{r}' \rho_{j,\boldsymbol{n}}(\boldsymbol{r}') = -q_j \qquad (6.6)$$

We note in passing that this approach follows in spirit the general definition of the Dirac δ-function via Gaussian distributions presented in Appendix B.6.2.

Adding the Gaussians to the original charge distribution in Eq. (6.4) yields

$$\rho_i^{(1)}(r') = \sum_{\{n\}}' \sum_{j=1}^{N} [q_j \delta(r' - r_j + n) + \rho_{j,n}(r')] \tag{6.7}$$

which may be perceived as a set of screened charges. Their potential decays much more rapidly than that of the original point charges. Indeed, as we demonstrate explicitly in Appendix F.1.1.1, the electrostatic potential corresponding to $\rho_i^{(1)}(r')$ is given by

$$\Phi^{(1)}(r_i) = \sum_{\{n\}}' \sum_{j=1}^{N} q_j \frac{\text{erfc}(\alpha|r_{ij} + n|)}{|r_{ij} + n|} \tag{6.8}$$

where erfc(y) is the complementary error function [11, 37] that decreases monotonically as y increases. Hence, its decay with increasing interparticle separation is controllable by the parameter α. In practice, α is usually chosen such that erfc(y) is essentially zero for separations larger than half the shortest side length of the central cell. In this case, only neighbors within the central cell, $(n = 0)$ have to be taken into account in the lattice summation, that is, in the summation over $\{n\}$ in Eq. (6.8).

As a next step we have to take care of the electrostatic potential contribution of the Gaussian charge clouds themselves,

$$\rho_i(r') - \rho_i^{(1)}(r') = -\sum_{\{n\}}' \sum_{j=1}^{N} \rho_{j,n}(r') \tag{6.9}$$

where the right side follows by substracting Eq. (6.7) from Eq. (6.4). We recall that the sum over lattice vectors in Eq. (6.9) excludes the term $j = i$ if $n = 0$, which is the Gaussian located at r_i where we wish to calculate the total electrostatic potential $\Phi(r_i)$.

In what follows it turns out to be convenient to first retain the so-called *self-term* in the sum over lattice vectors, which describes the interaction of a charge cloud with itself. Clearly, this contribution is unphysical and needs to be corrected for at the end of the derivation. We thus split the right side of Eq. (6.9) into two contributions, namely

$$
\rho^{(2)}(\boldsymbol{r}') = -\sum_{\{\boldsymbol{n}\}}\sum_{j=1}^{N}\rho_{j,\boldsymbol{n}}(\boldsymbol{r}')
$$

$$
= \sum_{\{\boldsymbol{n}\}}\sum_{j=1}^{N}q_j\left(\frac{\alpha}{\sqrt{\pi}}\right)^3\exp\left[-\alpha^2(\boldsymbol{r}'-\boldsymbol{r}_j+\boldsymbol{n})^2\right] \quad (6.10a)
$$

$$
\rho_i^{(3)}(\boldsymbol{r}') = \rho_{i,0}(\boldsymbol{r}')
$$

$$
= -q_i\left(\frac{\alpha}{\sqrt{\pi}}\right)^3\exp\left[-\alpha^2(\boldsymbol{r}'-\boldsymbol{r}_i)^2\right] \quad (6.10b)
$$

such that the lattice sum appearing in Eq. (6.10a) may now be carried out without any restriction. As a consequence, the charge distribution $\rho^{(2)}(\boldsymbol{r}')$ is independent of index i. We are thus dealing with a distribution of smooth, Gaussian charge clouds of total charge $+q_j$ which vary *periodically* in space. This, in turn, suggests carrying out the calculation of the corresponding contribution to the electrostatic potential, $\Phi^{(2)}(\boldsymbol{r}_i)$, in the *reciprocal* space spanned by lattice vectors $\boldsymbol{k} = (2\pi m_x/s_x, 2\pi m_y/s_y, 2\pi m_z/s_z)$ $(m_x, m_y, m_z \in \mathbb{Z})$. As shown explicitly in Appendices F.1.1.2 and F.1.1.3, the resulting potential is given by

$$
\Phi^{(2)}(\boldsymbol{r}_i) = \frac{4\pi}{V}\sum_{\boldsymbol{k}\neq\boldsymbol{0}}\sum_{j=1}^{N}\frac{q_j}{k^2}\exp\left[-\frac{k^2}{4\alpha^2}\right]\exp[-i\boldsymbol{k}\cdot\boldsymbol{r}_{ij}] + \Phi_{\mathrm{LR}}^{(2)}(\boldsymbol{r}_i) \quad (6.11)
$$

The first term on the right side of Eq. (6.11) represents contributions from the Fourier coefficients of $\Phi^{(2)}(\boldsymbol{r}_i)$ related to *non-zero* wavevectors $(\boldsymbol{k} \neq \boldsymbol{0})$. Each summand is weighted by a Gaussian of width $(2\alpha)^{-1}$, such that damping is controlled by the wave*number* $k = |\boldsymbol{k}|$. With increasing k the summands vanish increasingly rapidly, as long as α is finite and not too large. This is the great advantage of representing the potential from the Gaussian background in Fourier rather than real-space where the convergence would be much slower.

The last term on the right side of Eq. (6.11) is the so-called long-range potential $\Phi_{\mathrm{LR}}^{(2)}(\boldsymbol{r}_i)$ representing the Fourier coefficients for $\boldsymbol{k} = \boldsymbol{0}$ in the expansion in Eq. (6.11). As argued in Appendix F.1.1.3, $\Phi_{\mathrm{LR}}^{(2)}(\boldsymbol{r}_i)$ is sensitive to the specific boundary conditions employed in an actual computer simulation. Assuming that the *total* volume (central cell plus periodic replicas) is a large sphere surrounded by a dielectric continuum and characterized by a dielectric constant ϵ', the long-range contribution is given by

$$
\Phi_{\mathrm{LR}}^{(2)}(\boldsymbol{r}_i) = \frac{4\pi}{V}\frac{\boldsymbol{r}_i\cdot\boldsymbol{M}}{2\epsilon'+1} \quad (6.12)
$$

where

$$M = \sum_{j=1}^{N} q_j r_j \tag{6.13}$$

is the total dipole moment of the system.

Finally, we need to consider the potential related to the charge distribution $\rho_i^{(3)}(r')$ [see Eq. (6.10b)], which corrects the (unphysical) self-interaction included in $\Phi^{(2)}(r_i)$. As detailed in Appendix F.1.1.4,

$$\Phi^{(3)}(r_i) = -q_i \frac{2\alpha}{\sqrt{\pi}} \tag{6.14}$$

We are now in a position to write down an expression for the total potential energy of the three-dimensional Coulomb system [see Eq. (6.1)] within the Ewald formulation. Adding all contributions to the total electrostatic potential and inserting the result into Eq. (6.1) we obtain

$$U_C^{3d} = U_{CR}^{3d} + U_{CF}^{3d} + U_{CS}^{3d} + U_{CLR}^{3d} \tag{6.15}$$

On the right side of Eq. (6.15), U_{CR}^{3d} and U_{CF}^{3d} result from real- [see Eq. (6.8)] and Fourier-space ($k \neq 0$) contributions [see Eq. (6.11)] of the electrostatic potential, respectively, that is

$$
\begin{aligned}
U_{CR}^{3d} &= \frac{1}{2} \sum_{i=1}^{N} q_i \Phi^{(1)}(r_i) \\
&= \frac{1}{2} \sum_{i=1}^{N} \sum_{j=1}^{N} \sum_{\{n\}}{}' q_i q_j \frac{\mathrm{erfc}\,(\alpha|r_{ij} + n|)}{|r_{ij} + n|} \tag{6.16a}
\end{aligned}
$$

$$
\begin{aligned}
U_{CF}^{3d} &= \frac{1}{2} \sum_{i=1}^{N} q_i \left[\Phi^{(2)}(r_i) - \Phi_{LR}^{(2)}(r_i) \right] \\
&= \frac{2\pi}{V} \sum_{i=1}^{N} \sum_{j=1}^{N} \sum_{k \neq 0} \frac{q_i q_j}{k^2} \exp\left[-\frac{k^2}{4\alpha^2} \right] \exp[-ik \cdot r_{ij}] \tag{6.16b}
\end{aligned}
$$

The remaining terms on the right side of Eq. (6.15), U_{CLR}^{3d} and U_{CS}^{3d} follow from the long-range [see Eq. (6.12)] and self-contributions [see Eq. (6.14)] as

$$U_{CLR}^{3d} = \frac{1}{2} \sum_{i=1}^{N} q_i \Phi_{LR}^{(2)}(r_i) = \frac{2\pi}{V} \frac{1}{2\epsilon' + 1} M^2 \tag{6.17a}$$

$$U_{CS}^{3d} = \frac{1}{2} \sum_{i=1}^{N} q_i \Phi^{(3)}(r_i) = -\frac{\alpha}{\sqrt{\pi}} \sum_{i=1}^{N} q_i^2 \tag{6.17b}$$

where $M = |\boldsymbol{M}|$.

At this point some comments on the use of Eq. (6.15) in actual computer simulations seem to be appropriate. For simplicity, we consider a system in a cubic simulation box; that is, $s_x = s_y = s_z = s$. As mentioned [see discussion following Eq. (6.8)], the parameter α is usually chosen large enough such that only neighboring molecules within the central cell ($\boldsymbol{n} = \boldsymbol{0}$) need to be considered to compute the real-space contribution [see Eq. (6.16a)] of the total Ewald energy. In practice, typical values of the screening parameter are $\alpha s \approx 5 - 7$, corresponding to a cut-off of the real-space interactions at $r_c \approx s/2$. The resulting double sum is then evaluated with periodic boundary conditions in all spatial directions combined with the minimum image convention (see Section 5.2.2).

As far as the Fourier-space contribution to the Ewald sum for the configurational potential energy is concerned [see Eq. (6.16b)], the sum over wavevectors is truncated at a cut-off wavenumber $k_c = (2\pi/s)\sqrt{m_c^2}$ with $m_c^2 = \max\limits_{m_x,m_y,m_z} \left(m_x^2 + m_y^2 + m_z^2\right)$. Note, however, that the convergence of the Fourier part depends on α via the Gaussian damping function $\exp[-k^2/4\alpha^2]$. The larger is α (that is, the shorter is r_c), the larger m_c^2 needs to be chosen to make sure that terms with $|\boldsymbol{k}| > k_c$ can indeed be neglected. Thus, the challenge in applying Ewald summations is to find the right balance between the number of terms evaluated in real and Fourier-space, respectively. In practice, typical values of $m_c^2 \approx 30 - 60$, corresponding to a summation over $1000 - 1500$ different wave vectors (i.e., wave vectors not related by symmetry). We also note that the Fourier part can be rewritten more compactly as

$$U_{\mathrm{CF}}^{\mathrm{3d}} = \frac{2\pi}{V} \sum_{k \neq 0} \frac{1}{k^2} \exp\left[-\frac{k^2}{4\alpha^2}\right] \widetilde{a}(\boldsymbol{k})\widetilde{a}^*(\boldsymbol{k}) \tag{6.18}$$

where we have introduced the complex quantity

$$\widetilde{a}(\boldsymbol{k}) = \sum_{i=1}^{N} q_i \exp[-i\boldsymbol{k} \cdot \boldsymbol{r}_i] \tag{6.19}$$

and $\widetilde{a}^*(\boldsymbol{k})$ is the complex conjugate of $\widetilde{a}(\boldsymbol{k})$.

Equation (6.18) implies that the Fourier-space contribution of the energy can be evaluated for each wavevector as a product of two *single-particle sums*. This is of great importance in terms of computational speed, because for each wavevector its evaluation boils down to computing $2N$ terms [see Eqs. (6.18) and (6.19)] instead of N^2 terms as in Eq. (6.16b).

We finally note that most of the recent simulations of ionic systems are carried out under so-called "conducting" or "tin-foil" boundary conditions corresponding to the choice $\epsilon' = \infty$ in the long-range part of the Ewald energy [see Eq. (6.17a)]. With this choice, the long-range contribution can be completely neglected. On the other hand, with "vacuum" boundary conditions ($\epsilon' = 1$), the long-range term is nonzero and *positive*, implying that a situation characterized by $M \neq 0$ is associated with an energy penalty. This is consistent with our macroscopic considerations in Appendix F.1.1.3, where we emphasize that a polarized sphere in vacuum experiences a depolarizing field acting against the polarization within the sphere [see Eq. (F.42)].

Up to this point we have focused on the (total) energy related to (three-dimensional) ionic systems, which is particularly important in MC simulations. However, one may also be interested in performing MD simulations, where, unlike in MC, the evolution of the system is not governed by changes in the configurational potential *energy* (see Section 5.2) but rather by the *forces* determining the motion of the particles. Explicit expressions for the Coulombic contribution to the forces, as well as expressions for various components of the stress tensor (see Section 1.2.2) within the Ewald formulation are given in Appendix F.1.2.

6.2.2 From point charges to point dipoles

Based on our derivation of the Ewald energy of a Coulombic system, it is quite straightforward to formulate the analog for a system of point dipoles μ_i, $i = 1, \ldots, N$. Assuming, as before, a periodic replication of the central cell in all three spatial directions, the total energy of the bulk dipolar system is given by

$$U_{\mathrm{D}}^{\mathrm{3d}} = \frac{1}{2} \sum_{i=1}^{N} \sum_{j=1}^{N} \sum_{\{n\}}' u_{\mathrm{DD}} \left(r_{ij} + n, \mu_i, \mu_j \right) \tag{6.20}$$

where the dipole dipole interaction between two particles i and j is defined by

$$u_{\mathrm{DD}} \left(r_{ij}, \mu_i, \mu_j \right) = \frac{\mu_i \cdot \mu_j}{r_{ij}^3} - 3 \frac{(\mu_i \cdot r_{ij})(\mu_j \cdot r_{ij})}{r_{ij}^5} \tag{6.21}$$

The easiest derivation of an Ewald expression for the lattice sum in Eq. (6.20) parallel to the one in Eq. (6.15) for the ionic fluid is based on the observation that u_{DD} can be rewritten as

$$u_{\mathrm{DD}} \left(r_{ij}, \mu_i, \mu_j \right) = (\mu_i \cdot \nabla_i)(\mu_j \cdot \nabla_j) \Psi(r_{ij}) \tag{6.22}$$

where $\Psi(r_{ij})$ is the electrostatic potential of a unit point charge (i.e., a charge $q_i = +1$), namely

$$\Psi(r_{ij}) = \frac{1}{r_{ij}} \tag{6.23}$$

where $r_{ij} \equiv |\boldsymbol{r}_{ij}|$. Thus, from a formal point of view, the pair energy between a pair of *dipoles* can be derived from the Coulomb interaction potential between a pair of *charges*, $u_{\mathrm{CC}}(r_{ij}) = q_i q_j / r_{ij}$ through the transformation $q_i \to (\boldsymbol{\mu}_i \cdot \boldsymbol{\nabla}_i)$ and $q_j \to (\boldsymbol{\mu}_j \cdot \boldsymbol{\nabla}_j)$. The same applies, of course, to the total configurational potential energy in Eq. (6.20), which we can rewrite as

$$U_{\mathrm{D}}^{3\mathrm{d}} = \frac{1}{2} \sum_{i,j=1}^{N} {\sum_{\{\boldsymbol{n}\}}}' (\boldsymbol{\mu}_i \cdot \boldsymbol{\nabla}_i)(\boldsymbol{\mu}_j \cdot \boldsymbol{\nabla}_j) \Psi(|\boldsymbol{r}_{ij} + \boldsymbol{n}|) \qquad (6.24)$$

We now recall that the key idea in deriving the Ewald expression for Coulombic systems was to divide the charge distribution and the resulting electrostatic potential into several independently converging contributions. Explicit expressions for the Ψ-functions corresponding to real-space, Fourier, and long-range contributions to the Ewald expression for the Coulomb energy can easily be extracted from Eqs. (6.16) and (6.17a), yielding

$$\Psi_{\mathrm{R}}(r_{ij}) = \frac{\mathrm{erfc}(\alpha r_{ij})}{r_{ij}} \qquad (6.25\mathrm{a})$$

$$\Psi_{\mathrm{F}}(\boldsymbol{r}_{ij}) = \frac{4\pi}{V} \sum_{k \neq 0} \frac{1}{k^2} \exp\left[-\frac{k^2}{4\alpha^2}\right] \exp\left[-i\boldsymbol{k} \cdot \boldsymbol{r}_{ij}\right] \qquad (6.25\mathrm{b})$$

$$\Psi_{\mathrm{LR}}(\boldsymbol{r}_i, \boldsymbol{r}_j) = \frac{4\pi}{V} \frac{\boldsymbol{r}_i \cdot \boldsymbol{r}_j}{2\epsilon' + 1} \qquad (6.25\mathrm{c})$$

Because of Eq. (6.22) each of these functions generates a contribution to the Ewald expression for the total configurational potential energy of a dipolar system.[4] By complete analogy with the Coulombic case, we can thus write [see Eq. (6.15)]

$$U_{\mathrm{D}}^{3\mathrm{d}} = U_{\mathrm{DR}}^{3\mathrm{d}} + U_{\mathrm{DF}}^{3\mathrm{d}} + U_{\mathrm{DLR}}^{3\mathrm{d}} + U_{\mathrm{DS}}^{3\mathrm{d}} \qquad (6.26)$$

---

[4]The self-part plays a different role in this context and will be discussed later.

where the first three terms are given explicitly by

$$
\begin{aligned}
U_{\text{DR}}^{3d} &\equiv \frac{1}{2} \sum_{i=1}^{N} \sum_{j=1}^{N} \sum_{\{n\}}{}' (\boldsymbol{\mu}_i \cdot \nabla_i)(\boldsymbol{\mu}_j \cdot \nabla_j) \Psi_{\text{R}}(|\boldsymbol{r}_{ij} + \boldsymbol{n}|) \\
&= \frac{1}{2} \sum_{i=1}^{N} \sum_{j=1}^{N} \sum_{\{n\}}{}' \{(\boldsymbol{\mu}_i \cdot \boldsymbol{\mu}_j) B(|\boldsymbol{r}_{ij} + \boldsymbol{n}|, \alpha) \\
&\quad - [\boldsymbol{\mu}_i \cdot (\boldsymbol{r}_{ij} + \boldsymbol{n})][\boldsymbol{\mu}_j \cdot (\boldsymbol{r}_{ij} + \boldsymbol{n})] C(|\boldsymbol{r}_{ij} + \boldsymbol{n}|, \alpha)\} \quad (6.27a)
\end{aligned}
$$

$$
\begin{aligned}
U_{\text{DF}}^{3d} &\equiv \frac{1}{2} \sum_{i=1}^{N} \sum_{j=1}^{N} (\boldsymbol{\mu}_i \cdot \nabla_i)(\boldsymbol{\mu}_j \cdot \nabla_j) \Psi_{\text{F}}(\boldsymbol{r}_{ij}) \\
&= \frac{2\pi}{V} \sum_{\boldsymbol{k} \neq 0} \frac{1}{k^2} \exp\left(-\frac{k^2}{4\alpha^2}\right) \widetilde{M}(\boldsymbol{k}) \widetilde{M}^*(\boldsymbol{k}) \quad (6.27b)
\end{aligned}
$$

$$
\begin{aligned}
U_{\text{DLR}}^{3d} &\equiv \frac{1}{2} \sum_{i-1}^{N} \sum_{j=1}^{N} (\boldsymbol{\mu}_i \cdot \nabla_i)(\boldsymbol{\mu}_j \cdot \nabla_j) \Psi_{\text{LR}}(\boldsymbol{r}_i, \boldsymbol{r}_j) \\
&= \frac{2\pi}{2\epsilon' + 1} \frac{M^2}{V} \quad (6.27c)
\end{aligned}
$$

where the functions B and C are defined as [140]

$$
B(r, \alpha) \equiv \frac{1}{r^3} \left[\frac{2\alpha r}{\sqrt{\pi}} \exp(-\alpha^2 r^2) + \text{erfc}(\alpha r) \right] \quad (6.28a)
$$

$$
C(r, \alpha) \equiv -\frac{1}{r} \frac{dB}{dr} = \frac{1}{r^5} \left[\frac{2\alpha r}{\sqrt{\pi}} (3 + 2\alpha^2 r^2) \exp(-\alpha^2 r^2) + 3\text{erfc}(\alpha r) \right] \quad (6.28b)
$$

In addition, in Eq. (6.27b),

$$
\begin{aligned}
\widetilde{M}(\boldsymbol{k}) &= \sum_{i=1}^{N} (\boldsymbol{\mu}_i \cdot \boldsymbol{k}) \exp(-i\boldsymbol{k} \cdot \boldsymbol{r}_i) \\
&= \sum_{i=1}^{N} [(\boldsymbol{\mu}_i \cdot \boldsymbol{k}) \cos(\boldsymbol{k} \cdot \boldsymbol{r}_i) - i(\boldsymbol{\mu}_i \cdot \boldsymbol{k}) \sin(\boldsymbol{k} \cdot \boldsymbol{r}_i)] \\
&\equiv \text{Re}\widetilde{M}(\boldsymbol{k}) - i\text{Im}\widetilde{M}(\boldsymbol{k}) \quad (6.29)
\end{aligned}
$$

and in Eq. (6.27c),

$$
\boldsymbol{M} = \sum_{i=1}^{N} \boldsymbol{\mu}_i \quad (6.30)
$$

is again the total dipole moment of the many–particle system in the central cell and M is its magnitude. Similar to ionic systems, the long-range term vanishes for the choice $\epsilon' = \infty$, reflecting the fact that a conducting environment prevents formation of surface charges and thus suppresses a depolarizing field inside the central cell.

Finally, the self-part of the dipolar Ewald energy, $U_{\mathrm{DS}}^{\mathrm{3d}}$, cannot be derived by applying the Nabla operator as for the other terms, because the corresponding self-term for charged systems is already a constant. Specifically, one realizes from Eq. (6.17b) that the Coulomb self-energy can be rewritten as

$$U_{\mathrm{CS}}^{\mathrm{3d}} = \frac{1}{2}\sum_{i=1}^{N}\sum_{j=1}^{N}q_i q_j \Psi_{\mathrm{S}}\left(r_{ij}\right) \qquad (6.31a)$$

$$\Psi_{\mathrm{S}}\left(r_{ij}\right) = \Psi_{\mathrm{S}} = -\frac{2\alpha}{\sqrt{\pi}}\delta_{ij} \qquad (6.31b)$$

Thus, the transformation $q_i \rightarrow \boldsymbol{\mu}_i \cdot \boldsymbol{\nabla}_i$ would lead to the erroneous result $U_{\mathrm{DS}}^{\mathrm{3d}} = 0$. We defer the correct derivation of the self-term to Appendix F.2.1 where we show that

$$U_{\mathrm{DS}}^{\mathrm{3d}} = -\frac{2\alpha}{3\sqrt{\pi}}\sum_{i=1}^{N}\mu_i^2 \qquad (6.32)$$

Regarding the use of Eq. (6.26) in practice we note that the same comments made earlier apply here as well [see discussion after Eq. (6.17)]. A detailed discussion of optimal choices for the Ewald parameters α and m_c^2 for dipolar systems can be found in Refs. 243 and 244. Finally, readers who are interested in performing MD simulations of dipolar fluids are referred to Appendix F.2.2 where we present explicit expressions for forces and torques associated with the three–dimensional Ewald sum [see Eq. (6.26)]. Moreover, explicit expressions for various components of the stress tensor can be found in Appendix F.2.3.

6.3   Ewald summation for confined fluids

Having understood the concepts of Ewald summation techniques for three-dimensional bulk systems, we now turn to systems that are *finite* in at least one spatial dimension. We focus on a slab-like geometry where the fluid is confined by two plane parallel and structureless solid surfaces separated by a distance s_z along the z-axis of the coordinate system and of infinite extent in the x–y plane (see also Section 1.3.2). Hence, for the time being, we shall be dealing with the situation analyzed in Section 1.3.2 from a purely thermodynamic perspective where a rectangular fluid lamella of area $A = s_x s_y$ and height s_z is repeated periodically in the x- and y-directions but, obviously, not in the direction perpendicular to the walls. Note, that we implicitly assume that the fluid lamella considered here can also be deformed in the x- and y-directions by external agents and therefore the thermodynamic analysis of Section 1.3.2 does not directly apply here because we explicitly assume in that section that the lamella is only subject to a compressional strain in the direction normal to the solid surfaces.[5]

Not surprisingly, the reduction of spatial symmetry renders the treatment of long-range fluid fluid interactions significantly more involved than in the three-dimensional case. This concerns not only the derivation of appropriate (Ewald or other) expressions for energies, forces, and torques, but also the computational effort needed in actual calculations.

An additional issue here concerns the choice of appropriate boundary conditions at the solid surfaces. Quantitatively speaking, these can be characterized by another dielectric constant ϵ'', which, on account of the finiteness of the system in the $\pm z$-direction, may (and will) in general differ from ϵ' characterizing the infinite surroundings in the $x-y$ plane in which the fluid lamella is embedded. For example, a typical question in electrochemistry is to what extent "image charges" induced by an ionic fluid in a confining conducting (i.e., metallic) substrate ($\epsilon'' = \infty$) affect the structure of the confined fluid itself. Consequently, computer simulations of confined fluids with long-range interactions are still quite challenging, and there is an ongoing discussion of how the computational methods can be optimized [231, 238].

In light of this background we discuss in this section three different approaches to Ewald summation in slab geometries. The first and second of these methods are appropriate for fluids confined between *insulating* substrates (i.e., $\epsilon'' = 1$). They differ from one another in the mathematical rigor of their derivation. They are also quite different in the computational effort required to evaluate the final expressions in an actual computer simulation. The third approach presented in this section is capable of dealing with system confined by metallic solid surfaces (characterized by $\epsilon'' = \infty$).

6.3.1 The rigorous approach

Based on the ideas discussed in Section 6.2, various rigorous extensions to slab-like systems have been proposed [245–251]. Because these systems are

---

[5]A suitable adjustment of the thermodynamic description presented in Section 1.3.2 to the situation considered here is, however, both feasible and straightforward.

infinite in two dimensions, we begin by considering N charged particles in a central cell extended periodically along the x- and y- directions. Thus the the configurational potential energy is given by

$$U_{\mathrm{C}}^{\mathrm{2d}} = \frac{1}{2} \sum_{i=1}^{N} q_i \Phi(\boldsymbol{r}_i) = \frac{1}{2} \sum_{i=1}^{N} \sum_{j=1}^{N} \sum_{\{\boldsymbol{n}\}}{}' \frac{q_i q_j}{|\boldsymbol{r}_{ij} + \boldsymbol{n}|} \qquad (6.33)$$

where, in contrast to the corresponding expressions for bulk systems [see Eqs. (6.1) and (6.2)], the lattice vectors $\{\boldsymbol{n}\}$ are quasi two–dimensional, that is $\boldsymbol{n} = (\boldsymbol{n}_\parallel, 0)$ with $\boldsymbol{n}_\parallel = (n_x s_x, n_y s_y)$. Starting from Eq. (6.33), the corresponding Ewald formulation can be worked out in close analogy with the treatment detailed in Section 6.2 for three–dimensional (bulk) fluids.

As before the main step consists of rewriting the (δ-like) charge distribution $\rho(\boldsymbol{r})$ corresponding to Eq. (6.33) as a sum of three contributions [cf., Eq. (6.7), (6.10)] which are then considered separately. This procedure is sketched in Appendix F.3.1. One finally obtains

$$U_{\mathrm{C}}^{\mathrm{2d}} = U_{\mathrm{CR}}^{\mathrm{2d}} + U_{\mathrm{CF}}^{\mathrm{2d}} + U_{\mathrm{CS}}^{\mathrm{2d}} \qquad (6.34)$$

where the real-space contribution has the same form as before in the three-dimensional case and is thus given by the right side of Eq. (6.16a) taking, of course, $\boldsymbol{n} = (\boldsymbol{n}_\parallel, 0)$. The same holds for the self-part, which is given by Eq. (6.17b). The Fourier contribution, however, takes a more complicated form, namely

$$\begin{aligned}
U_{\mathrm{CF}}^{\mathrm{2d}} = {}& \frac{\pi}{2A} \sum_{i=1}^{N} \sum_{j=1}^{N} q_i q_j \sum_{\boldsymbol{k}_\parallel \neq \boldsymbol{0}} \frac{\exp[i\boldsymbol{k}_\parallel \cdot \boldsymbol{R}_{ij}]}{k_\parallel} f(k_\parallel, z_{ij}, \alpha) \\
& - \frac{\sqrt{\pi}}{A} \sum_{i=1}^{N} \sum_{j=1}^{N} q_i q_j \left[\frac{\exp\left(-\alpha^2 z_{ij}^2\right)}{\alpha} + \sqrt{\pi} z_{ij} \mathrm{erf}\left(\alpha z_{ij}\right) \right] \quad (6.35)
\end{aligned}$$

where $\boldsymbol{R}_{ij} = \boldsymbol{R}_i - \boldsymbol{R}_j$ and $\boldsymbol{R}_i = (x_i, y_i)$ is the projection of the position vector of particle i onto the x–y plane. Furthermore, $z_{ij} = z_i - z_j$ and $\boldsymbol{k}_\parallel = (2\pi m_x/s_x, 2\pi m_y/s_y)$ is a two-dimensional reciprocal lattice vector (i.e., $\exp[i\boldsymbol{k}_\parallel \cdot \boldsymbol{n}_\parallel] = 1$). Finally, $k_\parallel = |\boldsymbol{k}_\parallel|$ and

$$f(k_\parallel, z_{ij}, \alpha) = \exp[k_\parallel z_{ij}]\mathrm{erfc}\left(\frac{k_\parallel}{2\alpha} + \alpha z_{ij}\right) + \exp[-k_\parallel z_{ij}]\mathrm{erfc}\left(\frac{k_\parallel}{2\alpha} - \alpha z_{ij}\right) \qquad (6.36)$$

In terms of computer time, evaluation of the Fourier contribution to the total configurational potential energy, $U_{\mathrm{CF}}^{\mathrm{2d}}$, requires significantly more effort than in the three-dimensional case [see Eqs. (6.18) and (6.19) for comparison]. The reason is that the double sum over all pairs of particles in Eq. (6.35) cannot be rewritten as a product of two single-particle sums. The corresponding expression for a system of point dipoles is even more involved as one may verify from Eq. (F.115) given in Appendix F.3.1.2.

Thus, for both the ionic and the dipolar systems, the actual use of the rigorously derived Ewald summation for slab systems leads to a substantial increase in computer time. One way of dealing with this problem would be to employ precalculated tables [252] for potential energies (and forces) on a three-dimensional spatial grid amended by a suitable interpolation scheme. Another strategy is to employ approximate methods such as the one presented in the subsequent Section 6.3.2.

6.3.2   Slab-adapted Ewald summation

In the following we discuss a modification of the well-established three-dimensional Ewald sum suitable for simulations of systems of point charges or dipoles with a slab-like geometry [252–257] with insulating substrates (i.e., $\epsilon'' = 1$). Within this method the slab geometry is taken into account by incorporating a vacuum region along the z-direction in the basic cell. Specifically, we choose a tetragonal cell with lateral dimensions $s_x = s_y = s$ (area $A = s^2$), and perpendicular dimension $d_z = s_z + s_{vac} = \gamma s$, where s_z is the wall separation and s_{vac} is the width of the vacuum region that can be controlled via the dilatation factor γ. The entire cell with volume $V = s^2 d_z = \gamma s^3$ is replicated in *all three* spatial directions, which results in a system of parallel fluid slabs alternating with vacuum slabs in the z–direction (see Fig. 6.1).

The entire system is then placed into a medium of infinite dielectric constant that prevents formation of surface charges at the *outer* boundaries. Thus, we can employ the conventional three-dimensional Ewald sums [see Eqs. (6.15) and (6.26) for the Coulombic and dipolar case, respectively] with "tin-foil" boundary conditions (i.e., $\epsilon' = \infty$).

It is clear, however, that the slab-adapted version can mimic the desired single-slab situation only if interactions between periodic images of the slab are neglible, which rigorously requires $s_{vac} \to \infty$ (i.e., $\gamma \to \infty$). Because this can hardly be realized, we follow Yeh and Berkowitz [254] and supplement the Ewald energy by a correction term, implicitly assuming that s_{vac} is large enough so that the remaining interslab interactions can be handled on the basis of continuum theory. Suppose we have a system of N point charges, which is globally neutral (i.e., $\sum_{i=1}^{N} q_i = 0$) but carries a net dipole moment in the z-direction, $M_z = \sum_{i=1}^{N} q_i z_i$. Because of the periodic replication, M_z is mirrored by the neighboring slabs, which will therefore act as parallel

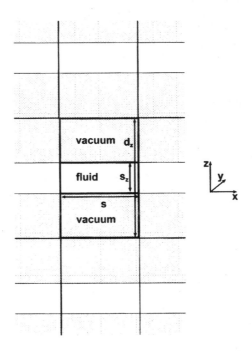

Figure 6.1: Sketch of the simulation cell (front view) employed in the slab-adapted three-dimensional Ewald sum.

capacitor plates with surface charges $P_z = M_z/V$. These charges generate a capacitor field, $\widetilde{\boldsymbol{E}} = \widetilde{E}_z \widehat{\boldsymbol{e}}_z$, within the central cell where

$$\widetilde{E}_z = 4\pi P_z \tag{6.37}$$

The corresponding contribution to the energy per volume follows as

$$\frac{\widetilde{U}_{\mathrm{C}}}{V} = -\frac{1}{2} P_z \widetilde{E}_z = -\frac{2\pi}{V^2} \sum_{i=1}^{N} \sum_{j=1}^{N} q_i q_j z_i z_j \tag{6.38}$$

The above considerations suggest considering $\widetilde{U}_{\mathrm{C}}$ as the energy contribution due to "interslab interactions" resulting from the periodic replication of the system in the z-direction. Clearly, these interslab interactions are an unwanted phenomenon. We therefore introduce an energy correction,

$$U_{C,c} = -\widetilde{U}_C = \frac{2\pi}{V} M_z^2 \tag{6.39}$$

which we add to the conventional three-dimensional Ewald sum for point charges with conducting boundaries (i.e., $\epsilon' = \infty$) to give

$$U_C^{\text{slab}} = U_C^{\text{3d}}\Big|_{\epsilon'=\infty} + U_{C,c} = U_C^{\text{3d}}\Big|_{\epsilon'=\infty} + \frac{2\pi}{V} M_z^2 \tag{6.40}$$

Equation (6.40) is our final expression for the energy of a charged system between insulating walls within the slab-adapted three-dimensional Ewald summation method. On the right side, the contribution $U_C^{\text{3d}}\big|_{\epsilon'=\infty}$ is defined by Eqs. (6.15), (6.16), and (6.17b).[6] The reader should also realize that, for the current system, the volume V appearing in the Fourier part of the energy [see Eq. (6.16b)] includes the vacuum space; that is, $V = s^2 d_z = \gamma s^3$. Moreover, here

$$\boldsymbol{k} = 2\pi \begin{pmatrix} m_x/s \\ m_y/s \\ m_z/\gamma s \end{pmatrix} \tag{6.41}$$

Because the correction term in Eq. (6.40) is proportional to M_z^2, the development of a net dipole moment in the z-direction leads to an increase in the total configurational potential energy. This is consistent with one's physical intuition if one recalls from elementary electrostatics that an infinitely extended slab polarized along the slab normal experiences a depolarizing field equal to $-4\pi P_z = -4\pi M_z/V$ acting against the polarization [242]. In fact, $U_{C,c}$ [see Eq. (6.39)] is nothing but the energy penalty due to the depolarizing field.

It is then straightforward to formulate a slab-adapted three-dimensional Ewald sum for a system of point dipoles. In this case,

$$P_z = \frac{1}{V} \sum_{i=1}^{N} \mu_{i,z} \tag{6.42}$$

and the energy correction compensating for interslab interactions follows from Eqs. (6.37)–(6.39) as

$$U_{D,c} = \frac{2\pi}{V} \sum_{i=1}^{N} \sum_{j=1}^{N} \mu_{i,z}\mu_{j,z} = \frac{2\pi}{V} M_z^2 \tag{6.43}$$

---

[6]Notice that the long-range part given in Eq. (6.17a) vanishes because of our current choice $\epsilon' = \infty$.

which has exactly the same form as for the corresponding ionic system [see Eq. (6.39)]. The total energy of the dipolar slab system is then calculated from

$$U_{\mathrm{D}}^{\mathrm{slab}} = U_{\mathrm{D}}^{\mathrm{3d}}\Big|_{\epsilon'=\infty} + U_{\mathrm{D,c}} = U_{\mathrm{D}}^{\mathrm{3d}}\Big|_{\epsilon'=\infty} + \frac{2\pi}{V} M_{\mathrm{z}}^2 \qquad (6.44)$$

where $U_{\mathrm{D}}^{\mathrm{3d}}$ is given in Eqs. (6.26)–(6.32) [noting that $U_{\mathrm{DLR}}^{\mathrm{3d}} = 0$ for our current choice $\epsilon' = \infty$; see Eq. (6.27c)]. Explicit expressions for the forces, torques, and the normal component of the stress tensor related to Eqs. (6.40) and (6.44), respectively, can be found in Appendix F.3.2.

In practice, most applications of the slab-adapted Ewald sum in three dimensions employ vacuum spaces that are three to five times thicker than the original substrate separation [252–257]. The energies obtained with the approximate Ewald sum then coincide almost perfectly with those of the rigorous method discussed in Section 6.3.1. For a systematic discussion of the errors involved in the slab-adapted version, we refer the interested reader to Ref. 256.

6.3.3 Reliability of the slab-adapted Ewald method

To demonstrate the reliability of the slab-adapted Ewald method introduced in the preceding Section 6.3.2, we present in the following results from lattice calculations [257]. Specifically, we consider a slab composed of dipolar spheres of diameter σ located at the sites of a face–centered cubic (fcc) lattice. The lattice vectors are $r = (\ell/2)\,(l_{\mathrm{x}}, l_{\mathrm{y}}, l_{\mathrm{z}})$, where ℓ is the lattice constant (fixed such that the reduced density $\rho\sigma^3 = 4\sigma^3/\ell^3 = 1.0$), and $\{l_\alpha\}$ ($\alpha = $ x,y,z) are integers with $l_{\mathrm{x}} + l_{\mathrm{y}} + l_{\mathrm{z}}$ even. An infinitely extended slab is then realized by setting $-\infty < l_{\mathrm{x}}, l_{\mathrm{y}} < \infty$ and $l_{\mathrm{z}} = 0, \ldots, n_{\mathrm{z}} - 1$ with n_{z} being the number of fcc layers in z-direction.

We evaluate the system's total dipolar energy for the following situations:

1. all dipoles point along the x-axis and

2. all dipoles point along the z-axis.

The latter, somewhat unphysical situation has been included to investigate the importance of the correction term given in Eq. (6.43). Clearly, the correction is irrelevant for case 1.

For both situations, we calculate the (dimensionless) dipolar energy per particle, $\sigma^3 U_{\mathrm{D}}/\mu^2 N$, via the slab-adapted three-dimensional Ewald sum[7] [see Eq. (6.44)] and with the rigorous Ewald method for dipolar slab systems

---

[7]Calculations were performed with the Ewald parameters $\alpha s = 7.0$ and $m_{\mathrm{c}}^2 = 80$.

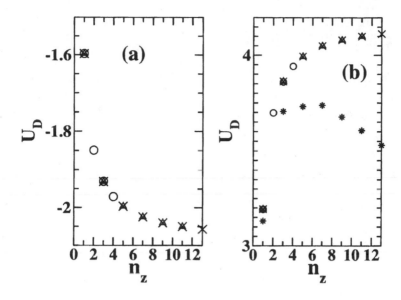

Figure 6.2: Dimensionless energy per particle for dipolar crystalline (fcc) slabs as a function of the number of lattice layers, assuming perfect order along the x-axis (a) and along the z-axis (b). Included are results from direct summation ($\bigcirc$), the rigorous Ewald sum for slab systems (Δ) [see Appendix F.3.1.2], and the slab-adapted three-dimensional Ewald sum ($\times$) [see Eq. (6.44)]. Part (b) additionally includes results from the latter method when the correction term [see Eq. (6.43)] is neglected ($*$).

as formulated in Appendix F.3.1.2. The results are compared in Fig. 6.2. Obviously, the two different Ewald methods are fully consistent both for cases 1 and 2. Figure 6.2(b) also includes results obtained with the slab-adapted three-dimensional Ewald sum but without the correction term given in Eq. (6.43). Clearly, neglecting this term causes erroneous results, although the dilatation factor controlling the space between two neighboring slabs has already been set to $\gamma = 10$. Thus, we conclude that it is essential to incorporate the correction to achieve consistency between the slab-adapted three-dimensional and the rigorous Ewald sum.

Also shown in Figs. 6.2 are results obtained via direct summation of the dipolar interactions. Using the fact that particles within a given layer contribute equally, the total configurational potential energy for case 2 can be cast as

$$U_{\mathrm{D}} = -\frac{1}{2} \sum_{i=1}^{N} \boldsymbol{\mu}_i \cdot \boldsymbol{E}(\boldsymbol{r}_i) = -\frac{\mu_z}{2} \sum_{i=1}^{N} E_z(\boldsymbol{r}_i) = -\frac{\mu_z}{2} \frac{N}{n_z} \sum_{l_z=0}^{n_z-1} E_z(l_z) \qquad (6.45)$$

where N/n_z is the number of dipoles per layer and $E_z(l_z)$ is the z-component of the field $\boldsymbol{E}$ acting on a dipole in lattice plane l_z. Without loss of generality we may assume that this dipole is located at a lattice site $\boldsymbol{r}_i = (l/2)\,(0,0,l_z)$. Then the field due to the other dipoles ($j \neq i$ with $\boldsymbol{\mu}_j = \mu_z \widehat{\boldsymbol{e}}_z$) located at $\boldsymbol{r}_j = (l/2)\,(l'_x, l'_y, l'_z)$ is given by

$$
\begin{aligned}
E_z(l_z) &= \mu_z \sum_{j \neq i}^{N} \left[\frac{3\,(z_j - z_i)^2}{r_{ij}^5} - \frac{1}{r_{ij}^3} \right] \\
&= \frac{8\mu_z}{l^3} \sum_{l'_x, l'_y = -\infty}^{\infty} \sum_{l'_z=0}^{n_z-1} \left[\frac{3\,(l_z - l'_z)^2}{\sqrt{l'^2_x + l'^2_y + (l_z - l'_z)^2}^{\,5}} \right. \\
&\qquad\qquad \left. - \frac{1}{\sqrt{l'^2_x + l'^2_y + (l_z - l'_z)^2}^{\,3}} \right]
\end{aligned}
\qquad (6.46)
$$

Similar expressions are obtained for case 1. Data plotted in Figs. 6.2 have been obtained by truncating the sums over l_x and l_y at ± 5000, which yields convergent results as long as $n_z \leq 4$. Comparing these results with those from the two Ewald methods, we conclude that not only the rigorous Ewald summation, but also the slab-adapted three-dimensional version provide quasi-exact results for the dipolar energy.

6.4 Insulating solid substrates

6.4.1 Dipolar interactions and normal stress

Having established the accuracy and reliability of the slab-adapted three-dimensional Ewald method, we present in this paragraph numerical results from GCEMC simulations (see Section 5.2.2) for a confined Stockmayer fluid.[8] The particles then interact with each other via both the long-range,

---

[8]The Stockmayer fluid consists of spherical particles with embedded (permanent) point dipoles.

anisotropic dipole dipole interactions u_{DD} given in Eq. (6.21) and the (spherical) LJ (12,6) potential given in Eq. (5.24). The Stockmayer fluid is frequently used to model polar molecular fluids with permanent electric dipole moments such as chloroform, for example [258, 259]. However, more recently, Stockmayer fluids have also been employed to model ferrofluids, which consist of (colloidal) particles with permanent *magnetic* dipole moments in carrier liquids like water or oil.

For technical reasons explained in Section 5.2.2, the GCEMC simulations have been performed using a slightly modified Stockmayer potential defined by

$$u\left(\mathbf{r}_{ij}, \boldsymbol{\mu}_i, \boldsymbol{\mu}_j\right) = u_{SR}\left(r_{ij}\right) + u_{DD}\left(\mathbf{r}_{ij}, \boldsymbol{\mu}_i, \boldsymbol{\mu}_j\right) \qquad (6.47)$$

where the short-range contribution is given in Eq. (5.39). In the calculations below we choose a cut-off radius of $r_c = 2.5$ for the short-range interactions.[9] Fluid molecules are confined by two planar solid walls separated by a distance s_z along the z-axis of the coordinate system and of infinite extent in the x–y plane. Assuming that fluid particles and atoms composing the wall interact via the LJ (12,6) potential given in Eq. (5.24) and averaging over the subspaces $-\infty < z \le -s_z/2$ and $s_z/2 \le z < \infty$ occupied by substrate particles, the fluid substrate potential follows as [cf. Eq. (5.71)]

$$\Phi^{[k]}\left(z; s_z\right) = \frac{4\pi}{45}\varepsilon\rho_s\sigma^3\left[\left(\frac{\sigma}{s_z/2 \pm z}\right)^9 - \left(\frac{\sigma}{s_z/2 \pm z}\right)^3\right] \qquad (6.48)$$

where the plus and the minus sign refer to the lower ($k = 1$) and upper ($k = 2$) wall, respectively. In Eq. (6.48), the reduced density of solid particles is set to $\rho_s\sigma^3 = 1$.

The GCEMC results presented in this section refer to Stockmayer fluids at a temperature $T = 1.60$ and dipole moment $m = 2.0$, which are typical for real polar molecular fluids [259]. The chemical potential is set equal to $\mu = 19.30$, so that the bulk fluid has an average mean density of $\bar{\rho} \simeq 0.6$. Keeping these parameters fixed, we investigated systems with substrate separations s_z in the range $1.7 \le s_z \le 10.0$.

In the following we focus on the normal stress, τ_{zz}, or rather on the disjoining pressure $f\left(s_z\right)$ defined in Eq. (5.57) for reasons given in Section 5.3.2. Numerical results for $f(s_z)$ are presented in Fig. 6.3, where the dipolar contribution to τ_{zz} has been calculated according to the microscopic expressions given in Eqs. (F.124)–(F.126) in Appendix F.3.2.

As before for confined fluids with short-range interaction potentials, the disjoining pressure $f\left(s_z\right)$ has damped oscillatory character with the oscillations vanishing when the confined systems become bulk-like at large wall

---

[9]Note that no such cut-off is applied to the long-range part of the potential.

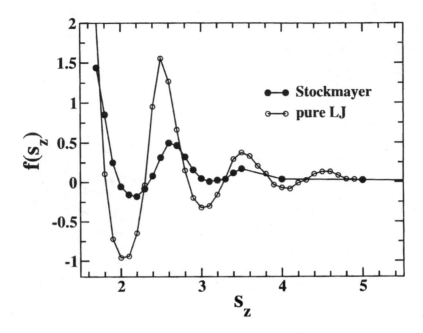

Figure 6.3: GCEMC results for the disjoining pressure $f(s_z)$ [see Eq. (5.57)] for a typical polar (Stockmayer) fluid as a function of the wall separation (data have been obtained at $T = 1.60$, $m = 2.0$, and $\mu = -19.30$ corresponding to an average bulk density $\bar{\rho} \simeq 0.6$). Also shown are corresponding results for an atomic LJ (12,6) fluid at the same temperature and bulk density.

separations s_z. This behavior agrees, at least on a qualitative level, with that observed for simpler systems interacting via spherically symmetric potentials only (see Section 5.3.3, Fig. 5.3). To illustrate that agreement, we have included in Fig. 6.3 GCEMC data for a pure LJ (12,6) fluid where the chemical potential has been fixed at a value chosen such that both the LJ (12,6) and the Stockmayer *bulk* fluids have the same average density (i.e., $\mu = -13.50$ for the corresponding LJ fluid). Upon decreasing the substrate separation from larger values, the function $f(s_z)$ first is essentially constant for both systems down to wall separations $s_z \approx 5\sigma$. Only at smaller separations does $f(s_z)$ exhibit pronounced oscillations with a period of roughly one molecular "diameter." One also sees that maxima and minima of $f(s_z)$ are more pronounced in the LJ (12,6) fluid where the positions of the extrema are slightly out of phase relative to $f(s_z)$ for the Stockmayer fluid. On the other hand, for both systems, the oscillations are accompanied by a step-like increase of the average adorption "rate" $\langle N \rangle / A$, which is shown in Fig. 6.4.

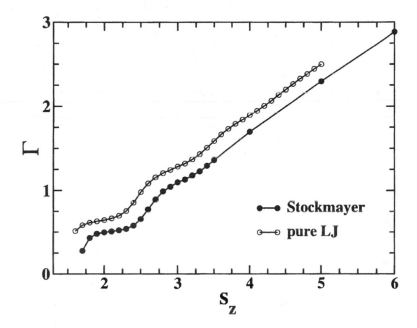

Figure 6.4: Adsorption rate of a Stockmayer fluid and a comparable LJ fluid as a function of the wall separation (parameters as in Fig. 6.3).

Based on previous studies on the relations between the oscillations in the disjoining pressure and structural changes (see Section 5.3.4), we interpret these features as fingerprints of stratification, which, to reiterate, is the tendency of fluid particles to arrange themselves in individual layers (strata) parallel to the confining walls. In this sense, maxima in the plot of $f(s_z)$ indicate the disappearance (formation) of complete layers of fluid molecules upon decreasing (increasing) s_z. At separations corresponding to minima of $f(s_z)$, on the other hand, particles are accommodated comfortably between the substrates. At other separations (particularly around the maxima), the fluid structure will be more or less frustrated (i.e., disordered).

Although this interpretation of the qualitative behavior holds for both the Stockmayer and the LJ (12,6) fluid, it is also interesting to compare the actual values of $f(s_z)$, which are indeed quite different for the two models considered. In particular, the oscillations in the LJ (12,6) fluid are characterized by much larger amplitudes, suggesting that the overall repulsion between the particles is stronger than in the dipolar system. In other words, it seems that even for strongly confined dipolar fluids, similar to what is found in bulk dipolar systems [260], the anisotropic dipolar interactions manifest themselves as a net *attraction*.

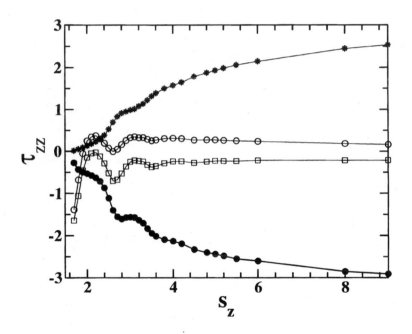

Figure 6.5: Contributions to the normal stress of a Stockmayer fluid as a function of the wall separation (parameters as in Fig. 6.3). ($\bullet$) τ_{zz}^{SR} ($\circ$) τ_{zz}^{FS}, ($*$) τ_{zz}^{D}, ($\square$) τ_{zz}.

That this is indeed the case can be seen from Fig. 6.5 where we have plotted separately the three contributions to the normal stress of the Stockmayer fluid,

$$\tau_{zz} = \tau_{zz}^{SR} + \tau_{zz}^{D} + \tau_{zz}^{FS} \tag{6.49}$$

It is seen that the oscillatory character of the normal stress is visible in all contributions to τ_{zz} at sufficiently small substrate separations. One also notes a difference in sign among the various individual contributions. For example, τ_{zz}^{D} is positive regardless of the substrate separation, indicating that the dipolar interactions tend to decrease the (magnitude of the) normal stress. Finally, in closing, it is worth noting that the features described above are not restricted to the specific thermodynamic conditions and not even to the specific model system we have considered here. Rather they should be perceived as generic features of confined dipolar liquids.

6.4.2 Orientational order in confined dipolar fluids

As a second application of the Ewald summation for systems with slab geometry we now consider strongly coupled confined dipolar liquids, that is, liquids where the dipolar interactions dominate the system's structure and phase behavior. Examples for such systems are ferrocolloidal films or confined molecular liquids with strong dipole moments and relatively weak van der Waals forces or hydrogen bonds [231].

The model we consider in the current section is similar to the Stockmayer fluid discussed in Section 6.4.1. However, here we replace the LJ interaction, which includes attractive dispersion interactions [see Eq. (5.24)], by the purely repulsive soft-sphere (SS) potential,

$$u_{SS}\left(r_{ij}\right) = 4\varepsilon \left(\frac{\sigma}{r_{ij}}\right)^{12} \tag{6.50}$$

In a similar spirit we neglect the attractive part of the fluid substrate potential introduced in Eq. (6.48). Choosing this somewhat minimalistic model has the advantage of permitting us to study more directly the interplay between spatial confinement and the long-range, anisotropic dipolar interactions. In fact, recent research on the bulk dipolar soft-sphere (DSS) (and related) model fluids demonstrated already that the specific properties of the dipolar interaction, combined with lack of long-range translational order can lead to new and unexpected physical behavior [260, 261]. In particular, it was shown that dense systems of dipolar spheres may develop spontaneous polarization, yielding a *liquid* state with long-range ferroelectric order [262–268]. Given the appearance of an orientationally ordered, yet liquid-like phase in bulk dipolar fluids, a main question addressed by us in Refs. 257 and 269, was whether spontaneous order of this type may also exist in nanoscopic slit-pores where it is not a priori clear whether confinement may support or inhibit ferroelectric order.

To this end we performed MC simulations in the mixed isostress isostrain ensemble introduced in Section 5.7 where a thermodynamic state is specified by the set $\{T, N, \tau_{\parallel}, s_{z0}\}$; that is, we are holding constant the number of molecules N rather than the chemical potential μ of the confined fluid. As an immediate consequence, the confined fluid can no longer be viewed as being in thermodynamic equilibrium with a bulk reservoir because their chemical potentials may (and very likely will) differ if the number of molecules in the confined system remains constant. Although this may seem to be a disadvantage with respect to situations encountered frequently in laboratory experiments, we nevertheless prefer this ensemble over the grand canonical one. The reason is that, for the confined DSS fluid, thermodynamic states of interest in the current context are expected to be relatively dense, such that transferring particles between the reservoir and the confined system might be very difficult.

This implies that in parallel laboratory experiments the exchange of matter between the confined fluid and a bulk reservoir is insignificant or that a bulk phase in the sense of the SFA experiment (see Section 5.3.1) may not even be present at all. Such a situation is realized, for example, in the SANS experiments described in Section 4.8.1 where a binary fluid mixture is confined to a nanoporous medium (i.e., pellets) without being in contact with a bulk reservoir.

The algorithm by which a (numerical representation of a) Markov chain is generated in this ensemble is discussed in detail in Section 5.7.7. However, for the DSS fluid, we need to also include random rotations of the molecules as part of the canonical substep of the algorithm in addition to their random displacement (see Section 5.2.2).

All simulations were performed for $T = 1.35$ and $m = 3.0$. These parameters have already been employed in earlier simulations of bulk DSS fluids for which long-range parallel order at sufficiently high densities or pressures was observed [262–265].

Our MC results for the confined dipolar fluids indicate that spontaneous order does indeed occur over a certain range of wall separations s_z. This can be seen from Figs. 6.6–6.8 for a system where $s_z = 7$. Specifically, in Fig. 6.6, we have plotted the global order parameter

$$P_1 = \frac{1}{N} \left| \sum_{i=1}^{N} \widehat{\boldsymbol{\mu}}_i \cdot \widehat{\boldsymbol{d}} \right| \tag{6.51}$$

as a function of the applied parallel pressure $P_{\parallel}$, where $\widehat{\boldsymbol{d}}$ is the global director. The latter is obtained from the simulations; the *instantaneous* value of $\widehat{\boldsymbol{d}}$ is the eigenvector corresponding to the largest eigenvalue of the order-parameter matrix [140]

$$\widehat{\mathbf{Q}} \equiv \frac{1}{2} \sum_{i=1}^{N} \left(3\widehat{\boldsymbol{\mu}}_i \cdot \widehat{\boldsymbol{\mu}}_i - \widehat{\mathbf{1}} \right) \qquad (6.52)$$

In the remainder of this section, we deliberately chose to deviate from our standard treatment of mechanical work in terms of *stresses* rather than *pressure* tensor elements. Even though we foster the former treatment throughout this book, the latter seems a bit more intuitive in the current context, especially with regard to existing literature on the bulk DSS fluid. However, we remind the reader that pressure tensor **P** and stress tensor $\boldsymbol{\tau}$ are trivially related through the relation $\mathbf{P} = -\boldsymbol{\tau}$.

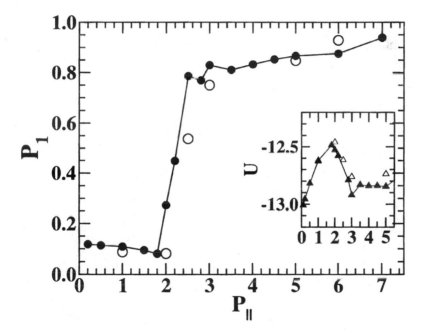

Figure 6.6: Order parameter P_1 and internal energy (inset) versus applied parallel pressure for a DSS fluid confined to a slit-pore system with wall separation $s_z = 7\sigma$. Results are obtained with $N = 256$ ($\bullet$, $\blacktriangle$) and $N = 500$ ($\circ$, $\triangle$) particles. Temperature and dipole moment are fixed at $T = 1.35$ and $m = 3.0$.

Starting from $P_{\parallel} = 0.1$ ($\tau_{\parallel} = -0.1$) and increasing (decreasing) the pressure (stress) up to $P_{\parallel} \approx 2.0$, the order parameter remains small at first, indicating that the system is globally isotropic over this pressure range. Upon further compression, however, P_1 increases sharply and for $P_\parallel > 2.6$ one observes fairly large values of the order parameter $P_1 \gtrsim 0.8$ that depend only weakly on the number of particles employed in the simulation. As all simulations were started from randomly oriented states, we conclude that states characterized by a large transverse pressure are indeed *spontaneously* polarized. They comport with a true ferroelectric phase.

The presence of an isotropic ferroelectric transition is also reflected by the total configurational potential energy per particle $\langle U \rangle / N$ plotted in the inset of Fig. 6.6. Increasing $P_\parallel$ from the initial smaller values, $\langle U \rangle / N$ first increases, but then begins to decrease at a transverse pressure of about $P_\parallel \simeq 2.0$ where P_1 begins to rise rather sharply. Clearly, the decrease of $\langle U \rangle / N$ can only be caused by the dipolar interactions, because the short-range fluid fluid and the fluid substrate potentials are purely repulsive.

As for "simple" fluids that have only translational degrees of freedom the structure of the confined DSS fluid is inhomogeneous on account of stratification (see Section 5.3.4). Because of the additional rotational degrees of freedom, however, the structure of the DSS fluid may be more complex as snapshots from the MC simulations in Fig. 6.7 illustrate. In the left part of that figure, a snapshot is presented for a globally isotropic system, whereas the right part shows a snapshot for an orientationally ordered phase. For the sake of clarity only molecules in one contact layer (i.e., the layers of molecules closest to one of the walls) are plotted.

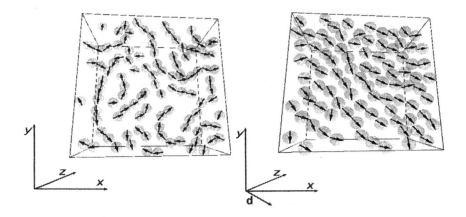

Figure 6.7: Left: Snapshot of the contact layer at $P_\parallel = 1.0$ (isotropic phase). Right: Snapshot of the contact layer at $P_\parallel = 5.0$ (spontaneously polarized phase). The thick arrow labeled by "d" denotes the direction of the global director in the x–y plane.

The globally isotropic state is characterized by the appearance of clusters and chains of particles with essentially random orientations, as expected for a dilute, strongly coupled dipolar fluid. In this regard, a main effect of spatial confinement is that the dipolar particles tend to form chains with in-plane rather than out-of-plane orientation. This anisotropy is even more pronounced in the orientationally ordered state, which is characterized by long, essentially straight chains that are aligned along a direction within the plane and parallel with the substrates. Closer inspection reveals that spheres forming neighboring chains tend to arrange themselves in an out of registry conformation. This way they avoid side side configurations where neighboring dipoles would prefer to be antiparallel rather than parallel as indicated by the snapshot plotted in the right part of Fig. 6.7. The structure perpendicular to the chains is generally more open, and an analysis of the corresponding in-plane correlation functions [257] shows that the ordered system still exhibits a liquid-like structure within the layers. This is similar to the confined fluid's bulk counterpart, which is also characterized by strongly anisotropic, yet short-range spatial correlations in the ferroelectric nematic state [262, 263, 265].

An important difference between the confined system and the bulk, however, concerns the thermodynamic conditions related to the *onset* of long-range parallel order. In fact, based on the data plotted in the two parts of Fig. 6.8, we conclude that in the confined system the onset of order occurs at somewhat lower pressures/densities, indicating that the walls promote rather than inhibit spontaneous orientational order. This result is, at least at first sight, rather surprising, because the substrates in the current system do not couple directly to the fluid particle dipole moments. A rationale for this shift of the onset of spontaneous order in the confined relative to the bulk fluid is offered in Ref. 257 where we basically employ entropic arguments.

As we argue in this latter work, the global director $\hat{d}$ can point in *any* direction on the unit sphere in the bulk. In confined systems, on the other hand, the presence of walls inhibits order in the normal direction, as can be understood by simple (macroscopic) energy considerations. Indeed, with insulating walls, order in the normal direction would lead to surface charges on the inner sides of the confining walls. These would in turn generate a demagnetizing/depolarizing field in a direction *opposite* to that of the total dipole moment of the fluid. Therefore, the director $\hat{d}$ is restricted to point in a direction *parallel* to the walls; within this $(x-y)$ plane, however, the orientation of $\hat{d}$ is then arbitrary. This effect is reflected by the snapshot in the bottom part of Fig. 6.7. Because of this restriction, which is absent in the bulk as we already mentioned, orientational fluctuations (and there-

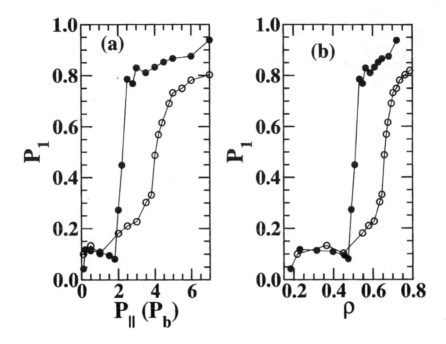

Figure 6.8: Order parameter P_1 as a function of the external parallel (bulk) pressure (a) and of the average density (b) for the confined system at $s_z = 7.0$ ($\bullet$) and the bulk system ($\circ$).

fore the orientational entropy) are supressed to a certain extent even in the isotropic phase and the ferroelectric transition occurs at significantly lower pressures/densities.

This argumentation is further supported by the predictions of a simple mean-field theory of the ferroelectric transition, which was originally presented in Ref. 257. Within this theory, we neglect any stratification (i.e., inhomogeneities of the local density) as well as any oscillations in the order parameter (which are indeed observed in the computer simulations). We also neglect nontrivial interparticle correlations. Our system can then be viewed as a system composed of N *uncorrelated* dipolar particles individually interacting with the mean field

$$\boldsymbol{E}^{\mathrm{MF}} = \rho P_1 \mu \left(\frac{4\pi}{3} - \frac{\pi}{2} \frac{\sigma}{s_z} \right) \widehat{\boldsymbol{e}}_x \qquad (6.53)$$

where the second term in parentheses is a correction of the bulk mean field ($\boldsymbol{E}^{\mathrm{MF}} = 4\pi\rho P_1 \mu \hat{\boldsymbol{e}}_x/3$) due to confinement [257]. Without loss of generality, we have set the director equal to the unit vector $\hat{\boldsymbol{e}}_x$ along the x-axis. Because particles do not interact, the Boltzmann factor weighting the individual dipole orientations (ω_i) is determined only by the field energy, $-\boldsymbol{\mu}(\omega_i) \cdot \boldsymbol{E}^{\mathrm{MF}}$. Consequently, P_1 can be obtained from the self-consistency relation

$$P_1 = \frac{\int d\omega \hat{\boldsymbol{\mu}}(\omega) \cdot \hat{\boldsymbol{e}}_x \exp\left[\beta\boldsymbol{\mu}(\omega) \cdot \boldsymbol{E}^{\mathrm{MF}}\right]}{\int d\omega \exp\left[\beta\boldsymbol{\mu}(\omega) \cdot \boldsymbol{E}^{\mathrm{MF}}\right]} \tag{6.54}$$

In Eq. (6.54), $\int d\omega$ represents an integration over *all* angular coordinates of the particles, i.e., $\omega = (\theta, \varphi)$. The computer simulation results (see Ref. 257) and the upper part of Fig. 6.7, however, suggest that dipole orientations along the z-axis are already highly disfavored in the isotropic phase. Within the mean-field theory, this phenomenon can only be formulated in a highly idealized manner, namely by assuming that the dipoles are *completely* restricted to point perpendicular to the walls (i.e., $\theta = \pi/2$). The self-consistency equation (6.54) then reduces to

$$P_1 = \frac{\int_0^{2\pi} d\varphi \cos\varphi \exp\left[\cos\varphi f(P_1)\right]}{\int_0^{2\pi} d\varphi \exp\left[\cos\varphi f(P_1)\right]} = \frac{I_1\left(f(P_1)\right)}{I_0\left(f(P_1)\right)} \tag{6.55}$$

where $f(P_1) = \beta\rho P_1 \mu^2 \left[4\pi/3 - \pi\sigma/(2s_z)\right]$ and $I_n(x)$ is the modified Bessel function of order n [141]. Expanding the right-hand side of Eq. (6.55) around $P_1 = 0$, we find that the system is polarized at densities above a critical density

$$\rho_{\mathrm{c,p}} = \frac{2k_B T}{\mu^2\left(4\pi/3 - \pi\sigma/2s_z\right)} \tag{6.56}$$

whereas for the corresponding bulk this density is [270, 271]

$$\rho_{\mathrm{c,b}} = \frac{3k_B T}{\mu^2 4\pi/3} \tag{6.57}$$

Comparing these two expressions, one sees that $\rho_{\mathrm{c,p}} < \rho_{\mathrm{c,b}}$ at all finite pore widths, in qualitative agreement to our MC results for $s_z = 7$. Quantitatively speaking, however, the mean-field theory turns out to be poor, as expected from corresponding bulk results for dipolar fluids [225]. This observation is not surprising in view of the strong orientational correlations between dipoles in the ferroelectric phase.

6.5   Conducting solid substrates

6.5.1   The boundary-value problem in electrostatics

So far our discussion of confined fluids with long-range electrostatic interactions has been restricted to the case of insulating walls characterized by
$\epsilon'' = 1$. However, in many cases, one might also want to consider dielectric interfaces ($\epsilon'' > 1$), which are important in biological systems such as
ion channels in proteins but also for self-assembled monolayers composed of
organic materials [272]. A special case in this context are conducting (i.e.,
metallic) interfaces for which $\epsilon'' = \infty$. They play a central role in electrochemical problems dealing with the structure and transport properties of
electrolytes close to metallic electrodes [273]. Another type of system where
conducting walls are essential are electrorheological fluids [274]. These are
colloidal dispersions of dielectric particles where structural and rheological
properties can be controlled by electric fields resulting from two metallic
electrodes [229].

From a theoretical point of view, the crucial difference between a system
confined by conducting substrates rather than insulating ones is that the
charges and dipoles in the original slablike system create "image charges"
and "image dipoles," respectively, within the confining metal [242]. These
images are, of course, merely theoretical constructs (rather than real charges
or dipoles) that allow one to solve the electrostatic boundary value problem.
To obtain this solution, one realizes that the Maxwell equations require the
electrostatic potential to be constant on the surface of the substrates; that
is, the tangential part of the field must vanish.

In practice, the existence of images implies that there are additional interactions to be taken into account such that the treatment of liquids in the
vicinity of conducting substrates seems even more complicated than the situations considered before (see Section 6.4). Surprisingly, however, the energy of
a system between conducting substrates can be mapped onto a problem with
three-dimensional periodicity [275, 276], which can subsequently be treated
by conventional Ewald summation methods such as the ones presented in
Section 6.2. In Section 6.5.2 (and the corresponding Appendix F.3.3), we
describe this mapping explicitly for systems of point charges. We then generalize our treatment to dipolar systems. This section closes with a discussion
of representative numerical results.

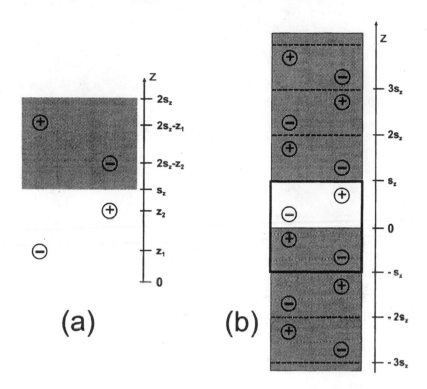

Figure 6.9: Sketch of the effect of conducting walls on two charged particles. (a) Presence of one conducting wall implies creation of one image charge per particle. (b) Two conducting walls yield an infinite number of images per particle, where one group of images has charges of the same sign as the original charge, whereas the other group is characterized by opposite charges. The structure in the z-direction can then be considered as an infinite periodic replication of the extended cell (original charges plus one set of images) marked by the thick frame.

6.5.2 Image charges in metals

We start by considering a single particle with charge q_i at some position r_i between two conducting (i.e., metallic) solid substrates of infinite thickness. To keep the notational burden to a minimum, we deviate from the remainder of this text in that we are assuming the substrate surfaces to be located at planes $z = 0$ and $z = s_z$ (instead of $z = \pm s_z/2$). According to the rules of electrostatics [242], which state that the tangential part of the electric field must vanish on conducting surfaces, the effect on *one* conducting solid surface (e.g., the upper one at $z = s_z$) consists of creating an image charge within the metal. As illustrated in Fig. 6.9(a), the position of the image is

$\boldsymbol{r}_i' = \boldsymbol{r}_i + 2\left(s_z - z_i\right)\widehat{\boldsymbol{e}}_z$, and its charge is

$$q_i' = -q_i \tag{6.58}$$

The total electrostatic potential in front of the substrate is given by the *sum* of the potentials caused by the original charge (q_i) plus that due to its image (q_i').

The presence of a *second* conducting substrate at $z = 0$ changes the situation drastically because not only the original particle but also its upper image are mirrored at the lower substrate such that they create image charges. These images located at positions $z < 0$ in turn induce new images in the upper substrate, and so forth ad infinitum [see Fig. 6.9 (b)]. Therefore, each charge generates an infinite number of images. The first group of images (which includes the one mentioned in the beginning) is characterized by charges $q_i' = -q_i$ located at

$$\boldsymbol{r}_i' = \boldsymbol{r}_i + 2\left(n_z s_z - z_i\right)\widetilde{\boldsymbol{e}}_z, \qquad n_z = 0, \pm 1, \pm 2, \ldots \tag{6.59}$$

The second group of images has the same charge as the original one; that is, $q_i'' = q_i$. These images are located at

$$\boldsymbol{r}_i'' = \boldsymbol{r}_i + 2n_z s_z \widehat{\boldsymbol{e}}_z, \qquad n_z = \pm 1, \pm 2, \ldots \tag{6.60}$$

Consider now N particles confined to a slit-pore with metallic substrate surfaces. The total configurational potential energy of this system is then obtained from

$$\widetilde{U}_{\mathrm{C}} = \frac{1}{2}\sum_{i=1}^{N} q_i \left[\Phi_{\mathrm{s}}\left(\boldsymbol{r}_i\right) + \Phi_{\mathrm{d}}\left(\boldsymbol{r}_i\right)\right] \tag{6.61}$$

where $\Phi_{\mathrm{s}}(\boldsymbol{r}_i)$ is the electrostatic potential arising from the images of particle i, whereas $\Phi_{\mathrm{d}}(\boldsymbol{r}_i)$ represents the contributions from particles $j \neq i$ and from their images. Using Eqs. (6.58)–(6.60), these potentials are given by

$$\Phi_{\mathrm{s}}\left(\boldsymbol{r}_i\right) = \sum_{n_z=-\infty}^{\infty}{}^{*} \frac{q_i}{|2n_z s_z \widehat{\boldsymbol{e}}_z|} - \sum_{n_z=-\infty}^{\infty} \frac{q_i}{|2\left(n_z s_z + z_i\right)\widehat{\boldsymbol{e}}_z|} \tag{6.62a}$$

$$\Phi_{\mathrm{d}}\left(\boldsymbol{r}_i\right) = \sum_{j\neq i}^{N}\left[\frac{q_j}{r_{ij}} + \sum_{n_z=-\infty}^{\infty}{}^{*} \frac{q_j}{|\boldsymbol{r}_{ij} + 2n_z s_z \widehat{\boldsymbol{e}}_z|}\right.$$

$$\left. - \sum_{n_z=-\infty}^{\infty} \frac{q_j}{|\boldsymbol{r}_{ij} + 2\left(n_z s_z + z_j\right)\widehat{\boldsymbol{e}}_z|}\right] \tag{6.62b}$$

where the asterisk is attached to the first summation sign in Eq. (6.62a) to indicate that the term with $n_z = 0$ is omitted. Moreover, in writing the summands, we used the fact that summation over n_z is equivalent to a summation over $-n_z$. Inserting Eq. (6.62a) into Eq. (6.61), we obtain after a straightforward rearrangement of terms

$$\widetilde{U}_C = \frac{1}{2} \sum_{i=1}^{N} \sum_{j=1}^{N} \left[\sum_{n_z=-\infty}^{\infty}{}' \frac{q_i q_j}{|\boldsymbol{r}_{ij} + 2n_z s_z \widehat{\boldsymbol{e}}_z|} - \sum_{n_z=-\infty}^{\infty} \frac{q_i q_j}{|\boldsymbol{r}_{ij} + 2\left(n_z s_z + z_j\right) \widehat{\boldsymbol{e}}_z|} \right]$$

(6.63)

where the prime attached to the first sum over n_z signifies that the term $n_z = 0$ is omitted only for $i = j$.

Finally, if the central cell comprising N particles is replicated along the x- and y- directions, we obtain a slab-like system confined between conducting walls. Introducing now three-dimensional lattice vectors

$$\overline{\boldsymbol{n}} = \begin{pmatrix} n_x s_x \\ n_y s_y \\ 2n_z s_z \end{pmatrix}$$

(6.64)

and replacing in Eq. (6.63) the sums over n_z by three-dimensional sums over the set of lattice vectors $\{\overline{\boldsymbol{n}}\}$, the total configurational potential energy of the system may be cast as

$$U_C = \frac{1}{2} \sum_{i=1}^{N} \sum_{j=1}^{N} \left[\sum_{\{\overline{\boldsymbol{n}}\}}{}' \frac{q_i q_j}{|\boldsymbol{r}_{ij} + \overline{\boldsymbol{n}}|} - \sum_{\{\overline{\boldsymbol{n}}\}} \frac{q_i q_j}{|\boldsymbol{r}_{ij} + \overline{\boldsymbol{n}} + 2z_j \widehat{\boldsymbol{e}}_z|} \right]$$

(6.65)

The lattice sums in Eq. (6.65) reflect the fact that the Coulombic system between conducting walls has, in a way, three-dimensional periodicity. The basic cell of this three-dimensional array contains the original cell with the N particles plus the first set of images, that is, the N images resulting from the presence of just the lower wall [see Fig. 6.9(b)]. In fact, as we show explicitly in Appendix F.3.3, the energy of the extended system with a total of $2N$ charges, $U_C^{\text{3d,ex}}$, is directly linked to U_C by the relation

$$U_C = \frac{1}{2} U_C^{\text{3d,ex}}$$

(6.66)

Thus, in a computer simulation with conducting interfaces, one only needs to calculate the energy (or forces) in the extended system, which turns out to be twice the original one. However, the current approach has the great advantage that it can take into account the three-dimensional periodicity of the extended system. Therefore, the conventional three-dimensional Ewald summation technique [see Eq. (6.15)] can be employed. As a consequence, simulations of systems between conducting interfaces are typically much faster than corresponding simulations between insulating substrates on account of the simplifying three-dimensional as opposed to the slab geometry.

6.5.3   Dipolar fluids

The above considerations for point charges can readily be generalized to dipolar systems between two conducting surfaces [277]. This follows again from the principles of elementary electrostatics, which tell us that each dipole $\boldsymbol{\mu}_i$ within the original basis cell creates two (infinitely large) groups of image dipoles. The first group is located at positions given in Eq. (6.59), where[10]

$$
\boldsymbol{\mu}_i' = \begin{pmatrix} -\mu_{i,\mathrm{x}} \\ -\mu_{i,\mathrm{y}} \\ \mu_{i,\mathrm{z}} \end{pmatrix}
\tag{6.67}
$$

The second group of images is located at positions defined by Eq. (6.60) and dipole moments $\boldsymbol{\mu}_i'' = \boldsymbol{\mu}_i$. Consider now N dipoles in the basic cell and replicate this cell in directions parallel to the walls. Then the total configurational potential energy can be written as (see Appendix F.3.3.2)

$$
\begin{aligned}
U_{\mathrm{D}} &= \frac{1}{2} \sum_{i=1}^{N} \sum_{j=1}^{N} \sum_{\{\overline{\boldsymbol{n}}\}}{}' \left\{ \frac{\boldsymbol{\mu}_i \cdot \boldsymbol{\mu}_j}{|\boldsymbol{r}_{ij} + \overline{\boldsymbol{n}}|^3} - \frac{3\left[\boldsymbol{\mu}_i \cdot (\boldsymbol{r}_{ij} + \overline{\boldsymbol{n}})\right]\left[\boldsymbol{\mu}_j \cdot (\boldsymbol{r}_{ij} + \bar{\boldsymbol{n}})\right]}{|\boldsymbol{r}_{ij} + \overline{\boldsymbol{n}}|^5} \right\} \\
&+ \frac{1}{2} \sum_{i=1}^{N} \sum_{j=1}^{N} \sum_{\{\overline{\boldsymbol{n}}\}} \left\{ \frac{\boldsymbol{\mu}_i \cdot \boldsymbol{\mu}_j'}{|\boldsymbol{r}_{ij} + \overline{\boldsymbol{n}} + 2z_j s_z \widehat{\boldsymbol{e}}_z|^3} \right. \\
&\left. - \frac{3\left[\boldsymbol{\mu}_i \cdot (\boldsymbol{r}_{ij} + \overline{\boldsymbol{n}} + 2z_j s_z \widehat{\boldsymbol{e}}_z)\right]\left[\boldsymbol{\mu}_j' \cdot (\boldsymbol{r}_{ij} + \overline{\boldsymbol{n}} + 2z_j s_z \widehat{\boldsymbol{e}}_z)\right]}{|\boldsymbol{r}_{ij} + \overline{\boldsymbol{n}} + 2z_j s_z \widehat{\boldsymbol{e}}_z|^5} \right\}
\end{aligned}
\tag{6.68}
$$

Finally, using essentially the same arguments as for the charged system, we can show that (see Appendix F.3.3.2)

$$
U_{\mathrm{D}} = \frac{1}{2} U_{\mathrm{D}}^{\mathrm{3d,ex}}
\tag{6.69}
$$

where $U_{\mathrm{D}}^{\mathrm{3d,ex}}$ is the total configurational potential energy of an extended system with a basis cell comprising the set of the N original dipoles plus the first set of image dipoles. Based on relation (6.69), we can again employ the conventional three-dimensional Ewald sum [see Eq. (6.26)] to calculate the energy of the slab system between conducting walls.

---

[10]To see this, consider the dipole as an arrangement of two charges of opposite sign and separated by some small distance and realize that each charge creates an image charge of opposite sign.

6.5.4   Metallic substrates and ferroelectricity

Given the appearance of spontaneous ferroelectric order in confined DSS fluids between insulating walls (see Section 6.4.2), it is interesting to consider the dependence of this phase transitions on the wall boundary conditions. To this end we have repeated the calculations described in Section 6.4.2 for a strongly coupled DSS fluid confined between two *conducting* walls, using Monte Carlo simulations in the $(N, s_z, P_{\parallel}, T)$ ensemble with $N = 500$ particles. The dipolar interactions were treated on the basis of Eq. (6.69). To compare with our previous results corresponding to the case of insulating walls (see Section 6.4.2), the reduced temperature, dipole moment, and wall separation have been set to the same values as before where $T = 1.35$, $m = 3.0$, and $s_z = 7$.

Numerical results for the order parameter P_1 as function of the applied transverse (i.e., parallel) pressure $P_{\parallel}$ are plotted in Fig. 6.10, where we have included corresponding data obtained with insulating walls (see Fig. 6.6) as a reference.

Clearly, the confined fluid between metallic walls does exhibit spontaneous ferroelectric ordering at sufficiently high pressures, as does its counterpart between insulating walls. Moreover, the pressure range in which the ferroelectric order develops upon compressing the fluid from the dilute limit is essentially the same for the two wall boundary conditions considered. Finally, inspection of the global director $\hat{\boldsymbol{d}}$ (which is a result of the simulations) indicates that metallic substrates (as do insulating ones) support ordering *parallel* to these substrates; that is, $\hat{\boldsymbol{d}}$ has directions within the x–y plane at all pressures considered. Thus, one would conclude that the dielectric contant characterizing the confining walls has only marginal influence on the phase transition at least at the wall separation considered here.

The same conclusion may be drawn from data for the total configurational potential energy plotted in Fig. 6.11. Upon increasing $P_{\parallel}$ from zero, the energy first rises in both systems as a consequence of the increasing repulsion between the particles. In this pressure range, the numerical values of U for metallic walls on the one hand, and insulating walls on the other hand, are essentially indistinguishable. Further compression then yields a sharp decrease of U that can be attributed to the decrease of dipolar energy due to orientational ordering (see Fig. 6.10). Within the ferroelectric phase the energies then increase again (upon increasing the pressure), with the values of U at high pressures being somewhat larger in the metallic case. On the other hand, local properties such as the density profiles and local order parameters turn out to be nearly identical [277].

Given the strong similarity of both the qualitative behavior and the ac-

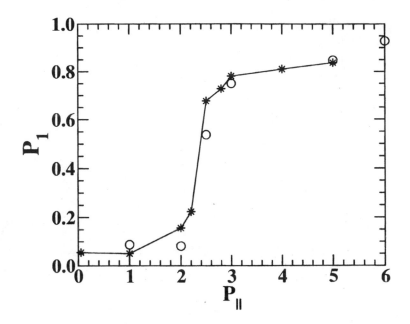

Figure 6.10: Order parameter P_1 as a function of the applied parallel pressure for a DSS fluid ($T = 1.35$, $m = 3.0$) confined between metallic walls with separation $s_z = 7$ (stars). Also shown are corresponding results for insulating walls (open circles) from Fig. 6.6 ($N = 500$).

tual thermophysical properties of the confined DSS fluid between metallic and insulating walls, one might wonder whether correct treatment of wall boundary conditions is important at all. One question appearing in this context concerns the influence of the substrate separation s_z. In particular, would we expect the same similarities (observed at $s_z = 7$) to also occur in more confined systems characterized by smaller values of s_z?

To get some insight into these questions we have performed various lattice calculations similar in spirit to those described in Section 6.3.3. Specifically, we have considered (infinitely extended) slabs composed of dipolar particles located at the sites of a face-centered cubic (fcc) lattice with (reduced) density $\rho_{\text{fcc}} = 1.0$. We have then employed the Ewald sum for dipolar systems between metallic walls [see Eq. (6.69)] to calculate the total dipolar energy U_{D} for various configurations characterized by perfect orientational order. Numerical results for U_{D} as a function of the number of lattice layers n_z are

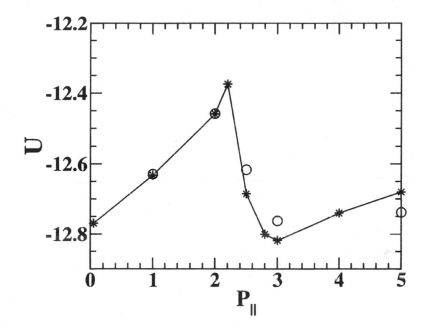

Figure 6.11: Total configurational energy U for DSS fluids between metallic (stars) and insulating (open circles) walls, respectively. Parameters are the same as in Fig. 6.10.

given in Fig. 6.12, where Part (a) compares the energies of a system oriented along the x-axis (i.e., $\hat{\boldsymbol{d}} \parallel \hat{\boldsymbol{x}}$) with those of a system oriented along the z-axis (i.e., $\hat{\boldsymbol{d}} \parallel \hat{\boldsymbol{z}}$).

It is observed that, regardless of the actual value of n_z, the energy U_D related to an ordering parallel to the walls is smaller than that related to perpendicular ordering. This explains why the MC simulations at $s_z = 7$ described above predict spontaneous ordering parallel to the walls. However, we also observe from Fig. 6.12(a) that the actual differences between the two ordering directions are large only for very thin films and decrease with increasing film thickness. This is in marked contrast to the corresponding energies for systems with insulating walls plotted in Fig. 6.2 revealing that *perpendicular* ordering (between insulating walls) is energetically unfavorable even for macroscopically thick slabs. We can understand these differences as a consequence of depolarizing fields that arise for perpendicular ordering

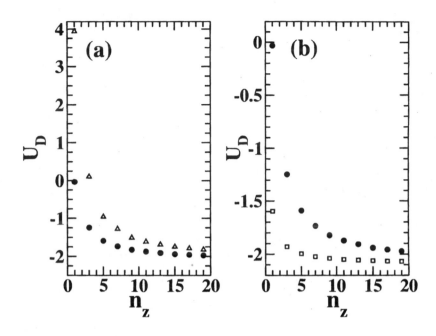

Figure 6.12: Dimensionless energy per particle for dipolar crystalline (fcc) slabs with perfect orientational order. Part (a) contains data for systems between metallic walls with order in the x–direction (solid circles) and the z–direction (open triangles), respectively. Part (b) compares data corresponding to metallic (solid circles) and insulating (open squares) walls for systems ordered in the x–direction.

between insulating walls but not for metallic walls.

Finally, we compare in Fig. 6.12 (b) the energies related to parallel ordering ($\hat{d} \parallel \hat{x}$) for the two wall boundary conditions considered. It is observed that the boundary conditions have a large effect only at very small values of n_z and become increasingly unimportant upon increasing n_z toward the bulk limit $n_z \to \infty$. This may explain why the energy values obtained in our MC simulations at s_z (which roughly corresponds to $n_z \approx 6$) within the ferroelectric phase are quite similar. Finally, the lattice energies plotted in Fig. 6.12(b) also reflect that the perfectly ordered system ($\hat{d} \parallel \hat{x}$) between insulating walls is generally characterized by smaller energy values (as compared with the metallic case), which is again consistent with our computer simulation results obtained at the largest pressure considered (see Fig. 6.11).

Chapter 7

Statistical mechanics of disordered confined fluids

7.1 Introductory remarks

So far we have considered only situations where the fluid is confined to a *single* pore. However, real porous solids often consist of an interconnected network of pores of various sizes and shapes [4]. Prominent examples of such *disordered* porous materials are mesoporous glasses such as Vycor and CPG (controlled pore glass), which are formed by spinodal decomposition of a mixture and subsequent removal of one component. Contrary to mesoporous glasses, which are characterized by relatively low porosity[1] of 30 to 60 percent, aerogels are formed by extremely dilute disordered networks of microscopic particles that occupy only a very small portion of the total volume (porosity 95 to 98 percent). An example is presented in Fig. 7.1. Mesoporous materials are of importance for a wide range of technical applications such as gas storage, separation processes, and heterogeneous catalysis and much progress has been made in the design, synthesis, and characterization of materials with novel properties [278, 279]. In this chapter, we are interested in the influence of such a *disordered* material on the structure and phase behavior of an adsorbed fluid.

Interest in this topic was stimulated by the intense experimental research on phase transitions in disordered media in the 1990s. The experimental studies have involved "simple" fluids such as ^{4}He or N_2 [131, 132, 280], fluid mixtures (e.g., isobutyric acid and water [115]), and anisotropic fluids such as nematic liquid crystals [281–284]. One main conclusion from these studies was that fluids in dilute aerogels can indeed undergo true phase transitions,

---

[1]The porosity is the volume fraction of the space available for the adsorbed fluid.

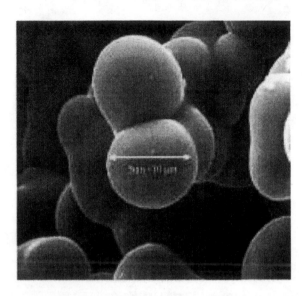

Figure 7.1: Inner structure of a carbon aerogel. From R. Emmerich, http://idw-online.de/public/pmid-42335/zeige-pm.html

whereas the existence of phase transitions in low-porosity mesoporous glasses still seems quite controversial [285, 286]. Therefore, most theoretical studies focus on the highly dilute case. Indeed, corresponding experiments have indicated that even extremely dilute media with porosities as high as 99.9 percent can alter the phase behavior of the adsorbed fluid drastically compared with the bulk.

Typical effects observed in systems with condensation and/or demixing phase transitions [115, 131, 132, 280] are shifts of the critical temperature toward significantly smaller values, an accompanying shift of the critical density (or composition), and a substantial narrowing of the coexistence curves. For nematic liquid crystals in silica aerogels, experiments have indicated that the isotropic nematic transition survives, but the long-range orientational (nematic) order occurring in the bulk is replaced by short-range or "quasi"-long range order [281–284]. One may therefore expect similar effects in other fluids displaying orientational order.

Compared with the large amount of experimental information, the theoretical understanding of fluids in highly dilute porous media like aerogels is far less developed. The challenge in this context is to understand the influence of the *quenched* (frozen) disorder realized by the nearly random aerogel network on the fluids properties. One of the earliest attempts to model this situation was a study by Brochard and de Gennes [287, 288], who suggested considering the adsorbed fluid as an experimental realization of the so-called random field Ising model (RFIM) [289]. The main idea here is that the local preferential attraction of the fluid by the solid surface within a pore, combined with the disordered character of the pore structure, induces a spatially random perturbation of the chemical potential. The latter can be represented as a local random magnetic field in the Ising picture.

A major drawback of the RFIM, however, is that it focuses entirely on the aspect of disorder, whereas confinement plays no role. To account for this problem, more recent theoretical studies, and computer simulations, of fluids in disordered media employ the concept of a quenched-annealed (QA) mixture [290, 291]. Here, the fluid molecules (the annealed species) equilibrate in a "matrix" consisting of particles quenched in a disordered configuration. Thus, QA models combine both disorder and confinement, the latter being guaranteed by the finite size of the matrix particles. In addition, preferential adsorption can be realized by assuming attractive (or other, more complex) interactions between fluid and matrix particles.

In what follows we first introduce in some more detail the concept of QA models and the resulting appearance of double averages. We then present the foundations of the so-called replica integral equation theory, a theoretical formalism appropriate for calculating two-particle correlation functions and thermodynamics quantities of QA systems, which are *homogeneous* on average. The last part of the chapter is devoted to applications of the replica integral equations, with an emphasis on fluids with long-range dipolar interactions.

7.2   Quenched-annealed models

In the framework of QA models, the disordered medium, such as the one depicted in Fig. 7.1, is modeled as a matrix consisting of N_m particles. The latter are frozen in place (*quenched*) according to a distribution $P\left(Q^{N_m}\right)$, where $Q^{N_m} = \{Q_1, \ldots, Q_{N_m}\}$ denotes the set of matrix particle coordinates. In the simplest case (e.g., hard-sphere matrices), these quenched variables are just the particle positions R_i. However, one may also consider the case of matrix particles with internal degrees of freedom, such as a charge or an orientation. In the latter case, the coordinates are $Q_i = (R_i, \Omega_i)$, with Ω_i being the set of Euler angles defining the particle orientation.

For the theoretical formalism to be described it is convenient to choose $P(Q^{N_m})$ as an equilibrium canonical distribution established at some temperature T_0,

$$P\left(\boldsymbol{Q}^{N_m}\right) = \frac{1}{Z_m} \exp\left[-\frac{U_{mm}\left(\boldsymbol{Q}^{N_m}\right)}{k_B T_0}\right] \tag{7.1}$$

where

$$U_{mm}\left(\boldsymbol{Q}^{N_m}\right) = \sum_{i=1}^{N_m-1} \sum_{j=i+1}^{N_m} u_{mm}\left(\boldsymbol{Q}_i, \boldsymbol{Q}_j\right) \tag{7.2}$$

is the configurational potential energy governing the distribution of the matrix particles (assuming pairwise additive interactions), and

$$Z_m = \int d\boldsymbol{Q}^{N_m} \exp\left[-\frac{U_{mm}\left(\boldsymbol{Q}^{N_m}\right)}{k_B T_0}\right] \tag{7.3}$$

is the corresponding configuration integral. The physical significance of T_0 is that of a quenching temperature, which is a temperature at which matrix particles in a given equilibrium configuration are suddenly "frozen" into their actual positions in that configuration. The notation $\int d\boldsymbol{Q}^{N_m} \ldots$ indicates an integration over the set of matrix particle coordinates. In writing Eq. (7.3) we have neglected combinatorial prefactors because matrix particles are not permitted to move, thus making them distinguishable through their spatial arrangement.

We now imagine that the free space left by the quenched matrix particles is occupied by a fluid of N_f *mobile* particles. The fluid particle coordinates, $\boldsymbol{q}_i$, are thus annealed variables that can equilibrate for a given configuration ("realization") of the matrix. Again, one may consider simple fluids with only translational degrees of freedom, where $\boldsymbol{q}_i = \boldsymbol{r}_i$. However, one may also consider anisotropic fluids for which $\boldsymbol{q}_i = (\boldsymbol{r}_i, \omega_i)$, where the set $\{\omega_i\}$ are Euler angles specifying the orientation. To complete the description of the QA model, one needs to specify the interactions between fluid particles and those between the fluid and the matrix. We assume again pair-wise additive interactions

$$U_{ff}\left(\boldsymbol{q}^{N_f}\right) = \sum_{i=1}^{N_f-1} \sum_{j=i+1}^{N_f} u_{ff}\left(\boldsymbol{q}_i, \boldsymbol{q}_j\right) \tag{7.4}$$

for the fluid fluid interactions where $\boldsymbol{q}^{N_f}$ is the set of fluid variables and

$$U_{fm}\left(\boldsymbol{q}^{N_f}, \boldsymbol{Q}^{N_m}\right) = \sum_{i=1}^{N_f} \sum_{j=1}^{N_m} u_{fm}\left(\boldsymbol{q}_i, \boldsymbol{Q}_j\right) \tag{7.5}$$

accounts for the fluid matrix interactions.

Because of the quenched nature of the matrix, the evaluation of an equilibrium property A of the adsorbed fluid representing, for example, its internal energy or pair correlation function is not at all straightforward. To see this, consider first the thermal average involving the fluid microscopic variables, q_i. For a given realization (i.e., configuration) Q^{N_m} of the matrix, and a given temperature T, this thermal average is defined as

$$\langle\ldots\rangle_{\boldsymbol{Q}} = \frac{1}{Z_{\boldsymbol{Q}}} \int d\boldsymbol{q}^{N_f} \exp\left\{-\frac{1}{k_B T}\left[U_{\mathrm{ff}}\left(\boldsymbol{q}^{N_f}\right) + U_{\mathrm{fm}}\left(\boldsymbol{q}^{N_f}, \boldsymbol{Q}^{N_m}\right)\right]\right\} \quad (7.6)$$

where

$$Z_{\boldsymbol{Q}} = \int d\boldsymbol{q}^{N_f} \exp\left\{-\frac{1}{k_B T}\left[U_{\mathrm{ff}}\left(\boldsymbol{q}^{N_f}\right) + U_{\mathrm{fm}}\left(\boldsymbol{q}^{N_f}, \boldsymbol{Q}^{N_m}\right)\right]\right\} \quad (7.7)$$

is the configurational integral of the fluid. Equations (7.6) and (7.7) have been formulated for an adsorbed fluid with a *fixed* number of particles (canonical ensemble), but they can be easily generalized to a grand canonical treatment where the fluid is coupled to a reservoir such that the particle number fluctuates around some average value (see Chapter 2). The latter situation is, in fact, relevant in many experiments of fluids in disordered aerogels [290]. In this chapter we concentrate on the canonical ensemble to keep the notation as compact as possible. Formulas relevant to a grand canonical description are given in Appendix G.1.

From a practical point of view, the thermal averages defined by Eq. (7.6) are not very meaningful as they depend on the specific realization of the matrix. Therefore one needs to supplement the thermal average by a "disorder average" over matrix configurations, yielding the double average

$$\left[\langle\ldots\rangle_{\boldsymbol{Q}}\right] = \int d\boldsymbol{Q}^{N_m} \langle\ldots\rangle_{\boldsymbol{Q}} P\left(\boldsymbol{Q}^{N_m}\right) \quad (7.8)$$

The problem in evaluating this double average is that the two sets of variables involved, $\boldsymbol{q}^{N_f}$ and $\boldsymbol{Q}^{N_m}$, are not treated on equal footing as in conventional statistical physics. Instead, as indicated by Eq. (7.8), the thermal average has to be performed *before* the disorder average is carried out. One way is to employ computer simulations where both types of averages can be directly evaluated. However, for complex interactions this procedure will be extremely time-consuming, especially at low temperatures where the number of matrix realizations required for the disorder average increases strongly (see, for example, Ref. 292).

7.3   The introduction of replicas

The appearance of double averages of the type just discussed is characteristic not only for the QA mixtures considered in this chapter, but also it is a generic feature of systems with quenched disorder. Prominent examples, which were extensively studied in the 1970s and 1980s, are spin glasses [293, 294] and random-field systems [289]. Work on these systems has established the so-called replica method, which allows one to circumvent the double averages by relating the original disordered system to an artificial, yet fully annealed "replicated" system [294]. Essentially the same methods can also be applied to QA mixtures, as first realized by Given and Stell [295–297].

To introduce the replica concept we consider an arbitrary physical quantity expressible in the form of Eq. (7.8), such as the internal energy of the adsorbed fluid

$$
\mathcal{U}_{\text{ff}} = \left[\left\langle \sum_{i=1}^{N_{\text{f}}-1} \sum_{j=i+1}^{N_{\text{f}}} u_{\text{ff}}\left(\boldsymbol{q}_i, \boldsymbol{q}_j\right) \right\rangle_{\boldsymbol{Q}} \right] \tag{7.9}
$$

Using Eqs. (7.6) and (7.8), the double average over one pair term appearing on the right side of Eq. (7.9), that is, for example, the term corresponding to $i = 1$, $j = 2$, can be written as

$$
\left[\left\langle u_{\text{ff}}\left(\boldsymbol{q}_1, \boldsymbol{q}_2\right) \right\rangle_{\boldsymbol{Q}} \right] = \int d\boldsymbol{Q}^{N_{\text{m}}} P\left(\boldsymbol{Q}^{N_{\text{m}}}\right) \frac{1}{Z_{\boldsymbol{Q}}}
$$
$$
\times \int d\boldsymbol{q}^{N_{\text{f}}} u_{\text{ff}}\left(\boldsymbol{q}_1, \boldsymbol{q}_2\right) \exp\left(-\frac{U_{\text{ff}} + U_{\text{fm}}}{k_{\text{B}}T}\right) \tag{7.10}
$$

We now multiply both the numerator and the denominator of the last term in the integrand by $Z_{\boldsymbol{Q}}^{n-1}$, where n is an arbitrary integer. Because $Z_{\boldsymbol{Q}}$ involves an integral over the N_{f} fluid particles coordinates [see Eq. (7.7)], this multiplication implies that we are introducing $n - 1$ copies (i.e., *replicas*) of the fluid particles. Assigning an arbitrary index α' to the variables $(\boldsymbol{q}_1, \boldsymbol{q}_2)$ appearing in the numerator of Eq. (7.10), and inserting Eq. (7.1) for $P\left(\boldsymbol{Q}^{N_{\text{m}}}\right)$, Eq. (7.10) may be recast as

$$
\left[\left\langle u_{\text{ff}}\left(\boldsymbol{q}_1, \boldsymbol{q}_2\right) \right\rangle_{\boldsymbol{Q}} \right] = \int d\boldsymbol{Q}^{N_{\text{m}}} \frac{1}{Z_{\text{m}} Z_{\boldsymbol{Q}}^{n}} \prod_{\alpha=1}^{n}
$$
$$
\times \int d\boldsymbol{q}_{\alpha}^{N_{\text{f}}} u_{\text{ff}}\left(\boldsymbol{q}_{1\alpha'}, \boldsymbol{q}_{2\alpha'}\right) \exp\left(-\frac{U_{\text{rep}}}{k_{\text{B}}T}\right) \tag{7.11}
$$

where we have used the notation $\int d\boldsymbol{q}_{\alpha}^{N_{\text{f}}} = \int d\boldsymbol{q}_{1\alpha} \int d\boldsymbol{q}_{2\alpha} \ldots \int d\boldsymbol{q}_{N_{\text{f}}\alpha}$ for the integrations related to the copy of the fluid with index α ($\alpha = 1, \ldots, n$).

Furthermore, U_{rep} appearing in Eq. (7.11) is given as

$$U_{\text{rep}} = \frac{T}{T_0} U_{\text{mm}}\left(\mathbf{Q}^{N_{\text{m}}}\right) + \sum_{\alpha=1}^{n} \left[U_{\text{ff}}\left(\mathbf{q}_{\alpha}^{N_{\text{f}}}\right) + U_{\text{fm}}\left(\mathbf{q}_{\alpha}^{N_{\text{f}}}, \mathbf{Q}^{N_{\text{m}}}\right) \right] \tag{7.12}$$

Inspection of Eq. (7.12) indicates that the replicated system introduced by the multiplication trick is a $(n + 1)$-component mixture composed of the matrix particles and altogether n identical copies (i.e., replicas) of the fluid particles. Each of the copies interacts with the matrix particles, but there are no interactions among different copies.

Returning to Eq. (7.11), one now uses the fact that the denominator appearing on the right-hand side, $Z_{\text{m}} Z_{\mathbf{Q}}^{n}$, which still depends on the realization $\mathbf{Q}^{N_{\text{m}}}$, becomes independent of $\mathbf{Q}^{N_{\text{m}}}$ in the limit $n \to 0$. More specifically, one has

$$\lim_{n \to 0} Z_{\text{m}} Z_{\mathbf{Q}}^{n} = Z_{\text{m}} \lim_{n \to 0} Z_{\mathbf{Q}}^{n} = Z_{\text{m}} \tag{7.13}$$

The same limiting value is obtained when taking the limit $n \to 0$ of the configuration integral of the replicated system;[2] that is if we consider

$$Z_{\text{rep}} = \int d\mathbf{Q}^{N_{\text{m}}} \prod_{\alpha=1}^{n} \int d\mathbf{q}_{\alpha}^{N_{\text{f}}} \exp\left(-\frac{U_{\text{rep}}}{k_B T}\right) \tag{7.14}$$

because

$$\lim_{n \to 0} Z_{\text{rep}} = \int d\mathbf{Q}^{N_{\text{m}}} \exp\left(-\frac{U_{\text{mm}}\left(\mathbf{Q}^{N_{\text{m}}}\right)}{k_B T_0}\right) = Z_{\text{m}} \tag{7.15}$$

where we have used the definition of U_{rep} given in Eq. (7.12)]. The equivalence of the two limits allows one to rewrite Eq. (7.11) as

$$\left[\langle u_{\text{ff}}\left(\mathbf{q}_1, \mathbf{q}_2\right)\rangle_{\mathbf{Q}}\right] = \lim_{n \to 0} \frac{1}{Z_{\text{rep}}} \int d\mathbf{Q}^{N_{\text{m}}} \prod_{\alpha=1}^{n} \int d\mathbf{q}_{\alpha}^{N_{\text{f}}} u_{\text{ff}}\left(\mathbf{q}_{1\alpha'}, \mathbf{q}_{2\alpha'}\right)$$
$$\times \exp\left(-\frac{U_{\text{rep}}}{k_B T}\right)$$
$$= \lim_{n \to 0} \langle u_{\text{ff}}\left(\mathbf{q}_{1\alpha'}, \mathbf{q}_{2\alpha'}\right)\rangle_{\text{rep}} \tag{7.16}$$

where $\langle \ldots \rangle_{\text{rep}}$ is a conventional canonical ensemble average in the replicated system. Combining Eqs. (7.16) and (7.9) one immediately obtains

$$\mathcal{U}_{\text{ff}} = \lim_{n \to 0} \left\langle \sum_{i=1}^{N_{\text{f}}-1} \sum_{j=i+1}^{N_{\text{f}}} u_{\text{ff}}\left(\mathbf{q}_{i\alpha'}, \mathbf{q}_{j\alpha'}\right) \right\rangle_{\text{rep}} = \lim_{n \to 0} \mathcal{U}_{\text{ff},\alpha'}^{\text{rep}} \tag{7.17}$$

---

[2]See Appendix G.1 for the corresponding formula in the grand canonical ensemble.

where $\mathcal{U}_{\mathrm{ff},\alpha'}^{\mathrm{rep}}$ is the internal energy of a subsystem formed by the fluid particles of copy α'.

Equation (7.17) is a representative example showing how one can calculate, in principle, a physical quantity of the QA mixture *without* evaluating the cumbersome double average. One first calculates the corresponding quantity in the replicated system. The latter contains more (that is, $n + 1$) components, but it is conceptionally easier to handle because *all* particles are mobile. This can be realized from Eq. (7.14), which shows that the coordinates of the matrix particles and the fluid particle replicas are treated on equal footing in the replicated system. By letting n then go to zero, one eventually arrives at the quantity of interest.

From a practical point of view, however, it is clear that the procedure described above is highly nontrivial: Apart from the necessity to deal with mixtures of $n+1$ components, the way to carry out the limit $n \to 0$ in practice is by no means straightforward. One method to deal with these difficulties is the replica integral equation formalism, which we will introduce Section 7.5. However, before doing this we first introduce the key concepts of the replica integral equations, which are the two-particle correlation functions of the QA system.

7.4 Correlation functions and fluctuations in the disordered fluid

From now on we focus on situations where the fluid adsorbed by a disordered matrix is both homogeneous and isotropic *after* averaging over different matrix configurations.[3] In such a situation, the fluid's singlet density is just a constant; that is,

$$
\begin{aligned}
\rho\left(\boldsymbol{q}\right) &= \left[\left\langle \sum_{i=1}^{N_{\mathrm{f}}} \delta\left(\boldsymbol{q} - \boldsymbol{q}_i\right) \right\rangle_Q\right] = \lim_{n \to 0} \left\langle \sum_{i=1}^{N_{\mathrm{f}}} \delta\left(\boldsymbol{q} - \boldsymbol{q}_{i\alpha'}\right) \right\rangle_{\mathrm{rep}} \\
&= \lim_{n \to 0} \overline{\rho}_{\alpha'}^{\mathrm{rep}} = \overline{\rho}
\end{aligned}
\tag{7.18}
$$

The expression after the first equal sign in Eq. (7.18) provides the statistical definition of the singlet density in the disordered system where $\delta\left(\boldsymbol{q} - \boldsymbol{q}_i\right) = \delta\left(\boldsymbol{r} - \boldsymbol{r}_i\right)$ for a simple fluid without internal degrees of freedom, whereas $\delta\left(\boldsymbol{q} - \boldsymbol{q}_i\right) = \delta\left(\boldsymbol{r} - \boldsymbol{r}_i\right)\delta\left(\omega - \omega_i\right)$ for anisotropic fluid particles. The second

---

[3]Note that the fluid structure for a *specific* realization will usually be highly inhomogeneous and/or anisotropic.

equal sign in Eq. (7.18) represents the link to the singlet density in the replicated system, which can found by using the same strategies introduced in Section 7.3. Finally, $\overline{\rho} = N_{\rm f}/V$ for a simple fluid, whereas $\overline{\rho} = (4\pi)^{-1}N_{\rm f}/V$ for anisotropic particles.

Given that the singlet density is just a constant, the most important quantities characterizing the local structure within the adsorbed fluid are the two-particle correlation functions. We start by considering the pair correlation function $g_{\rm ff}(\boldsymbol{q}, \boldsymbol{q}')$ between two fluid particles or, equivalently, the corresponding total correlation function $h_{\rm ff}(\boldsymbol{q}, \boldsymbol{q}') = g_{\rm ff}(\boldsymbol{q}, \boldsymbol{q}') - 1$. The statistical definition of the latter is given by the generalization of the corresponding definition for equilibrated systems [30],

$$\overline{\rho}^2 h_{\rm ff}(\boldsymbol{q}, \boldsymbol{q}') \equiv \left[\left\langle \sum_{i \neq j}^{N_{\rm f}} \delta(\boldsymbol{q} - \boldsymbol{q}_i)\, \delta(\boldsymbol{q}' - \boldsymbol{q}_j) \right\rangle_{\boldsymbol{Q}}\right] - \overline{\rho}^2 \qquad (7.19)$$

Treating the double average on the right-hand side as described in Section 7.3, one finds

$$
\begin{aligned}
\overline{\rho}^2 h_{\rm ff}(\boldsymbol{q}, \boldsymbol{q}') &= \lim_{n \to 0}\left[\left\langle \sum_{i \neq j}^{N_{\rm f}} \delta(\boldsymbol{q} - \boldsymbol{q}_{i\alpha'})\, \delta(\boldsymbol{q}' - \boldsymbol{q}_{j\alpha'}) \right\rangle_{\rm rep} - (\overline{\rho}_{\alpha'}^{\rm rep})^2\right] \\
&= \overline{\rho}^2 \lim_{n \to 0} h_{\alpha'\alpha'}^{\rm rep}(\boldsymbol{q}, \boldsymbol{q}') \qquad (7.20)
\end{aligned}
$$

where $h_{\alpha'\alpha'}^{\rm rep}$ is the total correlation function between two fluid particles of the *same* copy in the replicated system. A practical way to actually calculate the total correlation function as well as various other functions to be introduced below will be presented in Section 7.5. Here we note that the function $h_{\rm ff}(\boldsymbol{q}, \boldsymbol{q}')$ alone is already sufficient to calculate the internal (fluid–fluid) energy of the disordered fluid. In fact, combining Eqs. (7.9) and (7.19), one has

$$\mathcal{U}_{\rm ff} = \frac{\overline{\rho}^2}{2} \int \mathrm{d}\boldsymbol{q} \int \mathrm{d}\boldsymbol{q}'\, [h_{\rm ff}(\boldsymbol{q}, \boldsymbol{q}') + 1]\, u_{\rm ff}(\boldsymbol{q}, \boldsymbol{q}') \qquad (7.21)$$

by complete analogy with the corresponding bulk fluid relation [30].

The next correlation function we consider is characteristic for a quenched-annealed system in the sense that it vanishes for conventional, fully annealed fluids. This is the so-called blocked correlation function $h_{\rm b}(\boldsymbol{q}, \boldsymbol{q}')$ defined by

$$\overline{\rho}^2 h_{\rm b}(\boldsymbol{q}, \boldsymbol{q}') \equiv \left[\left\langle \sum_{i=1}^{N_{\rm f}} \delta(\boldsymbol{q} - \boldsymbol{q}_i) \right\rangle_{\boldsymbol{Q}} \left\langle \sum_{j=1}^{N_{\rm f}} \delta(\boldsymbol{q}' - \boldsymbol{q}_j) \right\rangle_{\boldsymbol{Q}}\right] - \overline{\rho}^2 \qquad (7.22)$$

For conventional fluids the outer (disorder) average of the first term on the right side is absent and each thermal average equals the singlet density. Thus, $h_b = 0$ for systems without quenched disorder. In the presence of disorder, on the other hand, the blocked correlation function is usually nonzero, because the singlet density for a particular realization, $\left\langle \sum_{i=1}^{N_f} \delta \left(q - q_i \right) \right\rangle_Q$, can be *inhomogeneous* and thus very different from its disorder average, $\overline{\rho}$. Thus, h_b can be interpreted as a measure of matrix-induced fluctuations of the local density. The relation of the blocked correlation function to the replicated system is somewhat different from the cases discussed before because of the appearance of two thermal averages superordinated by the disorder average in Eq. (7.22). The final result (see Appendix G.2 for a derivation) is given by

$$h_b \left(q, q' \right) = \lim_{n \to 0} h_{\alpha\beta}^{\text{rep}} \left(q, q' \right), \qquad \alpha \neq \beta \qquad (7.23)$$

where $h_{\alpha\beta}^{\text{rep}} \left(q, q' \right)$ is the total correlation function between two fluid particles of different copies in the replicated system.

The total and the blocked correlation function introduced above are already sufficient to describe the structure within the adsorbed fluid. However, to describe thermal fluctuations we need to introduce two additional correlation functions. The first one is the response function $G_{\text{ff}} \left(q, q' \right)$ defined as

$$
\begin{aligned}
G_{\text{ff}} \left(q, q' \right) &\equiv \left[\left\langle \sum_{i=1}^{N_f} \sum_{j=1}^{N_f} \delta \left(q - q_i \right) \delta \left(q' - q_j \right) \right\rangle_Q \right] \\
&\quad - \left[\left\langle \sum_{i=1}^{N_f} \delta \left(q - q_i \right) \right\rangle_Q \left\langle \sum_{j=1}^{N_f} \delta \left(q' - q_j \right) \right\rangle_Q \right] \\
&= \overline{\rho} \delta \left(q - q' \right) + \overline{\rho}^2 h_c \left(q, q' \right)
\end{aligned}
\qquad (7.24)
$$

where the second member of the equation defines the so-called "connected" correlation function, $h_c(q, q')$. Combining Eq. (7.24) with the definitions in Eqs. (7.19) and (7.22), one sees that the connected function is related to the total and blocked correlation function via

$$h_c \left(q, q' \right) = h_{\text{ff}} \left(q, q' \right) - h_b \left(q, q' \right) = \lim_{n \to 0} \left[h_{\alpha\alpha}^{\text{rep}} \left(q, q' \right) - h_{\alpha\beta}^{\text{rep}} \left(q, q' \right) \right] \quad (7.25)$$

where the far right side introduces the connection to the replicated system [see Eqs. (7.20) and (7.23)].

To see the importance of the connected and response function for thermal fluctuations, we present two examples. The first one concerns fluctuations of the number of fluid particles N_f in an adsorbed fluid coupled to a particle reservoir (grand canonical ensemble). These fluctuations are commonly measured by the isothermal compressibility, κ_T. A definition of this quantity within the framework of statistical thermodynamics of disordered systems is given by [298]

$$\kappa_T = \frac{1}{\bar{\rho} k_B T} \frac{\left[\langle N_f^2 \rangle_Q - \langle N_f \rangle_Q^2 \right]}{\left[\langle N_f \rangle_Q \right]} \tag{7.26}$$

where we remind the reader that $[\ldots]$ signifies the average over matrix representations[4] [see Eq. (7.8)]. Combining the above equation with Eq. (7.24), one obtains

$$\kappa_T \bar{\rho} k_B T = \frac{1}{V \bar{\rho}} \int d\boldsymbol{q} \int d\boldsymbol{q}' G_{ff} (\boldsymbol{q}, \boldsymbol{q}') = 1 + \frac{\bar{\rho}}{V} \int d\boldsymbol{q} \int d\boldsymbol{q}' h_c (\boldsymbol{q}, \boldsymbol{q}') \tag{7.27}$$

where the second line shows that it is the spatial (and angular) integral over the connected correlation function that determines the compressibility.

As a second example we consider the dielectric constant ϵ_D of a dipolar fluid, where each particle carries a permanent dipole moment, $\boldsymbol{\mu}$, and therefore $\boldsymbol{q} = (\boldsymbol{r}, \omega)$. The dielectric constant measures fluctuations of the *total* dipole moment $\boldsymbol{M} = \sum_i \boldsymbol{\mu}_i$ and is defined as [299]

$$\frac{(\epsilon_D - 1)(2\epsilon_D + 1)}{9\epsilon_D} = \frac{4\pi\rho}{9} \frac{1}{N_f k_D T} \left[\langle \boldsymbol{M} \cdot \boldsymbol{M} \rangle_Q - \langle \boldsymbol{M} \rangle_Q \cdot \langle \boldsymbol{M} \rangle_Q \right] \tag{7.28}$$

assuming the canonical ensemble. Writing the total moment as

$$\boldsymbol{M} = \int d\boldsymbol{r} \int d\omega \sum_{i=1}^{N_f} \delta (\boldsymbol{r} - \boldsymbol{r}_i) \delta (\omega - \omega_i) \boldsymbol{\mu} (\omega) \tag{7.29}$$

and employing Eq. (7.24), Eq. (7.28) can be cast as

$$\frac{(\epsilon_D - 1)(2\epsilon_D + 1)}{9\epsilon_D} = \frac{4\pi}{9k_B T} \int d\bar{\boldsymbol{r}} \int d\omega \int d\omega' \boldsymbol{\mu} (\omega) \cdot \boldsymbol{\mu} (\omega') G_{ff} (\bar{\boldsymbol{r}}, \omega, \omega')$$

$$= \frac{4\pi}{9k_B T} \left[1 + \bar{\rho} \int d\bar{\boldsymbol{r}} \int d\omega \int d\omega' \right.$$

$$\left. \times \boldsymbol{\mu} (\omega) \cdot \boldsymbol{\mu} (\omega') h_c (\bar{\boldsymbol{r}}, \omega, \omega') \right] \tag{7.30}$$

---

[4]See Eqs. (3.37) and (5.78) for corresponding expressions in the bulk and ordered porous media, respectively.

In writing the first line of Eq. (7.30) we used the fact that, in a system that is *on average* homogeneous and isotropic, correlation functions depend on the positions of the fluid molecules only via the separation *vector* $\bar{r} = r - r'$. This allows us to replace the double integral $\int dr \int dr' \ldots$ by $V \int d\bar{r} \ldots$ after carrying out the trivial integration over r'. The integral over orientations on the second line of Eq. (7.30) is a projection of the connected correlation function onto the scalar product $\mu(\omega) \cdot \mu(\omega')$. This quantity can be evaluated by using a rotationally invariant expansion of h_c [258].

Finally, we note that the (angle-averaged) structure factor of the adsorbed fluid $S(k)$, which is accessible in scattering experiments, contains both a blocked and a connected part. To see this we start from the expression

$$
S(k) = \frac{1}{N_f} \left[\left\langle \sum_{i=1}^{N_f} \sum_{j=1}^{N_f} \exp\left[ik \cdot (r_i - r_j)\right] \right\rangle_Q \right]
$$
$$
= 1 + \frac{\bar{\rho}}{V} \int dq \int dq' \exp\left[ik \cdot (r - r')\right] h_{\text{ff}}(q, q') \qquad (7.31)
$$

where on the second line we have used the statistical definition of the total fluid–fluid correlation function given in Eq. (7.19). The structure factor can be rewritten in terms of the Fourier transform of $h_{\text{ff}}(q, q')$ defined by

$$
\widetilde{h}_{\text{ff}}(k, \omega, \omega') = \int d\bar{r} \, h_{\text{ff}}(\bar{r}, \omega, \omega') \exp(ik \cdot \bar{r}) \qquad (7.32)
$$

where we assume a fluid with both translational and orientational degrees of freedom as the most general case. For a simple fluid the above relation simplifies to

$$
\widetilde{h}_{\text{ff}}(k) = \int d\bar{r} \, h_{\text{ff}}(\bar{r}) \exp(ik \cdot \bar{r}) = 4\pi \int d\bar{r} \bar{r}^2 h_{\text{ff}}(\bar{r}) \frac{\sin(k\bar{r})}{k\bar{r}} \qquad (7.33)
$$

as we argue in Appendix G.4 where k and $\bar{r}$ are the magnitude of the wavevector and the separation vector, respectively. Inserting Eq. (7.32) into Eq. (7.31) we obtain

$$
S(k) = 1 + \bar{\rho} \int d\omega \int d\omega' \widetilde{h}_{\text{ff}}(k, \omega, \omega')
$$
$$
= 1 + \bar{\rho} \int d\omega \int d\omega' \left[\widetilde{h}_c(k, \omega, \omega') + \widetilde{h}_b(k, \omega, \omega') \right] \qquad (7.34)
$$

where the second line results from the definition (7.25) of the connected correlation function. Equation (7.34) shows that indeed both the connected and the blocked correlations are involved in the structure factor of the adsorbed fluid. As a consequence, the long-wavelength (small angle) limit of the structure factor, $S(k \rightarrow 0)$, does *not* coincide with the (reduced) isothermal compressibility given in Eq. (7.27), which may be expressed as

$$\kappa_{\mathrm{T}} \overline{\rho} k_{\mathrm{B}} T = 1 + \overline{\rho} \int d\omega \int d\omega' \widetilde{h}_{\mathrm{c}} \left(\boldsymbol{k} \rightarrow \boldsymbol{0}, \omega, \omega' \right) \qquad (7.35)$$

This discrepancy is in contrast to conventional fluids where $S\left(k \rightarrow 0\right) = \kappa_{\mathrm{T}} \overline{\rho} k_{\mathrm{B}} T$ [30].

7.5 Integral equations

We now turn to the actual calculation of the correlation functions introduced in the preceding section. Our strategy is based on the fact that all particles in the multicomponent replicated system are mobile. This allows the application of standard liquid state approaches such as integral equation theories [30] as has first been realized by Given and Stell [295–297]. The only serious complication is the limit $n \rightarrow 0$ relating the replicated to the original disordered system [see, for example, Eq. (7.20)].

In this chapter we will deal with this problem by starting from integral equations for the $(n+1)$-component mixture and assuming then permutation symmetry between the replicas. Thereby the n-dependence in the equations becomes isolated, which finally allows us to take the limit $n \rightarrow 0$ relatively easily.

Of course, an implicit assumption of this procedure is that the permutation symmetry between the replica indices is *preserved* even for non-integer values in the range $n < 1$. Breaking the replica symmetry does indeed occur in several disordered systems with low-temperature glassy states [293, 294]. However, in this context, we are only interested in the description of the (high-temperature) fluid phase, where the assumption of preservation of replica symmetry for all n is reaonable.

7.5.1 Replica Ornstein-Zernike equations

At the core of any integral equation approach we have the (exact) Ornstein-Zernike (OZ) equation [300] relating the total correlation function(s) of a given fluid to the so-called direct correlation function(s). For the replicated system at hand, the OZ equation is that of a multicomponent mixture [30], namely

$$h_{ij}^{\text{rep}}\left(\boldsymbol{q}_1, \boldsymbol{q}_2\right) = c_{ij}^{\text{rep}}\left(\boldsymbol{q}_1, \boldsymbol{q}_2\right) + \sum_k \overline{\rho}_k \int d\boldsymbol{q}_3 h_{ik}^{\text{rep}}\left(\boldsymbol{q}_1, \boldsymbol{q}_3\right) c_{kj}^{\text{rep}}\left(\boldsymbol{q}_3, \boldsymbol{q}_2\right) \qquad (7.36)$$

where the component indices i, j, and k can assume $0, 1, \ldots, n$. Here 0 represents the mobile matrix particles, $1, \ldots, n$ the copies of fluid particles, and the c_{ij} are the corresponding direct correlation functions. The convolution integrals in Eq. (7.36) can be circumvented by introducing Fourier transforms [see Eq. (7.32)] of all correlation functions. The real-space OZ equation (7.36) then transforms into

$$\begin{aligned} \widetilde{h}_{ij}^{\text{rep}}\left(\boldsymbol{k}, \omega_1, \omega_2\right) &= \widetilde{c}_{ij}^{\text{rep}}\left(\boldsymbol{k}, \omega_1, \omega_2\right) \\ &+ \sum_k \overline{\rho}_k \widetilde{h}_{ik}^{\text{rep}}\left(\boldsymbol{k}, \omega_1, \omega_3\right) \otimes \widetilde{c}_{kj}^{\text{rep}}\left(\boldsymbol{k}, \omega_3, \omega_2\right) \qquad (7.37) \end{aligned}$$

where the symbol "$\otimes$" denotes both multiplication and an integral over the orientation of the third particle if that particle possesses orientational degrees of freedom. The important point about the Fourier-transformed OZ equation (7.37) is that it decouples with respect to $\boldsymbol{k}$.

In a next step we make use of the fact that the n copies of the fluid particles are *identical* (this is obvious from the introduction of the replicated system described in Section 7.3). As a consequence, there is permutation symmetry between the replica indices. This implies for the fluid fluid and fluid matrix/matrix fluid correlations

$$\begin{aligned} \widetilde{f}_{ii}^{\text{rep}} &= \widetilde{f}_{\alpha'\alpha'}^{\text{rep}} \\ \widetilde{f}_{0i}^{\text{rep}} &= \widetilde{f}_{0\alpha'}^{\text{rep}} \\ \widetilde{f}_{i0}^{\text{rep}} &= \widetilde{f}_{\alpha'0}^{\text{rep}}, \qquad \forall\, i = 1, \ldots, n \qquad (7.38) \end{aligned}$$

where $f = h$ or c and α' is an arbitrary replica index. Furthermore, the correlations between *different* fluid copies have the symmetry

$$\widetilde{f}_{i,j\neq i}^{\text{rep}} = \widetilde{f}_{\alpha',\beta'\neq\alpha'}^{\text{rep}}, \qquad \forall\, i, j = 1, \ldots, n \qquad (7.39)$$

Using relations (7.38) and (7.39) along with the symmetries of the singlet density, $\overline{\rho}_i = \overline{\rho}_{\alpha'}$, $i = 1, \ldots, n$, and writing the OZ equations [see Eq. (7.37)] separately for each type of correlation function, one obtains

$$\tilde{h}^{\text{rep}}_{00} = \tilde{c}^{\text{rep}}_{00} + \overline{\rho}_0 \tilde{h}^{\text{rep}}_{00} \otimes \tilde{c}^{\text{rep}}_{00} + n \overline{\rho}_{\alpha'} \tilde{h}^{\text{rep}}_{0\alpha'} \otimes \tilde{c}^{\text{rep}}_{\alpha'0} \tag{7.40a}$$

$$\tilde{h}^{\text{rep}}_{0\alpha'} = \tilde{c}^{\text{rep}}_{0\alpha'} + \overline{\rho}_0 \tilde{h}^{\text{rep}}_{00} \otimes \tilde{c}^{\text{rep}}_{0\alpha'}$$
$$+ \overline{\rho}_{\alpha'} \left[\tilde{h}^{\text{rep}}_{0\alpha'} \otimes \tilde{c}^{\text{rep}}_{\alpha'\alpha'} + (n-1) \tilde{h}^{\text{rep}}_{0\alpha'} \otimes \tilde{c}^{\text{rep}}_{\beta'\alpha'} \right] \tag{7.40b}$$

$$\tilde{h}^{\text{rep}}_{\alpha'0} = \tilde{c}^{\text{rep}}_{\alpha'0} + \overline{\rho}_0 \tilde{h}^{\text{rep}}_{\alpha'0} \otimes \tilde{c}^{\text{rep}}_{00}$$
$$+ \overline{\rho}_{\alpha'} \left[\tilde{h}^{\text{rep}}_{\alpha'\alpha'} \otimes \tilde{c}^{\text{rep}}_{\alpha'0} + (n-1) \tilde{h}^{\text{rep}}_{\alpha'\beta'} \otimes \tilde{c}^{\text{rep}}_{\alpha'0} \right] \tag{7.40c}$$

$$\tilde{h}^{\text{rep}}_{\alpha'\alpha'} = \tilde{c}^{\text{rep}}_{\alpha'\alpha'} + \overline{\rho}_0 \tilde{h}^{\text{rep}}_{\alpha'0} \otimes \tilde{c}^{\text{rep}}_{0\alpha'}$$
$$+ \overline{\rho}_{\alpha'} \left[\tilde{h}^{\text{rep}}_{\alpha'\alpha'} \otimes \tilde{c}^{\text{rep}}_{\alpha'\alpha'} + (n-1) \tilde{h}^{\text{rep}}_{\alpha'\beta'} \otimes \tilde{c}^{\text{rep}}_{\beta'\alpha'} \right] \tag{7.40d}$$

$$\tilde{h}^{\text{rep}}_{\alpha'\beta'} = \tilde{c}^{\text{rep}}_{\alpha'\beta'} + \overline{\rho}_0 \tilde{h}^{\text{rep}}_{\alpha'0} \otimes \tilde{c}^{\text{rep}}_{0\alpha'}$$
$$+ \overline{\rho}_{\alpha'} \left[\tilde{h}^{\text{rep}}_{\alpha'\alpha'} \otimes \tilde{c}^{\text{rep}}_{\alpha'\beta'} + \tilde{h}^{\text{rep}}_{\alpha'\beta'} \otimes \tilde{c}^{\text{rep}}_{\alpha'\alpha'} \right]$$
$$+ (n-2) \overline{\rho}_{\alpha'} \tilde{h}^{\text{rep}}_{\alpha'\beta'} \otimes \tilde{c}^{\text{rep}}_{\alpha'\beta'} \tag{7.40e}$$

where we have dropped the arguments of the correlation functions to emphasize the structure of the equations. Indeed, inspection of Eqs. (7.40) shows that the n-dependence of the correlation functions has now become *isolated*. This process allows us to perform the last step of our derivation, that is, the limit $n \to 0$ relating the replicated to the original disordered system. Using the definitions (7.20), (7.23), and (7.25) of the total, blocked, and connected correlation function, respectively, and introducing the notations $\tilde{h}_{\text{mm}} = \lim_{n \to 0} \tilde{h}^{\text{rep}}_{00}$, $\overline{\rho}_{\text{m}} = \lim_{n \to 0} \overline{\rho}_0$, $\tilde{h}_{\text{fm}} = \lim_{n \to 0} \tilde{h}^{\text{rep}}_{\alpha'0}$, and $\tilde{h}_{\text{mf}} = \lim_{n \to 0} \tilde{h}^{\text{rep}}_{0\alpha'}$, one obtains[5]

$$\tilde{h}_{\text{mm}} = \tilde{c}_{\text{mm}} + \overline{\rho}_{\text{m}} \tilde{h}_{\text{mm}} \otimes \tilde{c}_{\text{mm}} \tag{7.41a}$$

$$\tilde{h}_{\text{mf}} = \tilde{c}_{\text{mf}} + \overline{\rho}_{\text{m}} \tilde{h}_{\text{mm}} \otimes \tilde{c}_{\text{mf}} + \overline{\rho} \tilde{h}_{\text{mf}} \otimes \tilde{c}_{\text{c}} \tag{7.41b}$$

$$\tilde{h}_{\text{fm}} = \tilde{c}_{\text{fm}} + \overline{\rho}_{\text{m}} \tilde{h}_{\text{fm}} \otimes \tilde{c}_{\text{mm}} + \overline{\rho} \tilde{h}_{\text{c}} \otimes \tilde{c}_{\text{fm}} \tag{7.41c}$$

$$\tilde{h}_{\text{ff}} = \tilde{c}_{\text{ff}} + \overline{\rho}_{\text{m}} \tilde{h}_{\text{fm}} \otimes \tilde{c}_{\text{mf}}$$
$$+ \overline{\rho} \left[\tilde{h}_{\text{ff}} \otimes \tilde{c}_{\text{ff}} - \tilde{h}_{\text{b}} \otimes \tilde{c}_{\text{b}} \right] \tag{7.41d}$$

$$\tilde{h}_{\text{b}} = \tilde{c}_{\text{b}} + \overline{\rho}_{\text{m}} \tilde{h}_{\text{fm}} \otimes \tilde{c}_{\text{mf}} + \overline{\rho} \tilde{h}_{\text{ff}} \otimes \tilde{c}_{\text{b}}$$
$$+ \overline{\rho} \left[\tilde{h}_{\text{b}} \otimes \tilde{c}_{\text{ff}} - 2 \tilde{h}_{\text{b}} \otimes \tilde{c}_{\text{b}} \right] \tag{7.41e}$$

Finally, subtracting the last two of these equations from each other, one finds for the connected correlation function

$$\tilde{h}_{\text{c}} = \tilde{c}_{\text{c}} + \overline{\rho} \tilde{h}_{\text{c}} \otimes \tilde{c}_{\text{c}} \tag{7.42}$$

---

[5]The same notation is used for the various direct correlation functions.

Together Eqs. (7.41) and (7.42) form the replica-symmetric Ornstein-Zernike (RSOZ) equations first derived by Given and Stell in 1992 [295–297]. They are *exact* relationships, as are the OZ equations for conventional fluids. One specific feature of the RSOZ equations is the decoupling of the matrix structure from the fluid correlations [see Eq. (7.41a)]. This reflects that the matrix is quenched and thus not influenced by the structure of the adsorbed fluid. Consequently, the matrix correlations serve as *input* to the theory. In fact, one may even employ experimental data (e.g., from neutron scattering) for the matrix structure factor, which is related to the matrix correlations (assuming a matrix without rotational degrees of freedom) by $S_{mm}(k) \equiv 1 + \rho_m \tilde{h}_{mm}(k)$. The direct correlation function between matrix particles then follows from Eq. (7.41a), which implies $1 + \rho_m \tilde{h}_{mm}(k) = (1 - \rho_m \tilde{c}_{mm}(k))^{-1}$.

The above example already indicates that the practical solution of the whole set of RSOZ equations becomes particularly easy for simple QA systems where the correlations appearing in (7.41a)-(7.42) depend only on the wavenumber. For molecular fluids and matrices, on the other hand, the angle-dependence of the correlations can be handled by using rotationally invariant expansions. This procedure, which is outlined in Appendix G.3, results in a system of linear RSOZ equations for the correlation function coefficients.

7.5.2 Closure relationships

The RSOZ equations Eqs. (7.41) and (7.42) still involve both the total and the direct correlation functions. Therefore, appropriate closure expressions relating the correlation functions to the pair potentials are needed to calculate the correlation functions at given densities and temperatures. Typically, one uses standard closure expressions familiar from *bulk* liquid state theory [30]. One should note, however, that the performance of these closures for disordered fluids can clearly not be taken for granted. Instead, they need to be tested for each new model system under consideration.

In the following discussion we present as an example closure expressions appropriate for systems where both fluid and matrix particles are spherical and have fixed diameters ("hard cores") σ_f (fluid) and σ_m (matrix). The corresponding fluid fluid, fluid matrix, and matrix matrix interactions then contain (apart from other contributions) the hard-sphere(HS) potential

$$u_{HS}(r) = \begin{cases} 0, & r > \sigma_{f(fm,m)} \\ \infty, & r < \sigma_{f(fm,m)} \end{cases} \tag{7.43}$$

where $\sigma_{fm} = (\sigma_f + \sigma_m)/2$ and r is the separation of the particles.

Following the strategy described at the beginning of Section 7.5, we start by considering closure relations for the replicated system, employing the notation introduced in Eq. (7.36). The (exact) hard-core conditions can be written as

$$
\begin{array}{rlll}
c_{00}^{\text{rep}}\left(\boldsymbol{q}, \boldsymbol{q}'\right) & = & -1 - \eta_{00}^{\text{rep}}\left(\boldsymbol{q}, \boldsymbol{q}'\right), & r < \sigma_{\text{m}} & (7.44\text{a}) \\
c_{i0}^{\text{rep}}\left(\boldsymbol{q}, \boldsymbol{q}'\right) & = & -1 - \eta_{i0}^{\text{rep}}\left(\boldsymbol{q}, \boldsymbol{q}'\right), & r < \sigma_{\text{fm}} & (7.44\text{b}) \\
c_{0i}^{\text{rep}}\left(\boldsymbol{q}, \boldsymbol{q}'\right) & = & -1 - \eta_{0i}^{\text{rep}}\left(\boldsymbol{q}, \boldsymbol{q}'\right), & r < \sigma_{\text{fm}} & (7.44\text{c}) \\
c_{ii}^{\text{rep}}\left(\boldsymbol{q}, \boldsymbol{q}'\right) & = & -1 - \eta_{ii}^{\text{rep}}\left(\boldsymbol{q}, \boldsymbol{q}'\right), & r < \sigma_{\text{f}} & (7.44\text{d})
\end{array}
$$

where $i = 1, \ldots, n$. Furthermore, $\eta^{\text{rep}} = h^{\text{rep}} - c^{\text{rep}}$ with $h_{00(i0,0i,ii)}^{\text{rep}} = -1$ for the separation range considered, reflecting the fact that the corresponding pair correlation functions are zero. Note that there is no hard-core condition for the correlations between particles of different fluid copies, $(i \neq j)$ because these particles do *not* interact, as may be realized from the configurational potential energy of the replicated system in Eq. (7.12). We also note that each of the relations in Eq. (7.44) involves only one species of particles. Therefore, there is no explicit n-dependence and we can directly perform the limit $n \to 0$, yielding

$$
c_{\text{ff(fm,mm)}}\left(\boldsymbol{q}, \boldsymbol{q}'\right) = -1 - \eta_{\text{ff(fm,mm)}}\left(\boldsymbol{q}, \boldsymbol{q}'\right), \qquad r < \sigma_{\text{f(fm,m)}} \qquad (7.45)
$$

and $c_{\text{mf}}\left(\boldsymbol{q}, \boldsymbol{q}'\right) = c_{\text{fm}}\left(\boldsymbol{q}', \boldsymbol{q}\right)$.

For separations outside the hard core, the direct correlation functions have to be approximated. "Classic" closure approximations recently applied to QA models are the Percus-Yevick (PY) closure [301], the mean spherical approximation (MSA) [302], and the hypernetted chain (HNC) closure [30]. None of these relations, when formulated for the replicated system, contains any coupling between different species, and we can directly proceed to the limit $n \to 0$. The PY closure then implies

$$
\begin{array}{rll}
c_{\text{mm}}\left(\boldsymbol{q}, \boldsymbol{q}'\right) & = & \left[1 + h_{\text{mm}}\left(\boldsymbol{q}, \boldsymbol{q}'\right)\right] \\
& & \times \left\{1 - \exp\left[-\dfrac{u_{\text{mm}}\left(\boldsymbol{q}, \boldsymbol{q}'\right)}{k_{\text{B}} T_0}\right]\right\}, \quad r > \sigma_{\text{m}} \quad (7.46\text{a}) \\
c_{\text{ff(fm)}}\left(\boldsymbol{q}, \boldsymbol{q}'\right) & = & \left[1 + h_{\text{ff(fm)}}\left(\boldsymbol{q}, \boldsymbol{q}'\right)\right] \\
& & \times \left\{1 - \exp\left[-\dfrac{u_{\text{ff(fm)}}\left(\boldsymbol{q}, \boldsymbol{q}'\right)}{k_{\text{B}} T}\right]\right\}, \quad r > \sigma_{\text{f(fm)}} \quad (7.46\text{b}) \\
c_{\text{b}}\left(\boldsymbol{q}, \boldsymbol{q}'\right) & = & 0, \quad \forall r & (7.46\text{c})
\end{array}
$$

Note the appearance of the quenching temperature T_0 (instead of T) for the matrix correlations [first member of Eq. (7.46)]. This is a consequence of the prefactor T/T_0 in front of the matrix contribution to the configurational potential energy of the replicated system [see Eq. (7.12)].

The PY closure has frequently been applied to various hard-sphere QA models [303], for which the first relation in Eq. (7.46) reduces to

$$c_{\text{ff(fm,mm)}}\left(\boldsymbol{q},\boldsymbol{q}'\right) = 0 \qquad (7.47)$$

that is, outside the hard core. A drawback of the PY closure is that the blocked direct correlation function c_b is zero for *all* separations and that the blocked total correlation function h_b resulting from combining Eq. (7.46) with the RSOZ equations turns out to be very small. This finding is in contrast to results from computer simulations [304], where h_b has significant values especially at small separations [304].

Next, consider the MSA, defined by

$$c_{\text{mm}}\left(\boldsymbol{q},\boldsymbol{q}'\right) = -\frac{1}{k_{\text{B}}T_0}u_{\text{mm}}\left(\boldsymbol{q},\boldsymbol{q}'\right), \qquad r > \sigma_{\text{m}} \qquad (7.48a)$$

$$c_{\text{ff(fm)}}\left(\boldsymbol{q},\boldsymbol{q}'\right) = -\frac{1}{k_{\text{B}}T}u_{\text{ff(fm)}}\left(\boldsymbol{q},\boldsymbol{q}'\right), \qquad r > \sigma_{\text{f(fm)}} \qquad (7.48b)$$

$$c_{\text{b}}\left(\boldsymbol{q},\boldsymbol{q}'\right) = 0, \qquad \forall\, r \qquad (7.48c)$$

The MSA is linear in the pair potentials and has been applied to a variety of QA models with electrostatic interactions [305–307]. However, concerning the blocked correlations, the MSA has the same drawbacks as the PY closure (to which the MSA, in fact, reduces for pure hard-core models). As a final example, we present the HNC closure defined by

$$c_{\text{mm}}\left(\boldsymbol{q},\boldsymbol{q}'\right) = -\frac{1}{k_{\text{B}}T_0}u_{\text{mm}}\left(\boldsymbol{q},\boldsymbol{q}'\right) - \ln\left[1 + h_{\text{mm}}\left(\boldsymbol{q},\boldsymbol{q}'\right)\right]$$
$$+ h_{\text{mm}}\left(\boldsymbol{q},\boldsymbol{q}'\right), \qquad r > \sigma_{\text{m}} \qquad (7.49a)$$

$$c_{\text{ff(fm)}}\left(\boldsymbol{q},\boldsymbol{q}'\right) = -\frac{1}{k_{\text{B}}T}u_{\text{ff(fm)}}\left(\boldsymbol{q},\boldsymbol{q}'\right) - \ln\left[1 + h_{\text{ff(fm)}}\left(\boldsymbol{q},\boldsymbol{q}'\right)\right]$$
$$+ h_{\text{ff(fm)}}\left(\boldsymbol{q},\boldsymbol{q}'\right), \qquad r > \sigma_{\text{f(fm)}} \qquad (7.49b)$$

$$c_{\text{b}}\left(\boldsymbol{q},\boldsymbol{q}'\right) = -\ln\left[1 + h_{\text{b}}\left(\boldsymbol{q},\boldsymbol{q}'\right)\right] + h_{\text{b}}\left(\boldsymbol{q},\boldsymbol{q}'\right), \qquad \forall\, r \qquad (7.49c)$$

Clearly, the HNC yields nonzero blocked correlations. Moreover, it is a particularly successful approximation for long-range, electrostatic interactions appearing in ionic and dipolar QA models (see Section 7.7 for a discussion of specific applications).

7.6 Thermodynamics of the replicated fluid

So far we have focused on the calculation of correlation functions in the disordered system, from which one may extract structural features as well as susceptibilities (see Section 7.4). However, in the context of phase behavior, one may be also interested in thermodynamic quantities such as the free energy, pressure, or chemical potential. In equilibrium fluids, all of these quantities can be related (assuming pair-wise additive potentials) to the usual pair correlation function(s) accessible by integral equation theory [30]. Not surprisingly, these relations become much more involved for QA systems [308] and we note in advance that explicit expressions for thermodynamic quantities can only be obtained for specific model systems and closure approximations. In the following discussion, we therefore restrict ourselves to the relations between the thermodynamics of the original and the replicated system, and refer to the literature for specific systems.

We start by considering the free energy, which is defined by

$$-\frac{\mathcal{F}}{k_B T} = [\ln Z_Q] = \frac{1}{Z_m} \int dQ^{N_m} \exp\left[-\frac{U_{mm}(Q^{N_m})}{k_B T_0}\right] \ln Z_Q \qquad (7.50)$$

where Z_Q is given in Eq. (7.7) and we have used the definition of the disorder average (7.8) in writing the second line of Eq. (7.50). To introduce the replicated system, we employ the mathematical identity

$$\ln x = \lim_{n \to 0} \frac{d}{dn} x^n \qquad (7.51)$$

which can be proved by using the identity $x^n \equiv \exp[n \ln x]$ and a Taylor expansion of the exponential, which yields $x^n = 1 + n \ln x + \mathcal{O}(n^2)$. A derivative with respect to n, followed by taking the limit $n \to 0$, then leads to Eq. (7.51). Identifying with x in Eq. (7.51) the configuration integral Z_Q and inserting the latter relation into Eq. (7.50), one obtains

$$
\begin{aligned}
-\frac{\mathcal{F}}{k_B T} &= \frac{1}{Z_m} \lim_{n \to 0} \frac{d}{dn} \int dQ^{N_m} Z_Q^n \exp\left[-\frac{U_{mm}(Q^{N_m})}{k_B T_0}\right] \\
&= \frac{1}{Z_m} \lim_{n \to 0} \frac{d}{dn} \int dQ^{N_m} \prod_{\alpha=1}^{n} \int dq_\alpha^{N_f} \exp[-\beta U^{rep}] \\
&= \frac{1}{Z_m} \lim_{n \to 0} \frac{d}{dn} Z_{rep} \qquad (7.52)
\end{aligned}
$$

where we have used the definitions given in Eqs. (7.12) and (7.14) of the configurational potential energy and the configuration integral of the replicated system. Finally, introducing the free energy of this replicated system

$$\mathcal{F}_{rep} = -k_B T \ln Z_{rep} \qquad (7.53)$$

and using Eq. (7.15), one finds

$$\mathcal{F} = \lim_{n \to 0} \frac{d}{dn} \mathcal{F}_{rep} \qquad (7.54)$$

Equation (7.54) represents a practical guide to calculate the free energy provided that

1. the free energy of the (annealed) replicated system can be expressed by the correlation functions of the replicated system in a closed manner and that

2. the dependence of $\mathcal{F}_{rep}$ on n can be isolated such that the mathematical operation $\lim_{n \to 0} d/dn \ldots$ can indeed be performed.

Whether these requirements can be met depends on the model considered and on the closure relation involved for the calculation of the correlation functions. Examples for which Eq. (7.54) has actually been used pertain to the class of simple QA systems, that is, QA systems with no rotational degree of freedom where the interaction potentials contain a spherical hard-core contribution plus (at most) an attractive perturbation. For such systems, the free energy has been calculated on the basis of correlation functions in the mean spherical approximation (or an optimized random-phase approximation) [114, 298].

We now turn to the pressure P and the chemical potential μ_f of the adsorbed fluid. Considering first the $(n + 1)$-component replicated system one has

$$\begin{aligned}
\mathcal{F}_{rep} &= -P_{rep}V + \mu_m^{rep}N_m + \sum_{\alpha=1}^{n} \mu_{f,\alpha}^{rep}N_f \\
&= -P_{rep}V + \mu_m^{rep}N_m + n\mu_{f,\alpha'}^{rep}N_f \qquad (7.55)
\end{aligned}$$

where P_{rep} is the pressure of the replicated system, and $\mu_{f,\alpha}^{rep}$ and μ_m^{rep} are the chemical potentials of one fluid copy and that of the matrix particles, respectively. Also, we have used the symmetry between the fluid copies in writing the third term on the second line of Eq. (7.55). Combining Eqs. (7.55) and (7.54), we obtain

$$\mathcal{F} = \lim_{n \to 0} \frac{d}{dn} \left(-P_{rep} + \mu_m^{rep}\rho_m \right) V + \lim_{n \to 0} \mu_{f,\alpha'}^{rep}N_f \qquad (7.56)$$

Based on Eq. (7.56), one finds for the pressure P and the (fluid) chemical potential μ_f of the original system

$$P \equiv -\left(\frac{\partial \mathcal{F}}{\partial V}\right)_{T,N_f,\rho_m} \overset{(7.56)}{=} \lim_{n\to 0} \frac{d}{dn}\left(P_{rep} - \mu_m^{rep}\rho_m\right) \qquad (7.57a)$$

$$\mu_f \equiv \left(\frac{\partial \mathcal{F}}{\partial N_f}\right)_{T,V,\rho_m} \overset{(7.56)}{=} \lim_{n\to 0} \mu_{f,\alpha'}^{rep} \qquad (7.57b)$$

Equations (7.57a) and (7.57b) provide two ways to calculate the pressure and chemical potential. The first one is to perform the appropriate derivative of the free energy, assuming that the latter can be evaluated for all states of interest. The second, more direct way is to employ the relations to the pressure and chemical potentials in the replicated system. This second strategy is particularly useful for the calculation of μ_f because $\lim_{n\to 0} \mu_{f,\alpha'}^{rep}$ can be cast in closed form for a variety of model systems and closure relations, including the HNC approximation for molecular fluids [309, 310]. The pressure is more difficult because of the presence of the second, matrix-related term on the right side of Eq. (7.56).

Finally, we note without proof that both pressure and chemical potential can also be obtained by integrating the compressibility given in Eq. (7.27). Explicitly, one has [308]

$$\left(\frac{\partial P}{\partial (\rho k_B T)}\right)_{V,T,\rho_m} = \rho k_B T \kappa_T = \rho \left(\frac{\partial \mu_f}{\partial (\rho k_B T)}\right)_{V,T,\rho_m} \qquad (7.58)$$

For a derivation of these formulas we refer to Refs. 308 and 311.

7.7 Applications

7.7.1 Model systems

The earliest applications of the replica integral equation approach date back to the beginning of the 1990s. They focused on quite simple QA systems such as hard-sphere (HS) and LJ (12,6) fluids in HS matrices (see, for example, Refs. 4, 286, 290, 298, 303, 312, and 313 for reviews). From a technical point of view, these studies have shown that the replica integral equations yield accurate correlation functions compared with parallel computer simulation results [292, 303, 314, 315]. Moreover, concerning phase behavior, it turned out that the simple LJ (12,6) fluid in HS matrices already displays features also observed in experiments of fluids confined to aerogels [131, 132]. These features concern shifts of the vapor liquid critical temperature toward values significantly smaller than in the bulk, an accompanying increase of the critical density, and a narrowing of the coexistence curve.

Motivated by this success, a series of more recent replica integral equation studies has focused on the effects of more realistic features of both the adsorbed fluid and its interactions with matrix. Examples are studies of the influence of templated matrix materials [316], associating fluids [317], LJ mixtures [114, 311], and QA systems with ionic interactions [305, 306, 318–320]. However, until recently, only one study [309] has been available on QA systems with angle-dependent (specifically anisotropic steric) interactions.

In this chapter we discuss recent replica integral equation results for an adsorbed dipolar model fluid [307, 310, 321]. Specifically, we consider the so-called Stockmayer fluid consisting of spherical particles interacting with each other via both the (spherically symmetric) LJ (12,6) potential [see Eq. (5.24)] and long-range, anisotropic dipole dipole interactions generated by point dipoles $\boldsymbol{\mu}_i$ embedded in the center of the particles. The dipolar interaction is given by Eq. (6.21), and the potential energy between a pair of Stockmayer molecules is given by

$$
u_{\mathrm{ff}}\left(\boldsymbol{r}, \omega_1, \omega_2\right) = \begin{cases} u_{\mathrm{LJ}}\left(r\right) + u_{\mathrm{DD}}\left(\boldsymbol{r}, \omega_1, \omega_2\right), & r > \sigma_{\mathrm{f}} \\ \infty, & r < \sigma_{\mathrm{f}} \end{cases} \tag{7.59}
$$

The hard core in Eq. (7.59) has been imposed for numerical convenience. As a consequence, it is mainly the van der Waals–like attractive (rather than the repulsive) part of the LJ (12,6) potential ($\propto r^{-6}$) that contributes to the fluid fluid potential. The strength of the dipolar relative to the attractive LJ interactions is conveniently measured by the "reduced" (i.e., dimensionless) dipole moment $m = \mu/\sqrt{\varepsilon\sigma^3}$. Depending on this parameter, the Stockmayer fluid may serve as a simple model for polar molecular fluids [258, 259] (small m) or for ferrofluids [227, 228] (large m). Here we consider a system with dipole moment $m = 2$, which is a value typical for moderately polar molecular fluids [259] such as chloroform. For this value of m, GCEMC simulations have been presented in Section 6.4.1.

In what follows we discuss the phase behavior of the Stockmayer fluid in the presence of disordered matrices of increasing complexity. All results are based on a variant of the HNC equation [see Eq. (7.49)], which yields very good results for bulk dipolar fluids [268, 322]. Moreover, subsequent studies of dipolar hard-sphere (DHS) fluids [defined by Eq. (7.59) with $u_{\mathrm{LJ}} = 0$] in disordered matrices by Fernaud et al. [323, 324] have revealed a very good performance of the HNC closure compared with parallel computer simulation results. The integral equations are solved numerically with an iteration procedure. To handle the multiple angle-dependence of the correlations functions, one employs expansions in so-called rotational invariants, using the same steps that are well established for bulk dipolar fluids [268, 322]. In particular, the RSOZ equations can be formulated as a system of linear equations for the expansion coefficients. This is outlined in Appendix G.3. For further details concerning the numerical procedure, we refer to Refs. [310] and [307].

7.7.2 Dipolar fluids in simple matrices

The simplest disordered medium is realized by a matrix consisting of positionally quenched hard spheres. In this case, both the matrix matrix and the resulting fluid matrix interaction are given by Eq. (7.43). For simplicity we assume that matrix and fluid particles have the same diameter.

The influence of the HS matrix on the phase behavior of an adsorbed Stockmayer fluid is checked most easily by investigation of the *stability limits* (spinodals) of the isotropic high-temperature phase. Indeed, localization of true phase coexistence lines is significantly more difficult because of the lack of a closed expression for the pressure within the replica HNC approach.

Typical results for HS matrices of different porosity are displayed in Fig. 7.2 [310]. The S-like shape of the corresponding *bulk* curve reflects the presence of two types of phase transitions within the density range considered: a gas liquid transition appearing at low and intermediate fluid densities (corresponding to the hill in the spinodal) and an isotropic-to-ferroelectric (IF) transition occurring at higher densities. These RHNC predictions are consistent with computer simulation results on Stockmayer fluids. This concerns in particular the vapor liquid critical temperature, the RHNC estimate for which is very close to the corresponding GCEMC estimate [268]. Condensation/evaporation in the bulk is signaled by a divergence of the isothermal compressibility κ_T. For adsorbed fluids, we find clear evidence for a gas liquid transition only at very small matrix densities, e.g., at $\rho_m = 0.1$, which corresponds to a porosity typical for a silica aerogel (where gas liquid transitions have indeed been observed experimentally [131, 132]). In more dense matrices, the compressibility remains small for all temperatures considered, which is consistent with recent lattice fluid studies of Kierlik et al. [285, 286] questioning the existence of gas–liquid transitions in dense disordered porous materials regardless of the nature of the fluid.

These considerations motivated us to focus mainly on the dilute matrix case in what follows. In particular, one observes from Fig. 7.2 that the (gas liquid) critical point of this system is displaced toward lower temperature and lower density where the latter shift essentially disappears if additional attractions between fluid particles and the matrix are included in the model [310].

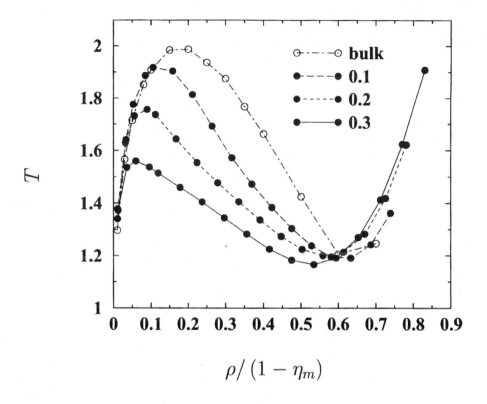

Figure 7.2: Replica HNC results for the temperatures corresponding to the stability limits of the homogeneous isotropic phase of confined Stockmayer fluids (and for the bulk) as a function of the renormalized fluid density ($\eta = \pi \rho \sigma^3 / 6$). Curves are labeled according to values of the reduced matrix density ρ_m.

These trends are similar to what is observed in simpler model fluids with purely spherically symmetric interactions [298, 313], which is to some extent expected because the gas liquid transition in Stockmayer fluids is mainly driven by the isotropic LJ (12,6) interactions underlying this model. We show in Ref. 307 that the main effects of HS matrices on the condensation can be reproduced when the dipolar model fluid is approximated by a fluid with angle-averaged dipolar interactions that are not only spherically symmetric but also short-ranged (they decay in proportion to r^{-6} for $r \to \infty$). This notion is particularly important for future simulation studies on adsorbed dipolar fluids.

At high fluid densities, bulk Stockmayer fluids exhibit an IF transition, which is signaled by a divergence of the dielectric constant ϵ_D [see Eq. (7.30)]. Results for ϵ_D are displayed in Fig. 7.3, which suggests that the IF transition

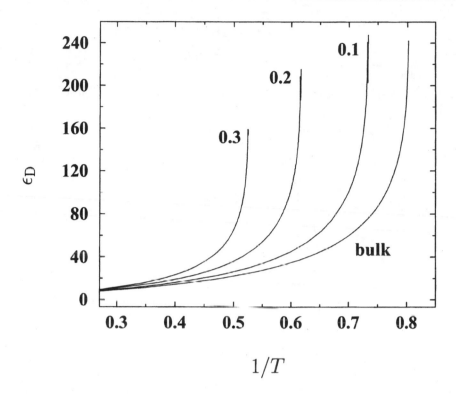

Figure 7.3: Dielectric constant ϵ_D versus (inverse) temperature for Stockmayer fluids at fixed fluid density $\rho = 0.7$. Curves are labeled according to values of the matrix density.

occurring in bulk Stockmayer fluids survives regardless of the matrix porosity [310]. Moreover, the transition temperatures indicated by the divergences of ϵ_D *increase* with increasing matrix density. To shed some light on this puzzling result, we performed a detailed study comparing partly quenched and fully equilibrated mixtures [310]. From this study it turns out that the shift of transition temperatures is mainly caused by the restricted volume accessible, which leads to an increased tendency of the fluid to form ferroelectric clusters in matrix-free regions of space. Similar conclusions have also been drawn later for a related model system, namely a DHS fluid in an HS matrix [323].

7.7.3 Dipolar fluids in complex matrices

More dramatic effects arise when the perturbation induced by the disordered matrix couples directly to the dipole moments of the fluid particles. Charged matrix particles provide an example. Their impact on a DHS fluid has been studied by Fernaud et al. [323]. They report a significant decrease of the dielectric constant and an enhanced tendency of dipoles to form aggregates at low densities. Another interesting case are dipolar fluid matrix interactions where each fluid particle "feels" both the dipole fields of its fluid neighbors and the additional dipole fields arising from the adsorbing medium.

Understanding the resulting interplay (or competition) between these interactions is relevant not only from an academic point of view but also from the perspective of adsorption processes in experimental systems. Indeed, dipolar fluid matrix interactions play a central role in purification processes such as liquid chromatography where polar liquids are adsorbed by disordered materials composed of molecules with polar headgroups [325].

The simplest matrix generating disordered dipolar fields consists of a system of DHS, which are quenched from an equilibrium fluid configuration at quenching temperature T_0. At this temperature, the coupling between two matrix particles is given by

$$\frac{1}{k_B T_0} u_{mm}\left(\boldsymbol{r}, \omega_1, \omega_2\right) = \begin{cases} u_{DD}\left(\boldsymbol{r}, \boldsymbol{\mu}_{m,1}, \boldsymbol{\mu}_{m,2}\right)/k_B T_0, & r > \sigma \\ \infty, & r < \sigma \end{cases} \quad (7.60)$$

where $u_{DD}\left(\boldsymbol{r}, \boldsymbol{\mu}_{m,1}, \boldsymbol{\mu}_{m,2}\right)$ is the dipolar interaction between two matrix particles. Throughout this work we assume the interaction strength $\mu_m^2/k_B T \sigma^3$ to have values such that the matrix is homogeneous and isotropic *on average*, that is, when averaged over different matrix configurations. Finally, the resulting fluid matrix coupling at temperature T is given by

$$\frac{1}{k_B T} u_{fm}\left(\boldsymbol{r}, \omega_1, \omega_2\right) = \begin{cases} u_{DD}\left(\boldsymbol{r}, \boldsymbol{\mu}_1, \boldsymbol{\mu}_{m,2}\right)/k_B T, & r > \sigma \\ \infty, & r < \sigma \end{cases} \quad (7.61)$$

The effect of the variation of μ_m^2 on the stability limits of a polar Stockmayer fluid, which implies a variation of the dipolar fluid matrix coupling, is illustrated in the upper part of Fig. 7.4. All results correspond to dilute matrices that do not suppress the gas liquid transition but lead to a significant shift of that transition. In particular, the critical temperature decreases with increasing μ_m, whereas the critical density increases. This characteristic effect is referred to as "preferential adsorption" in other contexts. The replica integral equation results thus demonstrate, at a microscopic level, that polar

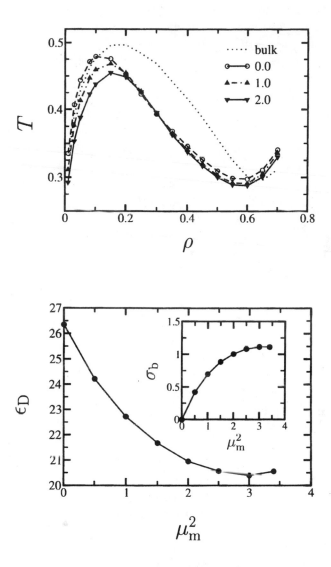

Figure 7.4: Top: Replica HNC predictions for the stability limits of the homogeneous isotropic phase of Stockmayer fluids adsorbed to disordered DHS matrices of density $\rho_m = 0.1$. Curves are labeled according to the reduced matrix dipole moment $\mu_m^2/k_B T_0 \sigma^3$ (the pure HS matrix corresponds to $\mu_m = 0$). Bottom: Dielectric constant of a dense adsorbed fluid as a function of the matrix dipole moment ($T = 0.5$, $\rho = 0.7$, $\rho_m = 0.1$). The inset shows the integrated blocking part of the dipole dipole correlation function.

interactions (arising, for example, from polar headgroups on the molecules) between a fluid and the disordered adsorbing material act essentially as a net attraction, as long as dilute or moderately dense fluids are considered. This notion is further supported by results for fluids with angle-averaged dipolar fluid matrix interactions as discussed in detail in Ref. [307].

At high fluid densities, the directional dependence of the dipolar fluid–matrix interactions dominates the properties of the adsorbed fluid. Specifically, one observes that even small values of μ_m^2 yield a pronounced decrease of the dielectric constant ϵ_D (see lower plot in Fig. 7.4). This result reflects the decreasing ability of the fluid to respond to an external field. Additional signatures of the disturbance of dielectric properties are the appearance of a blocked part of the dipole dipole correlation function (see inset in lower part of Fig. 7.4) and, more directly, the growth of magnitude and range of the dipolar correlation function between the fluid and the matrix (see Ref. [307] for details). Not surprisingly, these effects become particularly pronounced at low temperatures, where bulk Stockmayer fluids as well as Stockmayer fluids in neutral matrices exhibit an IF phase transition. A clear sign for such an instability is a divergence of the dielectric constant, which is observed only for relatively weak dipolar fluid matrix interactions, i.e., fluid matrix interactions that are significantly weaker than those between fluid particles.

This divergence suggests that larger fields destroy the ferroelectric ordering as one might indeed expect on physical grounds. However, even for the weakly disturbed systems, it seems likely that the frozen dipolar matrix fields influence the structure inside the low-temperature ferroelectric state. Based on work on related systems such as nematic liquid crystals in disordered silica matrices [281–284] (where the silica particles induce local ordering fields), one could imagine that the long-range ferroelectric ordering typical for bulk fluids is replaced by some type of short-range or "quasi"-long-range order characterized by a power-law decay of the correlation functions. In our opinion, a closer investigation of the nature of these low-temperature systems is presently beyond the replica-RHNC approach. This opinion is based on technical reasons because the OZ equations become too complex. In addition, more studies, and in particular simulation data, are required to test and improve the closure approximations under strongly coupled conditions (including a discussion of replica-symmetry breaking [294]).

Appendix A

Mathematical aspects of equilibrium thermodynamics

A.1 The trace of a matrix product

Consider two quadratic $n \times n$ matrices $\mathbf{A}$ and $\mathbf{B}$ with elements a_{ij} and b_{ij}, respectively. The trace of their product is defined as the sum of the diagonal components of matrix $\mathbf{C} = \mathbf{AB}$; that is,

$$
\begin{aligned}
\mathrm{Tr}\,\mathbf{C} &\equiv \sum_{i=1}^{n} C_{ii} = \mathrm{Tr}\,(\mathbf{AB}) \equiv \sum_{i=1}^{n}\sum_{j=1}^{n} a_{ij} b_{ji} = \sum_{j=1}^{n}\sum_{i=1}^{n} b_{ji} a_{ij} \\
&= \sum_{j=1}^{n} (\mathbf{BA})_{jj} \equiv \mathrm{Tr}\,(\mathbf{BA}) = \mathrm{Tr}\,\widetilde{\mathbf{C}}
\end{aligned}
\tag{A.1}
$$

where $\widetilde{\mathbf{C}} \equiv \mathbf{BA}$. Equation (A.1) also proves that, under the trace operation, matrices commute; that is, their order does not matter unlike for the matrix product itself where, of course,

$$
\mathbf{C} \neq \widetilde{\mathbf{C}}
\tag{A.2}
$$

holds in general. Exceptions are special cases where $\mathbf{A}$ and/or $\mathbf{B}$ represent the unit or zero matrices. Commutation of matrices under the trace operation are important in Section 1.5 in the context of Legendre transforms of thermodynamic potentials of confined phase.

A.2 Legendre transformation

A convenient way of switching between various sets of natural variables (see Section 1.5) is provided by the concept of Legendre transformation. We follow here the more comprehensive discussion presented in Chapter 5.2 of the excellent text by Callen [12] to which the interested reader is referred for more detail.

Suppose a function $f(x_1, x_2, \ldots, x_k) = f(\boldsymbol{x})$ exists such that the domain $D(f) \subseteq \mathbb{R}^k$. Hence, the point $(f, \boldsymbol{x})$ lies on a $(k+1)$-dimensional hypersurface (i.e., a surface in a $(k+2)$-dimensional space). The slope of this hypersurface in the ith direction is given by

$$f_i'(\boldsymbol{x}) = \left(\frac{\partial f}{\partial x_i} \right)_{\{\cdot\}\backslash x_i} \tag{A.3}$$

where we introduce the notation "$\{\cdot\}\backslash x_i$" to indicate that except for x_i all other $(k-1)$ variables are being held constant upon differentiating f. The goal then is to use some f_k' to specify f itself without sacrificing any information contained in the original representation of the hypersurface through $\boldsymbol{x}$.

In this venture the only problem is that an infinite number of functions $f(\boldsymbol{x}) + c$ with $c \in \mathbb{R}$ exist, all of which are giving rise to the same f_k'. In other words, the simple representation of f in terms of (some of) the $\{f_k'\}$ is not unique but ambiguous. A unique representation is, however, feasible by realizing that the hypersurface may equally well be represented by either the vector $\boldsymbol{x}$ satisfying the relation $f = f(\boldsymbol{x})$ or the envelope of tangent hyperplanes. The family of tangent hyperplanes may be characterized by the intercept of a new hyperplane g,

$$g(\boldsymbol{f}') = f(\boldsymbol{x}) - \boldsymbol{f}'^{\mathrm{T}}(\boldsymbol{x})\boldsymbol{x} \tag{A.4}$$

where $\boldsymbol{f}'^{\mathrm{T}} = (f_1', f_2', \ldots, f_k')$ is the transpose of the vector $\boldsymbol{f}$. Equation (A.4) is the basic identity of the Legendre transformation. Just like the original function selects a subset of points from $\mathbb{R}^k$, the intercept selects a subset of tangent planes. This equivalence is essentially an expression of duality between the more conventional *point* and the so-called Plücker[1] *line* geometry in multidimensional space.

The relation between f and g is bijective except for a sign difference. To realize the bijectivity, consider the differential

$$\mathrm{d}g = \mathrm{d}f - \boldsymbol{x}\,\mathrm{d}\boldsymbol{f}' - \boldsymbol{f}'\mathrm{d}\boldsymbol{x} = -\boldsymbol{x}\,\mathrm{d}\boldsymbol{f}' \tag{A.5}$$

---

[1]Julius Plücker (1801–1868), Professor of Mathematics and Physics at the Rheinische Friedrich–Wilhelms–Universität Bonn. He developed line geometry, which substitutes the straight line for the point as the basic geometrical unit in space.

because $df = f'dx$. The sign difference follows immediately from Eq. (A.5) because

$$-x_i = \left(\frac{\partial g}{\partial f_i'}\right)_{\{\cdot\}\backslash f_i'} \tag{A.6}$$

gives back the negative of the original set of variables. One should also realize that it is not necessary to replace the entire set of original variables by the new ones. Rather it is possible to choose an arbitrary subset of variables $\{x_{k'}\} \subseteq \{x_k\}$ and perform the Legendre transformation on it.

Last but not least, suppose $f(x_1, \ldots, x_k)$ exists such that f and its first and second partial derivatives are continuous in the neighborhood of a point $P = (x_1, \ldots, x_k) \in \mathbb{R}^k$. Under this presupposition, the theorem of Schwarz holds, which states that for the mixed second-order partial derivatives the order of differentiation is irrelevant; that is,

$$\left[\frac{\partial}{\partial x_i}\left(\frac{\partial f}{\partial x_j}\right)_{\{\cdot\}\backslash x_j}\right]_{\{\cdot\}\backslash x_i} = \left[\frac{\partial}{\partial x_j}\left(\frac{\partial f}{\partial x_i}\right)_{\{\cdot\}\backslash x_i}\right]_{\{\cdot\}\backslash x_j} \tag{A.7}$$

Applying Eq. (A.7) to thermodynamic state functions (instead of a general function f) gives rise to the celebrated Maxwell relations. They can be used to express certain quantities that are hard to measure or control in a laboratory experiment in terms of "mechanical" variables such as a set of stresses and strains and their temperature and density dependence.

A.3 Euler's theorem

When obtaining closed expressions for thermodynamic potentials, the concept of homogeneity of functions is of key importance. Suppose a function $f(x_1, x_2, \ldots, x_n)$ exists such that its domain $D(f) \subseteq \mathbb{R}^n$. This function is called *homogeneous of degree* k if it satisfies the equation

$$f(\lambda x_1, \lambda x_2, \ldots, \lambda x_n) = \lambda^k f(x_1, x_2, \ldots, x_n), \qquad \lambda \in \mathbb{R} > 0 \tag{A.8}$$

Differentiating both sides of Eq. (A.8) with respect to λ, we obtain Euler's theorem,

$$\sum_{i=1}^n \left(\frac{\partial f}{\partial(\lambda x_i)}\right)_{\{\cdot\}\backslash x_i} \frac{\partial(\lambda x_i)}{\partial \lambda} = \sum_{i=1}^n x_i \left(\frac{\partial f}{\partial(\lambda x_i)}\right)_{\{\cdot\}\backslash x_i} = k\lambda^{k-1} f \tag{A.9}$$

which holds for all $k \geq 1$. In particular, for the special case $k = 1$, we obtain from the previous expression

$$\sum_{i=1}^n x_i \left(\frac{\partial f}{\partial(\lambda x_i)}\right)_{\{\cdot\}\backslash x_i} = f(x_1, x_2, \ldots, x_n) \tag{A.10}$$

as a specialized Euler's theorem for homogeneous functions of degree 1.

Appendix B

Mathematical aspects of statistical thermodynamics

B.1 Stirling's approximation

Consider the function $\Gamma(x)$ defined through the expression

$$\Gamma(x) \equiv \int_0^\infty \exp(-t)\, t^{x-1} \mathrm{d}t \tag{B.1}$$

where it is easy to verify that $\Gamma(1) = 1$. It is also straightforward to show by partial integration that

$$\Gamma(x+1) = \int_0^\infty \exp(-t)\, t^x \mathrm{d}t = x\Gamma(x) \tag{B.2}$$

Thus, for $x = N \in \mathbb{N}$, we may write

$$\Gamma(N+1) = N! \tag{B.3}$$

by recursively applying Eq. (B.2).

Consider now the integral

$$N! = \int_0^\infty \exp(-t + t \ln N)\, \mathrm{d}t \equiv \int_0^\infty \exp[-Ng(t)]\, \mathrm{d}t \tag{B.4}$$

where it is clear from Eq. (B.1) the function $g(t)$ must have a maximum for some $t = t_{\max}$ because the integrand consists of a product of a monotonically decreasing [i.e., $\exp(-t)$] and increasing function (i.e., t^N), respectively.

Because of the definition of $g(t)$ given in Eq. (B.4), a straightforward calculation gives $t_{\max} = N$. Expanding $g(t)$ around the maximum at $g(N)$ in a Taylor series and retaining terms only up to second order, we find

$$g(t) \simeq g(N) + \frac{1}{2N^2}(t - N)^2 = 1 - \ln N + \frac{1}{2N^2}(t - N)^2 \qquad (B.5)$$

where the far right side follows from the definition of $g(t)$ in Eq. (B.4). From Eqs. (B.4) and (B.5), we find

$$
\begin{aligned}
N! \ &\simeq \ \exp(N \ln N - N) \int_0^\infty \exp\left(-\frac{(t - N)^2}{2N}\right) dt \\
&= \ \exp(N \ln N - N) \int_{-N}^\infty \exp\left(-\frac{\tilde{t}^2}{2N}\right) d\tilde{t} \\
&= \ \sqrt{2\pi N} \exp(N \ln N - N) \qquad\qquad\qquad (B.6)
\end{aligned}
$$

where we changed variables according to $t \to \tilde{t} = t - N$ between the first and the second lines and took the lower limit of integration to $-\infty$ because the integrand in the second line is extremely peaked around $\tilde{t} = 0$ for large N. Thus, in this limit, we have Stirling's approximation from the previous expression,

$$\ln N! = N \ln N - N + \mathcal{O}(\ln N) \qquad (B.7)$$

where the last term may be neglected in the limit $N \to \infty$. The reader may verify that, even for relatively small $N = 100$ [compared with the thermodynamic limit in statistical physical applications where $N = \mathcal{O}(10^{23})$ or even more], the value of $\ln N!$ calculated from Eq. (B.7) deviates by less than 1% from the exact result, which illustrates the power of Stirling's approximation.

B.2 Elements of function theory

B.2.1 The Cauchy-Riemann differential equations

Consider a complex function

$$
\begin{aligned}
f(z) \ &= \ u(z) + iv(z) & (B.8a) \\
z \ &= \ x + iy, \qquad z \in \mathbb{C}; x, y \in \mathbb{R} & (B.8b)
\end{aligned}
$$

where the real and imaginary parts are given by $u(z) = \mathrm{Re} f(z))$ and $v(z) = \mathrm{Im} f(z)$, respectively. We define the first derivative of $f(z)$ through

$$f'(z) \equiv \lim_{\Delta z \to 0} \frac{f(z + \Delta z) - f(z)}{\Delta z} \qquad (B.9)$$

which is *formally* equivalent to the definition of the first derivative of a real function. Functions $f(z)$ for which the above limit exists at some point z are called *differentiable*; functions for which this limit exists in some area B of the complex plane are called *analytic* or *regular*.

However, because of Eq. (B.8b) there is essentially an infinite number of ways to approach the point z by making Δz smaller. This nonuniqueness of the limit in Eq. (B.9) is a consequence of the fact that one is pursuing a path in the complex *plane* where an infinite degeneracy of such paths exists. For the limit in Eq. (B.9) to exist it is necessary that the specific path $\Delta z \to 0$ in the complex plane be irrelevant.

Let us consider two distinguished paths, one along the real and the other one along the imaginary axis of the complex plane. The first of these is characterized by $\Delta z = \Delta x$ and $\Delta y = 0$ so that we have from Eqs. (B.8b) and (B.9)

$$
\begin{aligned}
f'(z) &\equiv \lim_{\Delta z \to 0} \frac{f(z + \Delta z) - f(z)}{\Delta z} = \lim_{\Delta x \to 0} \frac{f(x + \Delta x + iy) - f(x + iy)}{\Delta x} \\
&= \lim_{\Delta x \to 0} \frac{u(x + \Delta x + iy) + iv(x + \Delta x + iy) - u(x + iy) - iv(x + iy)}{\Delta x} \\
&= \lim_{\Delta x \to 0} \frac{u(x + \Delta x + iy) - u(x + iy)}{\Delta x} \\
&\quad +i \lim_{\Delta x \to 0} \frac{v(x + \Delta x + iy) - v(x + iy)}{\Delta x} \\
&= \frac{\partial u}{\partial x} + i\frac{\partial v}{\partial x}
\end{aligned}
\tag{B.10}
$$

Along the second path in the complex plane we have $\Delta x = 0$ so that $\Delta z = i\Delta y$. By exactly the same reasoning one can then show that in this case

$$
f'(z) = \frac{1}{i}\frac{\partial u}{\partial y} + \frac{\partial v}{\partial y}
\tag{B.11}
$$

According to our supposition the two expressions in Eqs. (B.10) and (B.11) must be equal. For this equality to be reached, the real and imaginary parts of both expressions to be equal which gives rise to the so-called Cauchy-Riemann differential equations

$$
\frac{\partial u}{\partial x} = \frac{\partial v}{\partial y}
\tag{B.12a}
$$

$$
\frac{\partial u}{\partial y} = -\frac{\partial v}{\partial x}
\tag{B.12b}
$$

Theorem B.1 then follows.

Theorem B.1 *A function $f(z)$ defined over a domain $B(f) \subseteq \mathbb{C}$ is analytic if its partial derivatives with respect to x and y exist and the Cauchy–Riemann differential equations are satisfied.*

Proof B.1 *Consider*

$$\Delta f(z) = \Delta u(x, y) + i\Delta v(x, y) \tag{B.13}$$

where

$$\Delta u \equiv u(x + \Delta x, y + \Delta y) - u(x, y) = \frac{\partial u}{\partial x}\Delta x + \frac{\partial u}{\partial y}\Delta y \tag{B.14a}$$

$$\Delta u \equiv v(x + \Delta x, y + \Delta y) - v(x, y) = \frac{\partial v}{\partial x}\Delta x + \frac{\partial v}{\partial y}\Delta y \tag{B.14b}$$

so that

$$
\begin{aligned}
\Delta f(z) &= \left(\frac{\partial u}{\partial x} + i\frac{\partial v}{\partial x}\right)\Delta x + \left(\frac{\partial u}{\partial y} + i\frac{\partial v}{\partial y}\right)\Delta y \\
&= \left(\frac{\partial u}{\partial x} + i\frac{\partial v}{\partial x}\right)\underbrace{(\Delta x + i\Delta y)}_{\Delta z}
\end{aligned}
\tag{B.15}
$$

from which

$$f'(z) = \lim_{\Delta z \to 0} \frac{\Delta f}{\Delta z} = \lim_{\Delta z \to 0} \frac{f(z + \Delta z) - f(z)}{\Delta z} = \frac{\partial u}{\partial x} + i\frac{\partial v}{\partial x} \tag{B.16}$$

follows, which obviously is independent of the specific path $\Delta z \to 0$ in the complex plane, which completes the proof of Theorem B.1. q.e.d.

B.2.2   The method of steepest descent

Let us assume a complex function $f(z)$ exists [see Eqs. (B.8)] such that

1. $f(z)$ is analytic (see Appendix B.2.1) in a domain B of the complex plane.

2. $f(z) \to -\infty$ at the end of a path C, where C pertains to the domain B but is arbitrary because $f(z)$ is supposed to be analytic (see Theorem B.1).

With this function we now seek to evaluate the integral [cf. Eq. (2.22)]

$$\lim_{N \to \infty} I(N) \equiv \lim_{N \to \infty} \int_C dz \exp[Nf(z)] \tag{B.17}$$

Moreover, suppose a point $z = z_0$ exists such that $\mathrm{Re} f(z)$ assumes an extremum and $\mathrm{Im} f(z) = v(x_0, y_0) \equiv v_0 \simeq$ const. Because of the definition of $f(z)$ [see Eqs. (B.8)], this also implies that the integrand in Eq. (B.17) assumes a maximum at the point $z = z_0$. The necessary condition for an extremum of $\mathrm{Re} f(z)$ to exist may be stated more explicitly as

$$\left.\frac{\partial u}{\partial x}\right|_{z=z_0} = \left.\frac{\partial u}{\partial y}\right|_{z=z_0} = 0 \tag{B.18}$$

To find out what is the nature of the extremum, we consider

$$\left.\frac{\partial}{\partial x}\frac{\partial u}{\partial x}\right|_{z=z_0} = \left.\frac{\partial}{\partial x}\frac{\partial v}{\partial y}\right|_{z=z_0} \tag{B.19a}$$

$$\left.\frac{\partial}{\partial y}\frac{\partial u}{\partial y}\right|_{z=z_0} = -\left.\frac{\partial}{\partial y}\frac{\partial v}{\partial x}\right|_{z=z_0} \tag{B.19b}$$

where we obtain the right side by using the Cauchy-Riemann differential equations [see Eqs. (B.12)]. Because the order of differentiation is irrelevant, the previous expressions can be combined to give

$$\left.\frac{\partial^2 u}{\partial x^2}\right|_{z=z_0} + \left.\frac{\partial^2 u}{\partial y^2}\right|_{z=z_0} = 0 \tag{B.20}$$

which tells us that the curvature of the two-dimensional surface $u(x_0, y_0)$ along the real axis is equal to the curvature of $u(x_0, y_0)$ along the imaginary axis. In other words, the point $z = z_0$ is a saddle point.

To evaluate the integral in Eq. (B.17) we now specify the path C according to two criteria, namely

1. C should pass through the saddle point such that $u(x_0, y_0)$ becomes *maximum*.

2. Along C, $v(x_0, y_0) = v_0 \simeq$ const.

These conditions cause C to be the path of "steepest descent" from the saddle point. To achieve this result we need to establish a relation between the real and imaginary parts of $f(z)$.

Theorem B.2 $\mathrm{Re} f(z) = u(x, y)$ and $\mathrm{Im} f(z) = v(x, y)$ are related through

$$\nabla u(x, y) \cdot \nabla v(x, y) = 0 \tag{B.21}$$

Proof B.2 *Using the Cauchy-Riemann differential equations, one realizes that*

$$\frac{\partial u}{\partial x}\frac{\partial v}{\partial x} = \frac{\partial v}{\partial y}\frac{\partial v}{\partial x} = -\frac{\partial v}{\partial y}\frac{\partial u}{\partial y}$$

completing the proof. q.e.d.

Theorem B.3 *Consider a function $g(x, y)$ specifying a two-dimensional surface; then $\nabla g \perp g(x_0, y_0) = const$.*

Proof B.3

$$dg = \frac{\partial g}{\partial x}dx + \frac{\partial g}{\partial y}dy = \nabla g \cdot d\mathbf{r}$$

where $\mathbf{r}^T = (x, y)$ is a two-dimensional vector. For special cuts through the surface g satisfying $g(x_0, y_0) = const$, $dg = 0$, and therefore $\nabla g \cdot d\mathbf{r} = 0$ so that the vectors ∇g and $d\mathbf{r}$ must be orthogonal. q.e.d.

Because of Theorem B.3, ∇v is perpendicular to the line v_0. Theorem B.2, on the other hand, tells us that ∇u and ∇v are orthogonal, so that *any* line $v = const$ must also be *tangential* to ∇u. Thus, lines along which $v = const$ correspond to the steepest descent from the saddle.

Let us now expand $f(z)$ in a Taylor series around $z = z_0$ according to

$$f(z) = f(z_0) + \frac{1}{2}f''(z_0)(z - z_0)^2 + \mathcal{O}\left[(z - z_0)^3\right] \tag{B.22}$$

where we retain terms only up to second order in $z - z_0$ and the linear term is missing because of Eq. (B.18) and our supposition $v = v_0$ (i.e., $\partial v/\partial x = 0$) from which $f'(z_0) = 0$ immediately follows with the help of Theorem B.1. Inserting Eq. (B.22) into Eq. (B.17), we obtain

$$\begin{aligned} I(\mathsf{N}) &\simeq \exp[\mathsf{N}f(z_0)] \int_C dz \exp\left[\frac{1}{2}\mathsf{N}f''(z_0)(z - z_0)^2\right] \\ &= \exp[\mathsf{N}f(z_0)] \int_C dz \exp\left[-\frac{1}{2}\mathsf{N}|f''(z_0)|(z - z_0)^2\right] \end{aligned} \tag{B.23}$$

where the second line follows because $f(z)$ has its maximum at $z = z_0$, $v = v_0 = const$. Therefore $f''(z_0) < 0$ and is a real quantity. Let us also assume that in the immediate vicinity of z_0 the path C can be chosen so that it is parallel to the real axis and then $z - z_0 \approx x - x_0$. Under this assumption and because we are interested in the integral in the limit $N \rightarrow \infty$, we may replace $\int_C \ldots dz \rightarrow \int_{-\infty}^{\infty} \ldots dx$, which then gives

$$
\begin{aligned}
I\left(N\right) &\simeq \exp\left[Nf\left(x_0\right)\right] \int_{-\infty}^{\infty} dx \exp\left[-\frac{1}{2}N\left|f''\left(x_0\right)\right|\left(x-x_0\right)^2\right] \\
&= \exp\left[Nf\left(x_0\right)\right] \sqrt{-\frac{2\pi}{Nf''\left(x_0\right)}}
\end{aligned}
\tag{B.24}
$$

where Eq. (B.102) has also been used.

B.2.3 Gauß's integral theorem in two dimensions

To prove Theorem B.5 in the subsequent Appendix B.2.4, we first need to prove another theorem from Gauß. Suppose we are given two functions $f\left(x,y\right)$ and $g\left(x,y\right)$. We assume both $f\left(x,y\right)$ and $g\left(x,y\right)$ as well as their first derivatives to be continuous in a simply connected domain D where D is bounded by a piece-wise continuous curve C. Specifically we assume D to be represented by the set

$$
D = \left\{\left(x,y\right)\middle| a \leq x \leq b \vee \Psi_1\left(x\right) \leq y \leq \Psi_2\left(x\right)\right\} \subseteq \mathbb{R}^2
\tag{B.25}
$$

such that the contour C of D is described by the functions $\Psi_1\left(x\right)$ and $\Psi_2\left(x\right)$. Alternatively we may specify

$$
D = \left\{\left(x,y\right)\middle| c \leq y \leq d \vee \Phi_1\left(y\right) \leq x \leq \Phi_2\left(y\right)\right\} \subseteq \mathbb{R}^2
\tag{B.26}
$$

as indicated in Fig. B.1. Theorem B.4 by Gauß then asserts the following.

Theorem B.4

$$
\iint_D dxdy\left[\frac{\partial f\left(x,y\right)}{\partial x} + \frac{\partial g\left(x,y\right)}{\partial y}\right] = \int_C \left[f\left(x,y\right)dy - g\left(x,y\right)dx\right]
\tag{B.27}
$$

where it is important that the contour C is traversed in a counterclockwise fashion as indicated in Fig. B.1. Gauß's integral theorem can be proved as follows.

Proof B.4 *Consider first the integral [see Eq. (B.25)]*

$$
\iint_D dxdy\frac{\partial g\left(x,y\right)}{\partial y} = \int_a^b dx \int_{\Psi_1\left(x\right)}^{\Psi_2\left(x\right)} dy\frac{\partial g\left(x,y\right)}{\partial y}
\tag{B.28}
$$

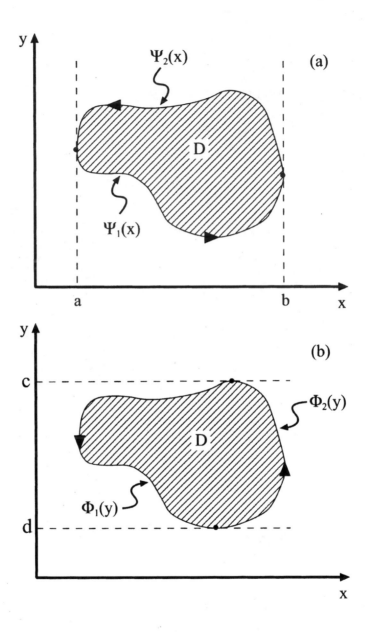

Figure B.1: Sketch of a simply connected domain D bounded by a piece-wise continuous curve C. The contour C is traversed in a counterclockwise fashion as indicated by the arrows. In (a) the contour is described by the functions $\Psi_1(x)$ and $\Psi_2(x)$ in the interval $\{x \mid a \leq x \leq b\}$, whereas in (b) the contour is specified alternatively by the functions $\Phi_1(y)$ and $\Phi_2(y)$ defined on the interval $\{y \mid c \leq y \leq d\}$.

Noticing that the function $g(x,y)$ is the antiderivative of the integrand on the right side of the previous expression, we have

$$\int_a^b dx \int_{\Psi_1(x)}^{\Psi_2(x)} dy \frac{\partial g(x,y)}{\partial y} = \int_a^b dx \left\{ g\left[x, \Psi_2(x)\right] - g\left[x, \Psi_1(x)\right] \right\}$$

$$= \int_a^b dx\, g\left[x, \Psi_2(x)\right] + \int_b^a dx\, g\left[x, \Psi_1(x)\right] \quad \text{(B.29)}$$

after interchanging the limits of integration on the second integral. Because the functions $\Psi_1(x)$ and $\Psi_2(x)$ describing the contour C appear as arguments of g, it is clear that the two conventional integrals appearing on the right side of the previous expression are equivalent to the line integral along C except for a sign difference because C is traversed in a clockwise rather than a counterclockwise fashion (see Fig. B.1). Hence, we have

$$\iint_D dx dy \frac{\partial g(x,y)}{\partial y} = -\int_C dx\, g(x,y) \quad \text{(B.30)}$$

By the same token we may write

$$\iint_D dx dy \frac{\partial f(x,y)}{\partial x} = \int_c^d dy\, f\left[\Phi_2(y), y\right] + \int_d^c dy\, f\left[\Phi_1(y), y\right] \quad \text{(B.31)}$$

because $f(x,y)$ is the antiderivative of $\partial f/\partial x$. As before the two conventional integrals appearing on the right side of the previous expression represent the line integral along C. However, in this case C is traversed in a counterclockwise fashion (see Fig. B.1) in accord with our original supposition, so that we may write

$$\iint_D dx dy \frac{\partial f(x,y)}{\partial x} = \int_C dy\, f(x,y) \quad \text{(B.32)}$$

Putting together Eqs. (B.30) and (B.32) yields Eq. (B.27) which completes the proof of Theorem B.4. q.e.d.

We note in passing that Gauß's theorem applies to domains in arbitrary dimensions. For example, in electrostatic or hydrodynamic problems, one is frequently confronted with the change of charge or mass density inside a three-dimensional volume. Using Gauß's theorem this change is equivalent to the net flux through the surface of this three-dimensional volume.

B.2.4   Cauchy integrals and the Laurent series

Consider a complex function $f(z)$ as defined in Eqs. (B.8).

Theorem B.5 *If $f(z)$ is analytic in some simply connected domain D, the integral over $f(z)$ along some closed path C pertaining to D vanishes.*

Proof B.5

$$\oint dz f(z) = \oint dz \left[u(x,y) + iv(x,y) \right] (dx + idy)$$

$$= \oint dx u(x,y) - \oint dy v(x,y) + i \oint dy u(x,y) + i \oint dx v(x,y)$$

Using Gauß's theorem [see Eq. (B.27)] we may rewrite the original integral as

$$\oint dz f(z) = - \iint\limits_{C} dx dy \left(\frac{\partial u}{\partial y} + \frac{\partial v}{\partial x} \right) + i \iint\limits_{C} dx dy \left(\frac{\partial u}{\partial x} - \frac{\partial v}{\partial y} \right)$$

As $f(z)$ is analytic, the Cauchy-Riemann differential equations are satisfied so that each of the two integrals above vanishes identically regardless of the specific choice of the path C. q.e.d.

Consider now a domain D, which is no longer simply connected. Such a situation arises if $f(z)$ is analytic everywhere in D except at some point $z = z_1$ where $f(z)$ is supposed to have a singularity (see Fig. B.2). Then, in fact, the integral around any closed path in D surrounding the singularity does not vanish but one may instead define the so-called residue

$$\text{Res} f(z_1) = \frac{1}{2\pi i} \oint dz f(z) \tag{B.33}$$

which vanishes in the absence of such a singularity according to Theorem B.5. However, similar to the proof of Theorem B.5 it can be shown that the precise path along which the residue is calculated is irrelevant for its value.

Suppose now $f(z)$ is analytic across a simply connected domain; then it is immediately clear that, if we pick a point $z = z_0$ in that domain, the quantity $f(z)/(z - z_0)$ will have a singularity at that point. Because of the above, the integral over $f(z)/(z - z_0)$ along any closed path surrounding $z = z_0$ will have some nonzero value that we seek to calculate. Because the closed path C surrounding $z = z_0$ is arbitrary, we take it to be a circle of

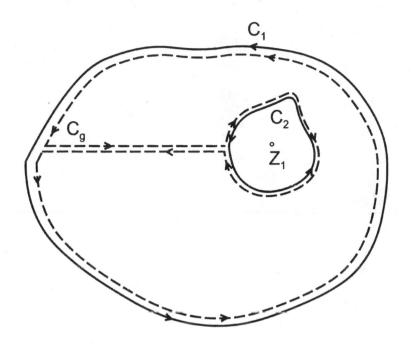

Figure B.2: Sketch of closed contours C_1, C_2, and C_g in the complex plane surrounding a singularity at z_1. Arrows indicate the direction in which the paths along the contours C_1, C_2, and C_g are traversed.

radius R around that point; that is, $z = z_0 + R \exp(it)$ or $\mathrm{d}z = iR \exp(it)\,\mathrm{d}t$. Thus, we obtain

$$
\oint \mathrm{d}z \frac{f(z)}{z - z_0} = \int_0^{2\pi} \mathrm{d}t \frac{f[z_0 + R \exp(it)]}{R \exp(it)} iR \exp(it)
$$

$$
= i \int_0^{2\pi} \mathrm{d}t\, f[z_0 + R \exp(it)] \overset{R=0}{=} 2\pi i f(z_0) \tag{B.34}
$$

where we set $R = 0$ in the last step because the path C surrounding $z = z_0$ is arbitrary and may therefore be taken to be a circle of radius $R = 0$ without loss of generality. With the transformation $z_0 \to z$ and $z_0 \to \zeta$ we recover Cauchy's integral formula from the previous expression,

$$
f(z) = \frac{1}{2\pi i} \oint \mathrm{d}\zeta \frac{f(\zeta)}{\zeta - z} \tag{B.35}
$$

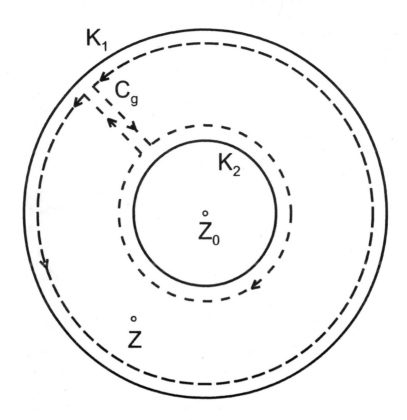

Figure B.3: Two circles K_1 and K_2 enclosing a point $z = z_0$. Also shown is a closed contour C_g in the complex plane where the arrows indicate the direction in which the path along the contour is traversed; z is a point enclosed by C_g.

Consider now the Cauchy integral (see Fig. B.3)

$$f(z) = \frac{1}{2\pi i}\oint_{C_g}d\zeta\frac{f(\zeta)}{\zeta - z} = \frac{1}{2\pi i}\left[\oint_{K_1}d\zeta\frac{f(\zeta)}{\zeta - z} - \oint_{K_1}d\zeta\frac{f(\zeta)}{\zeta - z}\right] \qquad (B.36)$$

where $f(z)$ is analytic everywhere in the domain surrounded by the curve C_g and K_1 and K_2 are two two circles as indicated in Fig. B.3. Outside the domain surrounded by the curve C_g and inside the circle K_2, $f(z)$ is supposed to have a singularity at a point $z = z_0$. The negative sign in front of the second integral is a consequence of the fact that along the second circle the integration is performed in a clockwise fashion. Because the direction along the paths along C_g connecting the circles K_1 and K_2 are traversed in opposite directions, one can argue that contributions to the integral along C_g from the connection between K_1 and K_2 vanish.

The first integral in Eq. (B.36) can be rewritten as

$$\oint_{K_1} d\zeta \frac{f(\zeta)}{\zeta - z} = \oint_{K_1} d\zeta \frac{f(\zeta)}{\zeta - z_0} \frac{1}{1 - \frac{z - z_0}{\zeta - z_0}} = \sum_{n=0}^{\infty} (z - z_0)^n \oint_{K_1} d\zeta \frac{f(\zeta)}{(\zeta - z_0)^{n+1}} \quad (B.37)$$

where we used the fact that

$$\sum_{n=0}^{\infty} x^n = \frac{1}{1 - x}, \qquad |x| < 1 \quad (B.38)$$

as one can verify by expanding the right side of Eq. (B.38) in a MacLaurin series. In a similar fashion we can rewrite the second integral in Eq. (B.36) as

$$
\begin{aligned}
-\oint_{K_2} d\zeta \frac{f(\zeta)}{\zeta - z} &= \oint_{K_2} d\zeta \frac{f(\zeta)}{z - z_0} \frac{1}{1 - \frac{\zeta - z_0}{z - z_0}} \\
&= \sum_{n=1}^{\infty} \frac{1}{(z - z_0)^n} \oint_{K_2} d\zeta f(\zeta) (\zeta - z_0)^{n-1} \\
&\stackrel{n'=-n}{=} \sum_{n'=-\infty}^{-1} (z - z_0)^{n'} \oint_{K_2} d\zeta \frac{f(\zeta)}{(\zeta - z_0)^{n'+1}} \\
&= \sum_{n=-\infty}^{-1} (z - z_0)^n \oint_{K_1} d\zeta \frac{f(\zeta)}{(\zeta - z_0)^{n+1}} \quad (B.39)
\end{aligned}
$$

where we used the fact that the curve surrounding the singularity is arbitrary and may therefore coincide with the curve K_1 in Eq. (B.37) without loss of generality. The two expressions in Eqs. (B.37) and (B.39) may thus be combined to the so-called Laurent series expansion

$$f(z) = \sum_{n=-\infty}^{\infty} a_n (z - z_0)^n \quad (B.40a)$$

$$a_n = \frac{1}{2\pi i} \oint d\zeta \frac{f(\zeta)}{(\zeta - z_0)^{n+1}} \quad (B.40b)$$

The Laurent expansion is very useful in analyzing the nature of singularities. However, a discussion of this aspect would go way beyond the scope of this Appendix. Therefore, we emphasize only the relation of the Laurent series to the residue introduced in Eq. (B.33). From the coefficients of the Laurent expansion, it is immediately obvious that

$$a_{-1} = \frac{1}{2\pi i} \oint d\zeta f(\zeta) \tag{B.41}$$

is identical with the expression in Eq. (B.33).

B.3   Lagrangian multipliers

The problem of finding extrema of a function $f(x_1, x_2, \ldots, x_m) \equiv f(\boldsymbol{x})$ subject to n constraints may be solved by using the method of Lagrangian multipliers. In the absence of such constraints the necessary condition for the existence of extrema may be stated as

$$df = \sum_{i=1}^{m} \left(\frac{\partial f}{\partial x_i} \right)_{\{\cdot\}\backslash x_i} dx_i = \boldsymbol{f}' \cdot d\boldsymbol{x} = 0 \tag{B.42}$$

from which

$$\boldsymbol{f}' = \boldsymbol{0} \tag{B.43}$$

follows because $d\boldsymbol{x}$ is arbitrary. If a set of n constraints specified by

$$\varphi_j(\boldsymbol{x}) = 0, \qquad j = 1, \ldots, n \tag{B.44}$$

has to be satisfied simultaneously, the variables $\{x_i\}$ are not independent anymore and Eq. (B.43) is no longer the correct necessary condition for the existence of these extrema.

However, consider a new function

$$F(\boldsymbol{x}) \equiv f(\boldsymbol{x}) + \underbrace{\sum_{j=1}^{n} \lambda_j \varphi_j(\boldsymbol{x})}_{=0, \text{ see Eq. (B.44)}} \tag{B.45}$$

where $\{\lambda_j\}$ is a set of n undetermined Lagrangian multipliers. The necessary condition for the existence of an extremum of F may be stated as

$$dF(\boldsymbol{x}) = \sum_{i=1}^{m} \left[\left(\frac{\partial f}{\partial x_i} \right)_{\{\cdot\}\backslash x_i} + \sum_{j=1}^{n} \lambda_j \left(\frac{\partial \varphi_j}{\partial x_i} \right)_{\{\cdot\}\backslash x_i} \right] dx_i = 0 \tag{B.46}$$

Notice that zeros of dF in Eq. (B.46) immediately satisfy the constraints specified in Eq. (B.44) unlike its counterpart df in Eq. (B.42). If we now specify the Lagrangian multipliers such that the term in brackets [...] vanishes identically, $d\boldsymbol{x}$ may assume arbitrary values and the necessary condition for the existence of an extremum can be restated as

$$\boldsymbol{\varphi}'\boldsymbol{\lambda} = -\boldsymbol{f}' \tag{B.47}$$

where the matrix $\boldsymbol{\varphi}'$ is given by

$$\boldsymbol{\varphi}' = \begin{pmatrix} (\partial\varphi_1/\partial x_1)_{\{\cdot\}\backslash x_1} & (\partial\varphi_2/\partial x_1)_{\{\cdot\}\backslash x_1} & \cdots & (\partial\varphi_n/\partial x_1)_{\{\cdot\}\backslash x_1} \\ (\partial\varphi_1/\partial x_2)_{\{\cdot\}\backslash x_2} & (\partial\varphi_2/\partial x_2)_{\{\cdot\}\backslash x_2} & \cdots & (\partial\varphi_n/\partial x_2)_{\{\cdot\}\backslash x_2} \\ \vdots & & \ddots & \vdots \\ (\partial\varphi_1/\partial x_m)_{\{\cdot\}\backslash x_m} & (\partial\varphi_2/\partial x_m)_{\{\cdot\}\backslash x_m} & \cdots & (\partial\varphi_n/\partial x_m)_{\{\cdot\}\backslash x_m} \end{pmatrix} \tag{B.48}$$

and $\boldsymbol{\lambda}^{\mathrm{T}} \equiv (\lambda_1, \lambda_2, \ldots, \lambda_n)$ is (the transpose of) a vector of n elements.

Whether the set of linear equations given in Eq. (B.47) has a solution (vector) $\boldsymbol{\lambda}$ depends on the ranks r and R of the functional matrices $\boldsymbol{\varphi}'$ and $\tilde{\boldsymbol{\varphi}}'$, respectively, which must satisfy

$$r = R \tag{B.49}$$

for such a solution to exist. Matrix $\tilde{\boldsymbol{\varphi}}'$ is defined as

$$\tilde{\boldsymbol{\varphi}}' \equiv \begin{pmatrix} (\partial\varphi_1/\partial x_1)_{\{\cdot\}\backslash x_1} & \cdots & (\partial\varphi_n/\partial x_1)_{\{\cdot\}\backslash x_1} & -(\partial f/\partial x_1)_{\{\cdot\}\backslash x_1} \\ (\partial\varphi_1/\partial x_2)_{\{\cdot\}\backslash x_2} & \cdots & (\partial\varphi_n/\partial x_2)_{\{\cdot\}\backslash x_2} & -(\partial f/\partial x_2)_{\{\cdot\}\backslash x_2} \\ \vdots & \ddots & & \vdots \\ (\partial\varphi_1/\partial x_m)_{\{\cdot\}\backslash x_m} & \cdots & (\partial\varphi_n/\partial x_m)_{\{\cdot\}\backslash x_m} & -(\partial f/\partial x_m)_{\{\cdot\}\backslash x_m} \end{pmatrix} \tag{B.50}$$

which we obtain by adding the vector $-\boldsymbol{f}$ as the rightmost column to the original matrix $\boldsymbol{\varphi}'$. Hence, as $\boldsymbol{\varphi}'$ is an $m \times n$ matrix, $\tilde{\boldsymbol{\varphi}}'$ is an $m \times (n+1)$ matrix. The latter can therefore either have rank $R = r + 1$ or $R = r$. In the former case it would be possible to form at least one nonvanishing subdeterminant of order $(n + 1) \times (n + 1)$ of the matrix $\tilde{\boldsymbol{\varphi}}'$ (assuming $m > n$). Hence, one needs to verify that this is impossible.

That $R = r$ for matrix $\tilde{\boldsymbol{\varphi}}'$ can be demonstrated if one employs the following properties of determinants, which we summarize without proof as follows:

1. Interchanging rows and columns leaves a determinant unaltered.

2. Multiplying a determinant by some number λ is equivalent to multiplying all elements in one of its rows (columns) by that number.

3. A determinant remains unchanged if one multiplies one of its rows (columns) by some number λ and adds it to another one.

4. A determinant vanishes if any two rows (columns) are identical.

Because of statement 1, all others hold for rows as well as for columns of a determinant.

Consider now the $(n+1) \times (n+1)$ subdeterminant of matrix $\widetilde{\varphi}'$ and multiply it by a product of numbers according to

$$
\prod_{j=1}^{n} \lambda_j
\begin{vmatrix}
\partial\varphi_1/\partial x_k & \cdots & \partial\varphi_n/\partial x_k & -\partial f/\partial x_k \\
\partial\varphi_1/\partial x_{k+1} & \cdots & \partial\varphi_n/\partial x_{k+1} & -\partial f/\partial x_{k+1} \\
\vdots & \ddots & & \vdots \\
\partial\varphi_1/\partial x_{k+n} & \cdots & \partial\varphi_n/\partial x_{k+n} & -\partial f/\partial x_{k+n} \\
\partial\varphi_1/\partial x_{k+n+1} & \cdots & \partial\varphi_n/\partial x_{k+n+1} & -\partial f/\partial x_{k+n+1}
\end{vmatrix}
$$

$$
=
\begin{vmatrix}
\lambda_1\left(\partial\varphi_1/\partial x_k\right) & \cdots & \lambda_n\left(\partial\varphi_n/\partial x_k\right) & -\partial f/\partial x_k \\
\lambda_1\left(\partial\varphi_1/\partial x_{k+1}\right) & \cdots & \lambda_n\left(\partial\varphi_n/\partial x_{k+1}\right) & -\partial f/\partial x_{k+1} \\
\vdots & \ddots & & \vdots \\
\lambda_1\left(\partial\varphi_1/\partial x_{k+n}\right) & \cdots & \lambda_n\left(\partial\varphi_n/\partial x_{k+n}\right) & -\partial f/\partial x_{k+n} \\
\lambda_1\left(\partial\varphi_1/\partial x_{k+n+1}\right) & \cdots & \lambda_n\left(\partial\varphi_n/\partial x_{k+n+1}\right) & -\partial f/\partial x_{k+n+1}
\end{vmatrix}
$$

$$\tag{B.51}$$

which follows with the aid of statement 2 above. From statement 3, it is then clear that if we add columns 1 to $n-1$ to the nth column the determinant remains unchanged. The nth column then contains elements

$$
\sum_{j=1}^{n} \lambda_j \left(\frac{\partial\varphi_j}{\partial x_l}\right)_{\{\cdot\}\backslash x_l}, \qquad l = k, \ldots, k+n+1
$$

However, because of Eq. (B.47) the elements of the nth column are identical to the ones forming column $n+1$ regardless of k such that all $(n+1) \times (n+1)$ subdeterminants of $\widetilde{\varphi}'$ vanish on account of statement 4 of our list of elementary properties of determinants. This, in turn, proves that the rank of matrix $\widetilde{\varphi}'$ is equal to the rank of matrix φ'.

If, in addition, one can show that $R = r = n$, a unique solution of Eq. (B.47) exists for the set of Lagrangian multipliers $\{\lambda_j\}$. However, whether the rank of the functional matrix φ' equals the number of constraints depends on the specific problem and needs to be investigated separately for each case under consideration.

B.4   Maximum term method

Consider the sum

$$S = \sum_{i=1}^{M} s_i \tag{B.52}$$

where we assume that all summands $s_i > 0$. Suppose a term $s_{\max}$ exists. Then the inequality

$$s_{\max} < S < M s_{\max} \tag{B.53}$$

must hold. Let now

$$s_{\max} = \mathcal{O}\left(e^M\right) \tag{B.54}$$

so that the inequality in Eq. (B.53) may be rephrased as

$$\mathcal{O}\left(M\right) < \ln S < \mathcal{O}\left(M\right) + \ln M \tag{B.55}$$

by taking the (natural) logarithm of all terms appearing in Eq. (B.53). Hence, in the limit $M \to \infty$, $\ln M$ becomes negligible compared with $\mathcal{O}\left(M\right)$. Therefore, in this limit, $\ln S$ is bounded from above and below by the same number $\mathcal{O}\left(M\right)$. Hence, we find

$$\ln S = \ln s_{\max} \tag{B.56}$$

meaning that the sum in Eq. (B.52) can be replaced by its maximum term.

B.5   Basis sets and the canonical ensemble partition function

B.5.1   Parseval's equation

Suppose we are given a complex function $f(x)$ defined on an interval $|a, b|$ and an orthonormal set of functions $\{\varphi_i(x)\}$ where $x \in \mathbb{R}$ is real. We may then expand $f(x)$ in terms of the functions $\{\varphi_i(x)\}$ according to

$$f(x) = \sum_{n=1}^{\infty} c_n \varphi_n(x) \tag{B.57}$$

where the expansion coefficients $\{c_n\} \in \mathbb{C}$ in general. From Eq. (B.57) we immediately have

$$\langle \varphi_m \mid f \rangle = \left\langle \varphi_m \left| \sum_{n=1}^{\infty} c_n \varphi_n \right. \right\rangle = \sum_{n=1}^{\infty} c_n \langle \varphi_m \mid \varphi_n \rangle \tag{B.58}$$

where we define the scalar product through

$$\langle g \,|\, h \rangle \equiv \int_a^b g^* (x) \, h (x) \, \mathrm{d}x \qquad (B.59)$$

and the asterisk is used to indicate the complex conjugate of the function $g(x)$. Because the functions $\{\varphi_i(x)\}$ are orthonormal, that is,

$$\langle \varphi_m \,|\, \varphi_n \rangle = \delta_{mn} \qquad (B.60)$$

where

$$\delta_{mn} \equiv \begin{cases} 0, & m \neq n \\ 1, & m = n \end{cases} \qquad (B.61)$$

is the Kronecker symbol, we have from Eq. (B.58)

$$c_n = \langle \varphi_n \,|\, f \rangle \qquad (B.62)$$

Consider next the function

$$g (x) \equiv f (x) - \sum_{n=0}^{N} c_n \varphi_n (x) \qquad (B.63)$$

and compute

$$\begin{aligned}
\vartheta &= \langle g \,|\, g \rangle \\
&= \langle f \,|\, f \rangle - \sum_{n=1}^{N} c_n^* \langle \varphi_n \,|\, f \rangle - \sum_{n=1}^{N} c_n \langle f \,|\, \varphi_n \rangle + \sum_{n=1}^{N} \sum_{m=1}^{N} c_n^* c_m \langle \varphi_n \,|\, \varphi_m \rangle \\
&= \langle f \,|\, f \rangle - \sum_{n=1}^{N} c_n^* c_n - \sum_{n=1}^{N} c_n c_n^* + \sum_{n=1}^{N} c_n^* c_n \\
&= \langle f \,|\, f \rangle - \sum_{n=1}^{N} c_n^* c_n \geq 0 \qquad (B.64)
\end{aligned}$$

In the derivation of Eq. (B.64) we used Eqs. (B.62), (B.60), and the fact that [see Eq. (B.62)] $c_n^* = \langle f \,|\, \varphi_n \rangle$. In the limit $N \to \infty$, Eq. (B.64) constitutes Bessel's inequality. If the equality holds in Eq. (B.64), we obtain Parseval's equation

$$\sum_n c_n^* c_n = \langle f \,|\, f \rangle \qquad (B.65)$$

A special case of Eq. (B.65) is obtained if the function f itself is normalized to 1; that is,

$$\sum_n c_n^* c_n = 1 \qquad (B.66)$$

B.5.2 Proof of Eq. (2.41)

To prove the validity of Eq. (2.41) we assume that the expansion of eigenfunctions in terms of a complete orthonormal basis exists [see Eq. (2.40)] so that we have

$$a_{ji} = \langle \psi_i | \phi_j \rangle \tag{B.67}$$

from Eq. (B.62). Let us furthermore assume that in addition to Eq. (2.40) the expansion

$$\psi_k = \sum_l b_{kl} \phi_l \tag{B.68}$$

is also possible where $\{b_{kl}\}$ is a set of constants similar to the expansion coefficients $\{a_{ji}\}$ in Eq. (2.40). From the discussion in Appendix B.5.1, it is immediately clear that

$$b_{kl} = \langle \phi_l | \psi_k \rangle = a_{lk}^* \tag{B.69}$$

where the far right side follows with the aid of Eq. (B.67) and from the fact that the functions $\{\phi_l\}$ form a complete orthonormal basis. Equation (B.69) shows that the coefficients $\{b_{kl}\}$ are related to the set of expansion coefficients $\{a_{kl}\}$ by taking the complex conjugate and reversing the order of the indices. Hence, it follows from Eq. (B.66) that

$$\sum_l b_{kl}^* b_{kl} = \sum_l a_{lk}^* a_{lk} = 1 \tag{B.70}$$

We now employ the expansion in Eq. (2.40) to replace the set of basis functions $\{\phi_i\}$ on the far right side of Eq. (2.41) so that

$$\left\langle \phi_i \left| \exp\left(-\lambda_2 \widehat{H}\right) \right| \phi_i \right\rangle = \sum_k \sum_l a_{ik}^* a_{il} \left\langle \psi_k \left| \exp\left(-\lambda_2 \widehat{H}\right) \right| \psi_l \right\rangle \tag{B.71}$$

Because the functions $\{\psi_l\}$ are eigenfunctions of $\widehat{H}$, we have [see Eq. (2.39)]

$$\exp\left(-\lambda_2 \widehat{H}\right) |\psi_l\rangle = \exp\left(-\lambda_2 E_l\right) |\psi_l\rangle \tag{B.72}$$

and therefore

$$\sum_k \sum_l a_{ik}^* a_{il} \left\langle \psi_k \left| \exp\left(-\lambda_2 \widehat{H}\right) \right| \psi_l \right\rangle = \sum_k \sum_l a_{ik}^* a_{il} \exp\left(-\lambda_2 E_l\right) \underbrace{\langle \psi_k | \psi_l \rangle}_{\delta_{kl}}$$

$$= \sum_l a_{il}^* a_{il} \exp\left(-\lambda_2 E_l\right) \tag{B.73}$$

Using this result and summing both sides of Eq. (B.71) over i, we find

$$\sum_i \left\langle \phi_i \left| \exp\left(-\lambda_2 \widehat{H}\right) \right| \phi_i \right\rangle = \sum_l \exp\left(-\lambda_2 E_l\right) \underbrace{\sum_i a_{il}^* a_{il}}_{=1}$$

$$= \sum_i \exp\left(-\lambda_2 E_i\right) = \mathcal{Q} \qquad (B.74)$$

where we changed the summation index on the second line, used Eq. (B.70), and the definition of $\mathcal{Q}$ in Eq. (2.38). Comparing Eq. (B.74) with Eq. (2.39) completes the proof of Eq. (2.41).

B.6 The classic limit of quantum statistics

B.6.1 The Dirac δ-function

To prove Eq. (2.97) in the subsequent Appendix B.6.2 we first need to introduce a "function"

$$\delta\left(x - b\right) \equiv \lim_{\alpha \to \infty} \sqrt{\frac{\alpha}{\pi}} \exp\left[-\alpha\left(x - b\right)^2\right] \qquad (B.75)$$

known as the Dirac δ-function[1] which we introduce here through the limiting process applied to Gaussian distributions where we realize from Eqs. (B.75) and (B.102) that

$$\int_{-\infty}^{\infty} dx\, \delta\left(x - b\right) = 1, \qquad \forall \alpha \qquad (B.76)$$

Note that $\delta\left(x - b\right)$ is not really a function in the ordinary sense because

$$\delta\left(x - b\right) = \begin{cases} 0, & x \neq b \\ \infty & x = b \end{cases} \qquad (B.77)$$

that is, the value of $\delta\left(x - b\right)$ is infinity and changes discontinuously at $x = b$. However, because of Eq. (B.76) the integral

$$\int_{-\infty}^{\infty} dx\, f\left(x\right) \delta\left(x - b\right) = \int_{b-\epsilon}^{b+\epsilon} dx\, f\left(x\right) \delta\left(x - b\right) \qquad (B.78)$$

---

[1]The "function" $\delta\left(x - b\right)$ was named after the English physicist Paul Adrien Maurice Dirac (1901–1984) in honor of his contributions to the development of quantum mechanics.

is a well-defined mathematical object where $\epsilon \ll 1$ is a vanishingly small quantity. We may change the limits of integration on the right side of Eq. (B.78) because $\delta(x - b)$ is nonzero only over a vanishingly small interval around b.

We may now argue on the basis of Fig. B.4 that the inequality

$$m(b - a) \leq \int_a^b dx\, f(x) \leq M(b - a) \tag{B.79}$$

will always be satisfied if $M = \max_{a \leq x \leq b} f(x)$ and $m = \min_{a \leq x \leq b} f(x)$ are the values of $f(x)$ at the *absolute* maximum and minimum in the interval $[a, b]$, respectively. Hence, some $\xi \in [a, b]$ will exist such that

$$\int_a^b dx\, f(x) = f(\xi) \int_a^b dx = f(\xi)(b - a) \tag{B.80}$$

represented by the hatched area in Fig. B.4.

Applying this logic to the integral on the right side of Eq. (B.78) we may write

$$\int_{b-\epsilon}^{b+\epsilon} dx\, f(x)\, \delta(x - b) = f(\xi) \underbrace{\int_{b-\epsilon}^{b+\epsilon} dx\, \delta(x - b)}_{=1} = f(\xi) \tag{B.81}$$

because of Eq. (B.76). Hence, taking the limit $\epsilon \to 0$ it follows from the previous expression and Eq. (B.78) that

$$\int_{-\infty}^{\infty} dx\, f(x)\, \delta(x - b) = f(b) \tag{B.82}$$

after replacing the integration limits by $\pm\infty$, which we may do on account of the "sharpness" of $\delta(x - b)$ [i.e., because of Eq. (B.77)]. By analogous reasoning it follows that in three dimensions

$$\int d\boldsymbol{r} f(\boldsymbol{r})\, \delta(\boldsymbol{r} - \boldsymbol{r}') = f(\boldsymbol{r}') \tag{B.83}$$

holds where in Cartesian coordinates

$$\delta(\boldsymbol{r} - \boldsymbol{r}') = \delta(x - x')\, \delta(y - y')\, \delta(z - z') \tag{B.84}$$

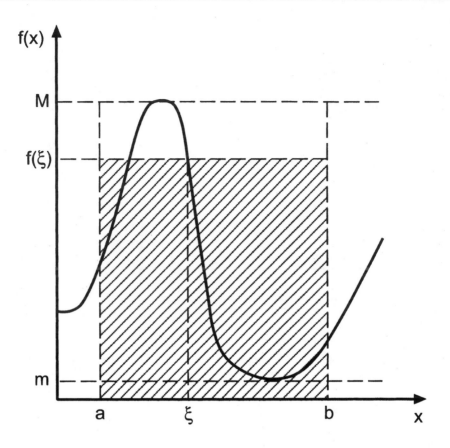

Figure B.4: Sketch of a nonmonotonic function $f(x)$ having an absolute maximum and minimum of height M and m, respectively. The hatched region corresponds to the area $f(\xi)(b-a)$.

An interesting relation is obtained by considering the Fourier transform of $f(r)$. By analogy with Eqs. (2.84) and (2.87) we may write

$$\widehat{f}(k) = \frac{1}{(2\pi)^3} \int dr' f(r') \exp(-ik \cdot r') \qquad \text{(B.85a)}$$

$$f(r') = \int dk \widehat{f}(k) \exp(ik \cdot r') \qquad \text{(B.85b)}$$

where the first equation constitutes the Fourier transform of $f(r)$ and the second one the inverse transformation. Inserting Eq. (B.85b) into Eq. (B.85a) we obtain

$$
\begin{aligned}
f(r') &= \frac{1}{(2\pi)^3} \iint dk dr f(r) \exp(-ik \cdot r) \exp(ik \cdot r') \\
&= \frac{1}{(2\pi)^3} \int dr f(r) \left\{ \int dk \exp[-ik \cdot (r - r')] \right\} \quad \text{(B.86)}
\end{aligned}
$$

Comparing the second line of the previous expression with Eq. (B.83) it is clear that

$$
\delta(r - r') = \frac{1}{(2\pi)^3} \int dk \exp[-ik \cdot (r - r')] \quad \text{(B.87)}
$$

is the Fourier representation of the Dirac δ-function.

B.6.2 Proof of the completeness relation

Let us now expand a complex function $f(r)$ in terms of some orthonormal set of functions $\{\varphi_i(r)\}$ [cf. Appendix B.5.1] according to

$$
f(r') = \sum_{i=1}^{\infty} c_i \varphi_i(r') \quad \text{(B.88)}
$$

where $\{c_i\} \in \mathbb{C}$ as in Eq. (B.57). As the basis $\{\varphi_i(r)\}$ is supposed to consist of orthonormal functions

$$
\int dr' \varphi_j^*(r') f(r') = \sum_{i=1}^{\infty} c_i \int dr' \varphi_j^*(r') \varphi_i(r') = \sum_{i=1}^{\infty} c_i \delta_{ij} = c_j \quad \text{(B.89)}
$$

where δ_{ij} denotes the Kronecker symbol. Upon substituting Eq. (B.89) back into the original expansion in Eq. (B.88), we obtain

$$
f(r') = \sum_{i=1}^{\infty} \left[\int dr \varphi_i^*(r) f(r) \right] \varphi_i(r') = \int dr \left[\sum_{i=1}^{\infty} \varphi_i^*(r) \varphi_i(r') \right] f(r)
$$
$$
\text{(B.90)}
$$

Comparing the right side of this expression with Eq. (B.82), we conclude that

$$
\delta(r - r') = \sum_{i=1}^{\infty} \varphi_i^*(r) \varphi_i(r') \quad \text{(B.91)}
$$

thus proving Eq. (2.97).

Finally, take as a specific example the eigenfunctions of the momentum operator (see Section 2.5) as a basis; that is

$$
\varphi_i(r') = \exp(ik_i \cdot r') \quad \text{(B.92)}
$$

where $\boldsymbol{k}_i$ is again taken to be the wave vector associated with the momentum of particle i. It then follows from Eq. (B.91) that

$$\delta\left(\boldsymbol{r} - \boldsymbol{r}'\right) = \sum_{i=1}^{\infty} \exp\left[-i\boldsymbol{k}_i \cdot \left(\boldsymbol{r} - \boldsymbol{r}'\right)\right] \tag{B.93}$$

Replacing in this expression the summation by an integration according to

$$\sum_{i=1}^{\infty} \ldots \rightarrow \frac{1}{(2\pi)^3} \int d\boldsymbol{k} \ldots$$

we recover the Fourier representation of the Dirac δ-function given above in Eq. (B.87). However, the reader should appreciate the fact that Eq. (B.91) is far more general than Eq. (B.93) because it holds for an *arbitrary* complete set of orthonormal functions. From that perspective it may thus be concluded that Eq. (B.93) is nothing but a statement about the completeness of the special basis represented by Eq. (B.92).

B.6.3   Quantum corrections due to wave-function symmetry

As we argued in Section 2.5.3, expressions of the form

$$\frac{1}{h^3} \int \exp\left(\frac{i\boldsymbol{p} \cdot \boldsymbol{r}}{\hbar}\right) \exp\left(-\frac{p^2}{2mk_\mathrm{B}T}\right) d\boldsymbol{p}$$

arise in the second term of Eq. (2.101) on account of permutations. To evaluate these integrals, we employ spherical polar coordinates and rewrite the previous expression more explicitly as

$$
\begin{aligned}
2\pi & \int_0^\pi \int_0^\infty \exp\left(-\frac{p^2}{2mk_\mathrm{B}T}\right) \exp\left(\frac{ipr\cos\varphi}{\hbar}\right) p^2 dp \sin\varphi d\varphi \\
= {} & 2\pi \int_0^\infty p^2 \exp\left(-\frac{p^2}{2mk_\mathrm{B}T}\right) dp \int_{-1}^1 \exp\left(\frac{iprx}{\hbar}\right) dx \\
= {} & \frac{4\pi\hbar}{r} \int_0^\infty p \exp\left(-\frac{p^2}{2mk_\mathrm{B}T}\right) \sin\left(\frac{pr}{\hbar}\right) dp \tag{B.94}
\end{aligned}
$$

where the factor 2π arises because we immediately carried out the integration over the angle ϑ in the first line of Eq. (B.94). On the second line we introduced the new variable $x = \cos\varphi$, which permits us to carry out the second integration over the angle φ. On the third line we finally use Euler's representation of complex numbers; that is,

$$\exp\left(\pm ix\right) = \cos\left(x\right) \pm i\sin\left(x\right) \tag{B.95}$$

The final expression in Eq. (B.94) can be simplified by partial integration where we note that

$$\int p\exp\left(-\frac{p^2}{2mk_{\mathrm{B}}T}\right)\mathrm{d}p = \frac{1}{2}\int \exp\left(-\frac{\tilde{p}}{2mk_{\mathrm{B}}T}\right)\mathrm{d}\tilde{p}$$
$$= -mk_{\mathrm{B}}T\exp\left(-\frac{p^2}{2mk_{\mathrm{B}}T}\right) + C \quad (B.96)$$

where we used the transformation $p \to \tilde{p} = p^2$ and C is an integration constant. Hence, partial integration of Eq. (B.94) leads to

$$\int\limits_0^\infty p\exp\left(-ap^2\right)\sin\left(py\right)\mathrm{d}p = \frac{y}{2a}\int\limits_0^\infty \exp\left(-ap^2\right)\cos\left(py\right)\mathrm{d}p \tag{B.97}$$

where we simplified the notation by introducing $a \equiv 1/2mk_{\mathrm{B}}T$ and $y = r/\hbar$.

Using again Eq. (B.95) we realize that the integral on the right side of Eq. (B.97) may be rewritten as

$$\int\limits_0^\infty \exp\left(-ap^2\right)\cos\left(py\right)\mathrm{d}p = \frac{1}{2}\exp\left(-\frac{y^2}{4a}\right)\left[\int\limits_{-iy/2a}^{\infty-iy/2a} \exp\left(-az^2\right)\mathrm{d}z\right.$$
$$\left. + \int\limits_{iy/2a}^{\infty+iy/2a} \exp\left(-az'^2\right)\mathrm{d}z'\right] \tag{B.98}$$

where we used the substitutions

$$z \equiv p - \frac{iy}{2a} \tag{B.99a}$$
$$z' \equiv p + \frac{iy}{2a} \tag{B.99b}$$

With the transformation $z' \to \tilde{z} = -z'$, the second integral in Eq. (B.98)

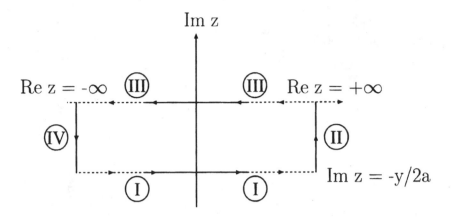

Figure B.5: Plot of the integration path in the complex plane used to evaluate the integral in Eq. (B.98). Individual portions of the integration path labeled I–IV refer to contributions from integrals given in Eq. (B.101).

may be rewritten so that its right side becomes

$$\frac{1}{2}\exp\left(-\frac{y^2}{4a}\right)\left[\int_{-iy/2a}^{\infty-iy/2a}\exp\left(-az^2\right)dz - \int_{-iy/2a}^{-\infty-iy/2a}\exp\left(-a\tilde{z}^2\right)d\tilde{z}\right]$$

$$= \frac{1}{2}\exp\left(-\frac{y^2}{4a}\right)\int_{-\infty-iy/2a}^{\infty-iy/2a}\exp\left(-az^2\right)dz \qquad (B.100)$$

after interchanging the integration limits on the second integral.

This last integral may be evaluated using function theoretical arguments. Consider a closed integration path in the complex plane as indicated in Fig. B.5. Integrating $\exp\left(-az^2\right)$ along this path we may write

$$\underbrace{\int_{-\infty-iy/2a}^{\infty-iy/2a}\exp\left(-az^2\right)dz}_{\text{Path I}} + \underbrace{\int_{\infty-iy/2a}^{\infty}\exp\left(-az^2\right)dz}_{\text{Path II}} + \underbrace{\int_{\infty}^{-\infty}\exp\left(-az^2\right)dz}_{\text{Path III}}$$

$$+ \underbrace{\int_{-\infty}^{-\infty-iy/2a}\exp\left(-az^2\right)dz = 0}_{\text{Path IV}} \qquad (B.101)$$

This integral must vanish because the integration path encloses an area in the complex plane in which the integrand does not have any poles according to Theorem B.5, which we proved in Appendix B.2.4 above. Moreover, we notice that integrals II and IV cancel each other because the integration is carried out parallel to the complex axis but the direction of the integration path is reversed between both integrals. This observation leaves us with integrals I and IV, which may be rearranged to give

$$
\int_{-\infty-iy/2a}^{\infty-iy/2a} \exp\left(-az^2\right) dz = \int_{-\infty}^{\infty} \exp\left(-az^2\right) dz = \sqrt{\frac{\pi}{a}}
\tag{B.102}
$$

Thus, from Eqs. (B.94), (B.97), (B.98), and (B.102) we conclude that

$$
\frac{1}{h^3} \int \exp\left(\frac{i\boldsymbol{p}\cdot\boldsymbol{r}}{\hbar}\right) \exp\left(-\frac{\boldsymbol{p}^2}{2mk_BT}\right) d\boldsymbol{p} = \frac{1}{\Lambda^3} \exp\left(-\frac{\pi r^2}{\Lambda^2}\right)
\tag{B.103}
$$

where we used the definitions of a and y given above as well as the expression for the de Broglie wavelength given in Eq. (2.103).

B.6.4 Quantum corrections to the Hamiltonian function

In Section 2.5.3 we derived the semiclassic expression for the canonical partition function [see Eq. (2.110)] based on the assumption that at sufficiently high temperatures we may replace the Hamiltonian operator by its classic analog, the Hamiltonian function [see Eq. (2.100)]. In this section we will sketch a more refined treatment of the semiclassic theory developed in Section 2.5 originally due to Hill and presented in detail in his classical work on statistical mechanics [326]. Because of Hill's clear and detailed exposition and because we need the final result mainly as a justification to treat confined fluids by means of classic statistical thermodynamics, we will just briefly outline the key ideas of Hill's treatment for reasons of completeness of the current work.

Our starting point is the expression

$$
\exp\left(-\beta\widehat{H}\right) \exp\left(i\boldsymbol{k}^N\cdot\boldsymbol{r}^N\right) = \exp\left(-\beta H\right) \exp\left(i\boldsymbol{k}^N\cdot\boldsymbol{r}^N\right) w\left(\boldsymbol{r}^N,\boldsymbol{p}^N;\beta\right)
\tag{B.104}
$$

where we use conventional notation $\beta \equiv 1/k_BT$ because in the end we wish to focus on the high-temperature limit $T^{-1} \to 0$. The key idea then is to determine the correction $w\left(\boldsymbol{r}^N,\boldsymbol{p}^N;\beta\right)$ such that in that limit the quantum mechanical analysis may be abandoned in favor of a classic one [see also Eq. (2.100)]. Defining in Eq. (B.104), $\exp\left(-\beta\widehat{H}\right)$ through its power-series expansion [see Eq. (2.39)], we may derive Bloch's equation,

$$\frac{\partial}{\partial\beta}\exp\left(-\beta\widehat{H}\right)\exp\left(i\boldsymbol{k}^N\cdot\boldsymbol{r}^N\right) = -\widehat{H}\exp\left(-\beta\widehat{H}\right)\exp\left(i\boldsymbol{k}^N\cdot\boldsymbol{r}^N\right) \quad (B.105)$$

Replacing in Bloch's equation the term $\exp\left(-\beta\widehat{H}\right)\exp\left(i\boldsymbol{k}^N\cdot\boldsymbol{r}^N\right)$ according to Eq. (B.104), we eventually arrive at a partial differential equation for the unknown function $w\left(\boldsymbol{r}^N,\boldsymbol{p}^N;\beta\right)$. This calculation requires a considerable amount of tedious yet straightforward algebraic manipulations detailed in Hill's book (see Eqs. (16.16)–(16.23) in Ref. 326). We shall therefore skip those steps of the derivation and just jump to the final result,

$$\begin{aligned}
\frac{\partial w}{\partial\beta} &= \frac{\hbar^2}{2m}\left[\boldsymbol{\nabla}^N\cdot\left(\boldsymbol{\nabla}^N w\right) - \beta\boldsymbol{\nabla}^N\cdot\left(\boldsymbol{\nabla}^N U\right) - 2\beta\left(\boldsymbol{\nabla}^N w\right)\cdot\left(\boldsymbol{\nabla}^N U\right)\right.\\
&\left.+\beta^2\left(\boldsymbol{\nabla}^N U\right)\cdot\left(\boldsymbol{\nabla}^N U\right)\right]\\
&+\frac{i\hbar}{m}\left[\left(\boldsymbol{\nabla}^N w\right)\cdot\boldsymbol{p}^N - \beta w\left(\boldsymbol{\nabla}^N U\right)\cdot\boldsymbol{p}^N\right] \equiv M\left(\beta\right) \quad (B.106)
\end{aligned}$$

where we introduce shorthand notation $\left(\boldsymbol{\nabla}^N\right)^{\mathrm{T}} \equiv \left(\nabla_1,\nabla_2,\ldots,\nabla_N\right)$ and $\nabla_i = \partial/\partial\boldsymbol{r}_i$. In deriving Eq. (B.106) we employed the space representation of the Hamiltonian operator [see Eq. (2.95)], and the fact that the classic Hamiltonian function can be split into kinetic- and potential-energy contributions according to Eq. (2.100). Terms proportional to $\boldsymbol{\nabla}^N$ in Eq. (B.106) arise from the kinetic part of $\widehat{H}$ applied to the product of terms on the right side of Eq. (B.104) (using, of course, the product rule of conventional calculus).

Equation (B.106) can be formally integrated to give

$$w\left(\boldsymbol{r}^N,\boldsymbol{p}^N;\beta\right) = 1 + \int\limits_0^\beta \mathrm{d}\beta' M\left(\beta';w\right) \quad (B.107)$$

satisfying the boundary condition $w\left(\boldsymbol{r}^N,\boldsymbol{k}^N;0\right) = 1$ such that in the limit of infinite temperature we recover Eq. (2.100) directly from Eq. (B.104). To solve the integral equation in Eq. (B.107), we expand $w\left(\boldsymbol{r}^N,\boldsymbol{p}^N;\beta\right)$ in the spirit of the Wenzel-Kramers-Brillouin (WKB) formalism and write

$$w\left(\boldsymbol{r}^N,\boldsymbol{p}^N;\beta\right) = \sum_{l=0}^\infty a_l\left(\boldsymbol{r}^N,\boldsymbol{p}^N;\beta\right)\hbar^l \quad (B.108)$$

where the set of unknown expansion functions $\{a_l\left(\boldsymbol{r}^N,\boldsymbol{p}^N;\beta\right)\}$ can be determined by inserting the *ansatz* into Eq. (B.107), which gives

$$
\begin{aligned}
\sum_{l=0}^{\infty} a_l \hbar^l &= 1 + \frac{1}{m}\sum_{l=0}^{\infty}\hbar^{l+2}\int_0^{\beta}\left[\boldsymbol{\nabla}^N\cdot\left(\boldsymbol{\nabla}^N a_l\right) - \beta' a_l \boldsymbol{\nabla}^N\cdot\left(\boldsymbol{\nabla}^N U\right)\right. \\
&\quad \left. -2\beta'\left(\boldsymbol{\nabla}^N a_l\right)\cdot\left(\boldsymbol{\nabla}^N U\right) + \beta'^2\left(\boldsymbol{\nabla}^N U\right)^2\right]\mathrm{d}\beta' \\
&\quad +\frac{i}{m}\sum_{l=0}^{\infty}\hbar^{l+1}\int_0^{\beta}\left[\left(\boldsymbol{\nabla}^N a_l\right)\cdot\boldsymbol{p}^N - \beta' a_l\left(\boldsymbol{\nabla}^N U\right)\cdot\boldsymbol{p}^N\right]\mathrm{d}\beta'
\end{aligned}
$$

$$(\text{B.109})$$

Equating in this expression terms of equal power in $\hbar$, one immediately finds that

$$
a_0\left(\boldsymbol{r}^N,\boldsymbol{p}^N;\beta\right) = 1 \tag{B.110}
$$

Inserting this result back into Eq. (B.109), the first-order function is obtained as

$$
a_1\left(\boldsymbol{r}^N,\boldsymbol{p}^N;\beta\right) = -\frac{i}{m}\left(\boldsymbol{\nabla}^N U\right)\cdot\boldsymbol{p}^N\int_0^{\beta}\mathrm{d}\beta'\beta' = -\frac{i\beta^2}{2m}\left(\boldsymbol{\nabla}^N U\right)\cdot\boldsymbol{p}^N \tag{B.111}
$$

which is no longer constant but depends on the configuration of molecules and their momenta. In a similar fashion the second-order function is obtained as

$$
\begin{aligned}
a_2\left(\boldsymbol{r}^N,\boldsymbol{p}^N;\beta\right) &= -\frac{\boldsymbol{\nabla}^N\cdot\left(\boldsymbol{\nabla}^N U\right)}{2m}\int_0^{\beta}\mathrm{d}\beta'\beta' + \frac{\left(\boldsymbol{\nabla}^N U\right)^2}{2m}\int_0^{\beta}\mathrm{d}\beta'\beta'^2 \\
&\quad +\frac{\left(\boldsymbol{p}^N\cdot\boldsymbol{\nabla}^N\right)U}{2m^2}\int_0^{\beta}\mathrm{d}\beta'\beta'^2 - \frac{\left[\left(\boldsymbol{\nabla}^N U\right)\cdot\boldsymbol{p}^N\right]^2}{2m^2}\int_0^{\beta}\mathrm{d}\beta'\beta'^3 \\
&= -\frac{1}{2m}\left[\frac{\beta^2}{2}\boldsymbol{\nabla}^N\cdot\left(\boldsymbol{\nabla}^N U\right) - \frac{\beta^3}{3}\left(\boldsymbol{\nabla}^N U\right)^2\right. \\
&\quad \left. -\frac{\beta^3}{3}\frac{\left(\boldsymbol{p}^N\cdot\boldsymbol{\nabla}^N\right)U}{m} + \frac{\beta^4}{4}\frac{\left[\left(\boldsymbol{\nabla}^N U\right)\cdot\boldsymbol{p}^N\right]^2}{4m}\right]
\end{aligned}
$$

$$(\text{B.112})$$

and so forth where the higher-order functions quickly become intractable on account of their rapidly increasing complexity.

On the basis of these first few terms in the expansion in Eq. (B.108), one eventually arrives at an improved approximation of Q in Eq. (2.110) if one replaces Eq. (2.100) by

$$\exp\left(-\frac{\widehat{H}}{k_B T}\right) \rightarrow \exp\left[-\frac{H\left(r^N, p^N\right)}{k_B T}\right]\left[1 + a_2\left(r^N, p^N; \beta\right)\hbar^2 + \ldots\right]$$

(B.113)

In writing Eq. (B.113) we have already neglected the term proportional to $a_1\left(r^N, p^N; \beta\right)$. Because this coefficient is linear in p^N, it will vanish upon integration over momentum subspace [see, for example, Eq. (2.111)]. From Eq. (B.113) one realizes that the correction

$$a_2\left(r^N, p^N; \beta\right)\hbar^2 = \mathcal{O}\left(\Lambda^2\right)$$

(B.114)

which depends also on derivatives of the total configurational potential energy [see Eq. (B.112)]. Hence, the correction vanishes for an ideal gas where by definition $U\left(r^N\right) \equiv 0$. In this latter case only the semiclassic correction due to symmetry properties of the wave function survives [see Eq. (2.110)].

Appendix C

Mathematical aspects of one-dimensional hard-rod fluids

C.1 Distance between mirror images

In this Appendix we will demonstrate that the distance $\overline{P_1 P_2}$ between a pair of points $P_1 \equiv r_1^{\mathrm{T}} = (z_1, z_2)$ and $P_2 \equiv r_2^{\mathrm{T}} (x_2, y_2)$ remains unaltered upon reflection at some straight line in the x–y plane. Consider the situation depicted in Fig. C.1 where the two points are separated by distances

$$r_{12} \equiv |r_2 - r_1| = \sqrt{(x_2 - x_1)^2 + (y_2 - y_1)^2} \tag{C.1a}$$

$$r_{12}' \equiv |r_2' - r_1'| = \sqrt{(x_2' - x_1')^2 + (y_2' - y_1')^2} \tag{C.1b}$$

We wish to demonstrate that the length of the line $\overline{P_1 P_2}$ remains unaltered if the points P_1 and P_2 are reflected at the line $y - -x$ as indicated in Fig. C.1.

In general the points P_i and P_i' are related through

$$r_i' = A(\alpha) r_i, \qquad i = 1, 2 \tag{C.2}$$

where the mapping $r_i \to r_i'$ is effected by the orthogonal transformation matrix

$$A(\alpha) = \begin{pmatrix} \cos\alpha & \sin\alpha \\ \sin\alpha & -\cos\alpha \end{pmatrix} \tag{C.3}$$

such that $\alpha/2$ is the angle between the straight line of reflection and the positive x-axis as shown in Fig. C.1 for the special case $\alpha = \frac{3}{2}\pi$. Hence, from

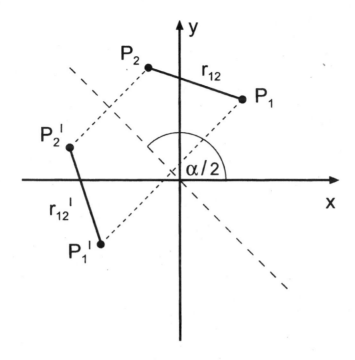

Figure C.1: Two points $P_1 = (x_1, y_1)$ and $P_2 (x_2, y_2)$ are transformed into corresponding points $P_1' (x_1', y_1')$ and $P_2' (x_2', y_2')$ through a reflection at the straight line $y = -x$ (- - -). The lines $\overline{P_1 P_2}$ and $\overline{P_1' P_2'}$ have lengths r_{12} and r_{12}', respectively (—).

Eqs. (C.2) and (C.3), it is immediately evident that

$$
\boldsymbol{r}_i' = \begin{pmatrix} x_i' \\ y_i' \end{pmatrix} = \begin{pmatrix} \cos\alpha & \sin\alpha \\ -\sin\alpha & \cos\alpha \end{pmatrix} \begin{pmatrix} x_i \\ y_i \end{pmatrix}
$$

$$
= \begin{pmatrix} x_i \cos\alpha + y_i \sin\alpha \\ -x_i \sin\alpha + y_i \cos\alpha \end{pmatrix}, \qquad i = 1, 2 \qquad \text{(C.4)}
$$

and therefore

$$
x_2' - x_1' = (x_2 - x_1)\cos\alpha + (y_2 - y_1)\sin\alpha \qquad \text{(C.5)}
$$
$$
y_2' - y_1' = (y_2 - y_1)\cos\alpha - (x_2 - x_1)\sin\alpha \qquad \text{(C.6)}
$$

Inserting this into Eq. (C.1b), we obtain

$$
r_{12}' = \sqrt{(x_2' - x_1')^2 + (x_2' - x_1')^2} = \sqrt{(x_2 - x_1)^2 + (y_2 - y_1)^2} = r_{12} \qquad \text{(C.7)}
$$

by comparison with Eq. (C.1a), which completes the proof. Hence, we conclude that any function $f(r_{12})$ depending only on the distance between a pair of points will equal $f(r'_{12})$ if the two points are reflected at *any* line in the x–y plane.

C.2   Integral transformations

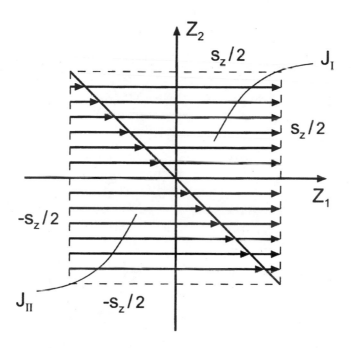

Figure C.2: Sketch of the domain J (see text) over which a function $f(z_{12})$ is to be integrated. The direction of integration is indicated by the arrows where subdomains J_I and J_{II} are defined in Eqs. (C.10).

Consider now the integral

$$I_2 = \int\limits_{-L/2}^{L/2} dz_2 \int\limits_{-L/2}^{L/2} dz_1 f(z_{12}) \tag{C.8}$$

where $z_{12} = |z_2 - z_1|$ such that $f(z_{12})$ is invariant under reflection along the line $z_2 = -z_1$ indicated in Fig. C.2. We split the domain

$$J \equiv \{(z_1, z_2)| - L/2 \le z_1 \le L/2 \vee -L/2 \le z_2 \le L/2\} \tag{C.9}$$

into two subdomains, namely,

$$J_I \equiv \{(z_1, z_2) | -L/2 \le z_2 \le L/2 \vee z_1 \ge -z_2\} \tag{C.10a}$$
$$J_{II} \equiv \{(z_1, z_2) | -L/2 \le z_2 \le L/2 \vee z_1 \le -z_2\} \tag{C.10b}$$

and write I_2 as a sum of two integrals according to

$$I_2 = \underbrace{\int_{-L/2}^{L/2} dz_2 \int_{-z_2}^{L/2} dz_1 f(z_{12})}_{I_I} + \underbrace{\int_{-L/2}^{L/2} dz_2 \int_{-L/2}^{-z_2} dz_1 f(z_{12})}_{I_{II}} \tag{C.11}$$

where for region of integration is J_I and J_{II}, respectively. Changing now in the first integral the order of integration, we may write

$$I_I = \int_{-L/2}^{L/2} dz_2 \int_{-z_2}^{L/2} dz_1 f(z_{12}) = \int_{-L/2}^{L/2} dz_1 \int_{-z_1}^{L/2} dz_2 f(z_{12}) \equiv I_I' \tag{C.12}$$

where the integration is still performed over the triangular area J_I.

Because $f(z_{12})$ does not change if we reflect each pair of points at the line $z_2 = -z_1$ according to the discussion in Appendix C.1, we may define the integral $I_{I'}$ alternatively over the subdomain J_{II} and write

$$I_I' = \int_{-L/2}^{L/2} dz_1 \int_{-z_1}^{L/2} dz_2 f(z_{12}) = \int_{-L/2}^{L/2} dz_1 \int_{-L/2}^{-z_1} dz_2 f(z_{12}) = \int_{-L/2}^{L/2} dz_2 \int_{-L/2}^{-z_2} dz_1 f(z_{12}) \tag{C.13}$$

where we used the fact that z_1 and z_2 may be treated as dummy indices and may therefore be interchanged. From Eq. (C.13) it is then clear that

$$I_I' = I_{II} \tag{C.14}$$

so that

$$I_2 = 2 I_{II} \tag{C.15}$$

follows without further ado. Hence, because our function $f(z_{12})$ depends only on the *distance* z_{12} between a pair of points rather than on z_1 and z_2, we see from Eqs. (C.11)–(C.13) that

$$I_2 = 2 \int_{-L/2}^{L/2} dz_2 \int_{-L/2}^{-z_2} dz_1 f(z_{12}) \tag{C.16}$$

We now employ this result to prove by complete induction that *in general* (i.e., for arbitrary $n \in \mathbb{N}$)

$$I_n = n! \int_{-L/2}^{L/2} dz_n \int_{-L/2}^{-z_n} dz_{n-1} \cdots \int_{-L/2}^{-z_2} dz_1 f \qquad (C.17)$$

Therefore, we *assume* that

$$
\begin{aligned}
I_{n\,1} &= \int_{-L/2}^{L/2} dz_{n-1} \int_{-L/2}^{L/2} dz_{n-2} \cdots \int_{-L/2}^{L/2} dz_1 f \\
&= (n-1)! \int_{-L/2}^{L/2} dz_{n-1} \int_{-L/2}^{-z_{n-1}} dz_{n-2} \cdots \int_{-L/2}^{-z_2} dz_1 f \qquad (C.18)
\end{aligned}
$$

where the function f is supposed to depend only on distances $z_{ij} = |z_j - z_i|$ between any pair of points i and j. From Eq. (C.18) we see that

$$I_n = \int_{-L/2}^{L/2} dz_n I_{n-1} = (n-1)! \int_{-L/2}^{L/2} dz_n \int_{-L/2}^{L/2} dz_{n-1} \int_{-L/2}^{-z_{n-1}} dz_{n-2} \cdots \int_{-L/2}^{-z_2} dz_1 f$$

$$(C.19)$$

which we may split into two terms according to

$$
\begin{aligned}
I_n &= (n-1)! \int_{-L/2}^{L/2} dz_n \int_{-L/2}^{-z_n} dz_{n-1} \int_{-L/2}^{-z_{n\,1}} dz_{n-2} \cdots \int_{-L/2}^{-z_2} dz_1 f \\
&+ (n-1)! \int_{-L/2}^{L/2} dz_n \int_{-z_n}^{L/2} dz_{n-1} \int_{-L/2}^{-z_{n-1}} dz_{n-2} \cdots \int_{-L/2}^{-z_2} dz_1 f \qquad (C.20)
\end{aligned}
$$

Consider now the second summand in Eq. (C.20) and interchange the order of integration between the first two of its integrals similar to the transformation between Eqs. (C.11) and (C.12). One obtains

$$\int_{-L/2}^{L/2} dz_{n-1} \int_{-z_{n-1}}^{L/2} dz_n \cdots = \int_{-L/2}^{L/2} dz_{n-1} \int_{-L/2}^{-z_{n-1}} dz_n \cdots \qquad (C.21)$$

where the right side follows by applying the same argument that led to Eq. (C.13). Inserting this expression into the second term of Eq. (C.20), we obtain

$$(n-1)! \int_{-L/2}^{L/2} dz_n \int_{-z_n}^{L/2} dz_{n-1} \int_{-L/2}^{-z_n 1} dz_{n-2} \dots \int_{-L/2}^{-z_2} dz_1 f$$

$$= (n-1)! \int_{-L/2}^{L/2} dz_{n-1} \int_{-L/2}^{-z_{n-1}} dz_n \int_{-L/2}^{-z_{n-1}} dz_{n-2} \dots \int_{-L/2}^{-z_2} dz_1 f \qquad (C.22)$$

We proceed by applying the same logic to the term

$$\dots \int_{-L/2}^{-z_{n-1}} dz_n \int_{-L/2}^{-z_{n-1}} dz_{n-2} \dots$$

which we split accordingly into two terms, namely

$$\dots \int_{-L/2}^{-z_{n-1}} dz_n \int_{-L/2}^{-z_n} dz_{n-2} \dots + \dots \int_{-L/2}^{-z_{n-1}} dz_n \int_{-z_n}^{-z_{n-1}} dz_{n-2} \dots$$

Applying again our transformation procedure to the second one of these, we realize that

$$\dots \int_{-L/2}^{-z_{n-1}} dz_n \int_{-z_n}^{-z_{n-1}} dz_{n-2} \dots = \dots \int_{-L/2}^{-z_{n-1}} dz_{n-2} \int_{-z_{n-2}}^{-z_{n-1}} dz_n \dots$$

$$= \dots \int_{-L/2}^{-z_{n-1}} dz_{n-2} \int_{-L/2}^{-z_{n-2}} dz_n \dots \qquad (C.23)$$

Putting this expression back into Eq. (C.22), we now arrive at

$$(n-1)! \int_{-L/2}^{L/2} dz_{n-1} \int_{-L/2}^{-z_n 1} dz_{n-2} \int_{-L/2}^{-z_n 2} dz_n \int_{-L/2}^{-z_n 2} dz_{n-3} \dots \int_{-L/2}^{-z_2} dz_1 f$$

so that we can apply identically the same procedure to the integral

$$\int_{-L/2}^{-z_{n-2}} dz_n \int_{-L/2}^{-z_{n-2}} dz_{n-3} \dots$$

at the next stage. Thus, one may continue until no pair of integrals has the same upper limit of integration. At this point, the original I_n has been decomposed into n identical terms and may therefore be rewritten as

$$I_n = n! \int\limits_{-L/2}^{L/2} \mathrm{d}z_n \int\limits_{-L/2}^{-z_n} \mathrm{d}z_{n-1} \ldots \int\limits_{-L/2}^{-z_2} \mathrm{d}z_1 f \qquad \text{(C.24)}$$

We therefore conclude that the form of I_{n-1} *hypothesized* in Eq. (C.18) implies Eq. (C.24) for I_n. Moreover, the case $n = 2$, proved explicitly above, is nothing but a special case of I_n as one realizes by comparing Eqs. (C.16) and (C.24) so that Eq. (C.24) completes our proof.

Finally, we stress that the decomposition of I_n according to Eq. (C.24) is based on the explicit assumption that the integrand f depends only on distances z_{ij} and not on the individual coordinates z_i and z_j as independent variables.

C.3 Gaussian density distribution

C.3.1 Limiting behavior of the probability density

In the grand canonical ensemble, the probability of finding N molecules in a (constant) volume V at some fixed temperature T may be cast as

$$P(N) = C \exp\left(\frac{\mu N}{k_{\mathrm{B}} T}\right) \mathcal{Q} \qquad \text{(C.25)}$$

where C is a normalization constant. Assuming that $P(N)$ becomes maximum for the most probable number of molecules $N = \overline{N}$, we may expand $\ln P(N)$ about this value in a Taylor series according to

$$
\begin{aligned}
\ln P(N) &= \ln P\left(\overline{N}\right) + \underbrace{\left(\frac{\partial \ln P}{\partial N}\right)_{\{\cdot\}\backslash N}\bigg|_{N=\overline{N}}}_{=0} \left(N - \overline{N}\right) \\
&\quad + \frac{1}{2!}\left(\frac{\partial^2 \ln P}{\partial N^2}\right)_{\{\cdot\}\backslash N}\bigg|_{N=\overline{N}} \left(N - \overline{N}\right)^2 + \mathcal{O}\left(N^3\right) \\
&\simeq \ln P\left(\overline{N}\right) + \frac{1}{2!}\left(\frac{\partial^2 \ln P}{\partial N^2}\right)_{\{\cdot\}\backslash N}\bigg|_{N=\overline{N}} \left(N - \overline{N}\right)^2 \qquad \text{(C.26)}
\end{aligned}
$$

where the first-order derivative vanishes because we are expanding $P(N)$ about its maximum. From Eq. (C.25) it is also clear that

$$\left(\frac{\partial^2 \ln P}{\partial N^2}\right)_{\{\cdot\}\backslash N} = \left(\frac{\partial^2 \ln Q}{\partial N^2}\right)_{\{\cdot\}\backslash N} \tag{C.27}$$

However, the fact that the first-order term in the Taylor expansion vanishes equips us with an additional relation, namely

$$-\frac{\mu}{k_B T}\bigg|_{N=\overline{N}} = \left(\frac{\partial \ln Q}{\partial N}\right)_{\{\cdot\}\backslash N}\bigg|_{N=\overline{N}} \tag{C.28}$$

so that we can rewrite Eq. (C.26) as

$$P(N) = P(\overline{N})\exp\left[-\frac{a}{2}\left(N-\overline{N}\right)^2\right] \tag{C.29a}$$

$$a \equiv \frac{1}{k_B T}\left(\frac{\partial \mu}{\partial N}\right)_{\{\cdot\}\backslash N}\bigg|_{N=\overline{N}} \tag{C.29b}$$

In addition, we require $P(N)$ to be properly normalized; that is,

$$\int_0^\infty P(\Delta N)\,dN = \int_{-\overline{N}}^\infty P(\Delta N)\,d(\Delta N) = \int_{-\infty}^\infty P(\Delta N)\,d(\Delta N) \overset{!}{=} 1 \tag{C.30}$$

where $\Delta N \equiv N - \overline{N}$. As $P(\Delta N)$ can be expected to be peaked fairly sharply at $\Delta N = 0$, and because $\overline{N} \to \infty$ in the thermodynamic limit, we may change the lower integration limit to $-\infty$ without affecting the integral. The normalization condition leads to [see Eq. (B.102)]

$$P(\overline{N}) = \sqrt{\frac{a}{2\pi}} \tag{C.31}$$

With the aid of $P(N)$ we can also calculate the average number of molecules via

$$\langle N \rangle = \int_{-\infty}^\infty N P(N)\,d\Delta N = \int_{-\infty}^\infty \left(\Delta N + \overline{N}\right) P(N)\,d\Delta N$$

$$= \underbrace{\int_{-\infty}^\infty \Delta N P(N)\,d\Delta N}_{=0} + \overline{N}\underbrace{\int_{-\infty}^\infty P(N)\,d\Delta N}_{=1} = \overline{N} \tag{C.32}$$

where the first integral vanishes because $P(N)$ is an even function of ΔN and the second integral is nothing but the normalization condition. To determine the constant a in Eq. (C.29b), we consider the so-called second central moment of $P(N)$; that is,

$$\langle (\Delta N)^2 \rangle \equiv \int_{-\infty}^{\infty} (\Delta N)^2 P(N) \, d\Delta N$$

$$= -\sqrt{\frac{2a}{\pi}} \frac{d}{da} \int_{-\infty}^{\infty} \exp\left[-\frac{a}{2}(N-\overline{N})^2\right] d\Delta N \qquad \text{(C.33)}$$

which is an identity as one easily verify by performing the differentiation of the integral and using the definition of $P(N)$ in Eq. (C.29a) together with Eq. (C.31). However, because of Eq. (B.102) the previous expression may be recast as

$$\langle (\Delta N)^2 \rangle = -\sqrt{\frac{2a}{\pi}} \frac{d}{da} \sqrt{\frac{2\pi}{a}} = \frac{1}{a} \qquad \text{(C.34)}$$

from which

$$\langle (\Delta N)^2 \rangle = \langle (N - \langle N \rangle)^2 \rangle = \langle N^2 \rangle - \langle N \rangle^2 = \frac{1}{a} = \sigma_N^2 \qquad \text{(C.35)}$$

follows with the aid of Eq. (2.75). Hence, we may rewrite Eq. (C.29a) as

$$P(N) = \frac{1}{\sqrt{2\pi}\sigma_N} \exp\left[-\frac{(N - \langle N \rangle)^2}{2\sigma_N^2}\right] \qquad \text{(C.36)}$$

To analyze the simulation data in Section (5.4.2) for the isothermal compressibility, it turns out to be more convenient to transform variables in this last expression according to $N, \langle N \rangle \to \rho, \overline{\rho}$ to obtain an analogous expression for the density distribution $P(\rho)$ in a slit-pore.

C.3.2 Moivre-Laplace approximation

An alternative route to Eq. (C.36) proceeds as follows. Suppose we take a system of constant volume at fixed T and μ. Let us perform an experiment on this system that allows us to determine the particle number N. As under conditions of fixed T and μ, but where N may vary, there is a certain probability p that the result of our measurement will be positive, that is, it will give *precisely* N. Clearly, the outcome of the experiment may also be negative; that is, there is a probability $1 - p$ that we will measure some other particle number but N. We now envision to repeat our experiment n times. In this sequence of n measurements m are assumed to be positive (i.e., they have as a result N particles in the system) and therefore $n - m$ measurements will yield a negative result. The probability to measure N in the *entire* sequence of n experiments is then given by

$$P_n\left(m\right) = \binom{n}{m} = p^m\left(1 - p\right)^{n-m} = \frac{n!}{m!\left(n - m\right)!}p^m\left(1 - p\right)^{n-m} \quad \text{(C.37)}$$

where the combinatorial factor arises because the precise sequence of experiments with positive and negative outcome does not matter in the sequence. Equation (C.37) represents the so-called Bernoulli distribution. With the aid of Eq. (B.6) we may rewrite the combinatorial factor as

$$\frac{n!}{m!\left(n - m\right)!} = \sqrt{\frac{n}{2\pi m\left(n - m\right)}}\left(\frac{n}{m}\right)^m\left(\frac{n}{n - m}\right)^{n-m} \quad \text{(C.38)}$$

such that Eq. (C.37) may be recast as

$$P_n\left(m\right) = \sqrt{\frac{n}{2\pi m\left(n - m\right)}}\left(\frac{np}{m}\right)^m\left(\frac{n\left(1 - p\right)}{n - m}\right)^{n-m} \quad \text{(C.39)}$$

We now seek an approximate expression for Eq. (C.38) valid in the limit of a very large number of experiments; that is, we focus on the limit $n \to \infty$. To derive this approximate expression we introduce

$$\zeta \equiv \frac{m - np}{\sqrt{np\left(1 - p\right)}} \quad \text{(C.40)}$$

as an auxiliary quantity for the subsequent derivation. Using the previous expression for ζ, it is straightforward to realize that

$$
\begin{aligned}
\ln\left(\frac{np}{m}\right)^m\left(\frac{n\left(1 - p\right)}{n - m}\right)^{n-m} &= -m\ln\frac{m}{np} - \left(n - m\right)\ln\frac{n - m}{n\left(1 - p\right)} \\
&= -\left(np + \zeta\sqrt{np\left(1 - p\right)}\right) \\
&\quad \times \ln\left(1 + \zeta\sqrt{\frac{1 - p}{np}}\right) \\
&\quad - \left[n\left(1 - p\right) - \zeta\sqrt{np\left(1 - p\right)}\right] \\
&\quad \times \ln\left(1 - \zeta\sqrt{\frac{p}{n\left(1 - p\right)}}\right) \quad \text{(C.41)}
\end{aligned}
$$

In the limit $n \to \infty$ we realize from the definition of ζ in Eq. (C.40) that we may approximate the logarithmic functions in the previous expression by expanding them in a Taylor series according to

$$\ln\left(1 + \zeta\sqrt{\frac{1-p}{np}}\right) \simeq \zeta\sqrt{\frac{1-p}{np}} - \frac{\zeta^2}{2}\frac{1-p}{np} + \dots \tag{C.42a}$$

$$\ln\left(1 - \zeta\sqrt{\frac{p}{n(1-p)}}\right) \simeq -\zeta\sqrt{\frac{p}{n(1-p)}} - \frac{\zeta^2}{2}\frac{p}{n(1-p)} - \dots \tag{C.42b}$$

which we truncate after the quadratic term. Inserting these expressions on the far right side of Eq. (C.41), we yield

$$-\left(np + \zeta\sqrt{np(1-p)}\right)\left(\zeta\sqrt{\frac{1-p}{np}} - \frac{\zeta^2}{2}\frac{1-p}{np}\right) =$$

$$-\zeta\sqrt{np(1-p)} - \frac{\zeta^2}{2}(1-p) + \frac{\zeta^3}{2}\sqrt{\frac{(1-p)^3}{p}}\frac{1}{\sqrt{n}} \tag{C.43}$$

$$-\left[n(1-p) - \zeta\sqrt{np(1-p)}\right]\left(-\zeta\sqrt{\frac{p}{n(1-p)}} - \frac{\zeta^2}{2}\frac{p}{n(1-p)}\right) =$$

$$\zeta\sqrt{np(1-p)} - \frac{\zeta^2}{2}p - \frac{\zeta^3}{2}\sqrt{\frac{p^3}{(1-p)}}\frac{1}{\sqrt{n}} \tag{C.44}$$

Neglecting in these expressions terms proportional to $1/\sqrt{n}$ (because we focus on the limit $n \to \infty$), we obtain from Eqs. (C.40), (C.41), and (C.43) the approximate relation

$$\ln\left(\frac{np}{m}\right)^m\left(\frac{n(1-p)}{n-m}\right)^{n-m} \simeq -\frac{\zeta^2}{2} = -\frac{1}{2}\frac{(m-np)^2}{np(1-p)} \tag{C.45}$$

Reinserting this approximation into the right side of Eq. (C.39), we finally arrive at the Moivre-Laplace approximation, namely

$$P_n(m) \simeq \sqrt{\frac{n}{2\pi m(n-m)}} \exp\left[-\frac{1}{2}\frac{(m-np)^2}{np(1-p)}\right] \tag{C.46}$$

To establish a connection between Eqs. (C.46) and (C.36), let us introduce

$$np = \langle N \rangle \tag{C.47a}$$

$$np(1-p) = \langle N^2 \rangle - \langle N \rangle^2 = \sigma_N^2 \tag{C.47b}$$

where we used the definition of the variance of the particle-number distribution introduced in Eq. (2.75). Moreover, as we defined at the outset $p = m/n$ as the probability of successful measurements (i.e., measurements that have as their result the desired number of molecules N), we may rewrite the prefactor of the exponential function in Eq. (C.46) as [see Eq. (C.47b)]

$$\frac{n}{m(n-m)} = \left[\frac{m}{n}n\left(1 - \frac{m}{n}\right)\right]^{-1} = \frac{1}{np(1-p)} = \frac{1}{\sigma_N^2} \qquad \text{(C.48)}$$

Hence, we recover Eq. (C.36) using in Eq. (C.46) the expressions given in Eq. (C.47) and (C.48). We emphasize that no specific form of a partition function was invoked in the current derivation of Eq. (C.36).

Appendix D

Mathematical aspects of mean-field theories

D.1 Van der Waals model for confined fluids

D.1.1 Evaluation of the double integral in Eq. (4.21)

Transforming variables in the double integral in Eq. (4.24) from $\{\boldsymbol{r}_1, \boldsymbol{r}_2\}$ to $\{\boldsymbol{r}_1, \boldsymbol{r}_{12}\}$, we have from the definition of $a_{\mathrm{p}}(\xi)$

$$
\begin{aligned}
a_p &= \frac{2\varepsilon_{\mathrm{ff}}\sigma_{\mathrm{ff}}^6}{A\left(s_{z0} - 2\sigma_{\mathrm{fw}}\right)} \int_V \mathrm{d}\boldsymbol{r}_1 \int_V \mathrm{d}\boldsymbol{r}_{12}\, g\left(r_{12}\right) r^{-6} \\
&= \frac{2\varepsilon_{\mathrm{ff}}\sigma_{\mathrm{ff}}^6}{s_{z0} - 2\sigma_{\mathrm{fw}}} \int_{\sigma_{\mathrm{fw}}}^{s_{z0}-\sigma_{\mathrm{fw}}} \mathrm{d}z_1 \int_{\widetilde{V}(z_1)} \mathrm{d}\boldsymbol{r}_{12}\, r_{12}^{-6} \\
&= \frac{4\pi\varepsilon_{\mathrm{ff}}\sigma_{\mathrm{ff}}^6}{s_{z0} - 2\sigma_{\mathrm{fw}}} \int_{\sigma_{\mathrm{fw}}}^{s_{z0}-\sigma_{\mathrm{fw}}} \mathrm{d}z_1 \left\{ \int \mathrm{d}z \int \mathrm{d}\rho\, \rho \left(z^2 + \rho^2\right)^{-3} \right\}_{\widetilde{V}(z_1)}
\end{aligned}
\tag{D.1}
$$

where $\widetilde{V}(z_1)$ denotes the z_1-dependent volume restricted by the hard cores of fluid molecules and by the hard walls. In the last line of Eq. (D.1), cylindrical coordinates z, ρ, and θ are introduced for convenience and we immediately carried out the one trivial integration over the angle $0 \le \theta < 2\pi$. We assume $s_{z0} > 2\left(\sigma_{\mathrm{fw}} + \sigma_{\mathrm{ff}}\right)$. Then the integration on z_1 breaks down into three ranges:

1. $\sigma_{\mathrm{fw}} < z_1 < \sigma_{\mathrm{fw}} + \sigma_{\mathrm{ff}}$,

2. $\sigma_{\mathrm{fw}} + \sigma_{\mathrm{ff}} < z_1 < s_{z0} - \left(\sigma_{\mathrm{fw}} + \sigma_{\mathrm{ff}}\right)$,

3. $s_{z0} - (\sigma_{fw} + \sigma_{ff}) < z_1 < s_{z0} - \sigma_{fw}$.

In turn, the integrations on z and ρ can be broken into either two or three regions

$$
\begin{aligned}
a_1 &= \frac{4\pi\varepsilon_{ff}\sigma_{ff}^6}{s_{z0} - 2\sigma_{fw}} \int_{\sigma_{fw}}^{\sigma_{fw}+\sigma_{ff}} dz_1 \Bigg\{ \int_{-z_1+\sigma_{fw}}^{\sigma_{ff}} dz \int_{\sqrt{\sigma_{ff}^2-z^2}}^{\infty} \frac{d\rho\,\rho}{(z^2+\rho^2)^3} \\
&\quad + \int_{\sigma_{ff}}^{s_{z0}-\sigma_{fw}-z_1} dz \int_0^{\infty} \frac{d\rho\,\rho}{(z^2+\rho^2)^3} \Bigg\} \\
&= \frac{4\pi\varepsilon_{ff}\sigma_{ff}^6}{s_{z0}-2\sigma_{fw}} \int_{\sigma_{fw}}^{\sigma_{fw}+\sigma_{ff}} dz_1 \Bigg\{ \frac{\sigma_f + z_1 - \sigma_{fw}}{4\sigma_{ff}^4} \\
&\quad - \frac{1}{12}\left[\frac{1}{(s_{z0}-\sigma_{fw}-z_1)^3} - \frac{1}{\sigma_{ff}^3} \right] \Bigg\} \\
&= \frac{4\pi\varepsilon_{ff}\sigma_{ff}^6}{s_{z0}-2\sigma_{fw}} \Bigg\{ \frac{11}{24\sigma_f^2} - \frac{1}{24}\left[\frac{1}{(s_{z0}-2\sigma_{fw}-\sigma_{ff})^2} - \frac{1}{(s_{z0}-2\sigma_{fw})^2} \right] \Bigg\}
\end{aligned}
$$
(D.2)

In a similar fashion, we obtain

$$
\begin{aligned}
a_2 &= \frac{4\pi\varepsilon_{ff}\sigma_{ff}^6}{s_{z0}-2\sigma_{fw}} \int_{\sigma_{fw}+\sigma_{ff}}^{s_{z0}-\sigma_{fw}-\sigma_{ff}} dz_1 \Bigg\{ \int_{-z_1+\sigma_{fw}}^{-\sigma_{ff}} dz \int_0^{\infty} \frac{d\rho\,\rho}{(z^2+\rho^2)^3} \\
&\quad + \int_{-\sigma_{ff}}^{\sigma_{ff}} dz \int_{\sqrt{\sigma_{ff}^2-z^2}}^{\infty} \frac{d\rho\,\rho}{(z^2+\rho^2)^3} + \int_{\sigma_{ff}}^{s_{z0}-z_1-\sigma_{fw}} dz \int_0^{\infty} \frac{d\rho\,\rho}{(z^2+\rho^2)^3} \Bigg\} \\
&= \frac{4\pi\varepsilon_{ff}\sigma_{ff}^6}{s_{z0}-2\sigma_{fw}} \int_{\sigma_{fw}+\sigma_{ff}}^{s_{z0}-\sigma_{fw}-\sigma_{ff}} dz_1 \Bigg\{ \frac{2}{3\sigma_{ff}^3} - \frac{1}{12(z_1-\sigma_{fw})^3} \\
&\quad - \frac{1}{12(s_{z0}-z_1-\sigma_{fw})^3} \Bigg\} \\
&= \frac{8\pi\varepsilon_{ff}\sigma_{ff}^6}{s_{z0}-2\sigma_{fw}} \Bigg\{ \frac{1}{3\sigma_{ff}^3} |s_{z0}-2\sigma_{fw}-2\sigma_{ff}| \\
&\quad + \frac{1}{24}\left[\frac{1}{(s_{z0}-2\sigma_{fw}-\sigma_{ff})^2} - \frac{1}{\sigma_{ff}^2} \right] \Bigg\}
\end{aligned}
$$
(D.3)

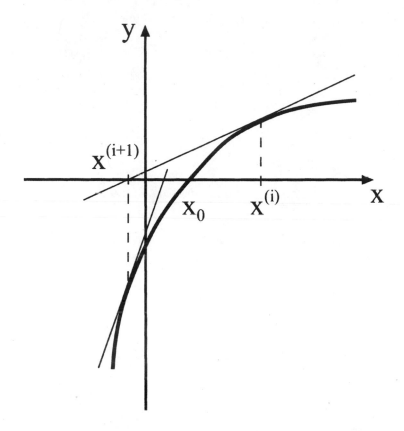

Figure D.1: Newton's iterative method to locate the zero of a function $y = f(x_0) = 0$ through a tangent construction (see text).

From Eqs. (D.2) and (D.3), we at last obtain

$$a_\text{p} = 2a_1 + a_2 = \frac{8\pi\varepsilon_\text{ff}\sigma_\text{ff}^3}{3} \frac{1}{s_{z0} - 2\sigma_\text{fw}} \left\{ s_{z0} - 2\sigma_\text{fw} - \frac{3\sigma_\text{ff}}{4} + \frac{\sigma_\text{ff}^3}{8\left(s_{z0} - 2\sigma_\text{fw}\right)^2} \right\}$$

(D.4)

Equation (4.24) follows immediately from Eqs. (D.4) and (4.25) and the definition of ξ.

D.1.2 Newton's method

To find the zero(s) of an analytic function $f(x)$ defined on a domain $D(f) \subseteq \mathbb{R}$, one may employ an interative technique due to Newton. Suppose an interval $[a, b]$ exists such that $f(x)$ is monotonic, continuous, and differentiable everywhere in $[a, b]$ (see Fig. D.1). Moreover assume a value $x_0 \in [a, b]$ such that $f(x_0) = 0$.

Consider now an *arbitrary* point $x^{(i)} \in [a, b] \neq x_0$. At this point the tangent to $f(x)$ is given by

$$g(x) = f\left(x^{(i)}\right) + f'\left(x^{(i)}\right)\left(x - x^{(i)}\right) \tag{D.5}$$

where $f'\left(x^{(i)}\right) \equiv \mathrm{d}f(x)/\mathrm{d}x\big|_{x=x^{(i)}}$. From this expression we may calculate the zero of the tangent (see Fig. D.1)

$$g\left(x^{(i+1)}\right) \overset{!}{=} 0 \longrightarrow x^{(i+1)} = x^{(i)} - \frac{f\left(x^{(i)}\right)}{f'\left(x^{(i)}\right)}, \qquad i = 1, \ldots, n \tag{D.6}$$

Equation (D.6) is the basis of Newton's iterative scheme. In the next iteration we replace $x^{(i)}$ on the right side of Eq. (D.6) by $x^{(i+1)}$ and obtain a new estimate $x^{(i+2)}$ until

$$\left|x^{(n+1)} - x^{(n)}\right| \lesssim \epsilon \tag{D.7}$$

where $\epsilon \ll 1$ is some predefined small number. Because of the properties of $f(x)$ specified at the outset, one obtains

$$x_0 \approx x^{(n+1)} \tag{D.8}$$

if the inequality stated in Eq. (D.7) is satisfied. The accuracy of this algorithm depends on whether $f'(x)$ is known analytically. An extension to functions depending on more than just a single variable is straightforward.

In applying Newton's method it is crucial that $f(x)$ is monotonic in the interval $[a, b]$. If this is not the case, an iterative solution of Eq. (D.6) may not converge to the correct value of x_0. In this case, Newton's method may, however, still work if the intial value $x^{(i)}$ is selected by an "educated guess"; that is, $x^{(i)}$ needs to be chosen sufficiently close to x_0 where $f(x)$ is still monotonic.

D.1.3 Maxwell's constraint

The densities of the coexisting gas

$$\rho_b^g(T_{0b}) \equiv \min\{\rho_{b1}, \rho_{b2}, \rho_{b3}\} \tag{D.9}$$

and liquid

$$\rho_b^l(T_{0b}) \equiv \max\{\rho_{b1}, \rho_{b2}, \rho_{b3}\} \tag{D.10}$$

must satisfy Maxwell's constraint [327]

$$0 \overset{!}{=} P\left(T_{0b}\right)\left(V^{g} - V^{l}\right) - \int\limits_{V^{l}}^{V^{g}} dV\, P\left(T_{0b}\right)$$

$$= P\left(T_{0b}\right)\frac{\rho_{b}^{l} - \rho_{b}^{g}}{\rho_{b}^{l}\rho_{b}^{g}} + k_{B}T_{0b}\ln\frac{\rho_{b}^{g}\left(1 - \rho_{b}^{l}b\right)}{\rho_{b}^{l}\left(1 - \rho_{b}^{g}b\right)} + a_{b}\left(\rho_{b}^{l} - \rho_{b}^{g}\right) \quad \text{(D.11)}$$

We assume the gas density $\rho_{b}^{g} \equiv \rho_{b}$ to be fixed; the density of coexisting liquid and the coexistence temperature are unknown. Therefore, we must solve Eqs. (4.43) and (D.11) *simultaneously* for T_{0b} and ρ_{b}^{l}. We solve the equations by the following procedure:

1. Compute $P\left(T_{i}\right)$ from Eq. (4.29) for fixed $\rho_{b}/\rho_{cb} \leq 1$ and initial temperature $T_{i}/T_{cb} = 1$.

2. Compute $\rho_{b}^{g}\left(T_{i}\right) \equiv \rho_{b}$ and $\rho_{b}^{l}\left(T_{i}\right)$ from Eq. (4.44) and use these to calculate

$$f\left(T_{i}\right) = P\left(T_{i}\right)\frac{\rho_{b}^{l} - \rho_{b}^{g}}{\rho_{b}^{l}\rho_{b}^{g}} + k_{B}T_{i}\ln\frac{\rho_{b}^{g}\left(1 - \rho_{b}^{l}b\right)}{\rho_{b}^{l}\left(1 - \rho_{b}^{g}b\right)} + a_{b}\left(\rho_{b}^{l} - \rho_{b}^{g}\right) \quad \text{(D.12)}$$

3. Obtain a new estimate for the coexistence temperature *via* Newton's method as $T_{i+1} = T_{i} - f\left(T_{i}\right)/f'\left(T_{i}\right)$ [see Eq. (D.6)], where

$$f' = k_{B}\frac{\rho_{b}^{l} - \rho_{b}^{g}}{\rho_{b}^{l}\left(1 - \rho_{b}^{g}b\right)} + k_{B}\ln\left[\frac{\rho_{b}^{g}\left(1 - \rho_{b}^{l}b\right)}{\rho_{b}^{l}\left(1 - \rho_{b}^{g}b\right)}\right] \quad \text{(D.13)}$$

is the (partial) derivative of f with respect to T.

4. Replace T_{i} by T_{i+1} and return to step 1.

The procedure is halted when $\left|f\left(T_{n}\right)\right| \leq 10^{-6}$. At the end it is sensible to check whether $\rho_{b}^{g}\left(T_{n}\right) = \rho_{b}$, as specified at the outset, and that $P\left(T_{n}, \rho_{b}^{g}\right) = P\left(T_{n}, \rho_{b}^{l}\right)$ for the coexisting phases, so that $T_{0b}\left(\rho_{b}\right) \equiv T_{n}$. Convergence is achieved after five or six iterations. The resulting bulk coexistence curve is plotted in Fig. 4.5. To illustrate the effect of confinement, we also plot in that same figure the pore coexistence curve for $s_{z} = 50$ determined by the procedure detailed above using, however, the corresponding equation of state for the pore fluid given in Eq. (4.28).

D.2 Lattice models

D.2.1 Numerical solution of Eq. (4.86)

To solve Eq. (4.86) we employ the Jacobi-Newton iteration technique, which proceeds iteratively in an alternating sequence of "local" and "global" minimization steps. Let $\rho_i^{k,l}$ be the local density at lattice site i in the kth local and the lth global minimization step. A *local* estimate for the corresponding minimum value of $\Omega^{k,l}$ is obtained via Newton's method [see Eq. (D.6)]; that is,

$$\rho_i^{k+1,l} = \rho_i^{k,l} - \frac{f\left(\rho_i^{k,l}\right)}{f'\left(\rho_i^{k,l}\right)}, \qquad i = 1, \ldots, \mathcal{N} \tag{D.14}$$

From Eqs. (4.83) we have for the functional derivative

$$f\left(\rho_i^{k,l}\right) \equiv \frac{\delta\Omega^{k,l}\left[\boldsymbol{\rho}^{k,l}\right]}{\delta\rho_i^{k,l}} = -k_{\mathrm{B}}T\ln\frac{\rho_i^{k,l}}{1-\rho_i^{k,l}} + \varepsilon_{\mathrm{ff}}\sum_{j\neq i}^{\nu(i)}\rho_j^{k,l} + \mu + \Phi_i \tag{D.15}$$

and therefore

$$f'\left(\rho_i^{k,l}\right) = -\frac{k_{\mathrm{B}}T}{\rho_i^{k,l}\left(1-\rho_i^{k,l}\right)} \tag{D.16}$$

It is important to realize that throughout each local minimization cycle the densities at nearest-neighbor sites of site i represented by the set $\left\{\rho_j^{k,l}\right\}$ remain fixed at the initial values assigned to them at the beginning of the local cycle. The iterative solution of Eq. (D.14) is halted if [see Eq. (D.7)]

$$\max_{i=1,\mathcal{N}}\left|\rho_i^{k+1,l} - \rho_i^{k,l}\right| \leq 10^{-7} \tag{D.17}$$

is satisfied, which is typically achieved in approximately two to three iterations. Local minimization is performed by visiting each lattice site i consecutively, and the local cycle ends once all sites have been considered.

"Global" minimization then involves updating the local density of the *entire* lattice according to

$$\rho_i^{0,l+1} = \rho_i^{k+1,l}, \qquad i = 1, \ldots, \mathcal{N} \tag{D.18}$$

thus providing new initial values for the next *local* minimization cycle [by setting $l + 1 \to l$ and returning to Eq. (D.14)]. Global minimization is carried out until [see Eq. (D.7)]

$$\max_{i=1,\mathcal{N}}\left|\rho_i^{0,l+1} - \rho_i^{0,l}\right| \leq 10^{-7} \tag{D.19}$$

which is achieved in roughly 100 steps.

Once local and global minimzation have converged for the current T and μ according to Eqs. (D.17) and (D.19), we may calculate the set of grand potentials $\{\Omega^\alpha\}$ for all morphologies from Eq. (4.83) and repeat the iterative procedure for a slightly different chemical potanial. Once the curves Ω^α versus μ have been obtained for all morphologies $\{M^\alpha\}$ and the given temperature T, we determine $\mu^{\alpha\beta}$ by numerically solving Eq. (1.76a). We then increase the temperature by a small amount δT and repeat the above procedure where some care has to be taken in the vicinity of critical points because the grand potentials for coexisting phases α and β become increasingly similar. Hence, we chose $\delta T = 2.5 \times 10^{-3}$ (in dimensionless units) as soon as the density difference $\Delta\rho^{\alpha\beta} \lesssim 0.05$.

To initiate the iterative scheme, suitable starting solutions for Eq. (D.14) are obviously required. These solutions are provided by the morphologies M^α at $T = 0$ for which $\mu^{\alpha\beta}$ can be calculated analytically from the expressions for Ω^α compiled in Table 4.1.

D.2.2 Binary mixtures

D.2.2.1 The mean-field Hamiltonian function

Here we work out expressions for the number N_{AA} (N_{BB}) of A–A (B–B) pairs. These pairs are directly connected sites, both of which are occupied by a molecule of species A (B). Likewise, expressions for N_{AB} (s) and the total number of molecules of species A and B at the solid substrate, N_{AW} (s) and N_{BW} (s), may be derived easily. The resulting expressions presented in Eqs. (4.118)–(4.121) contain terms that can be cast as

$$\sum_{k,l} s_{k,l} s_{k,l\pm1} = \sum_{k,l} s_{k,l}\rho_{l\pm1}m_{l\pm1} = n\sum_l \rho_l\rho_{l\pm1}m_l m_{l\pm1} \quad \text{(D.20a)}$$

$$\sum_{k,l} s_{k,l} s_{k,l\pm1}^2 = \sum_{k,l} s_{k,l}\rho_{l\pm1} = n\sum_l \rho_l\rho_{l\pm1}m_l \quad \text{(D.20b)}$$

$$\sum_{k,l} s_{k,l}^2 s_{k,l\pm1}^2 = \sum_{k,l} s_{k,l}^2\rho_{l\pm1} = n\sum_l \rho_l\rho_{l\pm1} \quad \text{(D.20c)}$$

$$\sum_{k,l} s_{k,l}^2 s_{k,l\pm1} = \sum_{k,l} s_{k,l}^2\rho_{l\pm1}m_{l\pm1} = n\sum_l \rho_l\rho_{l\pm1}m_{l\pm1} \quad \text{(D.20d)}$$

at the mean-field level. In addition,

$$\sum_{k,l} s_{k,l} \sum_{m \neq k}^{\nu(i)} s_{m,l} = 4 \sum_{k,l} s_{k,l} \rho_l m_l = 4n \sum_l \rho_l^2 m_l^2 \qquad \text{(D.21a)}$$

$$\sum_{k,l} s_{k,l} \sum_{m \neq k}^{\nu(i)} s_{m,l}^2 = 4 \sum_{k,l} s_{k,l} \rho_l = 4n \sum_l \rho_l^2 m_l \qquad \text{(D.21b)}$$

$$\sum_{k,l} s_{k,l}^2 \sum_{m \neq k}^{\nu(i)} s_{m,l}^2 = 4 \sum_{k,l} s_{k,l} \rho_l = 4n \sum_l \rho_l^2 \qquad \text{(D.21c)}$$

$$\sum_{k,l} s_{k,l}^2 \sum_{m \neq k}^{\nu(i)} s_{m,l} = 4 \sum_{k,l} s_{k,l}^2 \rho_l m_l = 4n \sum_l \rho_l^2 m_l \qquad \text{(D.21d)}$$

arise where the summation over the four nearest neighbors n.n. of lattice site k can be carried out explicitly if Eqs. (4.126) and (4.127) are invoked. Finally,

$$\sum_k s_{k,l} = n\rho_l m_l \qquad \text{(D.22a)}$$

$$\sum_k s_{k,l}^2 = n\rho_l \qquad \text{(D.22b)}$$

may be employed in Eqs. (2.4) and (2.5) of Ref. 84 to replace the sum over sites in lattice planes $l = 1, z$.

Hence, Eqs. (D.20) as well as (D.22) permit us to write down the mean-field expressions

$$N_{\mathrm{AA}}(\boldsymbol{\rho}, \boldsymbol{m}) = \frac{n}{8} \sum_{l=1}^{z} [\rho_l \rho_{l+1} (m_l + 1) (m_{l+1} + 1)$$
$$+ \rho_l \rho_{l-1} (m_l + 1) (m_{l-1} + 1) + 4\rho_l^2 (m_l + 1)^2] \quad \text{(D.23a)}$$

$$N_{\mathrm{BB}}(\boldsymbol{\rho}, \boldsymbol{m}) = \frac{n}{8} \sum_{l=1}^{z} [\rho_l \rho_{l+1} (m_l - 1) (m_{l+1} - 1)$$
$$+ \rho_l \rho_{l-1} (m_l - 1) (m_{l-1} - 1) + 4\rho_l^2 (m_l - 1)^2] \quad \text{(D.23b)}$$

$$N_{\mathrm{AB}}(\boldsymbol{\rho}, \boldsymbol{m}) = \frac{n}{4} [\rho_l \rho_{l+1} (1 - m_l m_{l+1}) + \rho_l \rho_{l-1} (1 - m_l m_{l-1})$$
$$+ 4\rho_l^2 (1 - m_l m_l)] \quad \text{(D.23c)}$$

$$N_{\mathrm{AW}}(\boldsymbol{\rho}, \boldsymbol{m}) = \frac{n}{2} [\rho_1 (1 + m_1) + \rho_z (1 + m_z)] \quad \text{(D.23d)}$$

$$N_{\mathrm{BW}}(\boldsymbol{\rho}, \boldsymbol{m}) = \frac{n}{2} [\rho_1 (1 - m_1) + \rho_z (1 - m_z)] \quad \text{(D.23e)}$$

which follow after somewhat tedious but straightforward algebraic manipulations.

Replacing now in Eq. (4.123) $N_{\alpha\beta}$ (s) and $N_{\alpha W}$ (s) ($\alpha, \beta = A, B$) by their mean-field counterparts given in Eqs. (D.23) eventually yields

$$
\begin{aligned}
H_{\mathrm{mf}}(\rho, m) &= \frac{\varepsilon n}{8} \sum_{l=1}^{n_z} \Bigg[\rho_l \rho_{l+1} (m_l + 1)(m_{l+1} + 1) \\
&\quad + \rho_l \rho_{l-1} (m_l + 1)(m_{l-1} + 1) + 4\rho_l \rho_l (m_l + 1)^2 \Bigg] \\
&\quad + \frac{\varepsilon \chi_B n}{8} \sum_{l=1}^{n_z} \Bigg[\rho_l \rho_{l+1} (m_l - 1)(m_{l+1} - 1) \\
&\quad + \rho_l \rho_{l-1} (m_l - 1)(m_{l-1} - 1) + 4\rho_l \rho_l (m_l - 1)^2 \Bigg] \\
&\quad + \frac{\varepsilon_{AB} n}{4} \sum_{l=1}^{n_z} \Bigg[\rho_l \rho_{l+1} (1 - m_l m_{l+1}) + \rho_l \rho_{l-1} (1 - m_l m_{l-1}) \\
&\quad + 4\rho_l \rho_l (1 - m_l m_l) \Bigg] \\
&\quad + \frac{\varepsilon_{AW} n}{2} [\rho_1 (1 + m_1) + \rho_z (1 + m_z)] \\
&\quad + \frac{\varepsilon_{AW} \chi_W n}{2} [\rho_1 (1 - m_1) + \rho_z (1 - m_z)] - n \sum_{l=1}^{n_z} \rho_l \mu \quad \text{(D.24)}
\end{aligned}
$$

as the mean-field analog of H given in Eq. (4.123). In Eq. (D.24) the first two sums account for the interaction between a pair of molecules of species A and B, respectively. Likewise, the third sum represents the contribution of A–B attractions to the mean-field Hamiltonian. The next two terms represent the interaction between a molecule of species A and B with the solid substrate and the last term couples the system to an (infinitely large) external reservoir of matter. Moreover, it is easy to verify that Eq. (D.24) degenerates to the expression

$$
\begin{aligned}
H_{\mathrm{mf}}(\rho) &= \frac{\varepsilon n}{2} \sum_{l=1}^{n_z} [\rho_l \rho_{l+1} + \rho_l \rho_{l-1} + 4\rho_l \rho_l] \\
&\quad + \varepsilon_W n [\rho_1 (1 - m_1) + \rho_z (1 - m_z)] - n \sum_{l=1}^{n_z} \rho_l \mu \quad \text{(D.25)}
\end{aligned}
$$

for a pure fluid ($m = 0$, $\chi_B = \chi_{AW} = 1$, $\varepsilon = \varepsilon_{AB}$) confined between chemically homogeneous substrates.

D.2.2.2 Derivation of Eqs. (4.132)

To derive Eqs. (4.132) we depart from Eq. (4.130), which may be written more explicitly as

$$
\Theta\left(\boldsymbol{n}, \boldsymbol{m}\right) = \prod_{l=1}^{z} n^{n} \left[\left(n - n_{l}\right)^{n - n_{l}} \left(n_{l}\frac{1 + m_{l}}{2}\right)^{n_{l}(1 + m_{l})/2} \right.
$$
$$
\left. \times \left(n_{l}\frac{1 - m_{l}}{2}\right)^{n_{l}(1 - m_{l})/2} \right]^{-1}
\tag{D.26}
$$

in the limit $n, \{n_{l}\} \to \infty$ using Stirling's approximation [see Eq. (B.7)]. Using [see Eqs. (4.126) and (4.127)]

$$
\rho_{l}^{A} = \rho_{l}\left(\frac{1 + m_{l}}{2}\right)
\tag{D.27a}
$$
$$
\rho_{l}^{B} = \rho_{l}\left(\frac{1 - m_{l}}{2}\right)
\tag{D.27b}
$$

so that Eq. (D.26) may be recast as

$$
\ln\Theta\left(\boldsymbol{\rho}, \boldsymbol{m}\right) = -n\sum_{l=1}^{z}\{\rho_{l}\ln\rho_{l} + \left(1 - \rho_{l}\right)\ln\left(1 - \rho_{l}\right) - \rho_{l}\ln 2
$$
$$
+ \frac{\rho_{l}}{2}\left[\left(1 + m_{l}\right)\ln\left(1 + m_{l}\right) + \left(1 - m_{l}\right)\ln\left(1 - m_{l}\right)\right]\}
\tag{D.28}
$$

Hence, together with Eqs. (4.131) and (D.24), Eq. (D.28) yields

$$
\omega\left(\boldsymbol{\rho}, \boldsymbol{m}; T, \mu\right) = f_{\text{int}}\left(\boldsymbol{\rho}, \boldsymbol{m}\right) + \frac{\varepsilon_{\text{AW}}}{2z}\left[\rho_{1}\left(m_{1} + 1\right) + \rho_{z}\left(m_{z} + 1\right)\right]
$$
$$
- \frac{\varepsilon_{\text{AW}}\chi_{\text{w}}}{2z}\left[\rho_{1}\left(m_{1} - 1\right) + \rho_{z}\left(m_{z} - 1\right)\right]
$$
$$
- \frac{1}{z}\sum_{l=1}^{z}\rho_{l}\mu
\tag{D.29}
$$

where the intrinsic free-energy density is defined as

$$
f_{\text{int}}\left(\boldsymbol{\rho}, \boldsymbol{m}\right) = u_{\text{AA}}\left(\boldsymbol{\rho}, \boldsymbol{m}\right) + u_{\text{BB}}\left(\boldsymbol{\rho}, \boldsymbol{m}\right) + u_{\text{AB}}\left(\boldsymbol{\rho}, \boldsymbol{m}\right) - Ts\left(\boldsymbol{\rho}, \boldsymbol{m}\right)
\tag{D.30}
$$

and the entropy density $s(\rho, m)$ as well as the various internal energies $u_{AA}(\rho, m)$, $u_{BB}(\rho, m)$, and $u_{AB}(\rho, m)$ are given by the expressions

$$
s = -\frac{k_B}{2z}\left[2\sum_{l=1}^{z}\{\rho_l \ln \rho_l + (1 - \rho_l)\ln(1 - \rho_l) - \rho_l \ln 2\}\right.
$$

$$
\left. + \sum_{l=1}^{z}\rho_l\{(1 + m_l)\ln(1 + m_l) + (1 - m_l)\ln(1 - m_l)\}\right] \quad \text{(D.31a)}
$$

$$
u_{AA} = \frac{\varepsilon}{8z}\sum_{l=1}^{z}\left[\rho_l\rho_{l+1}(m_l + 1)(m_{l+1} + 1)\right.
$$

$$
\left. + \rho_l\rho_{l-1}(m_l + 1)(m_{l-1} + 1) + 4\rho_l\rho_l(m_l + 1)^2\right] \quad \text{(D.31b)}
$$

$$
u_{BB} = \frac{\varepsilon\chi_b}{8z}\sum_{l=1}^{z}\left[\rho_l\rho_{l+1}(m_l - 1)(m_{l+1} - 1)\right.
$$

$$
\left. + \rho_l\rho_{l-1}(m_l - 1)(m_{l-1} - 1) + 4\rho_l\rho_l(m_l - 1)^2\right] \quad \text{(D.31c)}
$$

$$
u_{AB} = \frac{\varepsilon_{AB}}{4z}\sum_{l=1}^{z}\left[\rho_l\rho_{l+1}(1 - m_l m_{l+1})\right.
$$

$$
\left. + \rho_l\rho_{l-1}(1 - m_l m_{l-1}) + 4\rho_l\rho_l(1 - m_l m_l)\right] \quad \text{(D.31d)}
$$

In Eq. (D.31a) the factor proportional to $\ln 2$ accounts for the distinguishability of molecules of species A and B (i.e., their different "color"). Following Pini et al. the color of molecules may be given a more physical interpretation by interpreting it as a "spinlike variable in addition to translational degrees of freedom so that their mutual interaction depends both on their relative position and on their 'internal' state, namely whether the interacting particles belong to the same species or not" [328].

Differentiating Eqs. (D.29) according to Eqs. (4.132) we find after somewhat lengthy but straightforward algebraic manipulations our desired result,

$$
\begin{aligned}
zh_1^k &= -\mu^* + k_B T \left[\ln \frac{\rho_k}{1 - \rho_k} + \frac{1}{2} \left\{ (1 + m_k) \ln (1 + m_k) \right. \right. \\
&\qquad \left. \left. + (1 - m_k) \ln (1 - m_k) \right\} \right] \\
&\quad + \frac{\varepsilon}{4} (m_k + 1) \left[\rho_{k+1} (m_{k+1} + 1) + \rho_{k-1} (m_{k-1} + 1) + 4\rho_k (m_k + 1) \right] \\
&\quad + \frac{\varepsilon \chi_b}{4} (m_k - 1) \left[\rho_{k+1} (m_{k+1} - 1) + \rho_{k-1} (m_{k-1} - 1) + 4\rho_k (m_k - 1) \right] \\
&\quad + \frac{\varepsilon_{AB}}{2} \left[\rho_{k+1} (1 - m_k m_{k+1}) + \rho_{k-1} (1 - m_k m_{k-1}) + 4\rho_k \left(1 - m_k^2 \right) \right] \\
&\quad + \frac{\varepsilon_{AW}}{2} \left\{ \delta_{k1} \left[(1 + m_1) + \chi_w (1 - m_1) \right] \right. \\
&\qquad \left. + \delta_{kz} \left[(1 + m_z) + \chi_w (1 - m_z) \right] \right\}
\end{aligned}
\tag{D.32a}
$$

$$
\begin{aligned}
zh_2^k &= \frac{\rho_k}{2} \left[\beta^{-1} \ln \frac{1 + m_k}{1 - m_k} \right. \\
&\quad + \frac{\varepsilon}{2} \left[\rho_{k+1} (m_{k+1} + 1) + \rho_{k-1} (m_{k-1} + 1) + 4\rho_k (m_k + 1) \right] \\
&\quad + \frac{\varepsilon \chi_b}{2} \left[\rho_{k+1} (m_{k+1} - 1) + \rho_{k-1} (m_{k-1} - 1) + 4\rho_k (m_k - 1) \right] \\
&\quad - \varepsilon_{AB} \left[\rho_{k+1} m_{k+1} + \rho_{k-1} m_{k-1} + 4\rho_k m_k \right] \\
&\quad \left. + \varepsilon_{AW} (\delta_{k1} + \delta_{kz}) (1 - \chi_w) \right]
\end{aligned}
\tag{D.32b}
$$

where we also used Eqs. (D.31) and introduced $\mu^* \equiv \mu - \beta^{-1} \ln 2$, which takes into account the trivial contribution from the different "color" of molecules pertaining to one or the other species. In Eqs. (D.32), δ_{ij} is the Kronecker symbol. For the special case of a symmetric binary mixture ($\chi_B = 1$) confined between nonselective solid surfaces ($\chi_W = 1$), Eqs. (D.32) simplify considerably and we obtain

$$
\begin{aligned}
zh_1^k &= -\bar{\mu} + \beta^{-1}\left[\ln \frac{\rho_k}{1-\rho_k} \right. \\
&\quad \left. + \frac{1}{2}\left\{ (1+m_k)\ln(1+m_k) + (1-m_k)\ln(1-m_k) \right\} \right] \\
&\quad + \bar{\varepsilon}\,(\rho_{k+1} + \rho_{k-1} + 4\rho_k) \\
&\quad + \Delta\varepsilon\, m_k\,(\rho_{k+1}m_{k+1} + \rho_{k-1}m_{k-1} + 4\rho_k m_k) \\
&\quad + \varepsilon_{\mathrm{W}}\,(\delta_{k1} + \delta_{kz}) \qquad\qquad\qquad\qquad\qquad \text{(D.33a)} \\
zh_2^k &= \frac{\rho_k}{2}\left[\beta^{-1}\ln\frac{1+m_k}{1-m_k} \right. \\
&\quad \left. + 2\Delta\varepsilon\left\{ \rho_{k+1}m_{k+1} + \rho_{k-1}m_{k-1} + 4\rho_k m_k \right\} \right] \qquad \text{(D.33b)}
\end{aligned}
$$

where $\varepsilon_{\mathrm{W}} \equiv \varepsilon_{\mathrm{AW}}$ and

$$
\bar{\varepsilon} \equiv \frac{1}{2}\left(\varepsilon + \varepsilon_{\mathrm{AB}} \right) \qquad\qquad\qquad \text{(D.34a)}
$$

$$
\Delta\varepsilon \equiv \frac{1}{2}\left(\varepsilon - \varepsilon_{\mathrm{AB}} \right) \qquad\qquad\qquad \text{(D.34b)}
$$

D.2.2.3   Numerical solution of Eq. (4.143)

As we restrict ourselves to nearest-neighbor interactions, $h_{1,2}^k$ depend only on the set of variables $\{\rho_j, m_j \,|\, k-1 \le j \le k+1\}$ as one can verify from Eqs. (D.32). Hence, $\mathbf{D}$ in Eq. (4.142) has a band structure where all elements

$$
\frac{\partial h_{1,2}^i}{\partial \rho_j} = 0 \qquad\qquad\qquad\qquad \text{(D.35a)}
$$

$$
\frac{\partial h_{1,2}^i}{\partial m_j} = 0, \qquad \forall\, |i - j| \ge 2 \qquad\qquad \text{(D.35b)}
$$

Moreover, as $\omega\,(\mathcal{P})$ is continuous and differentiable, we have

$$
\frac{\partial h_1^k}{\partial m_l} = \frac{\partial^2 \omega}{\partial m_l \partial \rho_k} = \frac{\partial^2 \omega}{\partial \rho_k \partial m_l} = \frac{\partial h_2^k}{\partial \rho_k}, \qquad \forall k, l \qquad \text{(D.36)}
$$

that is, $\mathbf{D}$ is symmetric with respect to its main diagonal. As in this work we focus on planar, chemically homogeneous substrates, an additional symmetry exists for the local order parameters with respect to a (virtual) midplane on the lattice that might coincide with an actual lattice plane if z is odd. Therefore, if z is odd we conclude that

$$\rho_{\Phi-k} = \rho_{\Phi+k} \tag{D.37a}$$

$$m_{\Phi-k} = m_{\Phi+k}, \quad k = 1, \ldots, \Phi - 1 \tag{D.37b}$$

where $\Phi = (z + 1)/2$. If, on the other hand, z is even

$$\rho_{\Phi-k+1} = \rho_{\Phi+k} \tag{D.38a}$$

$$m_{\Phi-k+1} = m_{\Phi+k}, \quad k = 1, \ldots, \Phi \tag{D.38b}$$

where now, of course, $\Phi = z/2$.

To simplify the subsequent discussion we restrict ourselves to the case of even z where we note in passing that for sufficiently large z the distinction between odd and even numbers of lattice planes becomes irrelevant. Then we may reorganize the $2z$ elements of vectors $\boldsymbol{x}$, $\boldsymbol{f}$, and $\boldsymbol{\nabla}$ such that the resulting matrix $\mathbf{D}$ has point symmetry with respect to an inversion center. More specifically, we may express $\mathbf{D}$ formally as

$$\mathbf{D} = \begin{pmatrix} \widetilde{\mathbf{A}} & \widetilde{\mathbf{B}} \\ \mathbf{B} & \mathbf{A} \end{pmatrix} \tag{D.39}$$

where elements of the submatrices are related through

$$\widetilde{A}_{\Phi-k+1,\Phi-l+1} = A_{k,l} \tag{D.40a}$$

$$\widetilde{B}_{\Phi-k+1,\Phi-l+1} = B_{k,l}, \quad k,l = 1, \ldots, z \tag{D.40b}$$

It is then easy to verify that

$$\widetilde{\mathbf{B}} = \begin{pmatrix} 0 & & \cdots & & 0 \\ \vdots & \ddots & & & \vdots \\ \dfrac{\partial h_2^{\Phi+1}}{\partial \rho_\Phi} & \dfrac{\partial h_1^{\Phi+1}}{\partial \rho_\Phi} & 0 & \cdots & 0 \\ \dfrac{\partial h_2^{\Phi+1}}{\partial m_\Phi} & \dfrac{\partial h_1^{\Phi+1}}{\partial m_\Phi} & 0 & \cdots & 0 \end{pmatrix} \tag{D.41}$$

which contains only four nonzero elements. Similarly, the last two rows of submatrix $\widetilde{\mathbf{A}}$ can be cast as

$$\widetilde{\mathbf{A}} = \begin{pmatrix} \vdots & & & \cdots & & & & \\ 0 & \cdots & 0 & \dfrac{\partial h_1^{\Phi-1}}{\partial \rho_\Phi} & \dfrac{\partial h_2^{\Phi-1}}{\partial \rho_\Phi} & \dfrac{\partial h_1^{\Phi}}{\partial \rho_\Phi} & \dfrac{\partial h_2^{\Phi}}{\partial \rho_\Phi} \\ 0 & \cdots & 0 & \dfrac{\partial h_2^{\Phi-1}}{\partial m_\Phi} & \dfrac{\partial h_1^{\Phi-1}}{\partial m_\Phi} & \dfrac{\partial h_2^{\Phi}}{\partial m_\Phi} & \dfrac{\partial h_1^{\Phi}}{\partial m_\Phi} \end{pmatrix} \tag{D.42}$$

One then realizes that in $\widetilde{\mathbf{A}}$ the element

$$\frac{\partial h_2^\Phi}{\partial \rho_\Phi} = \frac{\partial^2 \omega}{\partial \rho_\Phi \partial m_\Phi} = \frac{\partial^2 \omega}{\partial m_\Phi \partial \rho_\Phi} \qquad (D.43)$$

appears, whereas in $\widetilde{\mathbf{B}}$ the conjugate element

$$\frac{\partial h_2^{\Phi+1}}{\partial \rho_\Phi} = \frac{\partial^2 \omega}{\partial \rho_\Phi \partial m_{\Phi+1}} = \frac{\partial^2 \omega}{\partial m_{\Phi+1} \partial \rho_\Phi} \qquad (D.44)$$

arises. Similar considerations apply to the pair of elements $\partial h_1^\Phi / \partial \rho_\Phi$ and $\partial h_1^{\Phi+1} / \partial \rho_\Phi$ as well as to the corresponding two pairs of elements on the last rows of $\widetilde{\mathbf{A}}$ and $\widetilde{\mathbf{B}}$. Moreover, it is apparent from symmetry properties stated in Eqs. (D.38) (z even) that both $\partial h_2^\Phi / \partial \rho_\Phi$ and $\partial h_2^{\Phi+1} / \partial \rho_\Phi$ are acting on the same element of the vector $\boldsymbol{x} - \boldsymbol{x}_0$ in Eq. (4.141) such that Eq. (D.39) can be recast as

$$\mathbf{D}' = \begin{pmatrix} \widetilde{\mathbf{A}}' & \mathbf{0} \\ \mathbf{0} & \mathbf{A}' \end{pmatrix} \qquad (D.45)$$

where $\mathbf{0}$ is the $z \times z$ zero matrix, $\widetilde{\mathbf{A}}'$ is identical with $\widetilde{\mathbf{A}}$ except for the last two elements in the two bottom rows, that is [see Eq. (D.42)],

$$\widetilde{\mathbf{A}}' = \begin{pmatrix} \vdots & & \cdots & & \\ 0 & \cdots & 0 & \frac{\partial h_2^{\Phi-1}}{\partial \rho_\Phi} & \frac{\partial h_2^{\Phi-1}}{\partial \rho_\Phi} & \left(\frac{\partial h_1^\Phi}{\partial \rho_\Phi} + \frac{\partial h_1^\Phi}{\partial \rho_{\Phi+1}}\right) & \left(\frac{\partial h_1^\Phi}{\partial m_\Phi} + \frac{\partial h_1^\Phi}{\partial m_{\Phi+1}}\right) \\ 0 & \cdots & 0 & \frac{\partial h_2^{\Phi-1}}{\partial m_\Phi} & \frac{\partial h_1^{\Phi-1}}{\partial m_\Phi} & \left(\frac{\partial h_2^\Phi}{\partial \rho_\Phi} + \frac{\partial h_2^\Phi}{\partial \rho_{\Phi+1}}\right) & \left(\frac{\partial h_2^\Phi}{\partial m_\Phi} + \frac{\partial h_2^\Phi}{\partial m_{\Phi+1}}\right) \end{pmatrix}$$
$$(D.46)$$

and the relation between the new submatrices $\widetilde{\mathbf{A}}'$ and $\mathbf{A}'$ is the same as that between $\widetilde{\mathbf{A}}$ and $\mathbf{A}$ [see Eq. (D.40a)].

Because of these symmetry considerations we may replace Eq. (4.143) by

$$\widetilde{\boldsymbol{x}}_{i+1} = -\widetilde{\mathbf{A}}'^{-1} \cdot \boldsymbol{f}(\widetilde{\boldsymbol{x}}_i) + \widetilde{\boldsymbol{x}}_i \equiv \delta\widetilde{\boldsymbol{x}}_i + \widetilde{\boldsymbol{x}}_i \qquad (D.47)$$

where the (transpose of the) z-dimensional vectors

$$\widetilde{\boldsymbol{x}}^{\mathrm{T}} = (\rho_1, m_1, \ldots, \rho_\Phi, m_\Phi) \qquad (D.48a)$$
$$\boldsymbol{f}^{\mathrm{T}}(\widetilde{\boldsymbol{x}}) = (h_1^1, h_2^1, \ldots, h_1^\Phi, h_2^\Phi) \qquad (D.48b)$$

and the $z \times z$ matrix $\widetilde{\mathbf{A}}'$ replaces the $2z \times 2z$ matrix $\mathbf{D}$, which considerably reduces the numerical efforts necessary in solving the original Eq. (4.141) iteratively.

In practice, starting from a suitable solution $\widetilde{\boldsymbol{x}}_0^\alpha$ for a given phase $\mathcal{P}^\alpha$, we solve Eq. (D.47) iteratively until $|\boldsymbol{f}(\widetilde{\boldsymbol{x}}_{i+1})| \leq 10^{-11}$, which requires typically

10^2–10^3 iterations. Under this condition, $\widetilde{x}_{i+1}$ is a(n approximate, numerical) solution of the equation

$$f(\widetilde{x}) = 0 \qquad (D.49)$$

In general, we are not only interested in solutions of Eq. (D.49) but also, more specifically, in those solutions satisfying Eq. (1.76a), which defines the chemical potential at coexistence $\mu^{\alpha\beta}$ between phases (i.e., morphologies) M^α and M^β for a given temperature T. To determine $\mu^{\alpha\beta}$ at a slightly different temperature $T' = T + \delta T$, we expand ω in a Taylor series around some chemical potential μ_i, say, so that

$$\omega^{\alpha,\beta}(\mu_{i+1}) = \omega^{\alpha,\beta}(\mu_i) + \left.\frac{d\omega^{\alpha,\beta}}{d\mu}\right|_{\mu=\mu_i} (\mu_i - \mu_{i+1}) + \mathcal{O}\left[(\mu_i - \mu_{i+1})^2\right]$$

$$\simeq \omega^{\alpha,\beta}(\mu_i) - \rho_i^{\alpha,\beta}(\mu_i - \mu_{i+1}), \qquad T' = \text{const} \qquad (D.50)$$

where we dropped all other arguments to ease the notational burden. The second line of Eq. (D.50) follows from

$$\frac{d\omega}{d\mu} = \sum_{l=1}^{z}\left(\frac{\partial\omega}{\partial\rho_l}\frac{d\rho_l}{d\mu} + \frac{\partial\omega}{\partial m_l}\frac{dm_l}{d\mu}\right) + \frac{\partial\omega}{\partial\mu} = \frac{\partial\omega}{\partial\mu} = -\frac{1}{z}\sum_{l=1}^{z}\rho_l = -\rho \qquad (D.51)$$

where Eqs. (4.132) and (1.78) have also been employed. Assuming Eq. (1.76a) to hold for μ_{i+1}, we can solve Eq. (D.50) for μ_{i+1} to obtain

$$\mu_{i+1} = \mu_i + \frac{\omega^\beta(\mu_i) - \omega^\alpha(\mu_i)}{\rho_i^\beta - \rho_i^\alpha} \qquad (D.52)$$

thus providing an iterative scheme to calculate the chemical potential at coexistence. This scheme may be initiated by setting initially $\mu_i = \mu_0^{\alpha\beta}$ at the previous temperature T_0 and calculating $\rho_i^{\alpha,\beta}$ from $\widetilde{x}^{\alpha,\beta}$ at that temperature.

However, $\widetilde{x}^{\alpha,\beta}$ will no longer be solutions of Eq. (D.49) at T' and μ_{i+1}. Therefore, we solve Eqs. (D.49) and (D.52) until $|\delta\mu| \equiv \mu_{i+1}^{\alpha\beta} - \mu_i^{\alpha\beta} \lesssim 10^{-11}$. Hence, for a given temperature, M^α and M^β coexist at a chemical potential $\mu^{\alpha\beta} \equiv \mu_{i+1}$. However, the associated $\omega^{\alpha\beta}$ does not necessarily correspond to the *absolute* but may instead represent only a *relative* minimum of the grand-potential density. If, on the other hand, for any pair α, β, the grand-potential density assumes its *global* minimum, M^α and M^β are *thermodynamically* stable phases at coexistence. The range of temperatures and chemical potentials over which this condition is satisfied defines the coexistence line $\mu_x^{\alpha\beta}(T)$ between M^α and M^β (see Section 1.7).

Appendix E

Mathematical aspects of Monte Carlo simulations

E.1 Stochastic processes

E.1.1 The Chapman-Kolmogoroff equation

Let $y(t)$ be a general random process, that is, a process incompletely determined at any given time t. Specific examples are discussed in the context of MC simulations in Section 5.2. The random process can be described by a set of probability distributions $\{P_n\}$ where, for example, $P_2(y_1t_1, y_2t_2)\,\mathrm{d}y_1\mathrm{d}y_2$ is the probability of finding y_1 in the interval $[y_1, y_1 + \mathrm{d}y_1]$ at time $t = t_1$ and in the interval $[y_2, y_2 + \mathrm{d}y_2]$ at another time $t = t_2$. Thus, the set $\{P_n\}$ forms a hierarchy of probability distributions describing $y(t)$ in increasingly greater detail the larger is n.

The simplest random process is completely stochastic so that one may write, for example,

$$P_2(y_1t_1, y_2t_2) = P_1(y_1t_1)\, P_1(y_2t_2) \tag{E.1}$$

However, here we are concerned with a slightly more complex random process known as a *Markov process*, characterized by the equation

$$P_2(y_1t_1, y_2t_2) = P_1(y_1t_1)\, K_1(y_1t_1 \mid y_2t_2) \tag{E.2}$$

where $K_1(y_1t_1 \mid y_2t_2)$ is the *conditional* probability of finding y in the interval $[y_2, y_2 + \mathrm{d}y_2]$ at time $t = t_2$ *provided* $y = y_1$ at an earlier time $t = t_1 < t_2$.

Some properties relevant to the current discussion are listed below:

1. $K_1(y_1t_1 \mid y_2t_2)$ is normalized; that is,

$$\int K_1(y_1t_1 \mid y_2t_2)\,\mathrm{d}y_2 = 1 \tag{E.3}$$

431

2. The conditional probability serves as some sort of *propagator* in that it controls the temporal evolution of $y(t)$ in the sense of

$$\int P_1(y_1 t_1) K_1(y_1 t_1 \mid y_2 t_2) \, \mathrm{d}y_1 = P_1(y_2 t_2) \qquad \text{(E.4)}$$

3. The probability distributions satisfy a stationarity condition. In particular,

$$
\begin{aligned}
P_1(y_1 t_1) &= P_1(y_1) & \text{(E.5a)} \\
P_2(y_1 t_1, y_2 t_2) &= P_2(y_1, y_2; t_2 - t_1) & \text{(E.5b)}
\end{aligned}
$$

4. Most importantly, Markov processes have a "one-step memory". That is, to find y in the interval $[y_n, y_n + \mathrm{d}y_n]$ at $t = t_n$ depends only on the realization $y = y_{n-1}$ at the immediately preceding time $t = t_{n-1}$ but is independent of all earlier realizations $y = y_m$ at times $t = t_m$ where $1 \leq m \leq n - 2$. Mathematically speaking this can be cast as

$$K_{n-1}(y_1 t_1, \dots, y_{n-1} t_{n-1} \mid y_n t_n) = K_1(y_{n-1} t_{n-1} \mid y_n t_n) \qquad \text{(E.6)}$$

Consider now

$$P_2(y_{n-2} t_{n-2}) = \int P_3(y_{n-2} t_{n-2}, y_{n-1} t_{n-1}, y_n t_n) \, \mathrm{d}y_{n-1} \qquad \text{(E.7)}$$

and assume that $y(t)$ is a Markov process. Then

$$
\begin{aligned}
P_3(y_{n-2} t_{n-2}, y_{n-1} t_{n-1}, y_n t_n) &= P_1(y_{n-2} t_{n-2}) K_1(y_{n-2} t_{n-2} \mid y_{n-1} t_{n-1}) \\
&\quad \times K_2(y_{n-2} t_{n-2}, y_{n-1} t_{n-1} \mid y_n t_n) \\
&\overset{4.}{=} P_1(y_{n-2} t_{n-2}) K_1(y_{n-2} t_{n-2} \mid y_{n-1} t_{n-1}) \\
&\quad \times K_1(y_{n-1} t_{n-1} \mid y_n t_n) \qquad \text{(E.8)}
\end{aligned}
$$

employing property 4. above. Because of Eqs. (E.2) and (E.8), Eq. (E.7) can be recast as

$$K_1(y_{n-2} t_{n-2} \mid y_n t_n) = \int K_1(y_{n-2} t_{n-2} \mid y_{n-1} t_{n-1}) K_1(y_{n-1} t_{n-1} \mid y_n t_n) \, \mathrm{d}y_{n-1}$$

$$\text{(E.9)}$$

Because of property 3 an alternative formulation of the previous expression is given by

$$K_1\left(y_{n-2} \mid y_n; t+\tau\right) = \int K_1\left(y_{n-2} \mid y_{n-1}; t\right) K_1\left(y_{n-1} \mid y_n; \tau\right) dy_{n-1} \quad \text{(E.10)}$$

where $t \equiv t_{n-1} - t_{n-2}$ and $\tau \equiv t_n - t_{n-1}$. Equation (E.10) is the celebrated Chapman-Kolmogoroff equation in the theory of stochastic processes.

E.1.2 The Principle of Detailed Balance

Suppose a small time interval τ_c exists such that, during τ_c, y_{n-1} changes without strongly affecting $K_1\left(y_{n-2} \mid y_n; t+\tau_c\right)$ in Eq. (E.10). In the limit $\tau_c \to 0$ we may then expand the left side of Eq. (E.10) in a Taylor series according to

$$K_1\left(y_{n-2} \mid y_n; t+\tau_c\right) = K_1\left(y_{n-2} \mid y_n; t\right) + \frac{\partial K_1\left(y_{n-2} \mid y_n; t\right)}{\partial t} \tau_c + \mathcal{O}\left(\tau_c^2\right) \quad \text{(E.11)}$$

Inserting this expression into Eq. (E.10), we find after rearranging terms

$$\frac{\partial K_1\left(y_{n-2} \mid y_n; t\right)}{\partial t} = \int K_1\left(y_{n-2} \mid y_{n-1}; t\right) \frac{K_1\left(y_{n-1} \mid y_n; \tau_c\right)}{\tau_c} dy_{n-1}$$
$$- \frac{1}{\tau_c} K_1\left(y_{n-2} \mid y_n; t\right) \quad \text{(E.12)}$$

At this point it is convenient to introduce the transition probability per time interval τ_c via

$$\phi\left(y_{n-1} \mid y_n\right) = \lim_{\tau \to \tau_c} \frac{1}{\tau} K_1\left(y_{n-1} \mid y_n; \tau\right) \quad \text{(E.13)}$$

satisfying

$$\int \phi\left(y_{n-1} \mid y_n\right) dy_{n-1} = \int \phi\left(y_n \mid y_{n-1}\right) dy_{n-1} \overset{1.}{=} \lim_{\tau \to \tau_c} \frac{1}{\tau} = \frac{1}{\tau_c} \quad \text{(E.14)}$$

where we also employed the fact that the conditional probability is normalized [see property 1, introduced in Section E.1.1]. Multiplying the left side of Eq. (E.12) by $P_1\left(y_{n-2} t_{n-2}\right)$ (see Section E.1.1) and integrating over y_{n-2}, we may write

$$\frac{\partial}{\partial t'} \int K_1\left(y_{n-2} \mid y_n; t' - t_{n-2}\right) P_1\left(y_{n-2} t_{n-2}\right) dy_{n-2}$$
$$\overset{3.}{=} \frac{\partial}{\partial t'} \int K_1\left(y_{n-2} t_{n-2} \mid y_n; t'\right) P_1\left(y_{n-2} t_{n-2}\right) dy_{n-2}$$
$$\overset{2.}{=} \frac{\partial P_1\left(y_n t'\right)}{\partial t'} \quad \text{(E.15)}$$

where we introduced a new variable $t' \equiv t + t_{n-2}$ and invoked properties 2 and 3 from Section E.1.1 as indicated. By a similar token we may recast the first term on the right side of Eq. (E.12) as

$$
\begin{aligned}
&\int \int K_1 \left(y_{n-2} \mid y_{n-1}; t' - t_{n-2}\right) P_1 \left(y_{n-2}t_{n-2}\right) \phi \left(y_{n-1} \mid y_n\right) dy_{n-2}dy_{n-1} \\
&\overset{3.}{=} \int \int K_1 \left(y_{n-2}t_{n-2} \mid y_{n-1}; t'\right) P_1 \left(y_{n-2}t_{n-2}\right) \phi \left(y_{n-1} \mid y_n\right) dy_{n-2}dy_{n-1} \\
&\overset{2.}{=} \int P_1 \left(y_{n-1}t'\right) \phi \left(y_{n-1} \mid y_n\right) dy_{n-1} \qquad (E.16)
\end{aligned}
$$

where we also used the definition of ϕ in Eq. (E.13). Treating the second term on the right site of Eq. (E.12) on equal footing, we obtain

$$
\begin{aligned}
&\frac{1}{\tau_c} \int K_1 \left(y_{n-2} \mid y_n; t' - t_{n-2}\right) P_1 \left(y_{n-2}t_{n-2}\right) dy_{n-2} \\
&\overset{3.}{=} \frac{1}{\tau_c} \int K_1 \left(y_{n-2}t_{n-2} \mid y_n; t'\right) P_1 \left(y_{n-2}t_{n-2}\right) dy_{n-2} \\
&\overset{2.}{=} \frac{1}{\tau_c} P \left(y_n t'\right) = \int \phi \left(y_n \mid y_{n-1}\right) P \left(y_n t'\right) dy_{n-1} \qquad (E.17)
\end{aligned}
$$

where in the last step we employed the definition of $1/\tau_c$ given by Eq. (E.14). Putting together Eqs. (E.15)–(E.17), we finally arrive at

$$
\frac{\partial P_1 \left(y_n t\right)}{\partial t} = \int \left[P_1 \left(y_{n-1}t\right) \phi \left(y_{n-1} \mid y_n\right) - \phi \left(y_n \mid y_{n-1}\right) P \left(y_n t\right)\right] dy_{n-1} \quad (E.18)
$$

(replacing the time variable trivially according to $t' \to t$). In the context of the discussion of MC simulations in Section (5.2), the stationary solution of Eq. (E.18),

$$
\frac{\partial P_1 \left(y_n t\right)}{\partial t} = 0 \qquad (E.19)
$$

is of particular interest because it leads to a special formulation of the *Principle of Detailed Balance*, namely

$$
\Pi \equiv \frac{P_1 \left(y_{n-1}t\right)}{P_1 \left(y_n t\right)} = \frac{\phi \left(y_n \mid y_{n-1}\right)}{\phi \left(y_{n-1} \mid y_n\right)} \qquad (E.20)
$$

where Π is the transition probability associated with the change $y_{n-1} \longleftrightarrow y_n$. Equation (E.20) reflects the condition of microscopic reversibility, which we already introduced in Eq. (E.14).

E.2   Chemically striped substrates

Because we are concerned in this tutorial with the effects of chemical heterogeneity at the nanoscale on the behavior of the confined film, we expect the details of the atomic structure not to matter greatly for our purpose. Therefore, we adopt a continuum representation of the interaction of a film molecule with the substrate, which we obtain by averaging the film substrate interaction potential over positions of substrate atoms in the x–y plane. The resulting continuum potential can be expressed as

$$
\Phi^{[k]}(x, z) = n_A \sum_{m=-\infty}^{\infty} \sum_{m'=0}^{\infty} \int_{-\infty}^{\infty} \mathrm{d}y' \Bigg\{ \int_{-s_x/2+ms_x}^{-d_s/2+ms_x} \mathrm{d}x' u_{fw}\left(|r - r'|\right)
$$

$$
+ \int_{-d_s/2+ms_x}^{d_s/2+ms_x} \mathrm{d}x' u_{fs}\left(|r - r'|\right)
$$

$$
+ \int_{d_s/2+ms_x}^{s_x/2+ms_x} \mathrm{d}x' u_{fw}\left(|r - r'|\right) \Bigg\} \tag{E.21}
$$

In Eq. (E.21), $n_A = 2/\ell^2$ is the areal density of the (100) plane of the fcc lattice. The position of a film molecule is denoted by r, and $r' = (x', y', z' = \pm s_z/2 \pm m'\delta_\ell)$ represents the position of a substrate atom, where "$-$" refers to the lower $(k = 1)$, "$+$" to the upper $(k = 2)$ substrate, and δ_ℓ is the spacing between successive crystallographic planes in the $\pm z$ direction. We note that, because all features of the substrate at the atomic scale have been washed out in $\Phi^{[k]}$, our continuum model cannot account properly for solid formation, which, as mentioned briefly at the beginning of Section 5.4, is strongly influenced by the atomic structure of the substrate.

By interchanging the order of integration and introducing the transformation

$$
\begin{aligned}
x' \to x'' &= x - x' \\
y' \to y'' &= y - y' \\
z' \to z'' &= z - (\pm s_z/2 \pm m'\delta_\ell)
\end{aligned} \tag{E.22}
$$

we can rewrite the integrals on the right side of Eq. (E.21) as

$$
\int_a^b \mathrm{d}x' \int_{-\infty}^{\infty} \mathrm{d}y' u\left(|r - r'|\right) = -4\epsilon \int_{x-a}^{x-b} \mathrm{d}x'' \int_{-\infty}^{\infty} \mathrm{d}y'' \left[\left(\frac{\sigma}{r''}\right)^{12} - \left(\frac{\sigma^2}{r''}\right)^6\right] \tag{E.23}
$$

where a and b refer to integration limits and u and ϵ correspond to u_{fs} and ϵ_{fs} or to u_{fw} and ϵ_{fw}, depending on a and b. The definite integral over y'' can be found in standard tabulations (see, for example, No. 60 in Ref. 141). Thus, Eq. (E.23) simplifies to

$$4\epsilon \int\limits_{x-a}^{x-b} dx'' \int\limits_{-\infty}^{\infty} dy'' \left[\left(\frac{\sigma^2}{x''^2 + y''^2 + z''^2} \right)^6 - \left(\frac{\sigma^2}{x''^2 + y''^2 + z''^2} \right)^3 \right] =$$

$$\frac{3\pi\epsilon\sigma}{2} \int\limits_{x-a}^{x-b} dx'' \left[I_1 \left(x'', z''; d_s, s_x, s_z \right) - I_2 \left(x'', z''; d_s, s_x, s_z \right) \right] \qquad \text{(E.24)}$$

where

$$I_1 \left(x'', z'' \right) \equiv \frac{21}{32} \sqrt{\left(\frac{\sigma^2}{R} \right)^{11}} \qquad \text{(E.25a)}$$

$$I_2 \left(x'', z'' \right) \equiv \sqrt{\left(\frac{\sigma^2}{R} \right)^{5}} \qquad \text{(E.25b)}$$

$$R \equiv x''^2 + z''^2 \qquad \text{(E.25c)}$$

The remaining integration over x'' can also be carried out analytically (see, for example, No. 244 in Ref. 141). A tiresome computation yields

$$\int\limits_{x-a}^{x-b} dx'' I_1 \left(x'', z'' \right) = \frac{21}{32} \int\limits_{x-a}^{x-b} dx'' \sqrt{\left(\frac{\sigma^2}{R} \right)^{11}}$$

$$\equiv \frac{21\sigma}{32} I_3 \left(x'', z''; d_s, s_x, s_z \right) \Bigg|_{x''=x-a}^{x''=x-b}$$

$$= \frac{21\sigma}{32} \frac{x'' \sigma^{10}}{9z''^2 \sqrt{R^9}} \left[1 + \frac{8}{7}S + \frac{48}{35}S^2 + \frac{64}{35}S^3 + \frac{128}{35}S^4 \right] \Bigg|_{x''=x-a}^{x''=x-b} \qquad \text{(E.26)}$$

and

$$
\int\limits_{x-a}^{x-b} dx'' I_2\left(x'', z''\right) = \int\limits_{x-a}^{x-b} dx'' \sqrt{\left(\frac{\sigma^2}{R}\right)^5}
$$

$$
\equiv \left. \sigma I_4\left(x'', z''; d_s, s_x, s_z\right) \right|_{x''=x-a}^{x''=x-b}
$$

$$
= \left. \sigma \frac{x''\sigma^4}{3z''^2\sqrt{R^3}} \left[1 + 2S\right] \right|_{x''=x-a}^{x''=x-b} \tag{E.27}
$$

where the dimensionless quantity S is given by

$$
S \equiv \frac{R}{z''^2} \tag{E.28}
$$

To simplify the expressions, we define the auxiliary function

$$
\Delta\left(x'', z''; d_s, s_x, s_z\right) \equiv \frac{21}{32} I_3\left(x'', z''; d_s, s_x, s_z\right) - I_4\left(x'', z''; d_s, s_x, s_z\right) \tag{E.29}
$$

E.3 Molecular expressions for stresses

E.3.1 Normal component of stress tensor

E.3.1.1 Virial expression

To evaluate the partial derivative of the configuration integral in Eq. (5.62), we employ an approach suggested by Hill [21] and transform coordinates according to [see also Eq. (5.7)]

$$
z_i \rightarrow \widetilde{z}_i = z_i/s_z \tag{E.30}
$$

This permits us to recast the previous expression as [see Eq. (5.69)]

$$
\tau_{zz} = -\frac{k_{\mathrm{B}}T}{A_{z0}\Xi_{\mathrm{cl}}} \sum_{N=0}^{\infty} \frac{1}{N!\Lambda^{3N}} \exp\left(\frac{\mu N}{k_{\mathrm{B}}T}\right)
$$

$$
\times \frac{\partial}{\partial s_z} \left[s_z^N \prod_{i=1}^{N} \int\limits_{-s_x/2}^{s_x/2} dx_i \int\limits_{-s_y/2}^{s_y/2} dy_i \int\limits_{-1/2}^{1/2} d\widetilde{z}_i \exp\left(-\frac{U}{k_{\mathrm{B}}T}\right) \right] \tag{E.31}
$$

where $U = U_{FF} + U_{FS}$. The partial derivative in Eq. (E.31) may be evaluated according to the product rule, that is, $y' = u'v + uv'$ $(y = uv)$, where in the current case

$$u \equiv s_z^N \tag{E.32a}$$

$$v \equiv \prod_{i=1}^{N} \int_{-s_x/2}^{s_x/2} \mathrm{d}x_i \int_{-s_y/2}^{s_y/2} \mathrm{d}y_i \int_{-1/2}^{1/2} \mathrm{d}\widetilde{z}_i \exp\left(-\frac{U}{k_BT}\right) \tag{E.32b}$$

It then follows that

$$-\frac{k_BT}{A_{z0}\Xi_{cl}} \sum_{N=0}^{\infty} \frac{N}{N!\Lambda^{3N}} \exp\left(\frac{\mu N}{k_BT}\right) s_z^{N-1}$$

$$\times \prod_{i=1}^{N} \int_{-s_x/2}^{s_x/2} \mathrm{d}x_i \int_{-s_y/2}^{s_y/2} \mathrm{d}y_i \int_{-1/2}^{1/2} \mathrm{d}\widetilde{z}_i \exp\left(-\frac{U}{k_BT}\right)$$

$$= -\frac{k_BT}{V\Xi_{cl}} \sum_{N=0}^{\infty} \frac{N}{N!\Lambda^{3N}} \exp\left(\frac{\mu N}{k_BT}\right)$$

$$\times \prod_{i=1}^{N} \int_{-s_x/2}^{s_x/2} \mathrm{d}x_i \int_{-s_y/2}^{s_y/2} \mathrm{d}y_i \int_{-s_z/2}^{s_z/2} \mathrm{d}z_i \exp\left(-\frac{U}{k_BT}\right)$$

$$= -\frac{k_BT}{V} \sum_{N=0}^{\infty} \int \mathrm{d}r^N Np\left(r^N; N\right) = -\frac{\langle N \rangle k_BT}{V} \equiv \tau_{zz}^{id} \tag{E.33}$$

where $V \equiv A_{z0}s_z$ and Eq. (5.6) have also been used. Equation (E.33) represents the ideal-gas contribution to the normal component of the stress tensor.

To compute the derivative v' we realize that $U = U_{FF} + U_{FS}$ depends on s_z because $U_{FS} \propto \Phi^{[k]}$ depends on $z_i = \widetilde{z}_i s_z$ and U_{FF} depends on [see Eq. (5.69)]

$$r_{ij} = \sqrt{(x_i - x_j)^2 + (y_i - y_j)^2 + s_z^2\left(\widetilde{z}_i - \widetilde{z}_j\right)^2} \tag{E.34}$$

Hence, we find from Eq. (E.31)

$$\frac{1}{A_{z0}\Xi_{cl}} \sum_{N=0}^{\infty} \frac{1}{N!\Lambda^{3N}} \exp\left(\frac{\mu N}{k_BT}\right) \int \mathrm{d}r^N \exp\left(-\frac{U}{k_BT}\right) \left(\frac{\partial U_{FF}}{\partial s_z} + \frac{\partial U_{FS}}{\partial s_z}\right)$$

after reverting the transformation $\tilde{z}_i \to z_i$ [see Eq. (E.30)] where

$$
\begin{aligned}
\frac{\partial U_{\mathrm{FF}}}{\partial s_z} &= \frac{1}{2} \frac{\partial}{\partial s_z} \sum_{i=1}^{N} \sum_{j\neq i=1}^{N} u_{\mathrm{ff}}(r_{ij}) \\
&= \frac{1}{2} \sum_{i=1}^{N} \sum_{j\neq i=1}^{N} u_{\mathrm{ff}}'(r_{ij}) \frac{\partial r_{ij}}{\partial s_z} \\
&= \frac{1}{2s_z} \sum_{i=1}^{N} \sum_{j\neq i=1}^{N} u_{\mathrm{ff}}'(r_{ij}) \frac{z_{ij}^2}{r_{ij}} \\
&= \frac{1}{2s_z} \sum_{i=1}^{N} \sum_{j\neq i=1}^{N} u_{\mathrm{ff}}'(r_{ij}) \frac{(\boldsymbol{r}_{ij} \cdot \widehat{\boldsymbol{e}}_z)^2}{r_{ij}} \equiv \frac{W_{\mathrm{zz}}}{2s_z}
\end{aligned}
\tag{E.35a}
$$

$$
\frac{\partial U_{\mathrm{FS}}}{\partial s_z} = \frac{1}{s_z} \sum_{k=1}^{2} \sum_{i=1}^{N} \left(z_i \pm \frac{s_z}{2} \right) F_{\mathrm{z},i}^{[k]}
\tag{E.35b}
$$

where

$$
F_{\mathrm{z},i}^{[k]} \equiv -\frac{\partial \Phi^{[k]}}{\partial z_i}
\tag{E.36}
$$

is the force exerted by particle i on substrate k and W_{zz} is Clausius' virial [21]. Equation (E.35b) follows by noting from Eq. (5.68) that

$$
\pm z_i = \frac{z_i''}{s_z} - \left(\frac{1}{2} + \frac{m' \delta_\ell}{s_z} \right)
\tag{E.37}
$$

so that

$$
\frac{\partial \tilde{z}_i}{\partial s_z} = -\frac{1}{s_z^2} \left(z_i \pm \frac{s_z}{2} \right)
\tag{E.38}
$$

and therefore

$$
\frac{\partial \Phi^{[k]}}{\partial s_z} = \frac{\partial \Phi^{[k]}}{\partial \tilde{z}_i} \frac{\partial \tilde{z}_i}{\partial s_z} = \frac{\partial \Phi^{[k]}}{\partial z_i} \frac{\partial z_i}{\partial \tilde{z}_i} \frac{\partial \tilde{z}_i}{\partial s_z} = \frac{F_{\mathrm{z},i}^{[k]}}{s_z} \left(z_i \pm \frac{s_z}{2} \right)
\tag{E.39}
$$

Hence we obtain

$$\frac{1}{A_{z0}\Xi_{\rm cl}} \sum_{N=0}^{\infty} \frac{1}{N!\Lambda^{3N}} \exp\left(\frac{\mu N}{k_{\rm B}T}\right) \int d\boldsymbol{r}^N \exp\left(-\frac{U}{k_{\rm B}T}\right) \frac{\partial U_{\rm FF}}{\partial s_z}$$

$$= \frac{1}{2V\Xi_{\rm cl}} \sum_{N=0}^{\infty} \frac{1}{N!\Lambda^{3N}} \exp\left(\frac{\mu N}{k_{\rm B}T}\right) \int d\boldsymbol{r}^N \exp\left(-\frac{U}{k_{\rm B}T}\right) W_{zz}$$

$$= \frac{1}{2V} \sum_{N=0}^{\infty} \int d\boldsymbol{r}^N W_{zz} p\left(\boldsymbol{r}^N; N\right) = \frac{\langle W_{zz}\rangle}{2V} \tag{E.40a}$$

$$\frac{1}{A_{z0}\Xi_{\rm cl}} \sum_{N=0}^{\infty} \frac{1}{N!\Lambda^{3N}} \exp\left(\frac{\mu N}{k_{\rm B}T}\right) \int d\boldsymbol{r}^N \exp\left(-\frac{U}{k_{\rm B}T}\right) \frac{\partial U_{\rm FS}}{\partial s_z}$$

$$= \frac{1}{V} \sum_{N=0}^{\infty} \int d\boldsymbol{r}^N \sum_{k=1}^{2} \sum_{i=1}^{N} \left(z_i \pm \frac{s_z}{2}\right) F_{z,i}^{[k]} p\left(\boldsymbol{r}^N; N\right)$$

$$= \frac{1}{V} \left\langle \sum_{k=1}^{2} \sum_{i=1}^{N} \left(z_i \pm \frac{s_z}{2}\right) F_{z,i}^{[k]} \right\rangle \equiv -\tau_{zz}^{\rm FS} \tag{E.40b}$$

E.3.1.2 Force expression

A different expression for T_{zz} can be obtained directly from Eq. (5.62) without the transformation of coordinates. Therefore, it is convenient to recast the configuration integral as [329]

$$Z = \int_{-s_z/2}^{s_z/2} dz_1 g_1\left(z_1\right) \tag{E.41}$$

where

$$g_1\left(z_1\right) \equiv \prod_{i=1}^{N} \int_{-s_x/2}^{s_x/2} dx_i \int_{-s_y/2}^{s_y/2} dy_i \prod_{j=2}^{N} \int_{-s_z/2}^{s_z/2} dz_j \exp\left(-\frac{U_{\rm FF} + U_{\rm FS}}{k_{\rm B}T}\right) \tag{E.42}$$

By applying Leibniz's rule for the differentiation of an integral [11]

$$\frac{\partial Z}{\partial s_z} = \int_{-s_z/2}^{s_z/2} dz_1 \frac{\partial g_1}{\partial s_z} + \frac{1}{2}\left[g_1\left(z_1 = s_z/2\right) + g_1\left(z_1 = -s_z/2\right)\right]$$

$$= \int_{-s_z/2}^{s_z/2} dz_1 \frac{\partial g_1}{\partial s_z} \tag{E.43}$$

because $g_1 (z_1 = \pm s_z/2) = 0$. The latter follows because

$$\lim_{z_1 \to \pm s_z/2} U_{FS} = \infty \tag{E.44}$$

which is a direct consequence of the divergence of the function $\Delta (x, z)$ in that limit [see Eqs. (5.68), (5.69), (E.25c), (E.26), (E.27), (E.28), and (E.29)]. Defining therefore

$$Z = \prod_{j=1}^{k} \int_{-s_z/2}^{s_z/2} dz_j g_k (z_1, \ldots, z_k) \tag{E.45a}$$

$$g_k (z_1, \ldots, z_k) \equiv \prod_{i=1}^{N} \int_{-s_x/2}^{s_x/2} dx_i \int_{-s_y/2}^{s_y/2} dy_i \prod_{j=k+1}^{N} \int_{-s_z/2}^{s_z/2} dz_j \exp \left(-\frac{U}{k_B T} \right) \tag{E.45b}$$

the above analysis may be repeated $N - 1$ times. We finally obtain

$$\begin{aligned} \tau_{zz} &= -\frac{1}{A_{z0} \Xi_{cl}} \sum_{N=0}^{\infty} \frac{1}{N! \Lambda^{3N}} \exp \left(\frac{\mu N}{k_B T} \right) \int dr^N \exp \left(-\frac{U}{k_B T} \right) \frac{\partial U_{FS}}{\partial s_z} \\ &= -\frac{1}{A_{z0} \Xi_{cl}} \sum_{N=0}^{\infty} \frac{1}{N! \Lambda^{3N}} \exp \left(\frac{\mu N}{k_B T} \right) \int dr^N \exp \left(-\frac{U}{k_B T} \right) \sum_{k=1}^{2} \sum_{i=1}^{N} \frac{\partial \Phi^{[k]}}{\partial s_z} \\ &= \frac{\left\langle \mathsf{F}_z^{[1]} \right\rangle - \left\langle \mathsf{F}_z^{[2]} \right\rangle}{2 A_{z0}} \end{aligned} \tag{E.46}$$

where

$$\mathsf{F}_z^{[k]} \equiv \sum_{i=1}^{N} F_{z,i}^{[k]}, \qquad k = 1, 2 \tag{E.47}$$

and we notice that

$$\frac{\partial \Phi^{[k]}}{\partial s_z} = \underbrace{\frac{\partial \Phi^{[k]}}{\partial z_i}}_{-F_{z,i}^{[k]}} \underbrace{\frac{\partial z_i}{\partial s_z}}_{\mp \frac{1}{2}} = \pm \frac{F_{z,i}^{[k]}}{2} \tag{E.48}$$

which follows directly from Eq. (5.68). Equation (E.46) constitutes the so-called force expression for the normal component of the stress tensor.

E.3.2   Shear stress

E.3.2.1   Virial expression

To derive a molecular expression for the shear stress, we begin by realizing
that [see Eq. (1.66)]

$$\tau_{xz} = \frac{1}{A_{x0}} \left(\frac{\partial \Omega}{\partial (\alpha s_{x0})} \right)_{\{\cdot\}\backslash \alpha s_{x0}} = -\frac{k_B T}{A_{z0}\Xi} \sum_{N=0}^{\infty} \exp\left(\frac{\mu N}{k_B T} \right) \frac{\partial Z}{\partial (\alpha s_{x0})} \quad (E.49)$$

where we also employed Eqs. (2.112) and (2.120). To proceed let us write
the configuration integral more explicitly as

$$Z = \prod_{i=1}^{N} \int_{-s_y/2}^{s_y/2} dy_i \int_{-s_z/2}^{s_z/2} dz_i \int_{-s_x/2+\alpha s_{x0}(2z_i+s_z)/2s_z}^{s_x/2+\alpha s_{x0}(2z_i+s_z)/2s_z} dx_i \exp\left(-\frac{U_{FF}+U_{FS}}{k_B T} \right) \quad (E.50)$$

which incorporates the effect of deforming the fluid lamella in the x-direction.
It is convenient to eliminate the shear-strain dependence of the integration
limits through the transformation

$$x_i \rightarrow x_i' \equiv x_i - \frac{\alpha s_{x0}(2z_i + s_z)}{2s_z} \quad (E.51)$$

so that we can cast the partial derivative of the configuration integral in
Eq. (E.49) as

$$\frac{\partial Z}{\partial (\alpha s_{x0})} = -\frac{1}{k_B T} \prod_{i=1}^{N} \int_{-s_y/2}^{s_y/2} dy_i \int_{-s_z/2}^{s_z/2} dz_i \int_{-s_x/2}^{s_x/2} dx_i' \left(\frac{\partial U_{FF}}{\partial (\alpha s_{x0})} + \frac{\partial U_{FS}}{\partial (\alpha s_{x0})} \right)$$

$$\times \exp\left(-\frac{U}{k_B T} \right) \quad (E.52)$$

The shear strain then appears only in the argument of $u_{ff}(r_{ij})$ via

$$r_{ij} = \sqrt{\left[x_i' - x_j' + \frac{\alpha s_{x0}}{2s_z}(z_i - z_j) \right]^2 + (y_i - y_j)^2 + (z_i - z_j)^2} \quad (E.53)$$

It is then an easy matter to apply the derivation detailed in Appendix E.3.1.1
to show that

$$\tau_{xz}^{id} = 0 \quad (E.54a)$$

$$\tau_{xz}^{FF} = \frac{\langle W_{zx} \rangle}{2V} \quad (E.54b)$$

where W_{zx} is defined analogously to W_{zz} in Eq. (E.40a). For the corresponding fluid substrate contribution to the shear stress, we find

$$\frac{\partial U_{FS}}{\partial (\alpha s_{x0})} = \sum_{k=1}^{2} \sum_{i=1}^{N} \frac{\partial \Phi^{[k]}}{\partial x_i} \frac{\partial x_i}{\partial (\alpha s_{x0})} = -\sum_{k=1}^{2} \sum_{i=1}^{N} F_{x,i}^{[k]} \frac{\partial x_i}{\partial (\alpha s_{x0})} \qquad (E.55)$$

where $F_{x,i}^{[k]}$ is the x-component of the force exerted by fluid molecule i on substrate k. Introducing [see Eq. (5.103)]

$$x_i'^{[1]} = x_i + \frac{\alpha s_{x0}}{2 s_z} (2 z_i + s_z) \qquad (E.56a)$$

$$x_i'^{[2]} = x_i + \frac{\alpha s_{x0}}{2 s_z} (2 z_i - s_z) \qquad (E.56b)$$

it is straightforward to verify from Eqs. (E.49), (E.52), (E.55), and (E.56) that

$$\tau_{xz}^{FS} = -\frac{1}{V} \sum_{N=0}^{\infty} \int dr^N \sum_{k=1}^{2} \sum_{i=1}^{N} \left(z_i \pm \frac{s_z}{2} \right) F_{x,i}^{[k]} (\widehat{x}_i, z_i) \, p \left(r^N ; N \right)$$

$$= -\frac{1}{V} \left\langle \sum_{k=1}^{2} \sum_{i=1}^{N} \left(z_i \pm \frac{s_z}{2} \right) F_{x,i}^{[k]} (\widehat{x}_i, z_i) \right\rangle \qquad (E.57)$$

where the probability density in the grand canonical ensemble is defined in Eq. (E.40a).

E.3.2.2 Force expression

Here we derive a molecular expression for the shear stress parallel to the "force" expression for the compressional stress derived in Appendix E.3.1.2. However, here we employ a slightly different definition of the auxiliary functions $\{g_i\}$, which we introduce via

$$Z = \prod_{i=1}^{N} \int_{-s_y/2}^{s_y/2} dy_i \int_{-s_z/2}^{s_z/2} dz_i \int_{-s_x/2 + \alpha s_{x0}(2z_i+s_z)/2s_z}^{s_x/2 + \alpha s_{x0}(2z_i+s_z)/2s_z} dx_1 g_1 (x_1, y_1, \ldots, y_N, z_1, \ldots, z_N)$$

$$(E.58)$$

where

$$g_1 (x_1, y_1, \ldots, y_N, z_1, \ldots, z_N) = \prod_{i=2}^{N} \int_{-s_x/2 + \alpha s_{x0}(2z_i+s_z)/2s_z}^{s_x/2 + \alpha s_{x0}(2z_i+s_z)/2s_z} dx_i \exp \left(-\frac{U}{k_B T} \right)$$

$$(E.59)$$

Hence, we may again apply Leibniz's rule for the differentiation of a parameter integral to obtain

$$
\frac{\partial Z}{\partial (\alpha s_{x0})} \propto \int_{-s_x/2 + \alpha s_{x0}(2z_1 + s_z)/2s_z}^{s_x/2 + \alpha s_{x0}(2z_1 + s_z)/2s_z} dx_1 \frac{\partial g_1}{\partial (\alpha s_{x0})}
$$

$$
+ \left[g_1 \left(x_1 = -s_x/2 + \alpha s_{x0} \left(2z_1 + s_z \right)/2s_z \right) \right.
$$

$$
\left. - g_1 \left(x_1 = s_x/2 + \alpha s_{x0} \left(2z_1 + s_z \right)/2s_z \right) \right] \frac{z_1}{s_z} \quad (E.60)
$$

where we dropped the arguments $y_1, \ldots, y_N$ and $z_1, \ldots, z_N$ to simplify the notation. The reader should note that these last two terms do not vanish separately as we argued in Eq. (E.43) because U does not necessarily diverge at $x_1 = \pm s_x/2 + \alpha s_{x0} \left(2z_1 + s_z \right)/2s_z$ but

$$
g_1 \left(x_1 = -s_x/2 + \alpha s_{x0} \left(2z_1 + s_z \right)/2s_z \right) = g_1 \left(x_1 = s_x/2 + \alpha s_{x0} \left(2z_1 + s_z \right)/2s_z \right)
$$
$$
(E.61)
$$

because U is periodic in x on account of periodic boundary conditions. Thus, as before in Section (E.3.1.2) we may repeat the above argument $N - 1$ times to finally arrive at

$$
\frac{\partial Z}{\partial (\alpha s_{x0})} = -\frac{1}{k_B T} \prod_{i=1}^{N} \int_{-s_y/2}^{s_y/2} dy_i \int_{-s_z/2}^{s_z/2} dz_i \int_{-s_x/2 + \alpha s_{x0}(2z_i + s_z)/2s_z}^{s_x/2 + \alpha s_{x0}(2z_i + s_z)/2s_z} dx_i \frac{\partial U_{FS}}{\partial (\alpha s_{x0})}
$$

$$
\times \exp \left(-\frac{U}{k_B T} \right) \quad (E.62)
$$

Because of Eq. (5.103) we may rewrite the partial derivative in the previous expression as

$$
\frac{\partial U_{FS}}{\partial (\alpha s_{x0})} = \frac{1}{2} \sum_{k=1}^{2} \sum_{i=1}^{N} \pm F_{x,i}^{[k]} \left(\widehat{x}_i, z_i \right) \equiv \frac{1}{2} \left(F_x^{[1]} - F_x^{[2]} \right) \equiv \overline{F}_x \quad (E.63)
$$

where $+ \leftrightarrow k = 1$ and $- \leftrightarrow k = 2$, respectively. In Eq. (E.63), $\overline{F}_x$ is the x-component of the instantaneous net force exerted by the confined fluid on the substrates.

E.3.3 Virial expression for the bulk pressure

By an approach similar to the one outlined in Section E.3.1.1, we may derive a molecular expression for the bulk pressure P_b following again the original derivation of Hill [21]. As we pointed out in Section 1.3.1, the bulk fluid is homogeneous and isotropic on account of the absence of any external fields. Hence, we may write Eq. (1.29) alternatively as

$$V = A_{x0}s_x = A_{y0}s_y = A_{z0}s_z \tag{E.64}$$

which reflects this symmetry of the bulk fluid. Moreover, we realize from Eqs. (1.31) and (1.60) that

$$P_b = -\left(\frac{\partial \Omega}{\partial V}\right) = k_B T \left(\frac{\partial \ln \Xi_{cl}}{\partial V}\right) \tag{E.65}$$

where we also employed Eq. (2.81).

From Eq. (E.64) we realize that

$$\frac{\partial}{\partial V} = \frac{1}{3}\left(\frac{1}{A_{x0}}\frac{\partial}{\partial s_x} + \frac{1}{A_{y0}}\frac{\partial}{\partial s_y} + \frac{1}{A_{z0}}\frac{\partial}{\partial s_z}\right) \tag{E.66}$$

Hence, by analogy with Eq. (E.31), we may write

$$P_b = \frac{k_B T}{3\Xi_{cl}}\sum_{N=0}^{\infty}\frac{1}{N!\Lambda^{3N}}\exp\left(\frac{\mu N}{k_B T}\right)\sum_{\alpha=1}^{3}\frac{1}{A_{\alpha 0}}\frac{\partial}{\partial s_\alpha}\left[s_\alpha^N\int_{-1/2}^{1/2}d\widetilde{r}^N\exp\left(-\frac{U_{FF}}{k_B T}\right)\right] \tag{E.67}$$

where $\alpha = 1, 2, 3 \longleftrightarrow x, y, z$ and

$$\int_{-1/2}^{1/2}d\widetilde{r}^N\ldots \equiv \prod_{i=1}^{N}\int_{-1/2}^{1/2}d\widetilde{x}_i\int_{-1/2}^{1/2}d\widetilde{y}_i\int_{-1/2}^{1/2}d\widetilde{z}_i\ldots \tag{E.68}$$

is introduced as convenient shorthand notation where we transformed to unit-cube coordinates via

$$\alpha_i \to \widetilde{\alpha}_i = \alpha_i/s_\alpha \tag{E.69}$$

By the same logic applied before to Eq. (E.31), it is a simple matter to show that

$$P_b = P_b^{id} + P_b^{co} = \frac{\langle N\rangle k_B T}{V} + P_b^{co} \tag{E.70}$$

where the ideal-gas contribution P_b^{id} follows if Eq. (E.64) is also employed. The configurational contribution to the bulk pressure, P_b^{co}, involves terms of the form [see Eq. (E.35a)]

$$\frac{\partial U_{FF}}{\partial s_\alpha} = \frac{W_{\alpha\alpha}}{2s_\alpha} \tag{E.71}$$

where

$$W_{\alpha\alpha} = \sum_{i=1}^{N} \sum_{j \neq i=1}^{N} u'_{\mathrm{ff}}(r_{ij}) \frac{(\boldsymbol{r}_{ij} \cdot \widehat{\boldsymbol{e}}_{\alpha})^2}{r_{ij}} \tag{E.72}$$

Hence, when summed over α and because of Eq. (E.64) these terms may be combined leading to

$$P_{\mathrm{b}}^{\mathrm{co}} = -\frac{\langle W \rangle}{6V} \tag{E.73}$$

where

$$W \equiv \sum_{i=1}^{N} \sum_{j \neq i=1}^{N} u'_{\mathrm{ff}}(r_{ij}) \, r_{ij} \tag{E.74}$$

Together Eqs. (E.70), (E.73), and (E.74) constitute the virial expression for the bulk pressure, which parallels Eq. (5.63) for the stress exerted by a *confined* fluid on the planar substrates of a slit-pore.

Appendix F

Mathematical aspects of Ewald summation

F.1 Three-dimensional Coulombic systems

F.1.1 Energy contributions in Ewald formulation

F.1.1.1 Real-space contribution

To derive Eq. (6.8) for the real-space part of the Ewald potential, we start from Eq. (6.7) for the set of screened charges and apply Poisson's formula [see Eq. (6.3)]. This gives

$$\Phi^{(1)}(r_i) = {\sum_{\{n\}}}' \sum_{j=1}^{N} \left(\frac{q_j}{|r_{ij} + n|} + \Phi_{j,n}(r_i) \right) \tag{F.1}$$

where the first term in parentheses is the usual Coulomb potential and

$$\Phi_{j,n}(r_i) \equiv \int dr' \frac{\rho_{j,n}(r')}{|r_i - r'|} = -q_j \left(\frac{\alpha}{\sqrt{\pi}} \right)^3 \int dr' \frac{\exp\left[-\alpha^2 (r' - r_j + n)^2\right]}{|r_i - r'|} \tag{F.2}$$

is the potential due to a Gaussian charge cloud (total charge $-q_j$) located at $r_j - n$ [see Eq. (6.5)]. To evaluate the integral on the far right side of Eq. (F.2), we transform variables according to $r' \to R = r' - r_j + n$, which gives

$$\Phi_{j,n}(r_i) = -q_j \left(\frac{\alpha}{\sqrt{\pi}} \right)^3 \int dR \frac{\exp\left[-\alpha^2 R^2\right]}{|r_{ij} + n - R|} \tag{F.3}$$

In Eq. (F.3) the integration is carried out over the entire three-dimensional space.

447

The most convenient way of doing this is to transform to spherical coordinates $R = |\boldsymbol{R}|$, θ, and φ, where θ and φ are the polar and azimuthal angles, respectively, associated with the orientation of $\boldsymbol{R}$ in a space-fixed coordinate system. We may split the integral over R into two contributions from regions characterized by the inequalities

$$R \ < \ |\boldsymbol{r}_{ij} + \boldsymbol{n}| \equiv x \tag{F.4a}$$

$$R \ > \ x \tag{F.4b}$$

This separation of the integral can be effected by using an expansion in terms of spherical harmonics $\{Y_l^m\}$[258] valid for arbitrary vectors $\boldsymbol{r}_1$ and $\boldsymbol{r}_2$[1]; that is,

$$|\boldsymbol{r}_1 - \boldsymbol{r}_2|^{-1} = \sum_{l=0}^{\infty} \sum_{m=-l}^{l} \frac{4\pi}{2l+1} \frac{r_<^l}{r_>^{l+1}} Y_{lm}^*(\theta_1, \varphi_1) Y_{lm}(\theta_2, \varphi_2) \tag{F.5}$$

where θ_i and φ_i are polar and azimuthal angles associated with vectors $\boldsymbol{r}_1$ and $\boldsymbol{r}_2$, respectively. Notice that the complex conjugate $Y_{lm}^* = Y_{l,-m}$. In Eq. (F.5), $r_<$ ($r_>$) is the magnitude of the smaller (larger) vector of the pair $\boldsymbol{r}_1$ and $\boldsymbol{r}_2$. Setting $\boldsymbol{r}_1 = \boldsymbol{R}$ and $\boldsymbol{r}_2 = \boldsymbol{r}_{ij} + \boldsymbol{n}$, and inserting Eq. (F.5) into Eq. (F.3), one realizes that only terms characterized by $l = m = 0$ (with $Y_{00} = 1/\sqrt{4\pi}$) survive because [258]

$$\int_0^{2\pi} d\varphi \int_{-1}^{1} d\cos\theta \, Y_{lm}(\theta, \varphi) = \sqrt{4\pi} \delta_{l,0} \delta_{m,0} \tag{F.6}$$

Equation (F.3) can therefore be rewritten as

$$\Phi_{j,n}(\boldsymbol{r}_i) = -4\pi q_j \left(\frac{\alpha}{\sqrt{\pi}}\right)^3 \left[\frac{1}{x} \int_0^x dR R^2 \exp\left(-\alpha^2 R^2\right) + \int_x^{\infty} dR R \exp\left(-\alpha^2 R^2\right)\right] \tag{F.7}$$

where the first integral appearing in brackets can be recast by using

$$\int_0^x dR R^2 \exp\left(-\alpha^2 R^2\right) = -\frac{1}{2\alpha} \frac{\partial}{\partial \alpha} \int_0^x dR \exp\left(-\alpha^2 R^2\right)$$

$$= -\frac{1}{2\alpha} \frac{\partial}{\partial \alpha} \left[\frac{1}{\alpha} \int_0^{\alpha x} du \exp\left(-u^2\right)\right]$$

$$= -\frac{1}{2\alpha} \frac{\partial}{\partial \alpha} \left[\frac{\sqrt{\pi}}{2\alpha} \mathrm{erf}(\alpha x)\right] \tag{F.8}$$

---

[1] See Eq. (3.70) in Ref. 242.

where we have employed the definition of the error function [11, 37, 330],

$$\operatorname{erf}(y) = \frac{2}{\sqrt{\pi}} \int_0^y du \exp\left(-u^2\right) \qquad \text{(F.9)}$$

to arrive at the third line of Eq. (F.8). The remaining partial derivative can be carried out by using

$$\frac{\partial}{\partial y}\operatorname{erf}(y) = \frac{2}{\sqrt{\pi}} \exp\left(-y^2\right) \qquad \text{(F.10)}$$

which follows immediately from the Eq. (F.9) and Leibniz's rule for the differentiation of a parameter integral [11, 330]. One finally obtains

$$\int_0^x dR\, R^2 \exp\left(-\alpha^2 R^2\right) - \frac{\sqrt{\pi}}{4\alpha^3}\operatorname{erf}(\alpha x) - \frac{1}{2\alpha^2}x\exp\left(-\alpha^2 x^2\right) \qquad \text{(F.11)}$$

The second integral in Eq. (F.7) gives

$$\int_x^\infty dR\, R\exp\left(-\alpha^2 R^2\right) = -\frac{1}{2\alpha^2}\int_x^\infty dR\frac{\partial}{\partial R}\exp\left(-\alpha^2 R^2\right) = \frac{1}{2\alpha^2}\exp\left(-\alpha^2 x^2\right) \qquad \text{(F.12)}$$

Inserting Eqs. (F.11) and (F.12) into Eq. (F.7) and replacing x by $|r_{ij} + n|$ the electrostatic potential at r_i due to one Gaussian located at $r_j - n \neq r_i$ reduces to

$$\Phi_{j,n}(r_i) = -q_j \frac{\operatorname{erf}(\alpha|r_{ij} + n|)}{|r_{ij} + n|} \qquad \text{(F.13)}$$

Finally, inserting Eq. (F.13) into our initial Eq. (F.1) and using the identity

$$1 - \operatorname{erf}(y) = \operatorname{erfc}(y) \qquad \text{(F.14)}$$

we eventually arrive at Eq. (6.8).

F.1.1.2 Fourier-space contribution for nonzero wavevectors

To evaluate the electrostatic potential $\Phi^{(2)}(r)$ [see Eq. (6.11)] from the periodic Gaussian charge distribution $\rho^{(2)}(r)$ [see Eq. (6.10a)], it is most convenient to start from Laplace's equation [242], which says that

$$\Delta\Phi^{(2)}(r) = -4\pi\rho^{(2)}(r) \qquad \text{(F.15)}$$

where $\Delta = \nabla \cdot \nabla$ is the Laplace operator. The Laplace equation is equivalent to Poisson's equation [see Eq. (6.3)] and follows directly from the first Maxwell equation of electrostatics,

$$\nabla \cdot \boldsymbol{E}(\boldsymbol{r}) = 4\pi\rho(\boldsymbol{r}) \tag{F.16}$$

using the definition

$$\boldsymbol{E}(\boldsymbol{r}) = -\nabla\Phi(\boldsymbol{r}) \tag{F.17}$$

for the electric field $\boldsymbol{E}$. From Eq. (F.15) it is evident that the Laplace equation is a second-order differential equation that can be solved conveniently in Fourier space. To this end, we expand the charge distribution and the corresponding potential according to the (discrete) Fourier series

$$\rho^{(2)}(\boldsymbol{r}) = \sum_{\{\boldsymbol{k}\}} \widetilde{\rho}^{(2)}(\boldsymbol{k}) \exp(i\boldsymbol{k} \cdot \boldsymbol{r}) \tag{F.18a}$$

$$\Phi^{(2)}(\boldsymbol{r}) = \sum_{\{\boldsymbol{k}\}} \widetilde{\Phi}^{(2)}(\boldsymbol{k}) \exp(i\boldsymbol{k} \cdot \boldsymbol{r}) \tag{F.18b}$$

where $\boldsymbol{k}$ is a vector of the reciprocal lattice related to the set of real-space lattice vectors $\{\boldsymbol{n}\}$ [see text above Eq. (6.11)], and the quantities $\widetilde{\rho}^{(2)}(\boldsymbol{k})$ and $\widetilde{\Phi}^{(2)}(\boldsymbol{k})$ are Fourier coefficients of the charge distribution and the potential, respectively. These Fourier coefficients can be obtained from the corresponding real-space quantities via

$$\widetilde{\rho}^{(2)}(\boldsymbol{k}) = \frac{1}{V_{\text{sys}}} \int d\boldsymbol{r} \rho^{(2)}(\boldsymbol{r}) \exp(-i\boldsymbol{k} \cdot \boldsymbol{r}) \tag{F.19a}$$

$$\widetilde{\Phi}^{(2)}(\boldsymbol{k}) = \frac{1}{V_{\text{sys}}} \int d\boldsymbol{r} \Phi^{(2)}(\boldsymbol{r}) \exp(-i\boldsymbol{k} \cdot \boldsymbol{r}) \tag{F.19b}$$

where V_{sys} is the volume of the *entire* system consisting of the basic cell plus its periodic replicas. Thus, $V_{\text{sys}} = V n_{\text{cell}}$, where n_{cell} is the total number of cells.

Inserting the expansions (F.18) into Eq. (F.15), we have

$$\begin{aligned} \Delta\Phi^{(2)}(\boldsymbol{r}) &= \sum_{\{\boldsymbol{k}\}} \widetilde{\Phi}^{(2)}(\boldsymbol{k}) \Delta \exp(i\boldsymbol{k} \cdot \boldsymbol{r}) \\ &= -\sum_{\{\boldsymbol{k}\}} k^2 \widetilde{\Phi}^{(2)}(\boldsymbol{k}) \exp(i\boldsymbol{k} \cdot \boldsymbol{r}) \\ &= -4\pi \sum_{\{\boldsymbol{k}\}} \widetilde{\rho}^{(2)}(\boldsymbol{k}) \exp(i\boldsymbol{k} \cdot \boldsymbol{r}) \end{aligned} \tag{F.20}$$

We now recall that Fourier expansions (F.18) are orthogonal expansions (see, e.g., Ref. 242). It follows that each summand on the second line of Eq. (F.20) has to be equal to its counterpart on the third line so that

$$\widetilde{\Phi}^{(2)}(\boldsymbol{k}) = \frac{4\pi}{k^2}\widetilde{\rho}^{(2)}(\boldsymbol{k}) \tag{F.21}$$

which is Laplace's equation in Fourier space. Thus, given the Fourier coefficients of the charge distribution (see below), we can easily calculate from Eq. (F.21) all Fourier coefficients of the corresponding potential, except its contribution at $\boldsymbol{k} = \boldsymbol{0}$, which will be discussed in the subsequent Appendix F.1.1.3. Replacing in Eq. (F.18b), $\widetilde{\Phi}^{(2)}(\boldsymbol{k})$ by the expression given in Eq. (F.21) permits us to calculate the desired potential $\Phi^{(2)}(\boldsymbol{r})$.

Having in mind this strategy we start by evaluating the Fourier coefficients of $\widetilde{\rho}^{(2)}(\boldsymbol{k})$. Inserting the explicit expression for $\rho^{(2)}(\boldsymbol{r})$ given in Eq. (6.10a) into Eq. (F.19a), we have

$$\widetilde{\rho}^{(2)}(\boldsymbol{k}) = \frac{1}{V_{\text{sys}}}\left(\frac{\alpha}{\sqrt{\pi}}\right)^3 \sum_{\{\boldsymbol{n}\}}\sum_{j=1}^{N} q_j \int d\boldsymbol{r}\, \exp\left[-i\boldsymbol{k}\cdot\boldsymbol{r} - \alpha^2(\boldsymbol{r} - \boldsymbol{r}_j + \boldsymbol{n})^2\right] \tag{F.22}$$

The spatial integral on the right side is a standard (three-dimensional) Gaussian integral and can be carried out analytically [330]. Using, in addition, the relation $\exp(-i\boldsymbol{k}\cdot\boldsymbol{n}) = 1$ (which *defines* $\boldsymbol{k}$ as a reciprocal lattice vector), one finds

$$
\begin{aligned}
\widetilde{\rho}^{(2)}(\boldsymbol{k}) &= \frac{1}{V_{\text{sys}}}\sum_{\{\boldsymbol{n}\}}\exp\left(-\frac{k^2}{4\alpha^2}\right)\sum_{j=1}^{N} q_j \exp(-i\boldsymbol{k}\cdot\boldsymbol{r}_j) \\
&= \frac{1}{V}\exp\left(-\frac{k^2}{4\alpha^2}\right)\sum_{j=1}^{N} q_j \exp(-i\boldsymbol{k}\cdot\boldsymbol{r}_j)
\end{aligned} \tag{F.23}
$$

where the second line has been obtained by employing the relation

$$\frac{1}{V_{\text{sys}}}\sum_{\{\boldsymbol{n}\}} 1 = \frac{1}{V_{\text{sys}}}n_{\text{cell}} = \frac{1}{V} \tag{F.24}$$

Inserting Eq. (F.23) into Eq. (F.21) then yields

$$\widetilde{\Phi}^{(2)}(\boldsymbol{k}) = \frac{1}{V}\frac{4\pi}{k^2}\exp\left[-\frac{k^2}{4\alpha^2}\right]\sum_{j=1}^{N} q_j \exp[-i\boldsymbol{k}\cdot\boldsymbol{r}_j], \quad \boldsymbol{k} \neq \boldsymbol{0} \tag{F.25}$$

Finally, inserting the coefficients (F.25) into Eq. (F.18b) together with $\boldsymbol{r} = \boldsymbol{r}_i$ gives the first term on the right side of Eq. (6.11), which is the contribution to the electrostatic potential $\Phi^{(2)}(\boldsymbol{r}_i)$. The missing "long-range" term related to the special case $\boldsymbol{k} = \boldsymbol{0}$ is discussed in the subsequent Appendix F.1.1.3.

F.1.1.3   Long-range contribution

Evaluation of the long-range part of the electrostatic potential, $\Phi_{LR}^{(2)}(\boldsymbol{r}_i)$, which results from the long-wavelength limit ($\boldsymbol{k} = \boldsymbol{0}$) of the corresponding Fourier expansion, is the "trickiest" part in the derivation of the Ewald expression for the electrostatic potential of a Coulombic system. The problem is immediately apparent from Laplace's equation in Fourier space [see Eq. (F.21)] which, when solved for $\widetilde{\Phi}^{(2)}(\boldsymbol{k})$ *directly* at $\boldsymbol{k} = \boldsymbol{0}$, yields a *divergent* result because of the factor $1/k^2$. Fortunately, we are not really interested in the value of $\widetilde{\Phi}^{(2)}(\boldsymbol{k})$ for $\boldsymbol{k} = \boldsymbol{0}$. To realize the irrelevance of the value of $\widetilde{\Phi}^{(2)}(\boldsymbol{k})$ at $\boldsymbol{k} = \boldsymbol{0}$, consider the corresponding energy contribution

$$U^{(2)}(\boldsymbol{0}) \equiv \frac{1}{2}\sum_{j=1}^{N} q_j \widetilde{\Phi}^{(2)}(\boldsymbol{0}) \tag{F.26}$$

where [see Eq. (F.19b)]

$$\widetilde{\Phi}^{(2)}(\boldsymbol{0}) = \frac{1}{V_{\text{sys}}}\int d\boldsymbol{r}\,\Phi^{(2)}(\boldsymbol{r}) \tag{F.27}$$

is the spatial integral over the potential which must be independent of (particle) index j. Now recall that we are dealing with a *globally neutral* system, meaning that $\sum_{j=1}^{N} q_j = 0$. Consequently, $U^{(2)}(\boldsymbol{0})$ vanishes regardless of the actual value of $\widetilde{\Phi}^{(2)}(\boldsymbol{0})$.

Thus, in the following discussion we focus on the *limit* $\boldsymbol{k} \to \boldsymbol{0}$ of the full product $\widetilde{\Phi}^{(2)}(\boldsymbol{k})\exp(i\boldsymbol{k}\cdot\boldsymbol{r})$ appearing in Eq. (F.18b). More explicitly, given that we are dealing with an isotropic system where the direction of the wavevector $\boldsymbol{k}$ should not matter, we consider the angle–averaged quantity

$$\begin{aligned}
\Phi_{LR}^{(2)}(\boldsymbol{r}) &= \frac{1}{4\pi}\lim_{k\to0}\int_{-1}^{1} d\cos\theta_{\boldsymbol{k}}\int_{0}^{2\pi} d\varphi_{\boldsymbol{k}}\,\widetilde{\Phi}^{(2)}(\boldsymbol{k})\exp(i\boldsymbol{k}\cdot\boldsymbol{r}) \\
&= \frac{1}{4\pi}\lim_{k\to0}\int d\omega_{\boldsymbol{k}}\,\widetilde{\Phi}^{(2)}(\boldsymbol{k})\exp(i\boldsymbol{k}\cdot\boldsymbol{r})
\end{aligned} \tag{F.28}$$

where $\theta_{\boldsymbol{k}}$ and $\varphi_{\boldsymbol{k}}$ are the angles specifying the orientation of $\boldsymbol{k}$ and $\omega_{\boldsymbol{k}} = (\theta_{\boldsymbol{k}}, \varphi_{\boldsymbol{k}})$.

To evaluate the right side of Eq. (F.28) we consider first the charge-density coefficients $\tilde{\rho}^{(2)}(\boldsymbol{k})$ that give rise to the potential $\Phi^{(2)}$ for small, but nonvanishing $\boldsymbol{k}$. Expanding these coefficients in a Taylor series around $\boldsymbol{k} = \boldsymbol{0}$ and using Laplace's equation [see Eq. (F.21)], we obtain

$$\widetilde{\Phi}^{(2)}(\boldsymbol{k}) = \frac{4\pi}{k^2}\left[\left.\tilde{\rho}^{(2)}\right|_{0} + \boldsymbol{k}\cdot\nabla_{\boldsymbol{k}}\left.\tilde{\rho}^{(2)}(\boldsymbol{k})\right|_{0} + \frac{1}{2}\boldsymbol{k}\boldsymbol{k}\cdot\nabla_{\boldsymbol{k}}\nabla_{\boldsymbol{k}}\left.\tilde{\rho}^{(2)}(\boldsymbol{k})\right|_{0} + \mathcal{O}\left(k^3\right)\right] \tag{F.29}$$

We now consider the lowest-order expansion coefficients of $\tilde{\rho}^{(2)}(\boldsymbol{k})$ appearing on the right side of Eq. (F.29). Using the general definition (F.19a) for the Fourier coefficients and performing the required derivatives, we obtain

$$\tilde{\rho}^{(2)}\big|_0 = \frac{1}{V_{\text{sys}}} \int d\boldsymbol{r}\, \rho^{(2)}(\boldsymbol{r}) = \frac{1}{V_{\text{sys}}} Q^{(2)} \tag{F.30a}$$

$$\nabla_{\boldsymbol{k}} \tilde{\rho}^{(2)}(\boldsymbol{k})\big|_0 = -\frac{i}{V_{\text{sys}}} \int d\boldsymbol{r}\, \boldsymbol{r}\rho^{(2)}(\boldsymbol{r}) = -\frac{i}{V_{\text{sys}}} \boldsymbol{P}^{(2)} \tag{F.30b}$$

$$\nabla_{\boldsymbol{k}} \nabla_{\boldsymbol{k}} \tilde{\rho}^{(2)}(\boldsymbol{k})\big|_0 = -\frac{1}{V_{\text{sys}}} \int d\boldsymbol{r}\, \boldsymbol{r}\boldsymbol{r}\rho^{(2)}(\boldsymbol{r}) = -\frac{1}{V_{\text{sys}}} \mathbf{A}^{(2)} \tag{F.30c}$$

The quantities on the right side of Eqs. (F.30) have a simple and lucid physical interpretation in terms of the multipole moments of the charge distribution $\rho^{(2)}(\boldsymbol{r})$ [242]. Indeed, $Q^{(2)}$ is nothing but the monopole moment, $\boldsymbol{P}^{(2)}$ is the dipole moment, and the second-rank tensor $\mathbf{A}^{(2)}$ is related closely to the quadrupole moment. Explicit expressions for these quantities can be easily obtained by inserting Eq. (6.10a) into Eqs. (F.30) and carrying out the (Gaussian) spatial integrals. For the monopole, this procedure gives

$$\frac{Q^{(2)}}{V_{\text{sys}}} = \left(\frac{\alpha}{\sqrt{\pi}}\right)^3 \frac{1}{V_{\text{sys}}} \sum_{\{\boldsymbol{n}\}} \sum_{j=1}^N q_j \int d\boldsymbol{r}\, \exp\left[-\alpha^2(\boldsymbol{r} - \boldsymbol{r}_j + \boldsymbol{n})^2\right]$$

$$= \frac{1}{V_{\text{sys}}} \sum_{\{\boldsymbol{n}\}} \sum_{j=1}^N q_j = \frac{1}{V} \sum_{j=1}^N q_j = 0 \tag{F.31}$$

where we used Eq. (F.24). Thus, the monopole moment vanishes due to the global charge neutrality of the system. The dipole moment of the charge distribution $\rho^{(2)}(\boldsymbol{r})$ coincides with that of the original delta-like distribution in Eq. (6.4); that is,

$$\frac{\boldsymbol{P}^{(2)}}{V_{\text{sys}}} = \left(\frac{\alpha}{\sqrt{\pi}}\right)^3 \frac{1}{V_{\text{sys}}} \sum_{\{\boldsymbol{n}\}} \sum_{j=1}^N q_j \int d\boldsymbol{r}\, \boldsymbol{r} \exp\left[-\alpha^2(\boldsymbol{r} - \boldsymbol{r}_j + \boldsymbol{n})^2\right]$$

$$= \frac{1}{V_{\text{sys}}} \sum_{\{\boldsymbol{n}\}} \sum_{j=1}^N q_j (\boldsymbol{r}_j - \boldsymbol{n}) = \frac{\boldsymbol{M}}{V} \tag{F.32}$$

In writing the last member of Eq. (F.32) we have used the definition $\boldsymbol{M} = \sum_{j=1}^N q_j \boldsymbol{r}_j$ for the total dipole moment of the central cell and the fact that each replicated cell has exactly the same total dipole moment. Using similar arguments we obtain for the cartesian components $\left(\mathbf{A}^{(2)}\right)_{kl}$ ($k, l = $ x, y, or z) of the second-rank tensor $\mathbf{A}^{(2)}$,

$$
\begin{aligned}
\frac{\left(\mathbf{A}^{(2)}\right)_{kl}}{V_{\text{sys}}} &= \left(\frac{\alpha}{\sqrt{\pi}}\right)^3 \frac{1}{V_{\text{sys}}} \sum_{\{n\}} \sum_{j=1}^{N} q_j \int d\mathbf{r} \, (\mathbf{r})_k \, (\mathbf{r})_l \exp\left[-\alpha^2 \left(\mathbf{r} - \mathbf{r}_j + \mathbf{n}\right)^2\right] \\
&= \frac{1}{V} \left[\frac{1}{2\alpha^2} \delta_{kl} + \sum_{j=1}^{N} q_j \left(\mathbf{r}_j\right)_k \left(\mathbf{r}_j\right)_l \left(1 - \delta_{kl}\right)\right] \\
&\equiv \frac{1}{V} \left(\mathbf{D}^{(2)}\right)_{kl}
\end{aligned}
\tag{F.33}
$$

We proceed by inserting the nonvanishing multipole moments defined in Eqs. (F.30)–(F.33) into the expansion in Eq. (F.29), which gives

$$
\widetilde{\Phi}^{(2)}(\mathbf{k}) = -i\frac{4\pi}{k^2 V} \mathbf{k} \cdot \mathbf{M} - \frac{2\pi}{k^2 V} \mathbf{k} \mathbf{D}^{(2)} \mathbf{k} + \mathcal{O}(k)
\tag{F.34}
$$

As we emphasized before we are interested in the long-wavelength limit of $\widetilde{\Phi}^{(2)}(\mathbf{k})$ times the phase factor $\exp(i\mathbf{k} \cdot \mathbf{r})$. Expanding the latter in a Taylor series around $\mathbf{k} = \mathbf{0}$, that is

$$
\exp(i\mathbf{k} \cdot \mathbf{r}) = 1 + i\mathbf{k} \cdot \mathbf{r} - \frac{1}{2}(\mathbf{k} \cdot \mathbf{r})^2 + \mathcal{O}(k^3)
\tag{F.35}
$$

and combining this expansion with Eq. (F.34), we obtain

$$
\begin{aligned}
\widetilde{\Phi}^{(2)}(\mathbf{k}) \exp(i\mathbf{k} \cdot \mathbf{r}) &= -i\frac{4\pi}{k^2 V} \mathbf{k} \cdot \mathbf{M} - \frac{2\pi}{k^2 V} \mathbf{k} \mathbf{D}^{(2)} \mathbf{k} \\
&\quad + \frac{4\pi}{k^2 V} (\mathbf{k} \cdot \mathbf{r}) (\mathbf{k} \cdot \mathbf{M}) - i(\mathbf{k} \cdot \mathbf{r}) \frac{2\pi}{k^2 V} \mathbf{k} \mathbf{D}^{(2)} \mathbf{k} \\
&\quad + \frac{1}{2}(\mathbf{k} \cdot \mathbf{r})^2 \left(i\frac{4\pi}{k^2 V} \mathbf{k} \cdot \mathbf{M} + \frac{2\pi}{k^2 V} \mathbf{k} \mathbf{D}^{(2)} \mathbf{k}\right) + \mathcal{O}(k^2)
\end{aligned}
\tag{F.36}
$$

We now consider separately the terms on the right side of Eq. (F.36), focusing on the question whether they contribute to the desired (angle-averaged) potential $\Phi_{\text{LR}}^{(2)}(\mathbf{r})$ [defined in Eq. (F.28)]. The first term depends on $1/k$ and may therefore seem to diverge as we take the limit $\mathbf{k} \to \mathbf{0}$. However, as this first term also contains $\mathbf{k} \cdot \mathbf{M}$, it vanishes already for nonvanishing $\mathbf{k}$ because of the angle average in Eq. (F.28). To see this result, we note that the scalar product of two arbitrary unit vectors $\hat{\mathbf{a}}$ and $\hat{\mathbf{b}}$ can be expressed in terms of spherical harmonics as [258]

$$
\hat{\mathbf{a}} \cdot \hat{\mathbf{b}} = \frac{4\pi}{3} \sum_{m=-1}^{1} Y_{1m}^*(\omega_{\mathbf{a}}) Y_{1m}(\omega_{\mathbf{b}})
\tag{F.37}
$$

Therefore,

$$\int d\omega_k k \cdot M = k \, |M| \frac{4\pi}{3} \sum_{m=-1}^{1} \int d\omega_k Y_{1m}^* (\omega_k) Y_{1m} (\omega_M) = 0, \qquad \text{(F.38)}$$

where we have also used Eq. (F.6).

The next term on the right side of Eq. (F.36) is constant in k and involves the product $k D^{(2)} k$, which does not immediately vanish if averaged over orientations. Nevertheless, we can safely neglect this term. The reason is that it is independent of the position of particle i, with the immediate consequence that the corresponding energy contribution vanishes due to the global charge neutrality of the system [see text below Eq. (F.26)].

The third term on the right side of Eq. (F.36) contains the product $(k \cdot r)(k \cdot M)$. It has an explicit positional dependence even after performing the orientational average. Indeed, expanding both scalar products according to Eq. (F.37) and using the orthogonality of spherical harmonics given by [258]

$$\int d\omega Y_{lm}^*(\omega) Y_{l'm'}(\omega) = \delta_{ll'} \delta_{mm'} \qquad \text{(F.39)}$$

we find

$$
\begin{aligned}
\frac{1}{k^2} \int d\omega_k (k \cdot r)(k \cdot M) &= |r| \, |M| \left(\frac{4\pi}{3} \right)^2 \sum_{m=-1}^{1} \sum_{m'=-1}^{1} \\
&\quad \times \int d\omega_k Y_{1m}^* (\omega_k) Y_{1m} (\omega_r) Y_{1m'}^* (\omega_k) Y_{1m'} (\omega_M) \\
&= |r| \, |M| \left(\frac{4\pi}{3} \right)^2 \sum_{m=-1}^{1} Y_{1m}^* (\omega_M) Y_{1m} (\omega_r) \\
&= \frac{4\pi}{3} r \cdot M \qquad \text{(F.40)}
\end{aligned}
$$

where the last line has been obtained by using Eq. (F.37) in reverse direction.

The subsequent terms on the right side of Eq. (F.36) can be ignored because they are at least proportional to k and therefore vanish in the limit $k \to 0$. Thus, the potential $\Phi_{LR}^{(2)}(r)$ reduces to [see Eq. (F.28)]

$$\Phi_{LR}^{(2)}(r) = \frac{1}{k^2 V} \int d\omega_k (k \cdot r)(k \cdot M) = \frac{4\pi}{3V} r \cdot M \qquad \text{(F.41)}$$

The above expression for the long-range part of the electrostatic potential is consistent with a well-known result from macroscopic electrostatics regarding the average electric field inside a large sphere containing an (arbitrary) charge distribution. This field is given by [242]

$$\overline{E} = -\frac{4\pi}{3}P \tag{F.42}$$

where P is the polarization of the sphere. Clearly, $\overline{E}$ is independent of the radius of the sphere. Moreover, it is constant within the sphere, implying that the corresponding electrostatic potential is given by

$$\overline{\Phi}(r) = -r \cdot \overline{E} = \frac{4\pi}{3}r \cdot P \tag{F.43}$$

We now recall that our system is represented by one unit cell that is replicated in all three spatial directions. Thus, we can indeed take our system to be a (macroscopically) large sphere. As a consequence, the quantity P can be identified with the quantity $P^{(2)}/V_{sys} = M/V$ appearing in Eqs. (F.30b) and (F.32). We therefore conclude that the potential $\overline{\Phi}(r)$ is identical with long-range potential $\Phi_{LR}^{(2)}(r)$ given in Eq. (F.41).

The above considerations are useful because they permit one to understand from a macroscopic perspective why a long-range contribution to the electrostatic potential should arise. Moreover, they are particularly helpful because they indicate a strategy to introduce different boundary conditions into the Ewald summation technique. Indeed, the physical picture to which Eqs. (F.41)–(F.43) correspond is that the (macroscopically) large sphere is surrounded by a vacuum. In this case, any polarization in the sphere will generate surface charges at the interface between the sphere and the vacuum, and these charges in turn generate the average (or depolarization) field given in Eq. (F.42). If, on the other hand, the sphere is surrounded by a dielectricum with dielectric constant ϵ', the average field inside the sphere has to be corrected by the so-called reaction field [331],

$$E_{RF} = \frac{2(\epsilon' - 1)}{2\epsilon' + 1}\frac{4\pi}{3}P \tag{F.44}$$

which, as expected, vanishes for the special case $\epsilon' = 1$ (i.e., in the vacuum). Combining Eqs. (F.42) and (F.44), the total average field inside the sphere then becomes

$$\overline{E} \rightarrow \overline{E} + E_{RF} = -\frac{4\pi}{2\epsilon' + 1}P = -\frac{4\pi}{2\epsilon' + 1}\frac{M}{V} \tag{F.45}$$

Inserting Eq. (F.45) into Eq. (F.43) and taking $r = r_i$, one obtains the final expression for the long-range contribution of the electrostatic potential given in Eq. (6.12).

F.1.1.4 Self-contribution

The self-part of the Ewald electrostatic potential given in Eq. (6.14) can be derived in a fashion similar to our derivation of the real-space contribution in Appendix F.1.1.1. Starting from Poisson's formula [see Eq. (6.3)] and inserting Eq. (6.10b) for the charge density $\rho^{(3)}(r')$, we have

$$\Phi^{(3)}(r_i) = \int dr' \frac{\rho^{(3)}(r')}{|r_i - r'|} = -q_i \left(\frac{\alpha}{\sqrt{\pi}}\right)^3 \int dr' \frac{\exp\left[-\alpha^2 (r' - r_i)^2\right]}{|r_i - r'|} \quad \text{(F.46)}$$

The three-dimensional integral on the far right side of Eq. (F.46) can be evaluated by transforming variables according to $r' \to R = r' - r_i$ followed by a transformation to polar coordinates. This gives [330]

$$\Phi^{(3)}(r_i) = -4\pi q_i \left(\frac{\alpha}{\sqrt{\pi}}\right)^3 \int\limits_0^\infty dR\, R \exp\left(-\alpha^2 R^2\right) \quad \text{(F.47)}$$

which can be easily be evaluated in closed form to give Eq. (6.14).

Finally, it seems worth noting that the self-part can also be derived directly from Eq. (F.13) representing the potential $\Phi_{j,n}(r_i)$ caused by a Gaussian located at $r_j - n \neq r_i$. Indeed, considering $\Phi_{j,n}$ at $n = 0$, one obtains

$$- \lim_{r_{ij} \to 0} q_i \frac{\text{erf}(\alpha r_{ij})}{r_{ij}} \quad - \quad - \lim_{r_{ij} \to 0} q_i \left[\frac{2\alpha}{\sqrt{\pi}} - \frac{2}{3\sqrt{\pi}}\alpha^3 r_{ij}^2 + \mathcal{O}\left(r_{ij}^4\right)\right]$$

$$= -q_i \frac{2\alpha}{\sqrt{\pi}} \overset{(6.14)}{=} \Phi^{(3)}(r_i) \quad \text{(F.48)}$$

in the limit $r_{ij} \to 0$ where we have used the first few terms in a Taylor expansion of the error function $\text{erf}(x)$ around $x = 0$ given by

$$\text{erf}(x) = \frac{2}{\sqrt{\pi}}x - \frac{2}{3\sqrt{\pi}}x^3 + \mathcal{O}(x^5) \quad \text{(F.49)}$$

F.1.2 Force and stress tensor components

F.1.2.1 Force components

Based on Eqs. (6.15)–(6.17b) we can also derive the corresponding expressions for the force acting on particle i,

$$F_{C,i}^{3d} = -q_i \nabla_i \Phi(r_i) \quad \text{(F.50)}$$

Because of Eqs. (6.16a) and (6.17a) we can split the total force into a sum of three individual contributions, namely

$$F_{C,i}^{3d} = F_{CR,i}^{3d} + F_{CF,i}^{3d} + F_{CLR,i}^{3d} \quad \text{(F.51)}$$

The reader should realize that the self-part makes no contribution because the summand in Eq. (6.17b) is independent of the coordinates of particle i. Considering the individual contributions to the total force separately, we obtain after straightforward differentiation

$$
\begin{aligned}
\boldsymbol{F}_{\mathrm{CR},i}^{\mathrm{3d}} &= -q_i \sum_{j=1}^{N} \sideset{}{'}\sum_{\boldsymbol{n}} q_j \nabla_{ij} \frac{\mathrm{erfc}\left(\alpha \left|\boldsymbol{r}_{ij} + \boldsymbol{n}\right|\right)}{\left|\boldsymbol{r}_{ij} + \boldsymbol{n}\right|} \\
&= q_i \sum_{j=1}^{N} \sideset{}{'}\sum_{\boldsymbol{n}} q_j \left\{ \frac{2\alpha}{\sqrt{\pi}} \exp\left[-\alpha^2 \left(\boldsymbol{r}_{ij} + \boldsymbol{n}\right)^2\right] \right. \\
&\qquad \left. + \frac{\mathrm{erfc}\left(\alpha \left|\boldsymbol{r}_{ij} + \boldsymbol{n}\right|\right)}{\left|\boldsymbol{r}_{ij} + \boldsymbol{n}\right|} \right\} \frac{\boldsymbol{r}_{ij} + \boldsymbol{n}}{\left|\boldsymbol{r}_{ij} + \boldsymbol{n}\right|^2} \quad (\mathrm{F}.52a) \\
\boldsymbol{F}_{\mathrm{CF},i}^{\mathrm{3d}} &= -4\pi q_i \sum_{j=1}^{N} q_j \nabla_{ij} \left[\sum_{\boldsymbol{k} \neq \boldsymbol{0}} \frac{1}{k^2 V} \exp\left(-\frac{k^2}{4\alpha^2}\right) \exp\left(-i\boldsymbol{k} \cdot \boldsymbol{r}_{ij}\right) \right] \\
&= 4\pi i q_i \sum_{j=1}^{N} q_j \sum_{\boldsymbol{k} \neq \boldsymbol{0}} \frac{\boldsymbol{k}}{k^2 V} \exp\left(-\frac{k^2}{4\alpha^2}\right) \exp\left(-i\boldsymbol{k} \cdot \boldsymbol{r}_{ij}\right) \\
&= 4\pi q_i \sum_{\boldsymbol{k} \neq \boldsymbol{0}} \frac{\boldsymbol{k}}{k^2 V} \exp\left(-\frac{k^2}{4\alpha^2}\right) \sum_{j=1}^{N} q_j \sin\left(\boldsymbol{k} \cdot \boldsymbol{r}_{ij}\right) \quad (\mathrm{F}.52b) \\
\boldsymbol{F}_{\mathrm{CLR},i}^{\mathrm{3d}} &= -q_i \nabla_i \left[\boldsymbol{r}_i \cdot \sum_{j=1}^{N} \frac{4\pi q_j \boldsymbol{r}_j}{V\left(2\epsilon' + 1\right)} \right] = -q_i \frac{4\pi \boldsymbol{M}}{V\left(2\epsilon' + 1\right)} \quad (\mathrm{F}.52c)
\end{aligned}
$$

In writing Eqs. (F.52a) and (F.52b) we have taken into account that the operator ∇_i appearing in the original force expression [see Eq. (F.50)] can be replaced by its counterpart ∇_{ij} with respect to the distance vector $\boldsymbol{r}_{ij} = \boldsymbol{r}_i - \boldsymbol{r}_j$ where, of course,

$$
\nabla_{ij} \equiv \frac{\partial}{\partial \boldsymbol{r}_{ij}} = \frac{\boldsymbol{r}_{ij}}{r_{ij}} \frac{\mathrm{d}}{\mathrm{d} r_{ij}} \quad (\mathrm{F}.53)
$$

and $r_{ij} = \left|\boldsymbol{r}_{ij}\right|$ also hold. In deriving Eq. (F.52a) we also used Eqs. (F.10) and (F.14). Moreover, the last line of Eq. (F.52b) has been obtained using

$$
i\boldsymbol{k} \exp\left(-i\boldsymbol{k} \cdot \boldsymbol{r}_{ij}\right) = i\boldsymbol{k} \cos\left(\boldsymbol{k} \cdot \boldsymbol{r}_{ij}\right) + \boldsymbol{k} \sin\left(\boldsymbol{k} \cdot \boldsymbol{r}_{ij}\right) \quad (\mathrm{F}.54)
$$

where the cosine term (contrary to sine term) changes sign upon inversion, that is, $\boldsymbol{k} \rightarrow -\boldsymbol{k}$, and therefore vanishes in the sum over *all* wavevectors.

F.1.2.2 Stress tensor components

By analogy with Appendix E.3 we derive molecular expressions for various (diagonal) components of the stress tensor $\tau_{\gamma\gamma}$ (γ = x, y, or z) by realizing that we may write

$$\tau_{\gamma\gamma,\mathrm{C}}^{\mathrm{3d}} = \tau_{\gamma\gamma}^{\mathrm{id}} + \frac{\langle W_{\gamma\gamma,\mathrm{C}}^{\mathrm{3d}} \rangle}{A_{\gamma 0}} \tag{F.55}$$

in the grand canonical ensemble where $\tau_{\gamma\gamma}^{\mathrm{id}}$ is given in Eq. (E.33). From the definition of the Clausius virial [see Eq. (E.35)] and Eq. (6.15) for the total configurational potential energy of the three-dimensional Coulomb system in Ewald formulation, we have

$$W_{\gamma\gamma,\mathrm{C}}^{\mathrm{3d}} = \frac{\partial U_{\mathrm{C}}^{\mathrm{3d}}}{\partial s_{\gamma}} = \frac{\partial U_{\mathrm{CR}}^{\mathrm{3d}}}{\partial s_{\gamma}} + \frac{\partial U_{\mathrm{CF}}^{\mathrm{3d}}}{\partial s_{\gamma}} + \frac{\partial U_{\mathrm{CLR}}^{\mathrm{3d}}}{\partial s_{\gamma}}, \qquad \gamma = \text{x, y, or z} \tag{F.56}$$

because $U_{\mathrm{Cs}}^{\mathrm{3d}}$ is a constant that does not depend on the actual configuration [see Eq. (6.17b)]. To evaluate the partial derivatives on the right side of Eq. (F.56), it turns out to be convenient to transform to unit-cube coordinates via

$$\boldsymbol{r}_i \to \tilde{\boldsymbol{r}}_i \equiv \begin{pmatrix} x_i/s_{\mathrm{x}} \\ y_i/s_{\mathrm{y}} \\ z_i/s_{\mathrm{z}} \end{pmatrix}, \qquad i = 1, \ldots, N \tag{F.57}$$

Consider the first term on the right side of Eq. (F.56). From Eq. (6.16a) we obtain

$$\begin{aligned}
\frac{\partial U_{\mathrm{CR}}^{\mathrm{3d}}}{\partial s_{\gamma}} &= \frac{1}{2} \sum_{i=1}^{N} \sum_{j=1}^{N} \sideset{}{'}\sum_{\{n\}} q_i q_j \frac{\partial}{\partial s_{\gamma}} \frac{\mathrm{erfc}\,(\alpha\,|\boldsymbol{r}_{ij} + \boldsymbol{n}|)}{|\boldsymbol{r}_{ij} + \boldsymbol{n}|} \\
&= -\frac{1}{2} \sum_{i=1}^{N} \sum_{j=1}^{N} \sideset{}{'}\sum_{\{n\}} q_i q_j \left\{ \frac{\mathrm{erfc}\,(\alpha\,|\boldsymbol{r}_{ij} + \boldsymbol{n}|)}{|\boldsymbol{r}_{ij} + \boldsymbol{n}|^2} \right. \\
&\quad \left. + \frac{2\alpha}{\sqrt{\pi}} \frac{1}{|\boldsymbol{r}_{ij} + \boldsymbol{n}|} \exp\left[-(\alpha\,|\boldsymbol{r}_{ij} + \boldsymbol{n}|)^2\right] \right\} \frac{\partial}{\partial s_{\gamma}} |\boldsymbol{r}_{ij} + \boldsymbol{n}| \tag{F.58}
\end{aligned}$$

which follows with the aid of Eqs. (F.10) and (F.14). We now notice that because of Eq. (F.57)

$$|\boldsymbol{r}_{ij} + \boldsymbol{n}| = \sqrt{s_{\mathrm{x}}^2 \left(\tilde{x}_{ij} + n_{\mathrm{x}}\right)^2 + s_{\mathrm{y}}^2 \left(\tilde{y}_{ij} + n_{\mathrm{y}}\right)^2 + s_{\mathrm{z}}^2 \left(\tilde{z}_{ij} + n_{\mathrm{z}}\right)^2} \tag{F.59}$$

In the previous expression we used the fact that the lattice vectors $\boldsymbol{n} = (n_{\mathrm{x}} s_{\mathrm{x}}, n_{\mathrm{y}} s_{\mathrm{y}}, n_{\mathrm{z}} s_{\mathrm{z}})$. Therefore,

$$\frac{\partial}{\partial s_{\gamma}} |\boldsymbol{r}_{ij} + \boldsymbol{n}| = \frac{s_{\gamma} \left(\gamma_{ij} + n_{\gamma}\right)^2}{|\boldsymbol{r}_{ij} + \boldsymbol{n}|} = \frac{1}{s_{\gamma}} \frac{[(\boldsymbol{r}_{ij} + \boldsymbol{n}) \cdot \hat{\boldsymbol{e}}_{\gamma}]^2}{|\boldsymbol{r}_{ij} + \boldsymbol{n}|} \tag{F.60}$$

where $\widehat{e}_\gamma$ is a unit vector in the γ-direction and $\gamma_{ij} = \widetilde{r}_{ij} \cdot \widehat{e}_\gamma$ so that

$$
\frac{\partial U_{\mathrm{CR}}^{\mathrm{3d}}}{\partial s_\gamma} = -\frac{1}{2s_\gamma} \sum_{i=1}^{N} \sum_{j=1}^{N} {\sum_{\{n\}}}' q_i q_j \left\{ \frac{\mathrm{erfc}\,(\alpha\,|r_{ij} + n|)}{|r_{ij} + n|} \right.
$$
$$
\left. + \frac{2\alpha}{\sqrt{\pi}} \exp\left[-(\alpha\,|r_{ij} + n|)^2\right] \right\} \frac{[(r_{ij} + n) \cdot \widehat{e}_\gamma]^2}{|r_{ij} + n|^2} \quad \text{(F.61)}
$$

follows without further ado.

Turning to the second term on the right side of Eqs. (F.56), we realize that $\widetilde{a}(k)$ is independent of $\{s_\gamma\}$ because each term in the sum [see Eq. (6.19)] can be written as

$$
\exp\left(-ik \cdot r_i\right) = \exp\left[-2\pi i \left(m_x \widetilde{x}_i + m_y \widetilde{y}_i + m_z \widetilde{z}_i\right)\right] \quad \text{(F.62)}
$$

where we used the definition of the wavevectors k [see text before Eq. (6.11)]. This leaves us with

$$
\frac{\partial U_{\mathrm{CF}}^{\mathrm{3d}}}{\partial s_\gamma} = -\frac{2\pi}{V} \sum_{k \neq 0} \frac{1}{k^2} \exp\left(-\frac{k^2}{4\alpha^2}\right) \left[\frac{1}{V}\frac{\partial V}{\partial s_\gamma} + \left(\frac{2}{k} + \frac{k}{2\alpha^2}\right) \frac{\partial k}{\partial s_\gamma}\right] |\widetilde{a}(k)|^2
$$
$$
= -\frac{2\pi}{V s_\gamma} \sum_{k \neq 0} \frac{1}{k^2} \exp\left(-\frac{k^2}{4\alpha^2}\right) \left[1 - \left(\frac{2}{k^2} + \frac{1}{2\alpha^2}\right)\right.
$$
$$
\left. \times (k \cdot \widehat{e}_\gamma)^2\, |\widetilde{a}(k)|^2 \right] \quad \text{(F.63)}
$$

from Eq. (6.18) where we have used the fact that

$$
V = A_{\gamma 0} s_\gamma \quad \text{(F.64)}
$$

and

$$
k = 2\pi \sqrt{\left(\frac{m_x}{s_x}\right)^2 + \left(\frac{m_y}{s_y}\right)^2 + \left(\frac{m_z}{s_z}\right)^2} \quad \text{(F.65)}
$$

from which

$$
\frac{\partial k}{\partial s_\gamma} = -\frac{2\pi}{\sqrt{(m_x/s_x)^2 + (m_y/s_y)^2 + (m_z/s_z)^2}} \frac{m_\gamma^2}{s_\gamma^3} = -\frac{(k \cdot \widehat{e}_\gamma)^2}{k} \frac{1}{s_\gamma} \quad \text{(F.66)}
$$

follows directly where

$$
k \cdot \widehat{e}_\gamma = \frac{2\pi m_\gamma}{s_\gamma} \quad \text{(F.67)}
$$

is the projection of the wavevector k onto the γ-axis (i.e., the γ-component of k).

To evaluate the third contribution to $W_{\gamma\gamma}$ in Eq. (F.56), we realize from Eq. (6.17a) that

$$\frac{\partial U_{\mathrm{CLR}}^{3\mathrm{d}}}{\partial s_\gamma} = \frac{2\pi}{2\epsilon' + 1} \frac{\partial}{\partial s_\gamma} \frac{M^2}{V} = \frac{2\pi}{2\epsilon' + 1} \left[\frac{2M_\gamma^2}{V s_\gamma} - \left(\frac{M}{V} \right)^2 \frac{\partial V}{\partial s_\gamma} \right] \qquad \text{(F.68)}$$

In writing the first term on the right side of Eq. (F.68) we introduced the projection of the total dipole moment M [see Eq. (6.13)] onto the γ-axis, namely

$$M_\gamma = M \cdot \hat{e}_\gamma = \sum_{i=1}^{N} q_i r_i \cdot \hat{e}_\gamma = \sum_{i=1}^{N} q_i s_\gamma \tilde{r}_i \cdot \hat{e}_\gamma \qquad \text{(F.69)}$$

Equation (F.68) may be rewritten to give

$$\frac{\partial U_{\mathrm{CLR}}^{3\mathrm{d}}}{\partial s_\gamma} = -\frac{1}{V s_\gamma} \frac{2\pi}{2\epsilon' + 1} \left[M^2 - 2 \left(M \cdot \hat{e}_\gamma \right)^2 \right] \qquad \text{(F.70)}$$

for the long-range contribution to $W_{\gamma\gamma,\mathrm{C}}^{3\mathrm{d}}$. Finally, putting all this together we have from Eqs. (F.55), (F.56), (F.61), (F.63), and (F.70) the somewhat lengthy expression

$$\begin{aligned}
\tau_{\gamma\gamma,\mathrm{C}}^{3\mathrm{d}} = \ & \tau_{\gamma\gamma}^{\mathrm{id}} - \frac{1}{2V} \left\langle \sum_{i=1}^{N} \sum_{j=1}^{N} {\sum_{\{n\}}}' q_i q_j \left\{ \frac{\mathrm{erfc}\left(\alpha \left| r_{ij} + n \right| \right)}{\left| r_{ij} + n \right|} \right. \right. \\
& + \left. \frac{2\alpha}{\sqrt{\pi}} \exp\left[-\left(\alpha \left| r_{ij} + n \right| \right)^2 \right] \right\} \frac{\left[\left(r_{ij} + n \right) \cdot \hat{e}_\gamma \right]^2}{\left| r_{ij} + n \right|^2} \bigg\rangle \\
& - \frac{2\pi}{V^2} \left\langle \sum_{k \neq 0} \frac{1}{k^2} \exp\left(-\frac{k^2}{4\alpha^2} \right) \left[1 - \left(\frac{2}{k^2} + \frac{1}{2\alpha^2} \right) \left(k \cdot \hat{e}_\gamma \right)^2 \right] \left| \tilde{a}\left(k \right) \right|^2 \right\rangle \\
& - \frac{1}{V^2} \frac{2\pi}{2\epsilon' + 1} \left\langle \left[M^2 - 2 \left(M \cdot \hat{e}_\gamma \right)^2 \right] \right\rangle, \qquad \gamma = \mathrm{x,\ y,\ or\ z} \qquad \text{(F.71)}
\end{aligned}$$

for the diagonal components of the stress tensor in a Coulombic bulk system.

F.2 Three-dimensional dipolar system

F.2.1 Self-energy

We now derive an expression for the self-contribution to the dipolar energy in Ewald formulation given in Eq. (6.32) by recalling that the corresponding

Coulombic contribution [see Eq. (6.17b)] results from the interaction of the charges q_i at r_i with the corresponding Gaussian charge clouds centered at r_i and representing a total charge of $-q_i$. Moreover, we have seen at the end of Appendix F.1.1.4 that for a given particle i the self-part of the electrostatic potential can be calculated from the potential generated by a Gaussian at r_j by taking the limit $r_{ij} \to 0$. Keeping this observation in mind and replacing the charges q_i by operators $\mu_i \cdot \nabla_i$ as suggested by Eq. (6.22), we find the following prescription to calculate the dipolar self-contribution

$$U_{\text{DSF}}^{\text{3d}} = -\frac{1}{2} \lim_{r_{ij} \to 0} \lim_{\mu_j \to \mu_i} \sum_{i=1}^{N} (\mu_i \cdot \nabla_i)(\mu_j \cdot \nabla_j)\left(\frac{\text{erf}(\alpha r_{ij})}{r_{ij}}\right) \qquad (F.72)$$

Approximating $\text{erf}(\alpha r_{ij})/r_{ij}$ by its Taylor expansion for small distances r_{ij} given in Eq. (F.48), we obtain

$$(\mu_i \cdot \nabla_i)(\mu_j \cdot \nabla_j)\left(\frac{2\alpha}{\sqrt{\pi}} - \frac{2\alpha^3}{3\sqrt{\pi}}r_{ij}^2\right) = \frac{4\alpha^3}{3\sqrt{\pi}}(\mu_i \cdot \mu_j) \qquad (F.73)$$

from which

$$U_{\text{DSF}}^{\text{3d}} = -\frac{2\alpha^3}{3\sqrt{\pi}} \sum_{i=1}^{N} \mu_i^2 \qquad (F.74)$$

follows immediately by inserting Eq. (F.73) into Eq. (F.72) and taking the double limit. Equation (F.74) is identical to Eq. (6.32).

In Eq. (F.73) we used the fact that

$$\nabla_i = \nabla_{ij} = -\nabla_j \qquad (F.75)$$

and Eq. (F.53). Therefore,

$$(\mu_i \cdot \nabla_i)(\mu_j \cdot \nabla_j) = -(\mu_i \cdot \widehat{e}_{ij})(\mu_j \cdot \widehat{e}_{ij})\frac{\text{d}^2}{\text{d}r_{ij}^2} = -(\mu_i \cdot \mu_j)\frac{\text{d}^2}{\text{d}r_{ij}^2} \qquad (F.76)$$

F.2.2   Force and torque

According to Eqs. (6.26)–(6.32) the total force on particle i can be expressed as a sum of two contributions, namely

$$F_{\text{D},i}^{\text{3d}} = F_{\text{DR},i}^{\text{3d}} + F_{\text{DF},i}^{\text{3d}} \qquad (F.77)$$

because both long-range contributions and self-contributions in Eqs. (6.27c) and (6.32) turn out to be independent of the position of particle i and therefore do not contribute to the force. From Eq. (6.27a) it follows that

$$
\begin{aligned}
\boldsymbol{F}_{\text{DR},i}^{\text{3d}} = -\nabla_{ij} \sum_{j=1}^{N} \sideset{}{'}\sum_{\{\boldsymbol{n}\}} &\{(\boldsymbol{\mu}_i \cdot \boldsymbol{\mu}_j) B(|\boldsymbol{r}_{ij} + \boldsymbol{n}|, \alpha) \\
&- [\boldsymbol{\mu}_i \cdot (\boldsymbol{r}_{ij} + \boldsymbol{n})][\boldsymbol{\mu}_j \cdot (\boldsymbol{r}_{ij} + \boldsymbol{n})] C(|\boldsymbol{r}_{ij} + \boldsymbol{n}|, \alpha)\}
\end{aligned} \quad (\text{F.78})
$$

where the functions B and C are defined in Eqs. (6.28a). Transforming variables according to $\boldsymbol{r}_{ij} \to \boldsymbol{r} = \boldsymbol{r}_{ij} + \boldsymbol{n}$, noting that $\nabla_{\boldsymbol{r}} = \nabla_{ij}$, and that

$$
\nabla_{\boldsymbol{r}} = \frac{\boldsymbol{r}}{r} \frac{\text{d}}{\text{d}r} \quad (\text{F.79})
$$

direct differentiation on the right side of Eq. (F.78) gives

$$
\begin{aligned}
\boldsymbol{F}_{\text{DR},i}^{\text{3d}} = \sum_{j=1}^{N} \sideset{}{'}\sum_{\{\boldsymbol{n}\}} &\{[(\boldsymbol{\mu}_i \cdot \boldsymbol{\mu}_j) \boldsymbol{r} + \boldsymbol{\mu}_i (\boldsymbol{\mu}_j \cdot \boldsymbol{r}) + \boldsymbol{\mu}_j (\boldsymbol{\mu}_i \cdot \boldsymbol{r})] C(r, \alpha) \\
&- (\boldsymbol{\mu}_i \cdot \boldsymbol{r})(\boldsymbol{\mu}_j \cdot \boldsymbol{r}) \boldsymbol{r} D(r, \alpha)\}
\end{aligned} \quad (\text{F.80})
$$

where the function $D(r, \alpha)$ is defined as [see Eqs. (6.28a)]

$$
\begin{aligned}
D(r, \alpha) &\equiv -\frac{1}{r} \frac{\text{d}C}{\text{d}r} = \frac{1}{r} \frac{\text{d}}{\text{d}r} \left(\frac{1}{r} \frac{\text{d}B}{\text{d}r} \right) \\
&= \frac{1}{r^7} \left[\frac{2\alpha r}{\sqrt{\pi}} (15 + 10\alpha^2 r^2 + 4\alpha^4 r^4) \exp(-\alpha^2 r^2) + 15 \text{erfc}(\alpha r) \right]
\end{aligned} \quad (\text{F.81})
$$

The Fourier-space contribution follows from Eq. (6.27b) as

$$
\boldsymbol{F}_{\text{DF},i}^{\text{3d}} = -\frac{4\pi}{V} \sum_{\boldsymbol{k} \neq 0} \frac{1}{k^2} \exp\left(-\frac{k^2}{4\alpha^2} \right) \sum_{j=1}^{N} \nabla_{ij} (\boldsymbol{\mu}_i \cdot \boldsymbol{k})(\boldsymbol{\mu}_j \cdot \boldsymbol{k}) \exp(-i\boldsymbol{k} \cdot \boldsymbol{r}_{ij}) \quad (\text{F.82})
$$

where we employed Eq. (6.29). Differentiating in Eq. (F.82) with respect to $\boldsymbol{r}_{ij}$ gives (see Appendix F.1.2 for the parallel derivation in the Coulombic case)

$$
\begin{aligned}
\boldsymbol{F}_{\text{DF},i}^{\text{3d}} &= \frac{4\pi}{V} \sum_{\boldsymbol{k} \neq 0} \frac{1}{k^2} \exp\left(-\frac{k^2}{4\alpha^2} \right) \sum_{j=1}^{N} i\boldsymbol{k} (\boldsymbol{\mu}_i \cdot \boldsymbol{k})(\boldsymbol{\mu}_j \cdot \boldsymbol{k}) \exp(-i\boldsymbol{k} \cdot \boldsymbol{r}_{ij}) \\
&= \frac{4\pi}{V} \sum_{\boldsymbol{k} \neq 0} \frac{1}{k^2} \exp\left(-\frac{k^2}{4\alpha^2} \right) \sum_{j=1}^{N} \boldsymbol{k} (\boldsymbol{\mu}_i \cdot \boldsymbol{k})(\boldsymbol{\mu}_j \cdot \boldsymbol{k}) \sin(\boldsymbol{k} \cdot \boldsymbol{r}_{ij}) \quad (\text{F.83})
\end{aligned}
$$

where the last line is obtained via Eq. (F.54). Finally, using $\sin(x - y) = \sin x \cos y - \cos x \sin y$ and the definitions of real and imaginary parts of $\widetilde{M}(\boldsymbol{k})$ given in Eq. (6.29) we obtain

$$
\begin{aligned}
\boldsymbol{F}_{\mathrm{DF},i}^{\mathrm{3d}} = \ &\frac{4\pi}{V} \sum_{\boldsymbol{k}\neq 0} \frac{1}{k^2} \exp\left(-\frac{k^2}{4\alpha^2}\right) \Big[\sin(\boldsymbol{k}\cdot\boldsymbol{r}_i)\,\mathrm{Re}\widetilde{M}(\boldsymbol{k}) \\
&- \cos(\boldsymbol{k}\cdot\boldsymbol{r}_i)\,\mathrm{Im}\widetilde{M}(\boldsymbol{k})\Big](\boldsymbol{\mu}_i\cdot\boldsymbol{k})\,\boldsymbol{k}
\end{aligned}
\tag{F.84}
$$

The torque acting on particle i is defined by [140]

$$
\boldsymbol{T}_{\mathrm{D},i}^{\mathrm{3d}} \equiv -\boldsymbol{\mu}_i \times \left(\nabla_{\boldsymbol{\mu}_i}\Phi_{\mathrm{D},i}^{\mathrm{3d}}\right)
\tag{F.85}
$$

where $\Phi_{\mathrm{D},i}^{\mathrm{3d}}$ is the energy of particle i. From Eqs. (6.27) we realize that $\Phi_{\mathrm{D},i}^{\mathrm{3d}}$ can be written as a sum of three terms, namely

$$
\Phi_{\mathrm{D},i}^{\mathrm{3d}} = \Phi_{\mathrm{DR},i}^{\mathrm{3d}} + \Phi_{\mathrm{DF},i}^{\mathrm{3d}} + \Phi_{\mathrm{DLR},i}^{\mathrm{3d}}
\tag{F.86}
$$

and the differentiation is performed with respect to $\boldsymbol{\mu}_i$. Referring back to Eqs. (6.26)–(6.32) we realize that

$$
\boldsymbol{T}_{\mathrm{D},i}^{\mathrm{3d}} = \boldsymbol{T}_{\mathrm{DR},i}^{\mathrm{3d}} + \boldsymbol{T}_{\mathrm{DF},i}^{\mathrm{3d}} + \boldsymbol{T}_{\mathrm{DLR},i}^{\mathrm{3d}}
\tag{F.87}
$$

where

$$
\boldsymbol{T}_{\mathrm{DR},i}^{\mathrm{3d}} = -\boldsymbol{\mu}_i \times \sum_{j=1}^{N}\sideset{}{'}\sum_{\boldsymbol{n}} \left[\boldsymbol{\mu}_j B(r,\alpha) - \boldsymbol{r}(\boldsymbol{\mu}_i\cdot\boldsymbol{r})\,C(r,\alpha)\right]
\tag{F.88a}
$$

$$
\begin{aligned}
\boldsymbol{T}_{\mathrm{DF},i}^{\mathrm{3d}} = \ &-\boldsymbol{\mu}_i \times \sum_{\boldsymbol{k}\neq 0} \frac{4\pi}{k^2 V} \exp\left(-\frac{k^2}{4\alpha^2}\right)\left[\boldsymbol{k} - \boldsymbol{\mu}_i(\boldsymbol{\mu}_i\cdot\boldsymbol{k})\right] \\
&\times \left[\cos(\boldsymbol{k}\cdot\boldsymbol{r}_i)\,\mathrm{Re}\widetilde{M}(\boldsymbol{k}) + \sin(\boldsymbol{k}\cdot\boldsymbol{r}_i)\,\mathrm{Im}\widetilde{M}(\boldsymbol{k})\right]
\end{aligned}
\tag{F.88b}
$$

$$
\boldsymbol{T}_{\mathrm{DLR},i}^{\mathrm{3d}} = -\boldsymbol{\mu}_i \times \left[\frac{4\pi}{2\epsilon'+1}\frac{\boldsymbol{M}}{V}\right]
\tag{F.88c}
$$

In Eq. (F.88a) we use again the shorthand notation $\boldsymbol{r} \equiv \boldsymbol{r}_{ij} + \boldsymbol{n}$ and $r = |\boldsymbol{r}|$ as before [see below Eq. (F.78)].

F.2.3   Stress tensor

As before in Section F.1.2.2 diagonal components of the stress tensor of a dipolar fluid can be obtained from the relation

$$
\tau_{\gamma\gamma,\mathrm{D}}^{\mathrm{3d}} = \tau_{\gamma\gamma}^{\mathrm{id}} + \frac{\left\langle W_{\gamma\gamma,\mathrm{D}}^{\mathrm{3d}}\right\rangle}{A_{\gamma 0}}
\tag{F.89}
$$

where the ideal-gas contribution $\tau_{\gamma\gamma}^{\text{id}}$ is given in Eq. (E.33), invoking again the grand canonical ensemble for convenience. By analogy with Eq. (F.56) we have

$$W_{\gamma\gamma,\text{D}}^{\text{3d}} = \frac{\partial U_{\text{D}}^{\text{3d}}}{\partial s_{\gamma}} = \frac{\partial U_{\text{DR}}^{\text{3d}}}{\partial s_{\gamma}} + \frac{\partial U_{\text{DF}}^{\text{3d}}}{\partial s_{\gamma}} + \frac{\partial U_{\text{DLR}}^{\text{3d}}}{\partial s_{\gamma}}, \qquad \gamma = \text{x, y, or z} \qquad \text{(F.90)}$$

Based on the same transformation to unit-cube coordinates employed before [see Eq. (F.57)], we realize that the dependence on s_{γ} is buried in the argument of the functions B and C [see Eqs. (6.28a)] and in the factors $\boldsymbol{\mu}_{i,j} \cdot (\boldsymbol{r}_{ij} + \boldsymbol{n})$ as far as $U_{\text{DR}}^{\text{3d}}$ is concerned. Differentiating these terms with respect to s_{γ}, it is easy to verify that terms of the form

$$\frac{1}{s_{\gamma}} \left(\boldsymbol{\mu}_{i(j)} \cdot \widehat{\boldsymbol{e}}_{\gamma} \right) \left[(\boldsymbol{r}_{ij} + \boldsymbol{n}) \cdot \widehat{\boldsymbol{e}}_{\gamma} \right]$$

arise where $\boldsymbol{\mu}_{i(j)}$ stands for either $\boldsymbol{\mu}_i$ or $\boldsymbol{\mu}_j$. Employing also the relation among the functions B, C, and D [see Eqs. (6.28a), (F.81)] as well as Eq. (F.60), it is a simple matter to show that

$$\begin{aligned}
\tau_{\gamma\gamma,\text{DR}}^{\text{3d}} = -\frac{1}{2V} \Bigg\langle \sum_{i=1}^{N} \sum_{j=1}^{N} {\sum_{\{n\}}}' & \left\{ (\boldsymbol{\mu}_i \cdot \boldsymbol{\mu}_j) (\boldsymbol{r} \cdot \widehat{\boldsymbol{e}}_{\gamma})^2 \right. \\
& + \left[(\boldsymbol{\mu}_i \cdot \widehat{\boldsymbol{e}}_{\gamma})(\boldsymbol{\mu}_j \cdot \boldsymbol{r}) + (\boldsymbol{\mu}_j \cdot \widehat{\boldsymbol{e}}_{\gamma})(\boldsymbol{\mu}_i \cdot \boldsymbol{r}) \right] (\boldsymbol{r}_{ij} \cdot \widehat{\boldsymbol{e}}_{\gamma}) \Big\} C(r, \alpha) \\
& + (\boldsymbol{\mu}_i \cdot \boldsymbol{r})(\boldsymbol{\mu}_j \cdot \boldsymbol{r})(\boldsymbol{r} \cdot \widehat{\boldsymbol{e}}_{\gamma})^2 D(r, \alpha) \Bigg\rangle
\end{aligned} \qquad \text{(F.91)}$$

where we again transformed variables according to $\boldsymbol{r}_{ij} \to \boldsymbol{r} = \boldsymbol{r}_{ij} + \boldsymbol{n}$ and $r = |\boldsymbol{r}|$.

Turning to the Fourier-space contribution next, we immediately see that $U_{\text{DF}}^{\text{3d}}$ contains a factor

$$\frac{1}{V} \frac{1}{k^2} \exp\left(-\frac{k^2}{4\alpha^2} \right)$$

that has already been considered in the derivation of $\tau_{\gamma\gamma,\text{CF}}^{\text{3d}}$ in Eq. (F.63). We are then left with a derivative of the function $\widetilde{M}(\boldsymbol{k})$ [see Eq. (6.29)] which depends on s_{γ} because of Eq. (F.66). Introducing

$$\widetilde{Q}(\boldsymbol{k}) \equiv \sum_{i=1}^{N} (\boldsymbol{\mu}_i \cdot \widehat{\boldsymbol{e}}_{\gamma})(\boldsymbol{k} \cdot \widehat{\boldsymbol{e}}_{\gamma}) \exp\left(i\boldsymbol{k} \cdot \boldsymbol{r}_i \right) \qquad \text{(F.92)}$$

one obtains

$$
\tau_{\gamma\gamma,\mathrm{DF}}^{\mathrm{3d}} = -\frac{2\pi}{V^2} \left\langle \sum_{\boldsymbol{k}\neq\boldsymbol{0}} \frac{1}{k^2} \exp\left(-\frac{k^2}{4\alpha^2}\right) \left\{ \left[1 - \left(\frac{2}{k^2} + \frac{1}{2\alpha^2}\right)\right] (\boldsymbol{k}\cdot\widehat{\boldsymbol{e}}_\gamma)^2 \right.\right.
$$
$$
\left.\left. \times \left|\widetilde{M}(\boldsymbol{k})\right|^2 + \widetilde{Q}(\boldsymbol{k})\,\widetilde{M}^*(\boldsymbol{k}) + \widetilde{Q}^*(\boldsymbol{k})\,\widetilde{M}(\boldsymbol{k}) \right\} \right\rangle \qquad (\text{F.93})
$$

Finally, the long-range contribution to the energy [see Eq. (6.27c)] gives rise to a stress contribution

$$
\tau_{\gamma\gamma,\mathrm{DLR}} = -\frac{1}{V^2}\frac{2\pi}{2\epsilon'+1}\left\langle M^2 \right\rangle \qquad (\text{F.94})
$$

The reader should appreciate the difference between the previous expression and the last term on the right side of Eq. (F.71). This difference arises because, for a dipolar system, $M_\gamma = \sum_{i=1}^{N} \boldsymbol{\mu}_i \cdot \widehat{\boldsymbol{e}}_\gamma$ is independent of s_γ, whereas for a Coulomb system, M_γ depends on s_γ as one can verify from Eq. (F.69). The diagonal component of the *total* stress tensor is then obtained by adding the three contributions given in Eqs. (F.91), (F.93), and (F.94) [see Eqs. (F.89) and (F.90)].

F.3   Slab geometry

F.3.1   Rigorous expressions

F.3.1.1   Point charges

To derive Eq. (6.34) for a system of point charges in slab geometry, we proceed in a fashion analogous to the one employed for bulk systems in Section 6.2.1 and Appendix F.1. In other words, we divide the original charge density related to Eq. (6.33)

$$
\rho_i(\boldsymbol{r}') = {\sum_{\boldsymbol{n}_\|}}' \sum_{j=1}^{N} q_j \delta(z'-z_j)\,\delta(\boldsymbol{R}'-\boldsymbol{R}_j+\boldsymbol{n}_\|) \qquad (\text{F.95})
$$

into three contributions corresponding to a set of screened charges $\rho_i^{(1)}(\boldsymbol{r}')$, a periodic set of charge clouds screening those original ones $\rho^{(2)}(\boldsymbol{r}')$, and a self-contribution $\rho_i^{(3)}(\boldsymbol{r}')$ describing the interaction of each charge cloud with itself. We choose the charge clouds to be spherical Gaussians[2] such that the

---

[2]See Ref. 248 for other choices.

three contributions of the charge distribution are

$$
\begin{aligned}
\rho_i^{(1)}\left(\boldsymbol{r}'\right) \;=\;& \rho_i\left(\boldsymbol{r}'\right) - \left(\frac{\alpha}{\sqrt{\pi}}\right)^3 {\sum_{\boldsymbol{n}_\parallel}}' \sum_{j=1}^{N} q_j \exp\left[-\alpha^2\left(z' - z_j\right)^2\right] \\
& \times \exp\left[-\alpha^2\left(\boldsymbol{R}' - \boldsymbol{R}_j + \boldsymbol{n}_\parallel\right)^2\right] \qquad\qquad\text{(F.96a)}
\end{aligned}
$$

$$
\begin{aligned}
\rho^{(2)}\left(\boldsymbol{r}'\right) \;=\;& \left(\frac{\alpha}{\sqrt{\pi}}\right)^3 \sum_{\boldsymbol{n}_\parallel} \sum_{j=1}^{N} q_j \exp\left[-\alpha^2\left(z' - z_j\right)^2\right] \\
& \times \exp\left[-\alpha^2\left(\boldsymbol{R}' - \boldsymbol{R}_j + \boldsymbol{n}_\parallel\right)^2\right] \qquad\qquad\text{(F.96b)}
\end{aligned}
$$

$$
\begin{aligned}
\rho_i^{(3)}\left(\boldsymbol{r}'\right) \;=\;& -q_i \left(\frac{\alpha}{\sqrt{\pi}}\right)^3 \exp\left[-\alpha^2\left(z' - z_i\right)^2\right] \\
& \times \exp\left[-\alpha^2\left(\boldsymbol{R}' - \boldsymbol{R}_i\right)^2\right] \qquad\qquad\text{(F.96c)}
\end{aligned}
$$

which is completely analogous to the bulk expressions given in Eqs. (6.7), (6.10a), and (6.10b), respectively. Thus, we can immediately write down expressions for the potentials related to $\rho_i^{(1)}\left(\boldsymbol{r}'\right)$ and $\rho_i^{(3)}\left(\boldsymbol{r}\prime\right)$ [see Eqs. (6.8) and (6.14)]; that is,

$$
\Phi^{(1)}\left(\boldsymbol{r}_i\right) \;=\; {\sum_{\boldsymbol{n}}}' \sum_{j=1}^{N} q_j \frac{\operatorname{erfc}\left(\alpha\left|\boldsymbol{r}_{ij} + \boldsymbol{n}\right|\right)}{\left|\boldsymbol{r}_{ij} + \boldsymbol{n}\right|} \qquad\qquad\text{(F.97a)}
$$

$$
\Phi^{(3)}\left(\boldsymbol{r}_i\right) \;=\; -q_i \frac{2\alpha}{\sqrt{\pi}}, \qquad\qquad\text{(F.97b)}
$$

where in Eq. (F.97a), $\boldsymbol{n} = \left(\boldsymbol{n}_\parallel, 0\right)$.

However, the potential $\Phi^{(2)}\left(\boldsymbol{r}'\right)$ related to $\rho^{(2)}\left(\boldsymbol{r}'\right)$ differs from its bulk counterpart [see Eq. (6.14)] because the basic simulation cell of the current slab system is repeated in only two (of the three) spatial dimensions. Nevertheless, we can still apply our basic strategy detailed in Appendix F.1.1.2 to find the explicit expression for $\Phi^{(2)}\left(\boldsymbol{r}'\right)$.

We start by expanding the potential in Fourier space according to

$$
\Phi^{(2)}\left(\boldsymbol{r}\right) = \frac{s_z}{2\pi} \sum_{\boldsymbol{k}_\parallel} \exp\left[i\boldsymbol{k}_\parallel \cdot \boldsymbol{R}\right] \int\limits_{-\infty}^{\infty} \mathrm{d}k_z \widetilde{\Phi}^{(2)}\left(\boldsymbol{k}\right) \exp\left[ik_z z\right] \qquad\text{(F.98)}
$$

where $\boldsymbol{k}_\parallel = \left(2\pi m_x/s_x, 2\pi m_y/s_y\right)$ are two-dimensional reciprocal lattice vectors (i.e., $\exp\left[i\boldsymbol{k}_\parallel \cdot \boldsymbol{n}_\parallel\right] = 1$), whereas the vectors $\boldsymbol{k}$ appearing as arguments of the coefficients $\widetilde{\Phi}^{(2)}\left(\boldsymbol{k}\right)$ are still three-dimensional. Equation (F.98) follows from its three-dimensional counterpart (F.19b) if we replace in the full sum $\sum_{\boldsymbol{k}} \cdots = \sum_{k_x} \sum_{k_y} \sum_{k_z} \cdots$ the partial summation over the discrete variable k_z by an integration, that is $\sum_{k_z} \cdots \to (\Delta k_z)^{-1} \int_{-\infty}^{\infty} dk_z \cdots$ with $\Delta k_z = 2\pi/s_z$. This is consistent with viewing the system in slab geometry as three–dimensional with a basic cell becoming *infinitely* large in the z-direction (i.e., $s_z \to \infty$) such that $\Delta k_z \to 0$.

The Fourier coefficients $\tilde{\Phi}^{(2)}(\boldsymbol{k})$ appearing in Eq. (F.98) are linked to the corresponding coefficients of the charge density via the Fourier-transformed Laplace equation [see Eq. (F.21)]

$$
\begin{aligned}
\tilde{\rho}^{(2)}(\boldsymbol{k}) &= \frac{1}{V_{\text{sys}}} \int d\boldsymbol{r} \exp\left(-i\boldsymbol{k} \cdot \boldsymbol{r}\right) \rho^{(2)}(\boldsymbol{r}) \\
&= \frac{1}{V} \exp\left(-\frac{k_\parallel^2 + k_z^2}{4\alpha^2}\right) \sum_{j=1}^{N} q_j \exp\left(-ik_z z_j\right) \exp\left(-i\boldsymbol{k}_\parallel \cdot \boldsymbol{R}_j\right)
\end{aligned}
$$

(F.99)

where we have inserted Eq. (F.96b) to obtain the second line of Eq. (F.99). Combining Eqs. (F.99) and (F.21) and inserting the resulting Fourier coefficients $\tilde{\Phi}^{(2)}(\boldsymbol{k})$ into the expansion in Eq. (F.98), we find

$$
\begin{aligned}
\Phi^{(2)}(\boldsymbol{r}) &= \frac{2}{A} \sum_{j=1}^{N} \sum_{\boldsymbol{k}_\parallel \neq 0} q_j \exp\left[i\boldsymbol{k}_\parallel \cdot (\boldsymbol{R} - \boldsymbol{R}_j)\right] \int_{-\infty}^{\infty} dk_z \frac{1}{k_\parallel^2 + k_z^2} \\
&\quad \times \exp\left[-\frac{k_\parallel^2 + k_z^2}{4\alpha^2} + ik_z(z - z_j)\right] + \Phi^{(2)}_{\boldsymbol{k}_\parallel \to 0}(z - z_j)
\end{aligned}
$$
(F.100)

where the sum in the first line is restricted to nonzero wavevectors $\boldsymbol{k}_\parallel$ and $\Phi^{(2)}_{\boldsymbol{k}_\parallel \to 0}$ contains contributions from the long-wavelength limit (see below). In Eq. (F.100), the integral over the continuous variable k_z gives [248]

$$
\int_{-\infty}^{\infty} dk_z \frac{1}{k_\parallel^2 + k_z^2} \exp\left[-\frac{k_\parallel^2 + k_z^2}{4\alpha^2} + ik_z(z - z_j)\right] = \frac{\pi}{2k_\parallel} f\left(k_\parallel, z - z_j, \alpha\right)
$$

(F.101)

where the function $f(k_\parallel, z - z_j, \alpha)$ has been defined in Eq. (6.36).

From Eqs. (F.100) and (F.101) it follows that $\Phi^{(2)}_{\boldsymbol{k}_\parallel \to 0}(z - z_j)$ is defined as

$$
\Phi^{(2)}_{\boldsymbol{k}_\parallel \to 0}(z - z_j) \equiv \frac{\pi}{Ak_\parallel} \sum_{j=1}^{N} \lim_{k_\parallel \to 0} f\left(k_\parallel, z - z_j, \alpha\right)
$$
(F.102)

Thus, we consider the behavior of the function $f(k_\parallel, z - z_j, \alpha)$ for small wavenumbers $k_\parallel$. To this end we perform a Taylor expansion of both the exponentials and the complementary error functions appearing on the right side of Eq. (6.36). In this expansion, the lowest-order Taylor coefficients of $\operatorname{erfc}(y) = 1 - \operatorname{erf}(y)$ follow immediately from Eq. (F.49). We also note the relation $\operatorname{erf}(-y) = -\operatorname{erf}(y)$, which follows from the definition of the error function in Eq. (F.9). Expanding $f(k_\parallel, z - z_j, \alpha)$ [see Eq. (6.36)] around $k_\parallel = 0$, we obtain

$$
\begin{aligned}
f(k_\parallel, z - z_j, \alpha) &= 2 - 2k_\parallel (z - z_j) \operatorname{erf}[\alpha(z - z_j)] \\
&\quad - \frac{2}{\alpha\sqrt{\pi}} k_\parallel \exp\left[-\alpha^2 (z - z_j)^2\right] + \mathcal{O}(k_\parallel^2) \quad \text{(F.103)}
\end{aligned}
$$

where we retain only linear terms in $k_\parallel$. Combining the previous expression with Eq. (F.102) yields

$$
\begin{aligned}
\Phi^{(2)}_{k_\parallel \to 0}(z - z_j) &= \frac{2}{A} \sum_{j-1}^{N} q_j \lim_{k_\parallel \to 0} \frac{\pi}{2k_\parallel} \Big\{ 2 - 2k_\parallel (z - z_j) \operatorname{erf}[\alpha(z - z_j)] \\
&\quad - \frac{2}{\alpha\sqrt{\pi}} k_\parallel \exp\left[-\alpha^2 (z - z_j)^2\right] + \mathcal{O}(k^2) \Big\} \quad \text{(F.104)}
\end{aligned}
$$

Inspecting the right side of Eq. (F.104) we see that the first term in parentheses is constant. This term is irrelevant because of global charge neutraliy (i.e., $\sum_{i=1}^{N} q_i = 0$). We therefore obtain from Eqs. (F.100) and (F.104) as a final expression for the potential from the set of Gaussians

$$
\begin{aligned}
\Phi^{(2)}(r) &= \frac{\pi}{A} \sum_{j=1}^{N} \sum_{k_\parallel \neq 0} q_j \frac{\exp\left[i k_\parallel \cdot (R - R_j)\right]}{k_\parallel} f(k_\parallel, z, \alpha) \\
&\quad - \frac{2\sqrt{\pi}}{A} \sum_{j=1}^{N} q_j \frac{\exp[-\alpha^2 (z - z_j)^2]}{\alpha} \\
&\quad - \frac{2\pi}{A} \sum_{j=1}^{N} q_j (z - z_j) \operatorname{erf}[\alpha(z - z_j)] \quad \text{(F.105)}
\end{aligned}
$$

The corresponding contribution to the energy is

$$
U_{\text{CF}}^{\text{2d}} = \frac{1}{2} \sum_{i=1}^{N} q_i \Phi^{(2)}(r_i) \quad \text{(F.106)}
$$

which coincides with Eq. (6.35).

F.3.1.2   Point dipoles

The rigorous Ewald sum for a slab-like system of point dipoles follows from the corresponding expression for Coulombic systems [see Eqs. (6.34) and (6.35)]. The derivation proceeds in a fashion similar to the one already discussed for bulk systems in Section 6.2.2. That is, we replace the charges q_i and q_j in the energy expressions by the operators $(\boldsymbol{\mu}_i \cdot \nabla_i)$ and $(\boldsymbol{\mu}_j \cdot \nabla_j)$. As a result,

$$U_{\mathrm{D}}^{2\mathrm{d}} = U_{\mathrm{DR}}^{2\mathrm{d}} + U_{\mathrm{DF}}^{2\mathrm{d}} + U_{\mathrm{DSF}}^{2\mathrm{d}} \tag{F.107}$$

where both real-space and self-part have the same form as in the three-dimensional case and are thus given by Eqs. (6.27a) and (6.32), respectively.

To evaluate the Fourier part, we start by considering the first sum on the right side of Eq. (6.35) involving nonzero wavevectors $\boldsymbol{k}_{\parallel} \neq \boldsymbol{0}$. For each pair i and j and each wavevector $\boldsymbol{k}_{\parallel}$, the replacement of the charge q_j by the operator $(\boldsymbol{\mu}_j \cdot \nabla_j)$ yields

$$
\begin{aligned}
&(\boldsymbol{\mu}_j \cdot \nabla_j) \frac{\exp\left(i\boldsymbol{k}_{\parallel} \cdot \boldsymbol{R}_{ij}\right)}{k_{\parallel}} f\left(k_{\parallel}, z_{ij}, \alpha\right) \\
&= -i\left(\boldsymbol{\mu}_j \cdot \boldsymbol{k}_{\parallel}\right) \frac{\exp\left(i\boldsymbol{k}_{\parallel} \cdot \boldsymbol{R}_{ij}\right)}{k_{\parallel}} f\left(k_{\parallel}, z_{ij}, \alpha\right) \\
&\quad + (\boldsymbol{\mu}_j \cdot \widehat{\boldsymbol{e}}_z) \frac{\exp\left(i\boldsymbol{k}_{\parallel} \cdot \boldsymbol{R}_{ij}\right)}{k_{\parallel}} d\left(k_{\parallel}, z_{ij}, \alpha\right)
\end{aligned}
\tag{F.108}
$$

where $\boldsymbol{R}_{ij} = \boldsymbol{R}_i - \boldsymbol{R}_j$, $z_{ij} = z_i - z_j$, the function $f(k_{\parallel}, z_{ij}, \alpha)$ is defined in Eq. (6.36), and

$$
\begin{aligned}
d\left(k_{\parallel}, z_{ij}, \alpha\right) = \frac{\partial}{\partial z_j} f\left(k_{\parallel}, z_{ij}, \alpha\right) &= k_{\parallel} \exp\left(-k_{\parallel} z_{ij}\right) \mathrm{erfc}\left(\frac{k_{\parallel}}{2\alpha} - \alpha z_{ij}\right) \\
&\quad - k_{\parallel} \exp\left(k_{\parallel} z_{ij}\right) \mathrm{erfc}\left(\frac{k_{\parallel}}{2\alpha} + \alpha z_{ij}\right)
\end{aligned}
\tag{F.109}
$$

Now we need to differentiate Eq. (F.108) one more time because of the second operator $(\boldsymbol{\mu}_i \cdot \nabla_i)$ replacing q_i in the original energy expression in Eq. (6.35), which yields

$$(\boldsymbol{\mu}_i \cdot \nabla_i)(\boldsymbol{\mu}_j \cdot \nabla_j) \frac{\exp(i\boldsymbol{k}_\| \cdot \boldsymbol{R}_{ij})}{k_\|} f(k_\|, z_{ij}, \alpha)$$

$$= (\boldsymbol{\mu}_i \cdot \boldsymbol{k}_\|)(\boldsymbol{\mu}_j \cdot \boldsymbol{k}_\|) \frac{\exp(i\boldsymbol{k}_\| \cdot \boldsymbol{R}_{ij})}{k_\|} f(k_\|, z_{ij}, \alpha)$$

$$+i(\boldsymbol{\mu}_i \cdot \hat{\boldsymbol{e}}_z)(\boldsymbol{\mu}_j \cdot \boldsymbol{k}_\|) \frac{\exp(i\boldsymbol{k}_\| \cdot \boldsymbol{R}_{ij})}{k_\|} d(k_\|, z_{ij}, \alpha)$$

$$+i(\boldsymbol{\mu}_j \cdot \hat{\boldsymbol{e}}_z)(\boldsymbol{\mu}_i \cdot \boldsymbol{k}_\|) \frac{\exp(i\boldsymbol{k}_\| \cdot \boldsymbol{R}_{ij})}{k_\|} d(k_\|, z_{ij}, \alpha)$$

$$+ (\boldsymbol{\mu}_i \cdot \hat{\boldsymbol{e}}_z)(\boldsymbol{\mu}_j \cdot \hat{\boldsymbol{e}}_z) \frac{\exp(i\boldsymbol{k}_\| \cdot \boldsymbol{R}_{ij})}{k_\|} e(k_\|, z_{ij}, \alpha) \qquad \text{(F.110)}$$

where we have also used [see Eq. (F.109)]

$$\frac{\partial}{\partial z_i} f(k_\|, z_{ij}, \alpha) = -\frac{\partial}{\partial z_j} f(k_\|, z_{ij}, \alpha) = -d(k_\|, z_{ij}, \alpha) \qquad \text{(F.111)}$$

and the function

$$c(k_\|, z_{ij}, \alpha) = \frac{\partial}{\partial z_i} d(k_\|, z_{ij}, \alpha)$$

$$= \frac{4\alpha k_\|}{\sqrt{\pi}} \exp\left(-\frac{k_\|^2}{4\alpha^2} - \alpha^2 z_{ij}^2\right)$$

$$-k_\|^2 \left[\exp(k_\| z_{ij}) \operatorname{erfc}\left(\frac{k_\|}{2\alpha} + \alpha z_{ij}\right)\right.$$

$$\left. + \exp(-k_\| z_{ij}) \operatorname{erfc}\left(\frac{k_\|}{2\alpha} - \alpha z_{ij}\right)\right] \qquad \text{(F.112)}$$

Keeping in mind that the total Fourier energy involves a sum over *all* nonzero wavevectors $\boldsymbol{k}_\| \neq \boldsymbol{0}$ [see Eq. (6.35)], we may employ symmetry arguments to simplify the above expressions. For example, both $(\boldsymbol{\mu}_i \cdot \boldsymbol{k}_\|)(\boldsymbol{\mu}_j \cdot \boldsymbol{k}_\|)$ and the function $f(k_\|, z_{ij}, \alpha)$ appearing on the first line of the right side of Eq. (F.110) are invariant against inversion of the wavevectors, that is to say, a replacement of $\boldsymbol{k}_\| \rightarrow$ by $-\boldsymbol{k}_\|$. Therefore, only the real part of $\exp(i\boldsymbol{k}_\| \cdot \boldsymbol{R}_{ij})$ [i.e., $\cos(\boldsymbol{k}_\| \cdot \boldsymbol{R}_{ij})$] will contribute to the sum over all wavevectors. The same is true for the fourth term involving the product $(\boldsymbol{\mu}_i \cdot \hat{\boldsymbol{e}}_z)(\boldsymbol{\mu}_j \cdot \hat{\boldsymbol{e}}_z)$ and the function $e(k_\|, z_{ij}, \alpha)$, which is again invariant against inversion of $\boldsymbol{k}_\|$. However, the second and third terms on the right side of Eq. (F.110) are linear in $\boldsymbol{k}_\parallel$ and therefore change sign upon inversion of the wavevector. As a consequence, we have

$$
i \sum_{\boldsymbol{k}_\parallel \neq 0} \left[(\boldsymbol{\mu}_i \cdot \widehat{\boldsymbol{e}}_z) (\boldsymbol{\mu}_j \cdot \boldsymbol{k}_\parallel) + (\boldsymbol{\mu}_j \cdot \widehat{\boldsymbol{e}}_z) (\boldsymbol{\mu}_i \cdot \boldsymbol{k}_\parallel) \right]
$$

$$
\times \frac{\exp\left(i\boldsymbol{k}_\parallel \cdot \boldsymbol{R}_{ij}\right)}{k_\parallel} d\left(k_\parallel, z_{ij}, \alpha\right)
$$

$$
= -\sum_{\boldsymbol{k}_\parallel \neq 0} \left[(\boldsymbol{\mu}_i \cdot \widehat{\boldsymbol{e}}_z) (\boldsymbol{\mu}_j \cdot \boldsymbol{k}_\parallel) + (\boldsymbol{\mu}_j \cdot \widehat{\boldsymbol{e}}_z) (\boldsymbol{\mu}_i \cdot \boldsymbol{k}_\parallel) \right]
$$

$$
\times \frac{\sin\left(i\boldsymbol{k}_\parallel \cdot \boldsymbol{R}_{ij}\right)}{k_\parallel} d\left(k_\parallel, z_{ij}, \alpha\right) \tag{F.113}
$$

We now consider the remaining (second) sum in the Coulomb Fourier energy [see the right side of Eq. (6.35)] involving a double sum over pairs of dipoles i and j. Replacing the products q_i and q_j by the operators $(\boldsymbol{\mu}_i \cdot \nabla_i)$ and $(\boldsymbol{\mu}_j \cdot \nabla_j)$, respectively, as before yields

$$
(\boldsymbol{\mu}_i \cdot \nabla_i)(\boldsymbol{\mu}_j \cdot \nabla_j) \left[\frac{\exp\left(-\alpha^2 z_{ij}^2\right)}{\alpha} + \sqrt{\pi} z_{ij} \mathrm{erf}\left(\alpha z_{ij}\right) \right]
$$

$$
= (\boldsymbol{\mu}_i \cdot \widehat{\boldsymbol{e}}_z)(\boldsymbol{\mu}_j \cdot \widehat{\boldsymbol{e}}_z) \frac{\exp\left(-\alpha^2 z_{ij}^2\right)}{\alpha}
$$

$$
\times \left(-4\alpha^2 z_{ij}^2 - 2\alpha + 4\alpha^2 z_{ij}^2\right)
$$

$$
= -2\alpha \exp\left(-\alpha^2 z_{ij}^2\right)(\boldsymbol{\mu}_i \cdot \widehat{\boldsymbol{e}}_z)(\boldsymbol{\mu}_j \cdot \widehat{\boldsymbol{e}}_z) \tag{F.114}
$$

Finally, putting all this together we arrive at the rigorous expression for the Fourier-space contribution to the total configurational energy of the dipolar system in slab geometry, namely

$$
\begin{aligned}
U_{\mathrm{DF}}^{\mathrm{2d}} = {} & \frac{\pi}{2A} \sum_{i=1}^{N} \sum_{j=1}^{N} \sum_{\boldsymbol{k}_\parallel \neq 0} \left\{ (\boldsymbol{\mu}_i \cdot \boldsymbol{k}_\parallel)(\boldsymbol{\mu}_j \cdot \boldsymbol{k}_\parallel) \frac{\cos\left(\boldsymbol{k}_\parallel \cdot \boldsymbol{R}_{ij}\right)}{k_\parallel} f(k_\parallel, z_{ij}, \alpha) \right. \\
& - \left[(\boldsymbol{\mu}_i \cdot \widehat{\boldsymbol{e}}_z)(\boldsymbol{\mu}_j \cdot \boldsymbol{k}_\parallel) + (\boldsymbol{\mu}_j \cdot \widehat{\boldsymbol{e}}_z)(\boldsymbol{\mu}_i \cdot \boldsymbol{k}_\parallel) \right] \\
& \times \frac{\sin\left(\boldsymbol{k}_\parallel \cdot \boldsymbol{R}_{ij}\right)}{k_\parallel} d\left(k_\parallel, z_{ij}, \alpha\right) \\
& \left. + (\boldsymbol{\mu}_i \cdot \widehat{\boldsymbol{e}}_z)(\boldsymbol{\mu}_j \cdot \widehat{\boldsymbol{e}}_z) \frac{\cos\left(\boldsymbol{k}_\parallel \cdot \boldsymbol{R}_{ij}\right)}{k_\parallel} e\left(k_\parallel, z_{ij}, \alpha\right) \right\} \\
& + \frac{\sqrt{\pi}}{A} \sum_{i=1}^{N} \sum_{j=1}^{N} 2\alpha \exp\left(-\alpha^2 z_{ij}^2\right)(\boldsymbol{\mu}_i \cdot \widehat{\boldsymbol{e}}_z)(\boldsymbol{\mu}_j \cdot \widehat{\boldsymbol{e}}_z) \tag{F.115}
\end{aligned}
$$

F.3.2   Force, torque, and stress in systems with slab geometry

F.3.2.1   Point charges

For ionic systems, the total Coulomb force acting on particle i within the slab-adapted three-dimensional Ewald sum [see Eq. (6.40)] can be cast as

$$F_{C,i}^{slab} = F_{CR,i}^{slab} + F_{CF,i}^{slab} + F_{C,c,i}^{slab} \tag{F.116}$$

where the first two contributions are identical to the corresponding ones in a truly three–dimensional system and are thus given by Eqs. (F.52a) and (F.52b), respectively. The last term in Eq. (F.116) arises from the correction term in the Ewald energy, $U_{C,c}$ [see Eq. (6.39)]. One obtains

$$F_{C,c,i}^{slab} = -q_i \frac{4\pi}{V} \sum_{j=1}^{N} q_j z_j \widehat{e}_z = -q_i \frac{4\pi}{V} M_z \widehat{e}_z \tag{F.117}$$

Regarding the stress tensor of the system, the (Coulomb) components corresponding to the two orthogonal directions parallel to the walls (i.e., $\gamma = x, y$) can be calculated exactly as in the three-dimensional case (see Appendix F.1.2.2). On the other hand, the normal component ($\gamma = z$) is given by

$$\tau_{C,zz}^{slab} = \tau_{CR,zz}^{slab} + \tau_{CF,zz}^{slab} + \tau_{C,c,zz}^{slab} \tag{F.118}$$

where only the real-space part $\tau_{CR,zz}^{slab} = \tau_{CR,zz}^{3d}$ [see Eq. (F.58)].

To evaluate the Fourier-space contribution, $\tau_{CF,zz}^{slab}$, we note that, because of the *artificial* elongation of the basis cell in z-direction, neither the wavevectors involved in the Fourier contribution to the Ewald energy [see Eq. (6.18) with the wavevectors given in Eq. (6.41)] nor the volume V depend on s_z. The vectors $r_i = (x_i, y_i, s_z \widetilde{z}_i)$, on the other hand, depend on s_z if we employ scaled z-coordinates as indicated. Differentiation thus yields

$$\tau_{CF,zz}^{slab} = \frac{1}{A_{z0}} \left\langle \frac{\partial U_{CF}^{3d}}{\partial s_z} \right\rangle = -\frac{1}{2 A_{z0} s_z} \frac{1}{V} \sum_{k \neq 0} \frac{4\pi}{k^2} \exp\left(-\frac{k^2}{4\alpha^2}\right)$$
$$\times \left\langle \widetilde{b}(k) \widetilde{a}^*(k) + \widetilde{b}^*(k) \widetilde{a}(k) \right\rangle \tag{F.119}$$

where the quantity $\widetilde{a}(k)$ is defined in Eq. (6.19) and

$$\widetilde{b}(k) = \sum_{i=1}^{N} ik_z z_i \exp(-ik \cdot r_i) \tag{F.120}$$

Finally, as the energy correction term given in Eq. (6.39) depends on s_z only through the z-components of the position vectors $\boldsymbol{r}_i$, the expression for the corresponding stress in the direction normal to the confining substrates follows as

$$\tau_{\mathrm{C,c,zz}} = \frac{1}{A_{z0}} \left\langle \frac{\partial \widetilde{U}_{\mathrm{C}}}{\partial s_z} \right\rangle = \frac{1}{A_{z0} s_z} \frac{4\pi}{V} \sum_{i=1}^{N} \sum_{j=1}^{N} q_i q_j z_i z_j \qquad (\mathrm{F.121})$$

F.3.2.2   Point dipoles

For dipolar particles, all force contributions within the slab-adapted three-dimensional Ewald sum coincide with those for truly three-dimensional systems discussed in Appendix F.2.2. This result arises because the correction term to the total dipolar energy [see Eqs. (6.44) and (6.43)] is independent of particle positions. There is, however, a contribution to the total torque that we need to consider separately. The total torque can be cast as

$$\boldsymbol{T}_{\mathrm{D},i}^{\mathrm{slab}} = \boldsymbol{T}_{\mathrm{DR},i}^{\mathrm{slab}} + \boldsymbol{T}_{\mathrm{DF},i}^{\mathrm{slab}} + \boldsymbol{T}_{\mathrm{D,c},i}^{\mathrm{slab}} \qquad (\mathrm{F.122})$$

where $\boldsymbol{T}_{\mathrm{DR},i}^{\mathrm{slab}}$ and $\boldsymbol{T}_{\mathrm{DF},i}^{\mathrm{slab}}$ are given by the bulk expressions [see Eqs. (F.88)], whereas

$$\boldsymbol{T}_{\mathrm{D,c},i}^{\mathrm{slab}} = -\boldsymbol{\mu}_i \times \left(\frac{4\pi}{V} M_z \widehat{\boldsymbol{e}}_z \right) \qquad (\mathrm{F.123})$$

Turning next to the stress tensor we realize that its normal component within the slab-adapted three-dimensional Ewald formalism can be written as a sum of two contributions, namely

$$\tau_{\mathrm{D,zz}}^{\mathrm{slab}} = \tau_{\mathrm{DR,zz}}^{\mathrm{slab}} + \tau_{\mathrm{DF,zz}}^{\mathrm{slab}} \qquad (\mathrm{F.124})$$

because the correction term to the total configurational potential energy in Eq. (6.43) does not depend on s_z (note that the V is the volume of the artificial cell *including* the vacuum space in the z-direction). The real-space part on the right side of Eq. (F.124) coincides with its three-dimensional analog given in Eq. (F.91). The Fourier part, $\tau_{\mathrm{DF,zz}}^{\mathrm{slab}}$, can be derived along the same lines already discussed below Eq. (F.118). We finally obtain with little ado the expression

$$\tau_{\mathrm{DF,zz}}^{\mathrm{slab}} = \frac{1}{A_{z0}} \left\langle \frac{\partial U_{\mathrm{CF}}^{\mathrm{3d}}}{\partial s_z} \right\rangle = -\frac{1}{2 A_{z0} s_z} \frac{1}{V} \sum_{\boldsymbol{k} \neq 0} \frac{4\pi}{k^2} \exp\left(-\frac{k^2}{4\alpha^2} \right)$$
$$\times \left\langle \widetilde{V}(\boldsymbol{k}) \widetilde{M}^*(\boldsymbol{k}) + \widetilde{V}^*(\boldsymbol{k}) \widetilde{M}(\boldsymbol{k}) \right\rangle$$
$$(\mathrm{F.125})$$

where the quantity $\widetilde{M}\left(\mathbf{k}\right)$ is defined in Eq. (6.29) and

$$\widetilde{V}\left(\mathbf{k}\right) = \sum_{i=1}^{N} \left(\boldsymbol{\mu}_i \cdot \mathbf{k}\right) i k_z z_i \exp\left(-i\mathbf{k} \cdot \mathbf{r}_i\right) \tag{F.126}$$

F.3.3   Metallic substrates

F.3.3.1   Point charges

Here we derive Eq. (6.66) linking the energy of a slab-like system of point charges between metallic walls to that of an extended system with three-dimensional periodicity.

The basic cell of the extended system contains N charges in the original cell plus the first set of images; that is, the N images resulting from the presence of just the lower wall. Positions and charges of these image particles are then given by the relations [see Eq. (6.59) with $n_z = 0$]

$$\mathbf{r}_{i+N} = \mathbf{r}_i - 2z_i\widehat{\mathbf{e}}_z, \quad i = 1, \ldots, N \tag{F.127a}$$
$$q_{i+N} = -q_i, \quad i = 1, \ldots, N \tag{F.127b}$$

Replicating the extended basic cell periodically in all three spatial directions, the total energy of the resulting system is given by

$$U_{\mathrm{C}}^{\mathrm{3d,ex}} = \frac{1}{2} \sum_{i=1}^{2N} \sum_{j=1}^{2N} \sideset{}{'}\sum_{\overline{\boldsymbol{n}}} \frac{q_i q_j}{|\mathbf{r}_{ij} + \overline{\boldsymbol{n}}|} \tag{F.128}$$

where the lattice vectors $\overline{\boldsymbol{n}}$ are specified in Eq. (6.64) and the prime at the sum indicates that the term related to $i = j$ is omitted for $\overline{\boldsymbol{n}} = \mathbf{0}$.

We now split the double sum in Eq. (F.128) into four terms containing

1. Particle particle contributions $\sum_{i=1}^{N} \sum_{j=1}^{N}$,

2. Image image contributions $\sum_{i=N+1}^{2N} \sum_{j=N+1}^{2N}$,

3. Particle image contributions $\sum_{i=1}^{N} \sum_{j=N+1}^{2N}$, and

4. Image particle contributions $\sum_{i=N+1}^{2N} \sum_{j=1}^{N}$.

Terms 1 and 2 give the same result as one may varify from the relations

$$q_{i+N}q_{j+N} = (-q_i)(-q_j) = q_i q_j \tag{F.129a}$$

$$\begin{aligned}
|\boldsymbol{r}_{i+N,j+N} + \overline{\boldsymbol{n}}| &= |\boldsymbol{r}_{ij} + \overline{\boldsymbol{n}} - 2z_{ij}\widehat{\boldsymbol{e}}_z| \\
&= \sqrt{(x_{ij} + s_x n_x)^2 + (y_{ij} + s_y n_y)^2 + (2s_z n_z - z_{ij})^2} \\
&= \sqrt{(x_{ij} + s_x n_x)^2 + (y_{ij} + s_y n_y)^2 + (z_{ij} - 2s_z n_z)^2} \\
&= |\boldsymbol{r}_{ij} + \overline{\boldsymbol{n}} - 4s_z n_z \widehat{\boldsymbol{e}}_z| \tag{F.129b}
\end{aligned}$$

and the fact that we sum in Eq. (F.128) over an infinite set of lattice vectors $\{\overline{\boldsymbol{n}}\}$ such that the term $4s_z n_z \widehat{\boldsymbol{e}}_z$ on the right side of Eqs. (F.129) is irrelevant. By similar reasoning, terms 3 and 4 in the above decomposition give equivalent results because of

$$q_i q_{j+N} = -q_i q_j = q_{i+N} q_j \tag{F.130a}$$

$$\begin{aligned}
|\boldsymbol{r}_{i,j+N} + \overline{\boldsymbol{n}}| &= |\boldsymbol{r}_{ij} + 2z_j \widehat{\boldsymbol{e}}_z + \overline{\boldsymbol{n}}| \\
&= \sqrt{(x_{ij} + s_x n_x)^2 + (y_{ij} + s_y n_y)^2 + (z_i + z_j + 2s_z n_z)^2} \\
&= \sqrt{(x_{ij} + s_x n_x)^2 + (y_{ij} + s_y n_y)^2 + (-z_i - z_j - 2s_z n_z)^2} \\
&= |\boldsymbol{r}_{i+N,j} + \overline{\boldsymbol{n}} - 4s_z n_z \widehat{\boldsymbol{e}}_z| \tag{F.130b}
\end{aligned}$$

Putting all this together, Eq. (F.128) can be rewritten as

$$U_{\mathrm{C}}^{3\mathrm{d,ex}} = \left[\sum_{i=1}^{N} \sum_{j=1}^{N} {\sum_{\overline{\boldsymbol{n}}}}' \frac{q_i q_j}{|\boldsymbol{r}_{ij} + \overline{\boldsymbol{n}}|} - \frac{q_i q_j}{|\boldsymbol{r}_{ij} + 2z_j \widehat{\boldsymbol{e}}_z + \overline{\boldsymbol{n}}|} \right] \tag{F.131}$$

We therefore see that $U_{\mathrm{C}}^{3\mathrm{d,ex}}$ in Eq. (F.131) for a three-dimensional system with the extended basis cell is indeed exactly twice the energy U_{C} given in Eq. (6.65).

F.3.3.2   Point dipoles

To derive Eq. (6.68) for the total energy of an *infinite* slab of dipolar particles between metallic substrates, we go one step back and consider a situation where the central cell comprising N particles has not yet been replicated in the x- and y-directions. The corresponding energy can be written as

$$\widetilde{U}_{\mathrm{D}} = -\frac{1}{2} \sum_{i=1}^{N} \boldsymbol{\mu}_i \cdot \left(\boldsymbol{E}_i^{\mathrm{self}} + \boldsymbol{E}_i^{\mathrm{dis}} \right) \tag{F.132}$$

where E_i^{self} and E_i^{dis} are the electrostatic fields arising from the images of particle i, on the one hand, and from the other particles j and their images, on the other hand. Using short-hand notation

$$e_i\left(\boldsymbol{\mu}_j, \boldsymbol{r}_j\right) = \frac{3\boldsymbol{r}_{ij}\left(\boldsymbol{\mu}_j \cdot \boldsymbol{r}_{ij}\right)}{r_{ij}^5} - \frac{\boldsymbol{\mu}_j}{r_{ij}^3} \qquad (F.133)$$

the fields E_i^{self} and E_i^{dis} follow as

$$E_i^{\text{self}} = \sum_{n_z=-\infty}^{\infty}{}^* e_i\left(\boldsymbol{\mu}_i, \boldsymbol{r}_i + 2n_z s_z \widehat{\boldsymbol{e}}_z\right) + \sum_{n_z=-\infty}^{\infty} e_i\left[\boldsymbol{\mu}_i', \boldsymbol{r}_i + 2\left(n_z s_z - z_i\right)\widehat{\boldsymbol{e}}_z\right]$$
$$(F.134a)$$

$$E_i^{\text{dis}} = \sum_{j\neq i}^{N}\left\{ e_i\left(\boldsymbol{\mu}_j, \boldsymbol{r}_j\right) + \sum_{n_z=-\infty}^{\infty}{}^* e_i\left(\boldsymbol{\mu}_j, \boldsymbol{r}_j + 2n_z s_z \widehat{\boldsymbol{e}}_z\right) \right.$$
$$\left. + \sum_{n_z=-\infty}^{\infty} e_i\left[\boldsymbol{\mu}_j', \boldsymbol{r}_j + 2\left(n_z s_z - z_i\right)\widehat{\boldsymbol{e}}_z\right] \right\} \qquad (F.134b)$$

where the asterisk attached to the sums indicates that terms corresponding to $n_z = 0$ have been omitted.

Replicating the original cell now in the x- and y- directions essentially implies that the sums over the integer variable n_z in Eq. (F.134a) have to be replaced by three-dimensional lattice sums over the vectors $\overline{\boldsymbol{n}}$ introduced in Eq. (6.64) [see the analogous procedure for charges described below Eq. (6.62a)]. Inserting the resulting field expressions into Eq. (F.132) and summarizing, one obtains Eq. (6.68) after some tedious but straightforward algebraic manipulations.

As a next step we now have to prove Eq. (6.69) where we again proceed as before in the Coulombic case. The basic cell of the extended dipolar system contains N dipoles in the original cell plus the first set of images, which are the N images resulting from the presence of just the lower wall. Positions and orientations of these N image particles are then given by

$$\boldsymbol{r}_{i+N} = \boldsymbol{r}_i - 2z_i \widehat{\boldsymbol{e}}_z, \quad i = 1,\ldots,N \qquad (F.135a)$$

$$\boldsymbol{\mu}_{i+N} = \begin{pmatrix} -\mu_{i,x} \\ -\mu_{i,y} \\ \mu_{i,z} \end{pmatrix}, \quad i = 1,\ldots,N \qquad (F.135b)$$

Replicating the basic cell periodically in all three spatial directions, we obtain for the total configurational potential energy the expression

$$
U_{\mathrm{D}}^{\mathrm{3d,ex}} = \frac{1}{2} \sum_{i=1}^{2N} \sum_{j=1}^{2N} \sum_{\overline{n}}' \left\{ \frac{\boldsymbol{\mu}_i \cdot \boldsymbol{\mu}_j}{|\boldsymbol{r}_{ij} + \overline{\boldsymbol{n}}|^3} - 3\frac{[\boldsymbol{\mu}_i \cdot (\boldsymbol{r}_{ij} + \overline{\boldsymbol{n}})][\boldsymbol{\mu}_j \cdot (\boldsymbol{r}_{ij} + \overline{\boldsymbol{n}})]}{|\boldsymbol{r}_{ij} + \overline{\boldsymbol{n}}|^5} \right\}
$$

$$(F.136)$$

where the lattice vectors $\bar{n}$ are specified in Eq. (6.64) and the prime at the lattice sum indicates that $i = j$ is omitted for $\bar{n} = 0$. Separating now the double sum in Eq. (F.136) into

1. Particle particle contributions $\sum_{i=1}^{N} \sum_{j=1}^{N}$,

2. Image image contributions $\sum_{i=N+1}^{2N} \sum_{j=N+1}^{2N}$,

3. Particle image contributions $\sum_{i=1}^{N} \sum_{j=N+1}^{2N}$, and

4. Image particle contributions $\sum_{i=N+1}^{2N} \sum_{j=1}^{N}$

one finds that terms 1 and 2 give the same result after performing the lattice sum. Indeed, one can easily show that

$$
\begin{aligned}
\boldsymbol{\mu}_{i+N} \cdot \boldsymbol{\mu}_{j+N} &= \boldsymbol{\mu}_i \cdot \boldsymbol{\mu}_j & (F.137a)\\
|\boldsymbol{r}_{i+N,j+N} + \overline{\boldsymbol{n}}| &= |\boldsymbol{r}_{ij} + \overline{\boldsymbol{n}} - 4n_z s_z \widehat{\boldsymbol{e}}_z| & (F.137b)
\end{aligned}
$$

where Eq. (F.137b) is identical with Eq. (F.129b) and

$$
\begin{aligned}
&[\boldsymbol{\mu}_{i+N} \cdot (\boldsymbol{r}_{i+N,j+N} + \overline{\boldsymbol{n}})][\boldsymbol{\mu}_{j+N} \cdot (\boldsymbol{r}_{i+N,j+N} + \overline{\boldsymbol{n}})]\\
=\ &[\boldsymbol{\mu}_i \cdot (\boldsymbol{r}_{ij} + \overline{\boldsymbol{n}} - 4n_z s_z \widehat{\boldsymbol{e}}_z)][\boldsymbol{\mu}_j \cdot (\boldsymbol{r}_{ij} + \overline{\boldsymbol{n}} - 4n_z s_z \widehat{\boldsymbol{e}}_z)] \quad (F.138)
\end{aligned}
$$

where we recall that the terms with $4n_z s_z \widehat{\boldsymbol{e}}_z$ are irrelevant because we sum over an infinite set of lattice vectors in Eq. (F.136). Moreover, terms 3 and 4 in the above decomposition are also equivalent because of the relations [see also Eq. (F.130b)]

$$
\begin{aligned}
\boldsymbol{\mu}_i \cdot \boldsymbol{\mu}_{j+N} &= \boldsymbol{\mu}_{i+N} \cdot \boldsymbol{\mu}_j & (F.139a)\\
|\boldsymbol{r}_{i,j+N} + \overline{\boldsymbol{n}}| &= |\boldsymbol{r}_{i+N,j} + \overline{\boldsymbol{n}} - 4n_z s_z \widehat{\boldsymbol{e}}_z| & (F.139b)
\end{aligned}
$$

and

$$
\begin{aligned}
&[\boldsymbol{\mu}_i \cdot (\boldsymbol{r}_{i,j+N} + \overline{\boldsymbol{n}})][\boldsymbol{\mu}_{j+N} \cdot (\boldsymbol{r}_{i,j+N} + \overline{\boldsymbol{n}})]\\
=\ &[\boldsymbol{\mu}_{i+N} \cdot (\boldsymbol{r}_{i+N,j} + \overline{\boldsymbol{n}} - 4n_z \widehat{\boldsymbol{e}}_z)][\boldsymbol{\mu}_j \cdot (\boldsymbol{r}_{i+N,j} + \overline{\boldsymbol{n}} - 4n_z \widehat{\boldsymbol{e}}_z)] \quad (F.140)
\end{aligned}
$$

We therefore see that the energy $U_{\mathrm{D}}^{\mathrm{3d,ex}}$ [see Eq. (F.136)] is indeed exactly twice the energy U_{D} given in Eq. (6.68).

Appendix G

Mathematical aspects of the replica formalism

G.1 Replica expressions in the grand canonical ensemble

For a fluid coupled to a reservoir with chemical potential μ, the thermal average at a given matrix realization (and fixed number of matrix particles) is given [instead of the canonical expression Eq. (7.6)] by

$$
\langle \ldots \rangle_Q = \frac{1}{\Xi_Q} \sum_{N_{\rm f}=0}^{\infty} \frac{z_{\rm f}^{N_{\rm f}}}{N_{\rm f}!} \int {\rm d}q^{N_{\rm f}} \ldots \exp\left[-\frac{U_{\rm ff}\left(q^{N_{\rm f}}\right) + U_{\rm fm}\left(q^{N_{\rm f}}, Q^{N_{\rm m}}\right)}{k_{\rm B}T} \right] \quad (G.1)
$$

where $z_{\rm f} = \exp\left(-\mu_{\rm f}/k_{\rm B}T\right)$ is the fugacity and

$$
\Xi_Q = \sum_{N_{\rm f}=0}^{\infty} \frac{z_{\rm f}^{N_{\rm f}}}{N_{\rm f}!} \int {\rm d}q^{N_{\rm f}} \ldots \exp\left[-\frac{U_{\rm ff}\left(q^{N_{\rm f}}\right) + U_{\rm fm}\left(q^{N_{\rm f}}, Q^{N_{\rm m}}\right)}{k_{\rm B}T} \right] \quad (G.2)
$$

is the matrix-dependent partition sum. Equation (7.8), on the other hand, for the disorder average remains unchanged. Moreover, the replicas can be introduced exactly as in the canonical ensemble discussed in Section 7.3. In particular, the configurational potential energy of the semi-grand canonical replicated system is the same as for the canonical system [see Eq. (7.12)]. The partition function of the replicated system in the grand canonical ensemble is defined by

$$
\Xi_{\rm rep} = \sum_{N_1, N_2, \ldots, N_n} \frac{z_{\rm f}^{N_1 + N_2 + \ldots + N_n}}{N_1! N_2! \ldots N_n!} \int {\rm d}Q^{N_{\rm m}} \prod_{\alpha=1}^{n} \int {\rm d}q_\alpha^{N_{\rm f}} \exp\left[-\frac{U^{\rm rep}}{k_{\rm B}T} \right] \quad (G.3)
$$

All relations given in Section 7.4 between correlations in the disordered and replicated system, respectively, remain unchanged. Finally, we consider the grand canonical potential, Ω, of the adsorbed fluid. Proceeding exactly as for the free energy in the canonical ensemble (see Section 7.6), one finds the analogous expression

$$\Omega = \lim_{n \to 0} \frac{\mathsf{d}}{\mathsf{d}n} \Omega_{\text{rep}} \tag{G.4}$$

where $\Omega_{\text{rep}} = -k_{\text{B}} T \ln \Xi_{\text{rep}}$.

G.2   Derivation of Eq. (7.23)

To derive Eq. (7.23), which relates the blocked correlation function, h_{b}, to the replicated system, we start from the statistical physical definition of the blocked function given in Eq. (7.22). On the right side of this equation, the double average over one of the pair terms, that is, for example, the term with $i = 1$, $j = 2$, can be written as

$$
\left[\langle A(\boldsymbol{q}_1) \rangle_{\boldsymbol{Q}} \, \langle B(\boldsymbol{q}_2) \rangle_{\boldsymbol{Q}} \right] = \int \mathsf{d} \boldsymbol{Q}^{N_{\text{m}}} P\left(\boldsymbol{Q}^{N_{\text{m}}}\right) \frac{1}{Z_{\boldsymbol{Q}}^2}
$$
$$
\times \int \mathsf{d} \boldsymbol{q}^{N_{\text{f}}} A(\boldsymbol{q}_1) \exp\left(-\frac{U_{\text{ff}} + U_{\text{fm}}}{k_{\text{B}} T} \right)
$$
$$
\times \int \mathsf{d} \boldsymbol{q}^{N_{\text{f}}} B(\boldsymbol{q}_2) \exp\left(-\frac{U_{\text{ff}} + U_{\text{fm}}}{k_{\text{B}} T} \right) \tag{G.5}
$$

where we employ the definitions of the thermal averages and the disorder average given in Eqs. (7.6) and (7.8), respectively, and $A(\boldsymbol{q}_1) \equiv \delta(\boldsymbol{q} - \boldsymbol{q}_1)$ and $B(\boldsymbol{q}_2) \equiv \delta(\boldsymbol{q}' - \boldsymbol{q}_2)$.

We may now introduce the replicated system by multiplying both the numerator and the denominator of the integrand in Eq. (G.5) by $Z_{\boldsymbol{Q}}^{n-2}$. The numerator then involves altogether n multiple integrals over the fluid particle coordinates, $q^{N_{\text{f}}}$. In other words, because of the multiplication by $Z_{\boldsymbol{Q}}^{n-2}$, we introduce $n - 2$ copies of the fluid particles *in addition* to the two already present. The already existing copies are represented by the two integrals $\int \mathsf{d} \boldsymbol{q}^{N_{\text{f}}} \ldots$ in Eq. (G.5). To these two copies we assign arbitrary, yet different indices α' and β'. The functions $A(\boldsymbol{q}_1)$ and $B(\boldsymbol{q}_2)$ then become functions of particle 1 of copy α' and particle 2 of copy β'. Inserting then Eq. (G.1) for

$P\left(\mathbf{Q}^{N_m}\right)$, Eq. (G.5) can be rewritten as

$$
\left[\langle A\left(\mathbf{q}_1\right)\rangle_{\mathbf{Q}} \langle B\left(\mathbf{q}_2\right)\rangle_{\mathbf{Q}}\right] = \int d\mathbf{Q}^{N_m} \frac{1}{Z_m Z_{\mathbf{Q}}^n} \prod_{\alpha=1}^{n}
$$
$$
\times \int d\mathbf{q}_\alpha^{N_f} A\left(\mathbf{q}_{1\alpha'}\right) B\left(\mathbf{q}_{2\beta'}\right) \exp\left(-\frac{U_{rep}}{k_B T}\right) \text{(G.6)}
$$

where we employ the definition of the configurational potential energy of the replicated system in Eq. (7.12). Finally, we use the same "trick" as described in Eq. (7.14) and identify the denominator in Eq. (G.6) with the partition function of the replicated system in the limit $n \to 0$. This identification gives

$$
\left[\langle A\left(\mathbf{q}_1\right)\rangle_{\mathbf{Q}} \langle B\left(\mathbf{q}_2\right)\rangle_{\mathbf{Q}}\right] = \lim_{n\to 0} \frac{1}{Z_{rep}} \int d\mathbf{Q}^{N_m} \prod_{\alpha=1}^{n} \int d\mathbf{q}_\alpha^{N_f} A\left(\mathbf{q}_{1\alpha'}\right) B\left(\mathbf{q}_{2\beta'}\right)
$$
$$
\times \exp\left[-\frac{U_{rep}}{k_B T}\right]
$$
$$
= \lim_{n\to 0} \langle A\left(\mathbf{q}_{1\alpha'}\right) B\left(\mathbf{q}_{2\beta'}\right)\rangle_{rep}, \qquad \alpha' \neq \beta' \qquad \text{(G.7)}
$$

Applying the relation in Eq. (G.7) to each pair term in the definition of the blocked correlation function [see Eq. (7.22)] and using Eq. (7.18) for the density, one finally arrives at Eq. (7.23).

G.3 Molecular fluids

The purpose of this section is to reformulate the RSOZ equations in $\mathbf{k}$-space given in Eqs. (7.41a)–(7.42) in an appropriate form for a molecular fluid in a molecular matrix such that the resulting expressions are still numerically tractable. We specialize to the case of linear molecules, such as spheres with dipole moments or ellipsoidal particles whose orientation can be specified by two angles $\omega = (\theta, \varphi)$. The correlation functions $\tilde{f}_\gamma(\mathbf{k}, \omega_1, \omega_2)$ (where $f = h$ or c, $\gamma = mm, mf, fm, ff, b$ or c) then depend on seven variables altogether. They describe the orientations of the pair of particles, that of the wavevector, and the wavenumber $k = |\mathbf{k}|$.

To handle these variables, we expand each correlation function in an angle-dependent basis set of rotational invariants [258]. Taking advantage of the fact that, in an globally isotropic system, the direction of the wavevector $\mathbf{k}$ does not matter, we choose $\mathbf{k}$ to be parallel to the z-axis of the space-fixed coordinate system. The resulting "$\mathbf{k}$-frame" expansion is then defined by [258]

$$
\tilde{f}_\gamma(k, \omega_1', \omega_2') = \sum_{l_1, l_2} \sum_{\chi} \tilde{f}_{\chi,\gamma}^{l_1 l_2}(k) \Psi_\chi^{l_1 l_2}(\omega_1', \omega_2') \qquad \text{(G.8)}
$$

where the invariants

$$\Psi_\chi^{l_1 l_2} = \frac{4\pi}{\sqrt{(2l_1 + 1)(2l_2 + 1)}} Y_{l_1,\chi}(\omega_1') Y_{l_2,-\chi}(\omega_2') \tag{G.9}$$

contain spherical harmonics $Y_{l,\chi}$ with $|\chi| \leq \min(l_1, l_2)$ and ω_1' and ω_2' are the dipole orientations measured with respect to the wavevector $\mathbf{k} \parallel \hat{\mathbf{e}}_z$. Invariance of the fluid against inversion of $\mathbf{k}$ implies that $\tilde{f}_{\chi,\gamma}^{l_1 l_2}(k) = \tilde{f}_{-\chi,\gamma}^{l_1 l_2}(k)$ for all correlations γ involved [258]. Moreover, exchange symmetry of the fluid–fluid and matrix–matrix potentials imposes the restriction $\tilde{f}_{\chi,\text{ff(c,b,mm)}}^{l_1 l_2}(k) = \tilde{f}_{\chi,\text{ff(c,b,mm)}}^{l_2 l_1}(k)$. Note, however, that this symmetry does not hold for the correlations involving fluid and matrix particles where $\tilde{f}_{\chi,\text{fm}}^{l_1 l_2}(k) = \tilde{f}_{\chi,\text{mf}}^{l_2 l_1}(k)$.

Inserting the $\mathbf{k}$-frame expansions (G.8) of the correlation functions into the RSOZ Eqs. (7.41a)–(7.42), the angular integral buried in the products $\tilde{f}_\gamma \otimes \tilde{f}_\delta$ can be easily performed by using the orthogonality of the spherical harmonics [258] [cf. Eq. (F.39)]. Introducing, for compactness, matrices $\tilde{\mathbf{F}}_\chi$ with elements

$$\left[\tilde{\mathbf{F}}_{\chi,\gamma}(k)\right]_{l_1 l_2} = \frac{1}{\sqrt{2l_1 + 1}} \frac{1}{\sqrt{2l_2 + 1}} \tilde{f}_{\chi,\gamma}^{l_1 l_2}(k) \tag{G.10}$$

we finally obtain

$$\tilde{\mathbf{H}}_{\chi,\text{mm}}(k) = \tilde{\mathbf{C}}_{\chi,\text{mm}}(k) + (-1)^\chi \rho_\text{m} \tilde{\mathbf{C}}_{\chi,\text{mm}}(k) \tilde{\mathbf{H}}_{\chi,\text{mm}}(k) \tag{G.11a}$$

$$\begin{aligned}\tilde{\mathbf{H}}_{\chi,\text{fm}}(k) &= \tilde{\mathbf{C}}_{\chi,\text{fm}}(k) + (-1)^\chi \rho_\text{m} \tilde{\mathbf{C}}_{\chi,\text{fm}}(k) \tilde{\mathbf{H}}_{\chi,\text{mm}}(k) \\ &\quad + (-1)^\chi \rho \tilde{\mathbf{C}}_{\chi,\text{c}}(k) \tilde{\mathbf{H}}_{\chi,\text{fm}}(k)\end{aligned} \tag{G.11b}$$

$$\begin{aligned}\tilde{\mathbf{H}}_{\chi,\text{ff}}(k) &= \tilde{\mathbf{C}}_{\chi,\text{ff}}(k) + (-1)^\chi \rho_\text{m} \tilde{\mathbf{C}}_{\chi,\text{fm}}(k) \tilde{\mathbf{H}}_{\chi,\text{mf}}(k) \\ &\quad + (-1)^\chi \rho \tilde{\mathbf{C}}_{\chi,\text{ff}}(k) \tilde{\mathbf{H}}_{\chi,\text{ff}}(k) \\ &\quad - (-1)^\chi \rho \tilde{\mathbf{C}}_{\chi,\text{b}}(k) \tilde{\mathbf{H}}_{\chi,\text{b}}(k)\end{aligned} \tag{G.11c}$$

$$\tilde{\mathbf{H}}_{\chi,\text{c}}(k) = \tilde{\mathbf{C}}_{\chi,\text{c}}(k) + (-1)^\chi \rho \tilde{\mathbf{C}}_{\chi,\text{c}}(k) \tilde{\mathbf{H}}_{\chi,\text{c}}(k) \tag{G.11d}$$

where $\rho = 4\pi\bar{\rho} = N_\text{f}/V$ and $\rho_\text{m} = 4\pi\bar{\rho}_\text{m} = N_\text{m}/V$.

We note that the RSOZ equations in Eqs. (G.11a)–(G.11d) are decoupled with respect to the wavenumber k and the angular index χ. In practice, the matrices $\tilde{\mathbf{H}}_\chi$ and $\tilde{\mathbf{C}}_\chi$ have finite dimensions due to a truncation of the rotationally invariant expansion in Eq. (G.8) at appropriate values of l_1 and l_2. The RSOZ Eqs. (G.11a)–(G.11d) can thus be solved by standard matrix inversion techniques.

G.4   Proof of Eq. (7.33)

Our starting point in this section is [see Eq. (7.33)]

$$\widetilde{h}_{\text{ff}}(k) = \int d\overline{r} h_{\text{ff}}(\overline{r}) \exp(ik\overline{r}\cos\vartheta) \tag{G.12}$$

where we used the definition of the scalar product of two vectors, namely $k \cdot \overline{r} = k\overline{r}\cos\vartheta$, where ϑ is the angle between k and $\overline{r}$. The reader should note that this form of Eq. (7.33) is a consequence of the fact that h_{ff} depends only on the magnitude (but not on the direction) of $\overline{r}$ on account of the isotropy and homogeneity of the system under consideration (see discussion in Section 7.4). For a fluid composed of molecules with rotational degrees of freedom, $h_{\text{ff}}(\overline{r}) \rightarrow h_{\text{ff}}(\overline{r}, \omega_1, \omega_2)$ and the integrals in Eq. (7.32) can only be carried out by employing the Rayleigh expansion of the phase factor $\exp(ik \cdot \overline{r})$ [258]. However, in the current case of a "simple" fluid, homogeneity and isotropy may be invoked conveniently to rewrite the previous expression more explicitly as

$$\widetilde{h}_{\text{ff}}(k) = \int_0^{2\pi} \int_0^{\pi} \int_0^{\infty} h_{\text{ff}}(\overline{r}) \exp(ik\overline{r}\cos\vartheta)\, d\varphi \sin\vartheta d\vartheta \overline{r}^2 d\overline{r} \tag{G.13}$$

using spherical polar coordinates, where the integration over φ can be carried out trivially yielding a factor of 2π.

The integral over ϑ can also be carried out in closed form because $h_{\text{ff}}(\overline{r})$ does not depend on this angle. Using the transformation $x \equiv \cos\vartheta$, we can rewrite the integral over ϑ in Eq. (G.13) as

$$\int_0^{\pi} \sin\vartheta d\vartheta \exp(ik\overline{r}\cos\vartheta) = -\int_1^{-1} dx \exp(ik\overline{r}x)$$

$$= -\frac{1}{ik\overline{r}}[\exp(-ik\overline{r}) - \exp(ik\overline{r})]$$

$$= \frac{2\sin(k\overline{r})}{k\overline{r}} \tag{G.14}$$

where we used Euler's identity $\exp(\pm ix) \equiv \cos x \pm i \sin x$ for complex numbers. Combining Eq. (G.14) with the integration over φ in Eq. (G.13), we arrive at the far right side of Eq. (7.33).

It is also instructive to consider the limit of vanishing wavenumber $k \rightarrow 0$. In this case

$$\lim_{k \to 0} \frac{\sin(k\overline{r})}{k\overline{r}} = \lim_{k \to 0} \cos(k\overline{r}) = 1 \tag{G.15}$$

which follows from de l'Hospital's rule for taking the limit of an undetermined expression of the form $\frac{0}{0}$. In this case Eq. (7.33) simplifies to

$$\widetilde{h}_{\mathrm{ff}}(0) = 4\pi \int\limits_0^\infty \mathrm{d}\bar{r}\,\bar{r}^2 h_{\mathrm{ff}}(\bar{r}) \tag{G.16}$$

G.5   Numerical solution of integral equations

In this appendix we present solution strategies for integral equations such as the ones discussed in Chapter 7. However, rather than considering the replica-version of integral equations introduced in Chapter 7 for disordered fluids we focus here on the equations related to a fully annealed binary mixture of particles without internal degrees of freedom. This simpler system can be considered as a generic case appropriate to introduce the basic steps of the numerical solution of integral equations. Indeed, based on the numerical solution of an annealed binary mixture, it is straightforward to generalize the algorithm for more complicated cases such as partly quenched (spherical) mixtures and even molecular (annealed or partly quenched) fluids, where rotationally invariant expansions of correlation functions lead to integral equations formally equivalent to those of multicomponent mixtures [258].

Specifically, we consider a model consisting of two species A and B of spherical particles with diameters σ_A and σ_B and isotropic interaction potentials $u_{\alpha\beta}(r)$. As a consequence, the two-point correlation functions depend only on the separation $r = |r_1 - r_2|$ between the particles. We are particularly interested in the total correlation function $h_{\alpha\beta}(r)$ or, equivalently, the pair correlation function $g_{\alpha\beta}(r) = h_{\alpha\beta}(r) + 1$ and the direct correlation function $c_{\alpha\beta}(r)$, which provide together a complete description of the structure and thermodynamics of the mixture [30]. These correlation functions are linked by the exact OZ equations [see also Eq. (7.36) for the general case of a multicomponent mixture]

$$h_{\alpha\beta}(r) = c_{\alpha\beta}(r) + \sum_\gamma \rho_\gamma \int \mathrm{d}r_3 h_{\alpha\gamma}(|r_1 - r_3|)\, c_{\gamma\beta}(|r_3 - r_2|) \tag{G.17}$$

As mentioned in Eq. (7.36), after introducing Fourier transforms of the correlation functions, the OZ equations simplifies to [see Eqs. (7.32) and (7.33)]

$$\widetilde{h}_{\alpha\beta}(k) = \widetilde{c}_{\alpha\beta}(k) + \sum_\gamma \rho_\gamma \widetilde{h}_{\alpha\gamma}(k)\, \widetilde{c}_{\gamma\beta}(k) \tag{G.18}$$

In addition, assuming particles with hard-core repulsive interactions, we have the exact core conditions [see Eqs. (7.44)]

$$c_{\alpha\beta}(r) = -1 - \eta_{\alpha\beta}(r), \qquad r < \sigma_{\alpha\beta} \qquad (G.19)$$

where $\eta_{\alpha\beta}(r) = h_{\alpha\beta}(r) - c_{\alpha\beta}(r)$ and $\sigma_{\alpha\beta} = (\sigma_\alpha + \sigma_\beta)/2$. For separations beyond the hard core we need an approximation, and we consider here as an examplary case the (nonlinear) hypernetted chain (HNC) closure introduced in Eqs. (7.49) for the case of disordered fluids.

Specializing our treatment to the current binary mixture, the HNC closure reads as follows:

$$c_{\alpha\beta}(r) = -\frac{u_{\alpha\beta}(r)}{k_B T} - \ln\left[1 + h_{\alpha\beta}(r)\right] + h_{\alpha\beta}(r), \qquad r > \sigma_{\alpha\beta} \qquad (G.20)$$

where we stress that these equations are decoupled with respect to the species indices. The same is true for the simpler Percus–Yevick (PY) and mean spherical approximation (MSA) closure approximations introduced in Eqs. (7.46) and (7.48).

The goal is now to solve Eqs. (G.18)–(G.20) together for given partial densities ρ_α and temperature T. Apart from some notable exceptions such as the PY equations for hard spheres [332–334] and hard sphere mixtures [335, 336] and the MSA for charged [337, 338] or dipolar hard spheres [339], the actual solution has to be done numerically. To this end, all correlation functions are discretized on a lattice with a typical grid width of $\Delta r = 0.02\sigma$ and typical lattice sizes of $N_l = 1000 - 2000$ points (corresponding to separations of $10 - 20\sigma$). Within this predescribed range the correlation functions can be calculated via an iteration procedure. The flow scheme representing one iterative cycle is depicted schematically in Fig. G.1.

1. Initiation of iterative procedure
   As a starting point of the iteration one defines an initial guess $c_{\alpha\beta}^{in}(r)$ for the direct correlation functions. Indeed, a good initial guess is particularly important for strongly interacting systems, that is, those systems at high densities and/or low temperatures. Most conveniently one may use the final result $c_{\alpha\beta}^{out}(r)$ of a previous solution at a "nearby" thermodynamic state with similar densities and/or temperature as an initial guess for the densities and temperatures of interest. As an alternative, which is particularly useful for hard–core systems, one may take the (PY) direct correlation functions [332–336] of the underlying hard-sphere system as an input for the current iteration. Finally, it seems worth mentioning that for weakly interacting systems (especially for soft-core systems at high temperatures) convergence can often also be achieved with the trivial choice $c_{\alpha\beta}^{in}(r) = 0$.

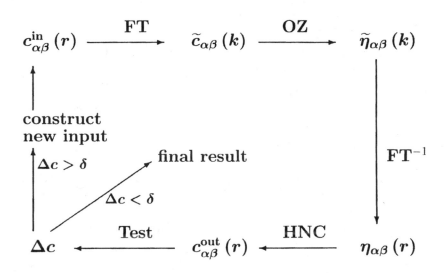

Figure G.1: Flow scheme of the numerical solution of the integral equations consisting of the OZ equations [see Eqs. (G.18) and (G.22)] together with the HNC closure [see Eqs. (G.19) and (G.20)]. The acronyms FT and FT^{-1} denote Fourier transformation and Fourier inversion, respectively; Δc is defined in Eq. (G.27); and δ is a threshold value set according to the desired accuracy.

2. Fourier transformation

   The next step consists of a Fourier transformation of the direct correlation functions according to the prescription [see also Eq. (7.33)]

   $$\widetilde{c}_{\alpha\beta}(k) = 4\pi \int dr \, r^2 \frac{\sin(kr)}{kr} c_{\alpha\beta}(r) \qquad (G.21)$$

   where the Fourier transforms $\{\widetilde{c}_{\alpha\beta}(k)\}$ are again discretized on a lattice with $N_k = N_l$ points and grid width $\Delta k = \pi/(N_l \Delta r)$. The actual Fourier transformation can be performed via fast Fourier transform techniques [175]. Some care has to be taken due to the discontinuity of the functions $\{c_{\alpha\beta}(r)\}$ at separations $r = \sigma_{\alpha\beta}$, which arise from the corresponding discontinuity of the pair potentials for hard-core systems. To circumvent this problem one usually computes the Fourier transform separately for the regions $r < \sigma_{\alpha\beta}$ and $r > \sigma_{\alpha\beta}$.

3. Solution of the OZ equation

   Having obtained the quantities $\widetilde{c}_{\alpha\beta}(k)$, we are now in the position to calculate the Fourier-transformed total correlation functions $\tilde{h}_{\alpha\beta}(k)$ by solving Eq. (G.18). However, for hard-core systems in particular, it is more convenient to consider instead the function $\tilde{\eta}_{\alpha\beta}(k) \equiv \tilde{h}_{\alpha\beta}(k) - \tilde{c}_{\alpha\beta}(k)$. These quantities are more suitable for the Fourier inversion (required to solve the closure relations) because the real-space functions $\eta_{\alpha\beta}(r)$ are continuous at the hard core (i.e., at $r = \sigma_{\alpha\beta}$). The total correlation functions $h_{\alpha\beta}(r)$, on the other hand, exhibit a discontinuity (as do the direct correlation functions). To calculate the quantities $\tilde{\eta}_{\alpha\beta}(k)$, we first rewrite Eq. (G.18) as

$$\tilde{\eta}_{\alpha\beta}(k) = \sum_{\gamma} \rho_{\gamma} \left[\tilde{\eta}_{\alpha\gamma}(k)\,\tilde{c}_{\gamma\beta}(k) + \tilde{c}_{\alpha\gamma}(k)\,\tilde{c}_{\gamma\beta}(k) \right] \qquad (G.22)$$

This latter equation may be rewritten more compactly by introducing the 2×2 matrices $\tilde{\mathbf{E}}(k)$ and $\tilde{\mathbf{C}}(k)$ with elements

$$\left[\tilde{\mathbf{E}}(k) \right]_{\alpha\beta} = \sqrt{\rho_{\alpha}\rho_{\beta}}\,\tilde{\eta}_{\alpha\beta}(k) \qquad (G.23a)$$

$$\left[\tilde{\mathbf{C}}(k) \right]_{\alpha\beta} = \sqrt{\rho_{\alpha}\rho_{\beta}}\,\tilde{c}_{\alpha\beta}(k) \qquad (G.23b)$$

Rewriting Eq. (G.22) in terms of $\tilde{\mathbf{E}}(k)$ and $\tilde{\mathbf{C}}(k)$ yields

$$\tilde{\mathbf{E}}(k) = \tilde{\mathbf{E}}(k)\,\tilde{\mathbf{C}}(k) + \tilde{\mathbf{C}}(k)\,\tilde{\mathbf{C}}(k) \qquad (G.24)$$

which may be solved for $\tilde{\mathbf{E}}(k)$ to give

$$\tilde{\mathbf{E}}(k) = \tilde{\mathbf{C}}(k)\,\tilde{\mathbf{C}}(k) \left[\mathbf{1} - \tilde{\mathbf{C}}(k) \right]^{-1} \qquad (G.25)$$

where $\mathbf{1}$ is the unit matrix. Equation (G.25) involves only 2×2 matrices. Thus, the inverse matrix $\left[\mathbf{1} - \tilde{\mathbf{C}}(k) \right]^{-1}$ can be calculated analytically so that $\tilde{\eta}_{\alpha\beta}(k)$ can easily be obtained from the elements of matrix $\tilde{\mathbf{E}}(k)$ according to its definition given in Eq. (G.23a).

4. Fourier inversion

Next we wish to calculate the (continuous) real-space functions $\eta_{\alpha\beta}(r)$. These functions simply follow from the Fourier inversion

$$\eta_{\alpha\beta}(r) = \frac{1}{2\pi^2} \int dk\, k^2 \frac{\sin(kr)}{kr}\,\tilde{\eta}_{\alpha\beta}(k) \qquad (G.26)$$

5. Solution of the closure expressions

Using the closure relationships Eq. (G.19) and (G.20) for separations below and above the hard core, we can now calculate the new direct correlation functions $c_{\alpha\beta}^{\text{out}}(r)$. We stress that the closure can be solved independently for each index combination $\alpha\beta$. Furthermore, note that the HNC closure $(r > \sigma_{\alpha\beta})$ involves the total correlation functions $h_{\alpha\beta}(r)$, which can be approximated by the current estimates $h_{\alpha\beta}(r) = c_{\alpha\beta}^{\text{in}}(r) + \eta_{\alpha\beta}(r)$.

6. Accuracy check

As a final step of an iteration cycle we have to test the quality of the solution using an appropriate self-consistency criterion. The choice of this criterion is not unique. However, a suitable quantity to be calculated in this context is

$$\Delta c \equiv \sqrt{\frac{1}{N_1} \sum_{i=1}^{N_1} \left[c_{\alpha\beta}^{\text{out}}(r_i) - c_{\alpha\beta}^{\text{in}}(r_i) \right]^2} \qquad (G.27)$$

If this quantity is smaller than some predescribed small number δ (e.g., $\delta = 10^{-6}$), one considers the solution $c_{\alpha\beta}^{\text{out}}(r)$ as being self-consistent and the iteration is stopped. If, on the contrary, $\Delta c > \delta$ the iteration is continued using appropriate input as described below.

7. Constructing new input

We have now reached a stage of our solution procedure where updated input is required for the next iteration cycle. Looking at this problem näively, it may seem that the most natural way of starting a new iteration consists of using the output functions $c_{\alpha\beta}^{\text{out}}(r)$ of the previous iteration n as an initial guess for the current iteration $n + 1$. However, in practice, this procedure often leads to an unstable iteration where Δc defined in Eq. (G.27) oscillates between subsequent iteration cycles or even increases. Several strategies to stabilize the iteration procedure have been proposed. Here "stable" means that Δc decreases monotonically as the number of iterations increases. The two most important strategies leading to a stable iteration procedure can be sketched as follows:

- Within the first method the new input of iteration $n+1$ is given as a linear interpolation between the input and the output functions of the previous iteration cycle; that is,

$$c_{\alpha\beta}^{\text{in},n+1}(r) = \alpha c_{\alpha\beta}^{\text{out},n}(r) + (1 - \alpha) c_{\alpha\beta}^{\text{in},n}(r), \qquad 0 < \alpha < 1 \quad (G.28)$$

The optimal choice of the parameter α has to be determined empirically. For typical calculations, $\alpha = 0.1 - 0.2$ and it takes of the order of 10^3 iterations to obtain satisfactory convergence [340].

- The second method is based on an idea of Ng and uses the "history" of the iteration procedure to construct the new input functions [341]. Specifically, one constructs new input by extrapolating the previous output from at least two or, in most cases, several previous iterations according to the prescription

$$
\begin{aligned}
c_{\alpha\beta}^{in,n+1}(r) &= [1 - \alpha_1(n) - \alpha_2(n)] \, c_{\alpha\beta}^{out,n}(r) \\
&\quad + \alpha_1(n) \, c_{\alpha\beta}^{out,n-1}(r) + \alpha_2(n) \, c_{\alpha\beta}^{out,n-2}(r) \, (G.29)
\end{aligned}
$$

where the prefactors $\alpha_1(n)$ and $\alpha_2(n)$ are determined in each cycle such that Δc given in Eq. (G.27) is minimized. The advantage of this method compared with the one described before is much faster convergence (typically only of the order of 10 iterations are required). At the same time, however, the procedure is more susceptible to numerical instabilities, implying that the quality of the input function has to be better. A more detailed discussion of this method can be found in the original paper of Ng [341].

Bibliography

[1] R. Lipowsky, *Curr. Opin. Colloids Interface Sci.*, **6**, 40 (2001).

[2] A. Cavallo, M. Müller, and K. Binder, *High Performance Computing in Science and Engineering*, Springer-Verlag, Berlin, 2005.

[3] C. Alba-Simionesco, B. Coasne, G. Dosseh, G. Dudziak, K. E. Gubbins, R. Radhakrishnan, and M. Sliwinska-Bartkowiak, *J. Phys.: Condens. Matter*, **18**, R15 (2006).

[4] L. D. Gelb, K. E. Gubbins, R. Radhakrishnan, and M. Sliwinska-Bartkowiak, *Rep. Prog. Phys.*, **62**, 1573 (1999).

[5] M. Schoen, *Computational Methods in Surface and Colloid Science*, Marcel Dekker, New York, 2000.

[6] E. H. Lieb and J. Yngvason, *Phys. Rep.*, **310**, 1 (1999).

[7] E. H. Lieb and J. Yngvason, *Phys. Rep.*, **310**, 669 (1999).

[8] S. G. Brush, *The Temperature of History. Phases of Science and Culture in the Nineteenth Century*, Burt Franklin, New York, 1977.

[9] Lord Kelvin, *Nature*, **1**, 551 (1870).

[10] J. Tyndall, *Heat Considered as a Mode of Motion*, Longmans and Green, London, 1863.

[11] G. Arfken, *Mathematical Methods for Physicists*, Academic Press, London, 1985.

[12] H. B. Callen, *Thermodynamics*, Wiley, New York, 1960.

[13] D. J. Evans and G. P. Morriss, *Statistical Mechanics of Nonequilibrium Liquids*, Academic Press, London, 1990.

[14] H. Goldstein, *Klassische Mechanik*, Aula-Verlag, Wiesbaden, 1991.

[15] H. E. Stanley, *Introduction to Phase Equilibria and Critical Phenomena*, Oxford University Press, Oxford, 1987.

[16] R. J. Baxter, *Exactly Solved Models in Statistical Mechanics*, Academic Press, London, 1990.

[17] D. A. McQuarrie, *Statistical Mechanics*, University Science Books, Sausalito, CA, 2000.

[18] R. H. Fowler, *Statistical Mechanics*, Cambridge University Press, Cambridge, UK, 1966.

[19] W. Nolting, *Grundkurs: Theoretische Physik, Vol. 6: Statistische Physik*, Verlag Zimmermann-Neufang, Ulmen, 1994.

[20] E. Schrödinger, *Statistical Thermodynamics*, Dover, New York, 1989.

[21] T. L. Hill, *An Introduction to Statistical Thermodynamics*, Dover, New York, 1986.

[22] J. G. Kirkwood, *J. Chem. Phys.*, **44**, 31 (1933).

[23] J. G. Kirkwood, *J. Chem. Phys.*, **45**, 116 (1934).

[24] E. Wigner, *Phys. Rev.*, **40**, 749 (1932).

[25] K. T. Vanderlick, H. T. Davis, and J. K. Percus, *J. Chem. Phys.*, **91**, 7136 (1989).

[26] H. T. Davis, *Statistical Mechanics of Phases, Interfaces, and Thin Films*, Wiley-VCH, New York, 1996.

[27] J. S. Rowlinson, *Proc. Roy. Soc. London A*, **402**, 67 (1985).

[28] D. A. McQuarrie and J. S. Rowlinson, *Mol. Phys.*, **60**, 977 (1987).

[29] J. W. Gibbs, *Elementary Principles in Statistical Mechanics Developed with Special Reference to the Rational Foundation of Thermodynamics*, Charles Scribner's Sons, New York, 1902.

[30] J. P. Hansen and I. R. McDonald, *Theory of Simple Liquids*, Academic Press, London, 1986.

[31] M. Thommes and G. H. Findenegg, *Langmuir*, **10**, 4270 (1994).

[32] G. H. Findenegg, M. Thommes, and T. Michalski, *Proceedings of the VIIIth European Symposium on Materials and Fluid Sciences in Microgravity*, ESA SP-333, 1992.

[33] D. Zhao, J. Feng, Q. Huo, N. Melosh, G. H. Frederickson, B. F. Chmelka, and G. D. Stucky, *Science*, **279**, 548 (1998).

[34] D. Zhao, Q. Huo, J. Feng, B. F. Chmelka, and G. D. Stucky, *J. Am. Chem. Soc.*, **120**, 6024 (1998).

[35] T. Hoffmann, D. Wallacher, P. Huber, R. Birringer, K. Knorr, A. Schreiber, and G. H. Findenegg, *Phys. Rev. B*, **72**, 064122 (2005).

[36] G. H. Findenegg, Private communication with S. H. L. Klapp (2006).

[37] M. R. Abramowitz and I. Stegun, *Handbook of Mathematical Functions*, National Bureau of Standards, Washington, D.C., 1965.

[38] M. Schoen, *Ber. Bunsenges. Phys. Chem.*, **100**, 1355 (1996).

[39] D. Nicholson and N. G. Parsonage, *Computer Simulation and the Statistical Mechanics of Adsorption*, Academic Press, London, 1982.

[40] M. Schoen, *Computer Simulations of Condensed Phases in Complex Geometries*, Springer, Berlin, 1993.

[41] D. J. Adams, *Mol. Phys.*, **29**, 307 (1975).

[42] L. A. Rowley, D. Nicholson, and N. G. Parsonage, *Mol. Phys.*, **31**, 365 (1976).

[43] J. E. Lane and T. H. Spurling, *Aust. J. Chem.*, **29**, 2103 (1980).

[44] I. K. Snook and W. van Megen, *J. Chem. Phys.*, **72**, 2907 (1980).

[45] M. Schoen, D. J. Diestler, and J. H. Cushman, *J. Chem. Phys.*, **87**, 5464 (1987).

[46] J. P. R. B. Walton and N. Quirke, *Mol. Simul.*, **2**, 361 (1989).

[47] R. Lustig, *Fluid Phase Equilib.*, **32**, 117 (1987).

[48] D. A. Lavis and G. M. Bell, *Statistical Mechanics of Lattice Systems: Closed Form and Exact Solutions*, Springer-Verlag, Berlin, 1999.

[49] D. W. Tolfree, *Rep. Prog. Phys.*, **61**, 313 (1998).

[50] F. Burmeister, C. Schäfle, B. Keilhofer, C. Bechinger, J. Boneberg, and P. Leiderer, *Adv. Mater.*, **10**, 495 (1998).

[51] A. Kumar and G. M. Whitesides, *Appl. Phys. Lett.*, **63**, 2002 (1993).

[52] P. Zeppenfeld, V. Diercks, R. David, F. Picaud, C. Ramseyer, and C. Girardet, *Phys. Rev. B*, **66**, 085414 (2002).

[53] C. Schäfle, C. Bechinger, B. Rinn, C. David, and P. Leiderer, *Phys. Rev. Lett.*, **83**, 5302 (1999).

[54] C. Schäfle, *Morphologie, Verdampfung und Kondensation von Flüssigkeiten auf benetzungsstrukturierten Oberflächen*, Ph.D. thesis, Universität Konstanz, Konstanz (2002).

[55] P. Lenz, C. Bechinger, C. Schäfle, P. Leiderer, and R. Lipowsky, *Langmuir*, **17**, 7814 (2001).

[56] S. Harkema, E. Schäffer, M. D. Morariu, and U. Steiner, *Langmuir*, **19**, 9714 (2003).

[57] D. Öner and T. J. McCarthy, *Langmuir*, **16**, 7777 (2000).

[58] H. B. Callen, *Thermodynamics and an Introduction to Thermostatics*, Wiley, New York, 1985.

[59] A. Isihara, *J. Phys. A*, **1**, 539 (1968).

[60] N. N. Bogoliubov, *Dokl. Akad. Nauk. SSSR*, **119**, 244 (1958).

[61] H. Bock, D. J. Diestler, and M. Schoen, *J. Phys.: Condens. Matter*, **13**, 4697 (2001).

[62] J. R. Silbermann, D. Woywod, and M. Schoen, *Phys. Rev. E*, **69**, 031606 (2004).

[63] M. Schoen and D. J. Diestler, *Phys. Rev. E*, **56**, 4427 (1997).

[64] H. Bock and M. Schoen, *Phys. Rev. E*, **59**, 4122 (1999).

[65] M. H. Adão, M. de Ruijter, M. Voué, and J. D. Coninck, *Phys. Rev. E*, **59**, 746 (1999).

[66] L. J. D. Frink and A. G. Salinger, *J. Chem. Phys.*, **110**, 5969 (1999).

[67] C. Bauer and S. Dietrich, *Phys. Rev. E*, **60**, 6919 (1999).

[68] M. Schneemilch, N. Quirke, and J. R. Henderson, *J. Chem. Phys.*, **118**, 816 (2003).

[69] M. Schneemilch and N. Quirke, *Molec. Simul.*, **29**, 685 (2003).

[70] A. Dupuis and J. M. Yeomans, *Fut. Gen. Compu. Sys.*, **20**, 993 (2004).

[71] N. Pesheva and J. de Coninck, *Phys. Rev. E*, **70**, 046102 (2004).

[72] P. Lenz and R. Lipowsky, *Phys. Rev. Lett.*, **80**, 1920 (1998).

[73] H. Gau, S. Herminghaus, P. Lenz, and R. Lipowsky, *Science*, **283**, 46 (1999).

[74] R. Lipowsky, P. Lenz, and P. S. Swain, *Colloids Surf. A*, **161**, 3 (2000).

[75] R. Lipowsky, *Interface Sci.*, **9**, 105 (2001).

[76] M. Brinkmann and R. Lipowsky, *J. Appl. Phys.*, **92**, 4296 (2002).

[77] W. F. P. Lenz and R. Lipowsky, *Europhys. Lett.*, **53**, 618 (2001).

[78] R. Lipowsky, M. Brinkmann, R. Dimova, T. Franke, J. Kierfeld, and X. Z. Zhang, *J. Phys.: Condens. Matter*, **17**, 537 (2005).

[79] R. Seemann, M. Brinkmann, E. J. Kramer, F. F. Lange, and R. Lipowsky, *Proc. Nat. Acad. Sci.*, **102**, 1848 (2005).

[80] M. Brinkmann, J. Kierfeld, and R. Lipowsky, *J. Phys. Math. Gen.*, **37**, 11547 (2004).

[81] A. Valencia and R. Lipowsky, *Langmuir*, **20**, 1986 (2004).

[82] G. Chmiel, A. Patrykiejew, W. Rzysko, and S. Sokolowski, *Mol. Phys.*, **83**, 19 (1994).

[83] C. Rascón and A. O. Parry, *J. Chem. Phys.*, **115**, 5258 (2001).

[84] D. Woywod and M. Schoen, *Phys. Rev. E*, **67**, 026122 (2003).

[85] D. Woywod and M. Schoen, *J. Phys.: Condens. Matter*, **16**, 4761 (2004).

[86] D. Woywod and M. Schoen, *Phys. Rev. E*, **73**, 011201 (2006).

[87] N. B. Wilding, F. Schmid, and P. Nielaba, *Phys. Rev. E*, **58**, 2201 (1998).

[88] P. H. van Konynenburg and R. L. Scott, *Philos. Trans. R. Soc. London Ser. A*, **298**, 495 (1980).

[89] A. S. Wightman, *Convexity in the Theory of Lattice Gases*, Princeton University Press, Princeton, NJ, 1979.

[90] D. A. Weitz, *MRS Bulletin*, **19**, 11 (1994).

[91] H. Greberg and G. N. Patey, *J. Chem. Phys*, **114**, 7128 (2001).

[92] D. Beysens and D. Estevé, *Phys. Rev. Lett.*, **54**, 2123 (1985).

[93] M. Sliwinska-Bartkowiak, S. L. Sowers, and K. E. Gubbins, *Langmuir*, **13**, 1182 (1997).

[94] S. Dietrich and M. Schick, *Phys. Rev. E*, **33**, 4952 (1986).

[95] A. J. Liu, D. J. Durian, E. Herbolzheimer, and S. Safran, *Phys. Rev. Lett.*, **65**, 1897 (1990).

[96] L. Monette, A. J. Liu, and G. S. Grest, *Phys. Rev. A*, **46**, 7664 (1992).

[97] S. Puri and K. Binder, *Phys. Rev. E*, **66**, 061602 (2002).

[98] M. Telo da Gama and R. Evans, *Mol. Phys.*, **48**, 687 (1983).

[99] Y. Fan, J. E. Finn, and P. A. Monson, *J. Chem. Phys.*, **99**, 8238 (1993).

[100] K. Binder, S. Puri, and H. L. Frisch, *Faraday Discuss.*, **112**, 103 (1999).

[101] F. Formisano and J. Teixera, *J. Phys.: Condens. Matter*, **12**, 351 (2000).

[102] F. Formisano and J. Teixera, *Eur. Phys. J. E*, **1**, 1 (2000).

[103] H. Furukawa, *Physica A*, **123**, 497 (1984).

[104] G. Porod, *Kolloid Z.*, **124**, 83 (1951).

[105] P. Debye, H. R. Anderson, and H. Brumberger, *J. Appl. Phys.*, **28**, 679 (1957).

[106] S. Schemmel, G. Rother, E. Eckerlebe, and G. H. Findenegg, *J. Chem. Phys.*, **122**, 244718 (2005).

[107] W. A. Steele, *The Interaction of Gases with Solid Surfaces*, Pergamon Press, Oxford, 1974.

[108] G. Rother, D. Woywod, M. Schoen, and G. H. Findenegg, *J. Chem. Phys.*, **120**, 11864 (2004).

[109] P. Gansen and D. Woermann, *J. Chem. Phys.*, **88**, 2655 (1984).

[110] C. M. Knobler, Private communication with M. Schoen (2006).

[111] P. Bryk, D. Henderson, and S. Sokołowski, *Langmuir*, **15**, 6026 (1999).

[112] J. Z. Tang and J. G. Harris, *J. Chem. Phys.*, **103**, 8201 (1995).

[113] L. D. Gelb and K. E. Gubbins, *Phys. Rev. E*, **56**, 3185 (1997).

[114] E. Schöll-Paschinger, D. Levesque, J.-J. Weis, and G. Kahl, *Phys. Rev. E*, **64**, 011502 (2001).

[115] Z. Zhuang, A. G. Casielles, and D. S. Cannell, *Phys. Rev. Lett.*, **77**, 2969 (1969).

[116] V. F. Sears, *Neutron News*, **3**, 26 (1992).

[117] M. M. Kohonen and H. K. Christenson, *J. Phys. Chem. B*, **106**, 6685 (2002).

[118] U. Wolff, *Phys. Rev. Lett.*, **60**, 1461 (1988).

[119] U. Wolff, *Phys. Rev. Lett.*, **62**, 361 (1989).

[120] D. J. Uzunov, *Introduction to the Theory of Critical Phenomena*, World Scientific, Singapore, 1993.

[121] D. Sornette, *Critical Phenomena in Natural Sciences*, Springer-Verlag, Berlin, 2004.

[122] M. Schoen, S. Hess, and D. J. Diestler, *Phys. Rev. E*, **2**, 2587 (1995).

[123] I. G. Sinai and V. Scheffer, *Introduction to Ergodic Theory*, Princeton University Press, Princeton, NJ, 1976.

[124] K. Petersen, *Ergodic Theory*, Cambridge University Press, Cambridge, UK, 1983.

[125] P. Walters, *An Introduction to Ergodic Theory*, Springer, New York, 2000.

[126] D. P. Landau and K. Binder, *Monte Carlo Simulations in Statistical Physics*, Cambridge University Press, Cambridge, UK, 2000.

[127] K. Binder and D. W. Heermann, *Monte Carlo Simulation in Statistical Physics*, Springer-Verlag, Heidelberg, 1988.

[128] G. V. Burgess, D. H. Everett, and S. Nutall, *Pure Appl. Chem.*, **61**, 1845 (1989).

[129] A. de Keizer, T. Michalski, and G. H. Findenegg, *Pure Appl. Chem.*, **63**, 1495 (1991).

[130] W. D. Machin, *Langmuir*, **10**, 1235 (1994).

[131] A. P. Y. Wong, S. B. Kim, W. I. Goldburg, and M. H. W. Chan, *Phys. Rev. Lett.*, **70**, 954 (1993).

[132] A. P. Y. Wong and M. H. W. Chan, *Phys. Rev. Lett.*, **65**, 2567 (1990).

[133] D. Tabor and R. H. S. Winterton, *Proc. Roy. Soc. London Ser. A*, **312**, 435 (1969).

[134] J. N. Israelachvili and D. Tabor, *Proc. Roy. Soc. London Ser. A*, **331**, 19 (1973).

[135] J. N. Israelachvili and P. M. McGuiggan, *Science*, **241**, 795 (1988).

[136] S. Granick, *Science*, **253**, 1374 (1991).

[137] G. Reiter, A. L. Demirel, J. Peanasky, L. L. Cai, and S. Granick, *J. Chem. Phys.*, **101**, 2606 (1994).

[138] N. Metropolis, A. W. Rosenbluth, M. N. Rosenbluth, A. H. Teller, and E. Teller, *J. Chem. Phys.*, **21**, 1087 (1953).

[139] W. W. Wood, *Physics of Simple Liquids*, North Holland, Amsterdam, 1968.

[140] M. P. Allen and D. J. Tildesley, *Computer Simulation of Liquids*, Clarendon, Oxford, 1987.

[141] I. N. Bronstein and K. A. Semendjajew, *Taschenbuch der Mathematik*, B. G. Teubner Verlagsgesellschaft, Stuttgart, 1991.

[142] N. B. Wilding and M. Schoen, *Phys. Rev. E*, **60**, 1081 (1999).

[143] J. N. Israelachvili, *Intermolecular and Surface Forces*, Academic Press, London, 1992.

[144] A. Tock, J. M. Georges, and L. Loubet, *J. Colloid Interface Sci.*, **126**, 150 (1988).

[145] B. Derjaguin, *Kolloid Z.*, **69**, 155 (1934).

[146] M. Gee, P. M. McGuiggan, J. N. Israelachvili, and A. M. Homola, *J. Chem. Phys.*, **93**, 1895 (1990).

[147] J. V. Alsten and S. Granick, *Phys. Rev. Lett.*, **61**, 2570 (1988).

[148] J. N. Israelachvili, P. M. McGuiggan, and A. M. Homola, *Science*, **240**, 189 (1988).

[149] J. Israelachvili, P. M. McGuiggan, M. Gee, A. M. Homola, M. Robbins, and P. Thompson, *J. Phys.: Condens. Matter*, **2**, SA 89 (1990).

[150] J. Klein and E. Kumacheva, *J. Chem. Phys.*, **108**, 6996 (1998).

[151] E. Kumacheva and J. Klein, *J. Chem. Phys.*, **108**, 7010 (1998).

[152] M. Schoen, T. Gruhn, and D. J. Diestler, *J. Chem. Phys.*, **109**, 301 (1998).

[153] S. Asakura and F. Oosawa, *J. Chem. Phys.*, **22**, 1255 (1954).

[154] M. Schmidt, *Computational Methods in Surface and Colloid Science*, Marcel Dekker, New York, 2000.

[155] B. Götzelmann, R. Evans, and S. Dietrich, *Phys. Rev. E*, **57**, 6785 (1998).

[156] P. Attard and J. L. Parker, *J. Phys. Chem.*, **96**, 5086 (1992).

[157] P. Attard and G. N. Patey, *J. Phys. Chem.*, **92**, 4970 (1990).

[158] A. Luzar, D. Bratko, and L. Blum, *J. Chem. Phys.*, **86**, 2955 (1987).

[159] R. Kjellander and S. Sarman, *Mol. Phys.*, **74**, 665 (1991).

[160] D. Henderson and M. Lozada-Cassou, *J. Colloid Interface Sci.*, **114**, 180 (1986).

[161] M. Lozada-Cassou and E. Diaz-Herrera, *J. Chem. Phys.*, **92**, 1194 (1990).

[162] W. A. Ducker, T. J. Senden, and R. M. Pashley, *Nature*, **353**, 239 (1991).

[163] D. Qu, J. S. Pedersen, S. Garnier, A. Laschewsky, H. Möhwald, and R. v. Klitzing, *Macromolecules*, **39**, 7364 (2006).

[164] S. H. L. Klapp, D. Qu, and R. v. Klitzing, *J. Phys. Chem. B*, **111**, 1296 (2007).

[165] I. K. Snook and D. Henderson, *J. Chem. Phys.*, **68**, 2134 (1978).

[166] V. Y. Antonchenko, V. V. Ilyin, N. N. Makovsky, A. N. Pavlov, and V. P. Sokhan, *Mol. Phys.*, **52**, 345 (1984).

[167] B. Götzelmann and S. Dietrich, *Phys. Rev. E*, **55**, 2993 (1997).

[168] R. Evans, J. R. Henderson, D. C. Hoyle, A. O. Parry, and Z. A. Sabeur, *Mol. Phys.*, **80**, 755 (1993).

[169] J. N. Israelachvili, *Acc. Chem. Res.*, **20**, 415 (1987).

[170] P. Bordarier, B. Rousseau, and A. H. Fuchs, *J. Chem. Phys.*, **106**, 7295 (1997).

[171] D. J. Diestler, M. Schoen, and J. H. Cushman, *Science*, **262**, 545 (1993).

[172] M. Schoen, D. J. Diestler, and J. H. Cushman, *J. Chem. Phys.*, **100**, 7707 (1994).

[173] M. Schoen, C. L. Rhykerd, Jr., D. J. Diestler, and J. H. Cushman, *Science*, **245**, 1223 (1989).

[174] C. L. Rhykerd, Jr., M. Schoen, and D. J. Diestler, *Nature*, **330**, 461 (1987).

[175] W. H. Press, B. P. Flannery, S. A. Teukolsky, and W. T. Vetterling, *Numerical Recipes in FORTRAN*, Cambridge University Press, Cambridge, UK, 1990.

[176] M. Schoen and S. Dietrich, *Phys. Rev. E*, **56**, 499 (1997).

[177] M. Gonzáles-Melchor, P. Orea, J. López-Lemus, F. Bresme, and J. Alejandre, *J. Chem. Phys.*, **122**, 094593 (2005).

[178] P. G. Watson, *Phase Transitions and Critical Phenomena*, Academic Press, London, 1972.

[179] F. Burmeister, C. Schäfle, T. Matthes, M. Böhmisch, J. Boneberg, and P. Leiderer, *Langmuir*, **13**, 2983 (1997).

[180] U. Drodofsky, J. Stuhler, T. Schulze, M. Drewsen, B. Brezger, T. Pfau, and J. Mlynek, *Appl. Phys.*, **65**, 755 (1997).

[181] J. Heier, E. J. Kramer, S. Walheim, and G. Krausch, *Macromolecules*, **30**, 6610 (1997).

[182] R. Garcia, M. Calleja, and F. Perez-Murano, *Appl. Phys. Lett.*, **72**, 2295 (1998).

[183] R. Zwanzig, *J. Chem. Phys.*, **22**, 1420 (1954).

[184] M. Schoen, J. C. L. Rhykerd, J. H. Cushman, and D. J. Diestler, *Mol. Phys.*, **66**, 1171 (1989).

[185] R. Evans, U. M. B. Marconi, and P. Tarazona, *J. Chem. Phys.*, **84**, 2376 (1986).

[186] B. N. Persson, *Phys. Rev. B*, **48**, 18140 (1993).

[187] P. A. Thompson and M. O. Robbins, *Science*, **250**, 792 (1990).

[188] P. A. Thompson and M. O. Robbins, *Phys. Rev. A*, **41**, 6830 (1990).

[189] M. Lupkowski and F. van Swol, *J. Chem. Phys.*, **95**, 1995 (1991).

[190] P. A. Thompson, G. S. Grest, and M. O. Robbins, *Phys. Rev. Lett.*, **68**, 3448 (1992).

[191] M. G. Rozman, M. Urbakh, and J. Klafter, *Phys. Rev. Lett.*, **77**, 683 (1996).

[192] M. G. Rozman, M. Urbakh, and J. Klafter, *Phys. Rev. E*, **54**, 6485 (1996).

[193] M. Schoen, D. J. Diestler, and J. H. Cushman, *Phys. Rev. B*, **47**, 5603 (1993).

[194] M. Schoen, *Mol. Simul.*, **17**, 369 (1996).

[195] P. Bordarier, B. Rousseau, and A. H. Fuchs, *Thin Solid Films*, **330**, 21 (1998).

[196] J. H. Weiner, *Statistical Mechanics of Elasticity*, Wiley, New York, 1983.

[197] A. Münster, *Statistical Thermodynamics*, Academic Press, London, 1969.

[198] P. Bordarier, M. Schoen, and A. H. Fuchs, *Phys. Rev. E*, **57**, 1621 (1998).

[199] P. W. Atkins and J. D. Paula, *Atkins' Physical Chemistry*, seventh edition, Oxford University Press, Oxford, 2002.

[200] R. Balian, *From Microphysics to Macrophysics. Methods and Applications of Statistical Physics*, volume 1, Springer-Verlag, Heidelberg, 1991.

[201] G. W. Castellan, *Physical Chemistry*, third edition, Addison-Wesley, New York, 1983.

[202] A. W. Adamson, *A Textbook of Physical Chemistry*, second edition, Academic Press, New York, 1979.

[203] M. K. Kamp, *Physical Chemistry. A Step-by-Step Approach*, Marcel Dekker, New York, 1979.

[204] J. P. Joule and W. Thomson, *Philos. Mag.*, **4**, 481 (1852).

[205] D. W. McClure, *Am. J. Phys.*, **39**, 288 (1971).

[206] D. S. Kothari, *Philos. Mag.*, **12**, 665 (1931).

[207] F. J. Ynduráin, *Relativistic Quantum Mechanics and Introduction to Field Theory*, Springer-Verlag, Berlin, 1996.

[208] D. V. Gogate, *Philos. Mag.*, **25**, 694 (1938).

[209] D. V. Gogate, *Philos. Mag.*, **26**, 166 (1938).

[210] P. T. Eubank, D. van Peursem, Y.-P. Chao, and D. Gupta, *A.I.Ch.E. J.*, **40**, 1580 (1994).

[211] J. B. Dence and D. J. Diestler, *Intermediate Physical Chemistry-Stationary Properties of Chemical Systems*, Wiley, New York, 1987.

[212] D. M. Heyes and C. T. Llaguno, *Chem. Phys.*, **168**, 61 (1992).

[213] T. R. Rybolt, *A.I.Ch.E. J.*, **35**, 2029 (1989).

[214] R. A. Pierotti and T. R. Rybolt, *J. Chem. Phys.*, **80**, 3826 (1984).

[215] J. H. Cole, D. H. Everett, C. T. Marshall, A. R. Paniego, J. C. Powl, and F. R. Reinoso, *J. Chem. Soc. Faraday Trans. I*, **70**, 2154 (1974).

[216] P. Wu and W. A. Little, *Cryogenics*, **24**, 415 (1984).

[217] W. A. Little, *Rev. Sci. Instrum.*, **55**, 661 (1984).

[218] M. Schoen and D. J. Diestler, *J. Chem. Phys.*, **109**, 5596 (1998).

[219] K. Binder, *Z. Phys. B*, **45**, 61 (1981).

[220] R. H. Swendsen and J.-S. Wang, *Phys. Rev. Lett.*, **58**, 86 (1987).

[221] F. Niedermeyer, *Phys. Rev. Lett.*, **61**, 2026 (1988).

[222] M. Seul and R. Wolfe, *Phys. Rev. Lett.*, **64**, 1903 (1990).

[223] G. J. Zarragoicoechea, *Mol. Phys.*, **96**, 1109 (1999).

[224] E. M. Hara, *Polyelectrolytes: Science and Technology*, Marcel Dekker, New York, 1993.

[225] S. H. L. Klapp, *J. Phys.: Condens. Matter*, **17**, R525 (2005).

[226] E. Blums, A. Cebers, and M. M. Maiorov, *Magnetic Fluids*, Walter de Gruyter, New York, 1997.

[227] S. Odenbach, *Ferrofluids, Magnetically Controllable Fluids and Their Applications*, Springer-Verlag, Berlin, 2002.

[228] R. E. Rosensweig, *Ferrohydrodynamics*, Cambridge University Press, Cambridge, UK, 1985.

[229] T. C. Halsey, *Electrorheological Fluids*, World Scientific, Singapore, 1992.

[230] C. Stubenrauch and R. von Klitzing, *J. Phys.: Condens. Matter*, **15**, R1197 (2003).

[231] J.-J. Weis, *J. Phys.: Condens. Matter*, **15**, S1471 (2003).

[232] J. A. Barker and R. O. Watts, *Mol. Phys.*, **26**, 789 (1973).

[233] D. J. Adams and E. M. Adams, *Mol. Phys.*, **42**, 907 (1981).

[234] J. W. Eastwood and R. W. Hockney, *J. Comp. Phys.*, **16**, 342 (1974).

[235] M. Deserno and C. Holm, *J. Chem. Phys.*, **109**, 7678 (1998).

[236] M. Deserno and C. Holm, *J. Chem. Phys.*, **109**, 7694 (1998).

[237] J. Lekner, *Physica*, **176**, 485 (1991).

[238] D. Frenkel and B. Smit, *Understanding Molecular Simulation*, Academic Press, San Diego, CA, 2002.

[239] S. W. de Leeuw, J. W. Perram, and E. R. Smith, *Proc. R. Soc. London Ser. A*, **373**, 27 (1980).

[240] S. W. de Leeuw, J. W. Perram, and E. R. Smith, *Proc. R. Soc. London Ser. A*, **373**, 56 (1980).

[241] S. W. de Leeuw, J. W. Perram, and E. R. Smith, *Proc. R. Soc. London Ser. A*, **373**, 177 (1983).

[242] J. D. Jackson, *Classical Electrodynamics*, Wiley, New York, 1999.

[243] P. G. Kusalik, *J. Chem. Phys.*, **93**, 3520 (1990).

[244] Z. Wang and C. Holm, *J. Chem. Phys.*, **115**, 6351 (2001).

[245] D. E. Parry, *Surf. Sci.*, **49**, 433 (1975).

[246] D. E. Parry, *Surf. Sci.*, **54**, 195 (1976).

[247] S. W. de Leeuw and J. W. Perram, *Mol. Phys.*, **37** (1979).

[248] D. M. Heyes and F. van Swol, *J. Chem. Phys.*, **75**, 5051 (1981).

[249] Y. J. Rhee, J. W. Hautman, and A. Rahman, *Phys. Rev. B*, **40**, 36 (1989).

[250] J. Hautman and M. L. Klein, *Mol. Phys.*, **92**, 379 (1991).

[251] A. Grzybowski, E. Gwozdz, and A. Brodka, *Phys. Rev. B*, **61**, 6706 (2000).

[252] E. Spohr, *J. Chem. Phys*, **107**, 6342 (1997).

[253] J. C. Shelley and G. N. Patey, *Mol. Phys.*, **88**, 385 (1996).

[254] I.-C. Yeh and M. L. Berkowitz, *J. Chem. Phys*, **111**, 3155 (1999).

[255] P. S. Crozier, R. L. Rowley, E. Spohr, and D. Henderson, *J. Chem. Phys.*, **112**, 9253 (2000).

[256] A. Arnold, J. de Joannis, and C. Holm, *J. Chem. Phys.*, **117**, 2496 (2002).

[257] S. H. L. Klapp and M. Schoen, *J. Chem. Phys.*, **117**, 8050 (2002).

[258] C. G. Gray and K. E. Gubbins, *Theory of Molecular Fluids*, Oxford University Press, London, 1984.

[259] M. E. van Leeuwen, *Fluid Phase Equilib.*, **99**, 1 (1994).

[260] P. I. C. Teixera, J. M. Tavares, and M. M. Telo da Gama, *J. Phys.: Condens. Matter*, **12**, R411 (2000).

[261] T. Tlusty, *Science*, **290**, 1328 (2000).

[262] D. Wei and G. N. Patey, *Phys. Rev. Lett.*, **68**, 2043 (1992).

[263] D. Wei and G. N. Patey, *Phys. Rev. A*, **46**, 7783 (1992).

[264] J.-J. Weis, D. Levesque, and G. J. Zarragoicoechea, *Phys. Rev. Lett.*, **69**, 913 (1992).

[265] J.-J. Weis and D. Levesque, *Phys. Rev. E*, **48**, 3728 (1993).

[266] B. Groh and S. Dietrich, *Phys. Rev. Lett.*, **72**, 2422 (1994).

[267] B. Groh and S. Dietrich, *Phys. Rev. E*, **50**, 3814 (1994).

[268] S. Klapp and F. Forstmann, *J. Chem. Phys.*, **106**, 9742 (1997).

[269] S. H. L. Klapp and M. Schoen, *J. Mol. Liq.*, **109**, 55 (2004).

[270] G. Ayton, M. J. P. Gingras, and G. N. Patey, *Phys. Rev. Lett.*, **75**, 2360 (1995).

[271] G. Ayton, M. J. P. Gingras, and G. N. Patey, *Phys. Rev. E.*, **56**, 562 (1997).

[272] V. Ballenegger and J. P. Hansen, *J. Chem. Phys.*, **122**, 114711 (2005).

[273] D. R. Berard, M. Kinoshita, X. Ye, and G. N. Patey, *J. Chem. Phys.*, **94**, 6271 (1994).

[274] G. L. Gulley and R. Tao, *Phys. Rev. E*, **56**, 4328 (1997).

[275] J. Hautman, J. W. Halley, and Y.-J. Rhee, *J. Chem. Phys.*, **91**, 467 (1989).

[276] J. W. Perram and M. A. Ratner, *J. Chem. Phys.*, **91**, 467 (1989).

[277] S. H. L. Klapp, *Mol. Simul.*, **32**, 609 (2006).

[278] T. J. Barton, L. M. Bull, W. G. Klemperer, D. A. Loy, B. McEnaney, M. Misono, P. A. Monson, G. Pez, G. W. Scherer, J. C. Vartulli, and O. M. Yaghi, *Chem. Mater.*, **11**, 2633 (1999).

[279] P. Selvam, S. K. Bhatia, and C. G. Sonwane, *Ind. Eng. Chem. Res.*, **40**, 3237 (2001).

[280] D. Tulimieri, J. Yoon, and M. H. W. Chan, *Phys. Rev. Lett.*, **82**, 121 (1999).

[281] L. Radzihovsky and J. Toner, *Phys. Rev. Lett.*, **79**, 4214 (1997).

[282] D. E. Feldman, *Phys. Rev. Lett.*, **84**, 4886 (2000).

[283] T. Bellini, M. Buscaglia, C. Chiccoli, F. Mantegazza, P. Pasini, and C. Zannoni, *Phys. Rev. Lett.*, **85**, 1008 (2000).

[284] T. Bellini, M. Buscaglia, C. Chiccoli, F. Mantegazza, P. Pasini, and C. Zannoni, *Phys. Rev. Lett.*, **88**, 245506 (2002).

[285] E. Kierlik, M. L. Rosinberg, G. Tarjus, and P. Vio, *Phys. Chem. Chem. Phys.*, **3**, 1201 (2001).

[286] E. Kierlik, P. A. Monson, L. Sarkisov, and G. Tarjus, *Phys. Rev. Lett.*, **87**, 055701 (2001).

[287] F. Brochard and P. G. de Gennes, *J. Phys. Lett.*, **44**, L785 (1983).

[288] P. G. de Gennes, *J. Phys. Lett.*, **44**, L785 (1984).

[289] T. Nattermann, *Spin Glasses and Random Fields*, World Scientific, Singapore, 1997.

[290] M. L. Rosinberg, *New Approaches to Problems in Liquid State Theory*, Kluwer, Dordrecht, the Netherlands, 1999.

[291] O. Pizio, *Computational Methods in Surface and Colloid Science*, Marcel Dekker, New York, 2000.

[292] L. Sarkisov and P. A. Monson, *Phys. Rev. E*, **61**, 7231 (2000).

[293] M. Mezard, G. Parisi, and M. A. Virasoro, *Spin Glasses and Beyond*, World Scientific, Singapore, 1987.

[294] K. Binder and A. P. Young, *Rev. Mod. Phys.*, **58**, 801 (1986).

[295] J. A. Given, *Phys. Rev. A*, **45**, 816 (1992).

[296] J. A. Given and G. Stell, *J. Chem. Phys.*, **97**, 4573 (1992).

[297] J. A. Given and G. Stell, *Physica A*, **209**, 495 (1994).

[298] E. Kierlik, M. L. Rosinberg, G. Tarjus, and P. A. Monson, *Phys. Chem. Chem. Phys.*, **106**, 264 (1997).

[299] S. H. L. Klapp and G. N. Patey, *J. Chem. Phys.*, **115**, 4718 (2001)

[300] L. S. Ornstein and F. Zernike, *Proc. Acad. Sci.*, **17**, 793 (1914).

[301] J. K. Percus and G. J. Yevick, *Phys. Rev.*, **110**, 1 (1958).

[302] J. L. Lebowitz and J. K. Percus, *Phys. Rev.*, **144**, 251 (1966).

[303] E. Lomba, J. Given, G. Stell, J.-J. Weis, and D. Levesque, *Phys. Rev. B*, **48**, 223 (1993).

[304] A. Meroni, D. Levesque, and J.-J. Weis, *J. Phys. Chem.*, **105**, 1101 (1996).

[305] B. Hribar, V. Vlachy, A. Trokhymchuk, and O. Pizio, *J. Chem. Phys.*, **107**, 6335 (1997).

[306] B. Hribar, O. Pizio, A. Trokhymchuk, and V. Vlachy, *J. Chem. Phys.*, **109**, 2480 (1998).

[307] C. Spöler and S. H. L. Klapp, *J. Chem. Phys.*, **121**, 9623 (2004).

[308] M. L. Rosinberg, G. Tarjus, and G. Stell, *J. Chem. Phys.*, **100**, 5172 (1994).

[309] M. Fernaud, E. Lomba, and J.-J. Weis, *Phys. Rev. E*, **64**, 051501 (2001).

[310] C. Spöler and S. H. L. Klapp, *J. Chem. Phys.*, **118**, 3628 (2003).

[311] E. Schöll-Paschinger, D. Levesque, G. Kahl, and J.-J. Weis, *Europhys. Lett.*, **55**, 178 (2001).

[312] D. M. Ford and E. Glandt, *Phys. Rev. E*, **50**, 1280 (1994).

[313] E. Kierlik, M. L. Rosinberg, G. Tarjus, and P. A. Monson, *J. Chem. Phys.*, **110**, 689 (1999).

[314] K. S. Page and P. A. Monson, *Phys. Rev. B*, **54**, 6557 (1996).

[315] M. Alvarez, D. Levesque, and J.-J. Weis, *Phys. Rev. B*, **60**, 5495 (1999).

[316] P. R. van Tassel, *Phys. Rev. E*, **60**, R25 (1999).

[317] A. Trokhymchuk, O. Pizio, M. Holovko, and S. Sokolowski, *J. Phys. Chem.*, **100**, 17004 (1996).

[318] B. Hribar, V. Vlachy, A. Trokhymchuk, and O. Pizio, *J. Phys. Chem. B*, **103**, 5361 (1999).

[319] B. Hribar, V. Vlachy, and O. Pizio, *J. Phys. Chem. B*, **104**, 4479 (2000).

[320] B. Hribar, V. Vlachy, and O. Pizio, *J. Phys. Chem. B*, **105**, 4727 (2001).

[321] C. Spöler and S. H. L. Klapp, *J. Chem. Phys.*, **120**, 6734 (2004).

[322] P. H. Fries and G. N. Patey, *J. Chem. Phys.*, **82**, 429 (1985).

[323] M. Fernaud, E. Lomba, J.-J. Weis, and D. Levesque, *Mol. Phys.*, **101**, 1721 (2003).

[324] M. Fernaud, E. Lomba, C. Martin, D. Levesque, and J.-J. Weis, *J. Chem. Phys.*, **119**, 364 (2003).

[325] A. C. Pierre and G. M. Panjok, *Chem. Rev.*, **102**, 4243 (2002).

[326] T. L. Hill, *Statistical Mechanics. Principles and Selected Applications*, Dover, New York, 1986.

[327] L. Reichl, *A Modern Course in Statistical Physics*, University of Texas Press, Austin, TX, 1980.

[328] D. Pini, M. Tau, A. Perola, and L. Reatto, *Phys. Rev. E*, **67**, 046116 (2003).

[329] M. Schoen, *Physica A*, **240**, 328 (1997).

[330] I. N. Bronstein and K. A. Semendjajew, *Taschenbuch der Mathematik*, Harri Deutsch, Frankfurt, Germany, 1981.

[331] H. Fröhlich, *Theory of Dielectrics*, Oxford University Press, Oxford, 1949.

[332] M. S. Wertheim, *Phys. Rev. Lett.*, **10**, 321 (1963).

[333] M. S. Wertheim, *J. Math. Phys.*, **5**, 643 (1964).

[334] E. Thiele, *J. Chem. Phys.*, **39**, 474 (1963).

[335] E. W. Grundke and D. Henderson, *Mol. Phys.*, **24**, 269 (1972).

[336] L. L. Lee and D. Levesque, *Mol. Phys.*, **26**, 1351 (1973).

[337] E. Waisman and J. L. Lebowitz, *J. Chem. Phys.*, **56**, 3086 (1972).

[338] E. Waisman and J. L. Lebowitz, *J. Chem. Phys.*, **56**, 3093 (1972).

[339] M. S. Wertheim, *J. Chem. Phys.*, **55**, 4291 (1971).

[340] A. A. Broyles, *J. Chem. Phys.*, **33**, 456 (1960).

[341] K.-C. Ng, *J. Chem. Phys.*, **61**, 2680 (1974).

Index